Gauge Theories in the Twentieth Century

Gauge Theories in the Twentieth Century

Editor

John C. Taylor

University of Cambridge

ICP

Imperial College Press

Published by

Imperial College Press
57 Shelton Street
Covent Garden
London WC2H 9HE

Distributed by

World Scientific Publishing Co. Pte. Ltd.
P O Box 128, Farrer Road, Singapore 912805
USA office: Suite 1B, 1060 Main Street, River Edge, NJ 07661
UK office: 57 Shelton Street, Covent Garden, London WC2H 9HE

British Library Cataloguing-in-Publication Data
A catalogue record for this book is available from the British Library.

The editor and the publisher would like to thank the following for their assistance and their permission to reproduce the articles found in this volume:

American Institute of Physics (*JETP Lett.*), American Physical Society (*J. Math. Phys.*, *Phys. Rev.*, *Phys. Rev. Lett.*), Elsevier Science Publishers B. V. (*Nucl. Phys.*, *Phys. Lett.*), Kluwer Academic Publishers, Polish Academy of Sciences (*Acta Phys. Pol.*), Royal Society, Springer-Verlag (*Z. Phys.*)

GAUGE THEORIES IN THE TWENTIETH CENTURY

ISBN 1-86094-281-4
ISBN 1-86094-282-2 (pbk)

Printed in Singapore.

Contents

Preface

My intention in this volume is to bring together some of the key papers in the development of gauge theories in physics, mainly from the 1920s to the 1980s. Inevitably, some equally important papers will have been omitted. The emphasis is on principles rather than applications. The volume is not intended to be a serious work in the history of science; still less is it a textbook of physics. Most of the ideas in the papers, so far as they have proved to be correct, are to be found in modern textbooks, such as [1]. Some first-hand accounts of the history are to be found in [2].

I am very grateful to Michael Atiyah, Tom Kibble, David Olive and Roger Phillips for their encouragement and advice. I also thank Ron Shaw for information supplied to me. None of these people is to blame for deficiencies which remain in this volume. I am grateful to David Grellscheid, Sakura Schafer-Nameki and Fabian Wagner for translating the German papers (though they are not responsible for any errors introduced by my editing).

1 Gauge invariance in electromagnetism

The early history of gauge theories has been beautifully recounted by Lochlainn O'Raifeartaigh in *The Dawning of Gauge Theory* [3], to which I will make frequent reference. Some of the papers reprinted here also appear in O'Raifeartaigh's book.

In classical theory, Maxwell's equations and the equations of motion of charged particles can be formulated in terms of the electric and magnetic fields, so gauge invariance plays no essential role (unless one wishes to write an action for the charged particle). But Maxwell in his *Treatise* did use the electromagnetic potentials. Paper 1.1 shows him, in modern terms, making a gauge transformation to get into the Coulomb gauge.

As is well known, the word *gauge* (*Eich* in German) originates with Weyl's attempt [4] (an English translation appears in [3]) in 1918 to unify (classical) electromagnetism with Einstein's theory of gravity. In his theory, under parallel transfer along a path C, a vector was assumed to undergo a change of length by a (path-dependent!) factor

$$\exp\left[\gamma \int_C A \cdot dx\right], \tag{1}$$

where A_μ is the 4-vector of the electromagnetic potentials and γ is some scalar coefficient. The idea did not work. Paper 1.2 is a chapter from an English translation of Weyl's book *Space-Time-Matter* (1922 edition), accompanied by a preface written in 1950 which explains how Weyl abandoned the theory.

In 1922, on the basis of examples of periodic orbits, Schrödinger suggested that the constant γ should be chosen to be imaginary, so that (1) becomes

$$\exp\left[\frac{-ie}{\hbar} \int_C A \cdot dx\right]. \tag{2}$$

An English translation of part of this paper is included in [3]. Schrödinger says, "I do not dare to say whether this would make any sense in the context of Weyl geometry."

In 1927, Fritz London published a "quantum mechanical interpretation of Weyl's theory" (paper 1.3, English translation 1.3e). London argues that the wave function ψ can, in some sense, be identified with a length scale l, transforming according to (2), at least with the proviso that the path C in should be chosen to be the particle's path in a semi-classical approximation. In his very last equation, London transforms the electromagnetic potentials by an imaginary gauge transformation. It is curious, in view of Klein's work to be mentioned next, that London is led to a sort of five-dimensional formalism, with proper time as the fifth coordinate.

The other route to an appreciation of the role of gauge transformations in quantum theory began from another attempt to find a geometrical interpretation for electromagnetism. In 1921, Kaluza (English translation in [3]) proposed a five-dimensional space-time, with the metric components $g_{5\mu}$ $(\mu \neq 5)$[1] related to the electromagnetic potentials A_μ (the interpretation of g_{55} is a moot point). Oskar Klein, in 1926 (paper 1.4, English translation 1.4e), wrote the five-dimensional relativistic Schrödinger equation,[2] and made the ansatz

[1] 0 is used instead of 5 by Kaluza, but this is confusing to modern readers.

[2] What we would now call the Klein-Gordon equation — the non-relativistic Schrödinger equation appeared only later in 1926.

of periodic dependence on x_5. This amounts to compactifying the fifth dimension as a circle. The electric charge is interpreted in terms of the angular momentum of the motion round the circle. Gauge transformations are just space-time-dependent transformations of the coordinate x_5.

Something formally very similar was done independently by Fock and published in 1927 (paper 1.5, English translation 1.5e). At the end of Section 1, Fock states very explicitly how the wave function behaves under gauge transformations.

In the Kaluza-Klein approach, both neutral and charged particles move classically on geodesics in five-dimensional space-time, but the geodesics followed by charged particles wrap around the x_5 circle while those for neutral particles do not.

From a present day standpoint, we are interested in these researches because they introduced gauge invariance into quantum theory. But the authors had ambitious aims: the unification of electromagnetism with Einstein's theory of gravity, and the derivation of the wave equation of quantum theory.

The physical significance of the electromagnetic potentials in quantum theory was only appreciated in 1959, by Aharanov and Bohm (paper 1.6).[3] An electron traversing paths *outside* a region of magnetic field can exhibit interference effects. Gauge transformations cannot in general make the phase factor (2) equal to 1 along *all* paths. The predicted effect was observed soon after, first by Chambers in 1960 (paper 1.7).

2 Non-abelian gauge theories

Non-abelian gauge theories were discovered by Yang and Mills, who published their work in 1954 (paper 2.1), and independently by Shaw (unpublished thesis, paper 2.2). Both Yang and Mills and Shaw were motivated by the idea of making isotopic spin invariance into a local, gauge invariance. Both Yang and Mills and Shaw worried whether the spin 1 particles would have to have zero mass.[4]

Shaw raises the question whether one of the gauge fields might be identified as the photon. He does this in the context of a now obsolete approximate strong interaction symmetry using the group $O(4)$.

O'Raifeartaigh in [3] recounts the history in more detail, and shows how both O. Klein (in 1938) and Pauli (in 1953) came close to discovering non-abelian gauge theories. Both used modifications of the Kaluza-Klein idea, with, in Pauli's case, a six-dimensional space-time.

In the non-abelian case, there is a generalization of the phase factor (2):

$$P\exp\left[-\frac{ig}{\hbar}\int_C A\cdot dx\right], \tag{3}$$

[3]A similar idea had appeared in a paper of 1949, *The Refractive Index in Electron Optics and the Principles of Dynamics*, by Ehrenberg and Siday [5].

[4]Yang and Mills remark that no Ward identity analogous to $k^\mu G_{\mu\nu}(k) = 0$, which in perturbative QED guarantees zero mass for the photon, had been proved. We now know 't Hooft's identity $k^\mu k^\nu G_{\mu\nu} = 0$ would have the same effect (paper 6.3).

where P denotes the matrix ordering of the (Hermitian) matrix gauge potential A along the path C. The quantity (3) is sometimes called the *gauge phase factor* (as in papers 3.2 and 13.1).

3 Gravity as a gauge theory

At about the same time as Yang, Mills and Shaw, Utiyama was working on a general theory of local gauge invariance [6] (reprinted in [3]). Utiyama cast gravity as a gauge theory as follows. His approach was to gauge the six-parameter group of Lorentz transformation. He began with a flat space-time, but uses general curvilinear coordinates u^μ. He defines

$$h_k^\mu = \frac{\partial u^\mu}{\partial x^k}, \tag{3}$$

where the x^k are Cartesian coordinates. He then goes on to regard the h_k^μ as *independent* quantities, and shows that this requires the original space to be curved. The analogue of the field in Yang-Mills theory is the curvature tensor, and an invariant action can be constructed therefrom (together with the determinant $\det h_k^\mu$).

Kibble in 1960 (paper 3.1) looked at this question again. Unlike Utiyama, he took the 10-parameter group of inhomogeneous Lorentz transformations (Poincaré group) as the group to gauge (that is, to make its parameters depend on position in space-time). The gravitational field variables are the 16 vierbein components h_k^μ and the 24 connection components A_μ^{ij}. The first order, Palatini form of gravity equations is obtained. Kibble works formally in a flat "background" space-time, but the quantities

$$g_{\mu\nu} = h_\mu^k h_{k\nu}$$

may be interpreted as the metric of a Riemannian space-time.

In these formulations, the analogy between Yang-Mills theory and gravitation is certainly not complete: in Yang-Mills there is no analogue of the vierbein.

A paper of Yang in 1974 (paper 3.2) also addresses this subject. Yang does not introduce vierbeins, and he gauges the 16-parameter group of general linear transformations. From his point of view, the natural gauge theory is not Einstein's gravity.

The similarity between gravity and Yang-Mills theory was used by Feynman (paper 6.1).

4 Gauge invariance and superconductivity

The role of gauge invariance in superconductivity was a matter of discussion for some years. In the phenomenological theory, the (approximate) London equation

$$\nabla \wedge \mathbf{j} = -\frac{1}{\lambda^2}\mathbf{B} \tag{4}$$

(with $\mathbf{j}$ the current, $\mathbf{B}$ the magnetic field and λ the penetration depth) implies in general (although the original application was restricted to the transverse part of (5)) that

$$\mathbf{j} = \frac{1}{\lambda^2}(\nabla\phi - \mathbf{A}) \tag{5}$$

for some scalar field ϕ. Gauge invariance demands that, under a gauge transformation where

$$\mathbf{A} \to \mathbf{A} + \nabla\omega\,, \tag{6}$$

$$\phi \to \phi + \omega\,. \tag{7}$$

Much of the discussion hinged on the interpretation of the field ϕ, which is connected with longitudinal plasma oscillations (compression waves in the superconducting electrons).

In the Ginzburg-Landau [7] model for the superconducting phase transition, ϕ is the phase of the order parameter.

In the BCS microscopic theory, there is a condensate of Cooper pairs, and

$$\exp\left[\frac{2ie\phi}{\hbar}\right] \tag{8}$$

is the (position-dependent) phase factor of this condensate. The fact that the condensate "chooses" a phase like this, is now regarded as an example of spontaneous[5] breaking of gauge invariance. A particularly bold modern exposition of this point of view can be found in Section 21.6 of Weinberg's textbook [1].

The original BCS theory accounted for the transverse part of the current $\mathbf{j}$ in (6) only. Several authors (for example Nambu in paper 4.1) extended the theory in a gauge-invariant way, so as to include the longitudinal part.

The Josephson effect (paper 4.2) is an example of the physical importance of the phase ϕ. Across a narrow insulating junction between two superconductors, a current flows which has a periodic dependence on the phase difference. If a potential difference is maintained between the two superconductors, the phase difference depends linearly on time, and hence the current oscillates.

Normally, gauge invariance implies that the photon must have zero mass. In the relativistic case, for example, the photon self-energy must have the form

$$(p_\mu p_\nu - p^2\eta_{\mu\nu})\Pi(p^2)\,, \tag{9}$$

apparently vanishing[6] at $p^2 = 0$. Schwinger, in 1962 (paper 4.3), argued that the zero mass could be avoided if, by non-perturbative effects, Π developed a pole at $p^2 = 0$. In the same year, Anderson (paper 4.4) pointed out that just such an effect takes place in superconductors. The spontaneous symmetry breaking by the Cooper pair vacuum condensate might have been expected to entail a massless Goldstone boson; but the Goldstone field is just the ϕ in (5), and, because of gauge invariance, has no physical significance independently of the electromagnetic potential.

5 Spontaneous symmetry breaking and particle physics

In 1964, two short letters were published showing how gauge fields can get mass due to spontaneous symmetry breaking, by Englert and Brout (paper 5.1) and by Higgs (paper 5.2).

[5] "Spontaneous symmetry breaking", or sometimes "vacuum symmetry breaking", is said to occur when the ground state of a system does not have the invariance of the action which describes it.

[6] In abelian theory, the $p_\mu p_\nu$ term (9) makes no contribution, because of current conservation.

They gave simple models in which some components of a multiplet of "elementary" scalar fields acquire a vacuum expectation value, that is, undergo Bose-Einstein condensation in the vacuum. The choice of component determines the symmetry breaking. The other components of the scalar multiplet are swallowed by gauge fields, providing the longitudinal polarization states necessary for spin 1 particles with mass.

Both sets of authors refer to previous work, particularly related to superconductivity. Higgs refers to Anderson; Englert and Brout refer to Nambu and Jona-Lasinio [8]. Neither papers directly refer to the Ginzburg-Landau model of superconductivity [7], although their field theories are essentially relativistic versions of this.

Higgs draws attention to the partial multiplets of massive scalar particles, now called Higgs particles, which accompany the gauge particles. These are quanta of the fields which get vacuum expectation values.

In the same year, 1964, Guralnik, Hagen and Kibble (paper 5.3) discussed further the quantum theory of the spontaneous symmetry breaking. In particular they showed how Goldstone's theorem [9], which shows that symmetry breaking requires massless quanta, is evaded in the case of gauge theories. In a paper of 1967, Kibble (paper 5.4) gave a fuller account of symmetry breaking in non-abelian gauge theories. In particular, he showed that the number of vector fields which acquire mass is equal to the number of symmetries which are broken.

Electroweak theory [10] is based upon the breaking of of a gauged $SU(2) \times U(1)$ symmetry down to the electromagnetic $U(1)$ group, giving three heavy vector particles and (at least) one Higgs scalar. But this is not the subject of the present volume.

A collection of original papers about dynamical gauge symmetry breaking appears in [11].

6 Gauge-fixing in non-abelian gauge theories

In order to calculate with quantum gauge field theories (at least by perturbation theory), one needs to fix a gauge: otherwise there are no unique potentials, and no unique propagator. For example, one may add a "gauge-fixing term" to the actions, so that the photon propagator becomes

$$\frac{-\eta_{\mu\nu}}{k^2 + i\epsilon}, \tag{10}$$

$\eta_{\mu\nu}$ being the Lorentz metric. But this propagator has poles with four different polarization states: the two physical transverse ones, but also a longitudinal one (parallel to $\mathbf{k}$) and a time-like one which has the wrong sign. In QED, it is possible to show that the latter two unwanted poles give cancelling contributions (because the current is divergenceless).

But in non-abelian theories this cancellation does not happen. This was appreciated by Feynman in 1963 (reprinted in paper 6.1, together with a discussion involving other experts, like B. de Witt). Feynman analyzed perturbation theory diagrams, and found he needed to include virtual "ghost" diagrams with an extra minus sign for each closed ghost loop.[7]

[7]This minus sign may be produced by making the ghost fields anti-commuting quantities, like fermion fields, even though they are scalars (or vectors in the case of gravity).

These were necessary in order to complete the cancellation between longitudinal and time-like poles. Feynman was really concerned with quantum gravity, but he used Yang-Mills theory as a simpler model.

In 1967, Faddeev and Popov (paper 6.2) gave a closed-form derivation of the ghost contributions, by manipulating the Feynman path integral. The ghosts are necessary in order to cancel the Jacobian of the transformation from gauge variation to the gauge-fixing function (which is trivial only in the abelian case). Similar conclusions were reached by de Witt [12].

In gauge theories with spontaneous symmetry breaking, there is a question what types of gauge-fixing are convenient to use. One may use a physical (unitary) gauge in which the massive vector particles have propagators

$$-\frac{\eta_{\mu\nu} - \frac{k_\mu k_\nu}{M^2}}{k^2 - M^2 + i\epsilon}. \tag{11}$$

But then the analysis of divergences by power-counting is complicated, and renormalizability is not obvious.[8] In 1971, 't Hooft (paper 6.3)[9] introduced a class of gauges in which the $1/M^2$ in (11) no longer appears, and power-counting works just as in QED. Much of the importance of spontaneous symmetry breaking of non-abelian gauge theories is as a method to generate renormalizable theories of charged spin 1 particles (as required in electroweak physics).

It is interesting to know whether there are choices of gauge-fixing (necessarily not explicitly Lorentz invariant) in which the only dynamical fields are physical ones and in which a Hamiltonian exists; so that unitarity is obvious. In QED, an example is the Coulomb gauge, in which the longitudinal part of the potential is chosen to be zero and the time-like part is eliminated. But the generalization of the Coulomb gauge to non-abelian theories runs into complications and difficulties [14]. Another example of a physical gauge is to choose one of the Lorentz components of each gauge field to be zero. This introduces into perturbation theory denominators like $1/n \cdot k$ (where n is a unit vector), and the Feynman integrals are not obviously well-defined.

The existence of such physical gauges, where canonical quantization can be applied, is an important question of principle. Some well-known textbooks begin their account of gauge theories by starting in such a gauge, and from it derive covariant gauges (with ghosts).

In 1978, Gribov (paper 6.4) observed that gauge-fixing conditions, involving say $\partial_\mu A^\mu$ or $\partial_i A_i$, may not fix the gauge uniquely. There may be "Gribov copies": different values of A related by gauge transformations and with the same value of the gauge-fixing function. Perturbation theory is unaffected by this observation, but there are potentially very important non-perturbative effects, as discussed in the first place by Gribov in the paper reprinted here.

[8] The renormalizability of field theories with massive vector mesons had been extensively investigated by M. Veltman in the late 1960s. The methods then developed influenced the later work of 't Hooft (paper 6.3) and 't Hooft and Veltman [16] on spontaneously broken gauge theories.

[9] For more of 't Hooft's many important contributions to gauge theories, see [13].

7 Gauge identities and unitarity

In QED perturbation theory, gauge-fixing destroys gauge invariance; but indirect consequences of the original invariance remain, and are essential to the consistency, especially the unitarity of the calculation. These consequences are the Ward-Takahashi identities [15]. Renormalization must be done in a way consistent with them (although this must happen automatically if a gauge-invariant regularization is used). As an example, the identities ensure that particles with equal bare charges also have equal renormalized charges.

In non-abelian theories, the corresponding identities are more complex. The first such identities were derived by 't Hooft in paper 6.3, but more general gauge identities were soon derived [16, 17]. These identities express the invariance of the action (including the gauge-fixing term) under certain non-linear transformations which mix ordinary fields and ghost fields (that is, mix commuting quantities with anti-commuting ones, as in supersymmetry). This symmetry is called BRST symmetry after the work of Becchi, Rouet and Stora (see the article reprinted as paper 7.1) and Tyutin. BRST symmetry is nilpotent, in the sense that, if Q is the operator which generates the symmetry, then $Q^2 = 0$. If the physical states[10] are assumed to be those annihilated by Q, then unitarity is guaranteed.

8 Asymptotic freedom

The inventors of non-abelian gauge theories had in mind the strong interactions, because of isotopic spin invariance. Sakurai, in 1960 [18], attempted to construct a gauge theory of strong interactions. Some of the key steps in the 1960s towards the theory of QCD were: the postulate of the existence (in some sense) of quarks [19], the idea of colour [20], evidence from sum rules etc. for the presence of free partons within hadrons [21], and, of course, the experiments on high-energy electron-nucleon scattering being done at SLAC.

The decisive discovery, which explained all this, was that of asymptotic freedom, independently in 1973 by Gross and Wilczek (paper 8.1), and by Politzer (paper 8.2). In QED, and in most four-dimensional field theories, there is an effective charge, depending on distance, which increases at small distances (the bare charge being defined in the limit where the distance approaches zero). This effect is sometimes explained intuitively as being due to "shielding" of the charge at larger distances by virtual charges in the vacuum, in analogy with condensed matter physics.[11] For unbroken non-abelian gauge theories (with not too many fermions or scalars) the opposite happens: the effective charge decreases at small distances (or at high energies).

The discovery of asymptotic freedom justified the use of perturbation theory in QCD, given also a "factorization theorem". The latter factorizes cross-sections into a short-distance part, to which perturbation theory is applicable, and a long-distance part ("structure functions") which have to be taken from experiment. It is essential that the short-distance part should be insensitive to low energies.[12]

[10] As opposed to states containing arbitrary longitudinal and time-like polarizations, and ghosts.

[11] In fact, this simple analogy is valid only for virtual particles which are bosons and are spinless [22].

[12] The cancellation of infra-red divergences, which is well-known in QED, occurs also in QCD, but only up to terms suppressed by inverse powers of the energy [23].

9 Monopoles and vortex lines

In 1931, Dirac (paper 9.1) proved that the existence of magnetic monopoles was consistent with quantum theory only if all pole strengths (m) and electric charges (e) satisfy the quantization condition

$$em = \frac{\hbar}{2} \times \text{integer}\,. \tag{12}$$

The reason is that, although no everywhere continuous magnetic vector potential $\mathbf{A}$ exists in the presence of a magnetic monopole, vector potentials do exist defined in different regions which, where they overlap, are connected by gauge transformations $\omega(\mathbf{x})$ satisfying

$$\exp\left(\frac{ie\omega}{\hbar}\right) = 1\,.$$

Then an everywhere continuous wave function for the electrically charged particle does exist.

In 1974, independently, 't Hooft (paper 9.2) and Polyakov (paper 9.3) showed that, in certain spontaneously broken non-abelian gauge theories, static classical solutions of the field equations exist which, viewed from large distances, are magnetic monopoles (obeying Dirac's quantization condition (12), of course). Such monopole solutions do not exist in the minimum standard model, but they do exist in some broken grand unified theories (of strong and electroweak interactions). These monopoles would be very heavy, but might have been produced during phase changes in the cooling early universe.

't Hooft-Polyakov monopoles are classical solutions (solitons) which must be stable for topological reasons. Somewhat similar topological structures had been discovered in 1973 by Nielsen and Olesen (paper 9.4). These were string-like, that is, indefinitely extended in one dimension. They are field theory analogues of the quantized flux tubes in superconductors.

10 Non-perturbative approaches

Asymptotic freedom justifies perturbative calculations of short-distance effects in QCD, but long-distance effects, such as the explanation of confinement and the particle spectrum, are beyond its reach. One of the most popular non-perturbative approaches has been to approximate the field theory by replacing space-time (or possibly only space) by a discrete lattice. For this to be a good approximation, the lattice spacing must be sufficiently small. One may then use a strong-coupling approximation, or a numerical simulation. One of the seminal papers, by Wilson in 1974, is reprinted as paper 10.1. Among other things, Wilson gives an example of a gauge-invariant action on a lattice, and proposes a mathematical criterion for confinement using the vacuum-expectation value of the Wilson loop operator (the operator in (3) with C a closed loop).

In a non-abelian gauge theory based on a semi-simple group, the group is necessarily compact. In the case of QED, the group may be taken to be $U(1)$, which is compact, or its covering group, which is not. In perturbation theory, this distinction has no force. In lattice formulations, however, compact and non-compact theories are quite different. Paper 10.2 is a discussion by Polyakov (1975) of the implications of compactness for confinement.

An alternative expansion to ordinary perturbation theory ('t Hooft, paper 10.3) results from taking the gauge group to be $SU(N)$ with N large (but Ng^2 remaining finite). The leading contribution comes from the sum of all planar graphs.

11 Instantons and vacuum structure

In 1975, Belavin, Polyakov, Schwartz and Tyupkin (paper 11.1) discovered *instantons*, that is, in four-dimensional Euclidean[13] field theory, self-dual or anti-self-dual (meaning $F^* = \pm F$, where $F^*_{12} = F_{34}$ etc.) Yang-Mills fields whose potentials approach pure gauge values at infinity. Such fields automatically obey the field equations. They are topologically stable because the gauge function at infinity is non-trivially mapped onto the 3-sphere at infinity, the degree of the mapping being defined by an integer

$$\nu = \frac{1}{64\pi^2} \int d^4x F_{ij,a} F^*_{ij,a} \tag{13}$$

(where a is a group index). Here the integrand is equal to a divergence, and the integral would vanish if A tended to zero at infinity faster than $1/\sqrt{x^2}$.

Instantons have a finite value of the (Euclidean) action, proportional to (13) and to $1/g^2$ (where g is the coupling strength). The contribution to the path integral from field configurations near the instanton is proportional:

$$\exp\left(\frac{-8\pi^2|\nu|}{g^2}\right). \tag{14}$$

Such contributions are beyond the reach of perturbation theory. In QCD, the effective value of g decreases with the relevant scale, and so (14) is small for small instantons but not for large ones.

If, then, field configurations with non-zero values of ν are relevant to the Euclidean path integral, how are they to be weighted? This question was answered in 1976 by Jackiw and Rebbi (paper 11.2) and by Callan, Dashen and Gross (paper 11.3). They must be weighted by a phase factor

$$\exp(i\theta\nu) \tag{15}$$

in the path integral, where θ is a real number (called θ by both groups of authors) which characterizes the theory (somewhat like another coupling strength).

In electroweak theory, instantons predict tiny, exponentially suppressed, lepton- and baryon-number violating effects.

In QCD, instantons solve the "$U(1)$ problem", showing how an anomaly means that there need be no light isosinglet pseudoscalar particle (analogous to the pions) [24]. But the phase (15) breaks P and CP invariance, and hence, for a reason not fully understood, θ must be very small [25].

[13] Euclidean field theory is relevant because properties of Lorentz field theory can be recovered from it, including, in particular, tunnelling probabilities.

12 Three-dimensional gauge fields and topological actions

In three-dimensional[14] non-abelian gauge theories with an odd number of fermions, an anomaly introduces (Redlich, 1984, reprinted as paper 12.1) an effective, parity violating action (α is a constant)

$$\frac{\alpha}{8\pi^2}\int d^3x\epsilon^{\lambda\mu\nu}\mathrm{tr}\left[\frac{1}{2}A_\lambda F_{\mu\nu}-\frac{1}{3}A_\lambda A_\mu A_\nu\right]. \tag{16}$$

This (Chern-Simons) term [26] is topological, in the sense that no metric in the 3-space is required in its definition. It is invariant under gauge transformations for which the gauge function maps trivially onto the sphere at infinity; but for gauge transformations which have a winding number w it changes by $\alpha w/g^2$. The condition that the exponential of $i/\hbar$ times (16) (occurring in the path integral) should change by a multiple of 2π enforces a quantization condition on α.

13 Gauge theories and mathematics

By a remarkable coincidence, not long before the discovery by physicists of non-abelian gauge theories, the relevant mathematical formalism — that of fibre bundles — had been invented by mathematicians. In applications to physics, the base of the bundle is space-time (or some part of it) and the gauge group acts on the fibre. A field is a section of a bundle. The gauge potential is a connection. The gauge phase factor (3) is a parallel displacement, and these constitute the holonomy group of the bundle.

The mathematics becomes relevant when the bundle is not trivial, for instance in the case of a magnetic monopole.

Yang (paper 13.1) treats some physical situations using a mathematical point of view.

Although mathematicians invented bundles, they did not have the Yang-Mills equations. Some important mathematics about the classification of 4-manifolds has been derived using these equations [27].

References

[1] S. Weinberg, *The Quantum Theory of Fields*, Cambridge, Cambridge University Press (1996).

[2] *History of Original Ideas and Basic Discoveries in Particle Physics*, edited by H.B. Newman and T. Ypsilantis, Plenum Press, New York and London (1996).

[3] L. O'Raifeartaigh, *The Dawning of Gauge Theories*, Princeton, Princeton University Press (1997).

[4] H. Weyl, *Sitzungber. Preus. Akad. Berlin* (1918) 465.

[14]Three-dimensional field theories are relevant to several situations in condensed matter physics, and also to quantum field theory at very high temperatures.

[5] W.E. Ehrenberg and R.E. Siday, *Proc. Roy. Soc.* **B62** (1949) 8.

[6] R. Utiyama, *Phys. Rev.* **101** (1956) 1597.

[7] V.L. Ginzburg and L.D. Landau, *J. Expt. Theor. Phys. (USSR)* **20** (1950) 1064.

[8] Y. Nambu and G. Jona-Lasinio, *Phys. Rev.* **122** (1961) 345.

[9] J. Goldstone, A. Salam and S. Weinberg, *Phys. Rev.* **127** (1962) 965.

[10] S. Weinberg, *Phys. Rev. Lett.* **19** (1967) 1264; A. Salam, in *Elementary Particles Theory*, edited by N. Svartholm, Stockholm, Almqvist Forlag AB (1968) 367.

[11] *Dynamical Gauge Symmetry Breaking*, edited by E. Farhi and R. Jackiw, World Scientific, Singapore (1982).

[12] B. de Witt, *Phys. Rev.* **160** (1967) 1113, **162** (1967) 1195, **163** (1967) 1239.

[13] G. 't Hooft, *Under the Spell of the Gauge Principle*, Vol. 19 of "Advanced series in mathematical physics", World Scientific, Singapore (1994).

[14] N.H. Christ and T.D. Lee, *Phys. Rev.* **D22** (1980) 939; P. Doust and J.C. Taylor, *Phys. Lett.* **197** (1987) 232.

[15] J.C. Ward, *Phys. Rev.* **78** (1950) 182; Y. Takahashi, *Nuovo Cim.* **6** (1957) 371.

[16] G. 't Hooft and M.T. Veltman, *Nucl. Phys.* **B44** (1972) 189, **B50** (1972) 318.

[17] J.C. Taylor, *Nucl. Phys.* **B33** (1971) 436; A.A. Slavnov, *Teor. i Mat. Fiz.* **10** (1972) 99; B.W. Lee and J. Zinn-Justin, *Phys. Rev.* **D5** (1972) 3121, 3137, 3155, and **D7** (1972) 1049.

[18] J.J. Sakurai, *Ann. Phys.* **11** (1960) 1.

[19] M. Gell-Mann, *Phys. Lett.* **8** (1964) 214; G. Zweig, CERN Report No. TH401 (1964).

[20] M.Y. Han and Y. Nambu, *Phys. Rev.* **139B** (1965) 1006; O.W. Greenberg, *Phys. Rev. Lett.* **13** (1964) 598.

[21] J. Bjorken, *Phys. Rev.* **179** (1969) 1547; C.G. Callan and D.J. Gross, *Phys. Rev. Lett.* **22** (1968) 156; R.P. Feynman, *Phys. Rev. Lett.* **23** (1969) 1415.

[22] R.J. Hughes, *Nucl. Phys.* **B186** (1981) 376.

[23] R. Doria, J. Frenkel and J.C. Taylor, *Nucl. Phys.* **B168** (1980) 93.

[24] G. 't Hooft, *Phys. Rev.* **D14** (1976) 3432.

[25] R.D. Peccei and H. Quinn, *Phys. Rev.* **D16** (1977) 1791; S. Weinberg, *Phys. Rev. Lett.* **40** (1978) 223; F. Wilczek, *Phys. Rev. Lett.* **40** (1978) 279.

[26] S. Deser, R. Jackiw and S. Templeton, *Ann. Phys.* **281** (2000) 409.

[27] S. Donaldson, *J. Diff. Geom.* **18** (1983) 269.

From: J.E. Maxwell, *A Treatise on Electricity and Magnetism* (1891), 3rd ed. (Dover, New York, 1954). Article 616.

615.] These may be regarded as the principal relations among the quantities we have been considering. They may be combined so as to eliminate some of these quantities, but our object at present is not to obtain compactness in the mathematical formulae, but to express every relation of which we have any knowledge. To eliminate a quantity which expresses a useful idea would be rather a loss than a gain in this stage of our enquiry.

There is one result, however, which we may obtain by combining equations (A) and (E), and which is of very great importance.

If we suppose that no magnets exist in the field except in the form of electric circuits, the distinction which we have hitherto maintained between the magnetic force and the magnetic induction vanishes, because it is only in magnetized matter that these quantities differ from each other.

According to Ampère's hypothesis, which will be explained in Art. 833, the properties of what we call magnetized matter are due to molecular electric circuits, so that it is only when we regard the substance in large masses that our theory of magnetization is applicable, and if our mathematical methods are supposed capable of taking account of what goes on within the individual molecules, they will discover nothing but electric circuits, and we shall find the magnetic force and the magnetic induction everywhere identical. In order, however, to be able to make use of the electrostatic or of the electromagnetic system of measurement at pleasure we shall retain the coefficient μ, remembering that its value is unity in the electromagnetic system.

616.] The components of the magnetic induction are by equations (A), Art. 591,

$$\left.\begin{aligned} a &= \frac{dH}{dy} - \frac{dG}{dz}, \\ b &= \frac{dF}{dz} - \frac{dH}{dx}, \\ c &= \frac{dG}{dx} - \frac{dF}{dy}. \end{aligned}\right\}$$

The components of the electric current are by equations (E), Art. 607,

$$\left.\begin{aligned} 4\pi u &= \frac{d\gamma}{dy} - \frac{d\beta}{dz}, \\ 4\pi v &= \frac{d\alpha}{dz} - \frac{d\gamma}{dx}, \\ 4\pi w &= \frac{d\beta}{dx} - \frac{d\alpha}{dy}. \end{aligned}\right\}$$

According to our hypothesis a, b, c are identical with $\mu\alpha, \mu\beta, \mu\gamma$ respectively. We therefore obtain

$$4\pi\mu u = \frac{d^2 G}{dx\,dy} - \frac{d^2 F}{dy^2} - \frac{d^2 F}{dz^2} + \frac{d^2 H}{dz\,dx}. \tag{1}$$

If we write
$$J = \frac{dF}{dx} + \frac{dG}{dy} + \frac{dH}{dz}, \tag{2}$$

and *
$$\nabla^2 = -\left(\frac{d^2}{dx^2} + \frac{d^2}{dy^2} + \frac{d^2}{dz^2}\right), \tag{3}$$

we may write equation (1),

$$4\pi\mu u = \frac{dJ}{dx} + \nabla^2 F.$$

Similarly,

$$\left.\begin{aligned} 4\pi\mu v &= \frac{dJ}{dy} + \nabla^2 G, \\ 4\pi\mu w &= \frac{dJ}{dz} + \nabla^2 H. \end{aligned}\right\} \tag{4}$$

If we write
$$\left.\begin{aligned} F' &= \frac{1}{\mu}\iiint \frac{u}{r}\,dx\,dy\,dz, \\ G' &= \frac{1}{\mu}\iiint \frac{v}{r}\,dx\,dy\,dz, \\ H' &= \frac{1}{\mu}\iiint \frac{w}{r}\,dx\,dy\,dz, \end{aligned}\right\} \tag{5}$$

$$\chi = \frac{4\pi}{\mu}\iiint \frac{J}{r}\,dx\,dy\,dz, \tag{6}$$

where r is the distance of the given point from the element xyz, and the integrations are to be extended over all space, then

$$\left.\begin{aligned} F &= F' + \frac{d\chi}{dx}, \\ G &= G' + \frac{d\chi}{dy}, \\ H &= H' + \frac{d\chi}{dz}. \end{aligned}\right\} \tag{7}$$

The quantity χ disappears from the equations (A), and it is not related to any physical phenomenon. If we suppose it to be zero everywhere, J will also be zero everywhere, and equations (5), omitting the accents, will give the true values of the components of $\mathfrak{A}$.

* The negative sign is employed here in order to make our expressions consistent with those in which Quaternions are employed.

From: H. Weyl, *Space–Time–Matter* (1922), 4th ed., English transl. H.L. Brose (Dover, 1930). Preface and para. 35.

PREFACE TO THE FIRST AMERICAN PRINTING

THIS translation is made from the fourth edition of RAUM ZEIT MATERIE which was published in 1921. Relativity theory as expounded in this book deals with the space-time aspect of classical physics. Thus, the book's contents are comparatively little affected by the stormy development of quantum physics during the last three decades. This fact, aside from the public's demand, may justify its re-issue after so long a time. Of course, had the author to re-write the book today, he would take into account certain events that have modified the situation in the intervening years. I mention three such points.

(1) The principle of general relativity had resulted above all in a new theory of the gravitational field. While it was not difficult to adapt also Maxwell's equations of the electromagnetic field to this principle, it proved insufficient to reach the goal at which classical field physics is aiming: a unified field theory deriving all forces of nature from one common structure of the world and one uniquely determined law of action. In the last two of its 36 sections, my book describes an attempt to attain this goal by a new principle which I called gauge invariance (Eichinvarianz). This attempt has failed. There holds, as we know now, a principle of gauge invariance in nature; but it does not connect the electromagnetic potentials φ_i , as I had assumed, with Einstein's gravitational potentials g_{ik} , but ties them to the four

components of the wave field ψ by which Schrödinger and Dirac taught us to represent the electron. For this and the following points, compare my book, GRUPPENTHEORIE UND QUANTENMECHANIK, Leipzig 1928, 2nd ed. 1931, the article, "Elektron und Gravitation" in *Zeitschr. f. Physik 56*, 1929, p. 330, and my Rouse Ball lecture "Geometry and Physics" in *Naturwissenschaften 19*, 1931, pp. 49–58. Of course, one could not have guessed this before the "electronic field" ψ was discovered by quantum mechanics! Since then, however, a unitary field theory, so it seems to me, should encompass at least these three fields: electromagnetic, gravitational and electronic. Ultimately the wave fields of other elementary particles will have to be included too—unless quantum physics succeeds in interpreting them all as different quantum states of one particle.

(2) Quite a number of unified field theories have sprung up in the meantime. They are all based on mathematical speculation and, as far as I can see, none has had a conspicuous success. Kaluza's five-dimensional theory, particularly in the garb of projective relativity, has been investigated and extended by several authors. The most recent attempts by Schrödinger and by Einstein combine Eddington's idea of an affine field theory with that of dropping the requirement of symmetry for the metric tensor g_{ik} and the components Γ^i_{kl} of the affine connection. Of the extensive literature I mention here only: E. Schrödinger, "The Final Affine Field Laws," in *Proc. Roy. Irish Ac.* (A) *51*, pp. 163–171, 205–216; *52*, pp. 1–9 (1947/48); A. Einstein, THE MEANING OF RELATIVITY, 3rd ed., Princeton, N. J., 1949, Appendix II; Schrödinger's book, SPACE-TIME STRUCTURE announced by Macmillan, and my lecture "50 Jahre Relativitätstheorie" at the first post-war meeting of the Gessellschaft deutscher Naturforscher und Aerzte (Munich, Oct. 1950, to be published soon). One non-speculative development which deserves mention is Einstein's mixed metric-affine formulation in which both the g_{ik} and the Γ^i_{kl} are taken as quantities capable of independent virtual variation; cf. Einstein, Sitzungsber. *Preuss. Ak. Wissensch.* 1925, p. 414; H. Weyl, *Phys. Review 77*, 1950, pp. 699–701.

(3) A new development began for relativity theory after 1925 with its absorption into quantum physics. The first great success was scored by Dirac's quantum mechanical equations of the electron, which introduced a new sort of quantities, the spinors,

besides the vectors and tensors into our physical theories. See Dirac's book, THE PRINCIPLES OF QUANTUM MECHANICS, 3rd ed., Oxford, Clarendon Press, 1947. The generally relativistic formulation of these equations offered no serious difficulties. But difficulties of the gravest kind turned up when one passed from one electron or photon to the interaction among an indeterminate number of such particles. In spite of several promising advances a final solution of this problem is not yet in sight and may well require a deep modification of the foundations of quantum mechanics, such as would account in the same basic manner for the elementary electric charge e as relativity theory and our present quantum mechanics account for c and h.

Zurich, October 1950

HERMANN WEYL

indeterminate. This edge, which appears as a two-dimensional configuration in the original co-ordinates is, therefore, three-dimensional in the new co-ordinates; it is the cylinder erected in the direction of the t-axis over the equator $z = 0$ of the sphere (65). The question arises whether it is the first or the second co-ordinate system that serves to represent the whole world in a regular manner. In the former case the world would not be static as a whole, and the absence of matter in it would be in agreement with physical laws; de Sitter argues from this assumption (*vide* note 32). In the latter case we have a static world that cannot exist without a mass-horizon; this assumption, which we have treated more fully, is favoured by Einstein.

§ 35. The Metrical Structure of the World as the Origin of Electromagnetic Phenomena *

We now aim at a final synthesis. To be able to characterise the physical state of the world at a certain point of it by means of numbers we must not only refer the neighbourhood of this point to a co-ordinate system but we must also fix on certain units of measure. We wish to achieve just as fundamental a point of view with regard to this second circumstance as is secured for the first one, namely, the arbitrariness of the co-ordinate system, by the Einstein Theory that was described in the preceding paragraph. This idea, when applied to geometry and the conception of distance (in Chapter II) after the step from Euclidean to Riemann geometry had been taken, effected the final entrance into the realm of infinitesimal geometry. Removing every vestige of ideas of "action at a distance," let us assume that world-geometry is of this kind; we then find that the metrical structure of the world, besides being dependent on the quadratic form (1), is also dependent on a linear differential form $\phi_i \, dx_i$.

Just as the step which led from the special to the general theory of relativity, so this extension affects immediately only the world-geometrical foundation of physics. Newtonian mechanics, as also the special theory of relativity, assumed that uniform translation is a unique state of motion of a set of vector axes, and hence that the position of the axes at one moment determines their position in all other moments. But this is incompatible with the intuitive principle of the **relativity of motion.** This principle could be satisfied, if facts are not to be violated drastically, only by maintaining the conception of **infinitesimal** parallel displacement of a vector set of axes; but we found ourselves obliged to regard the

* *Vide* note 33.

affine relationship, which determines this displacement, as something physically real that depends physically on the states of matter ("**guiding field**"). The properties of *gravitation* known from experience, particularly the equality of inertial and gravitational mass, teach us, finally, that gravitation is already contained in the guiding field besides inertia. And thus the general theory of relativity gained a significance which extended beyond its original important bearing on **world-geometry** to a significance which is specifically *physical*. The same certainty that characterises the relativity of motion accompanies the principle of the **relativity of magnitude.** We must not let our courage fail in maintaining this principle, according to which the size of a body at one moment does not determine its size at another, in spite of the existence of rigid bodies.* But, unless we are to come into violent conflict with fundamental facts, this principle cannot be maintained without retaining the conception of *infinitesimal* congruent transformation; that is, we shall have to assign to the world besides its *measure-determination* at every point also a *metrical relationship*. Now this is not to be regarded as revealing a "geometrical" property which belongs to the world as a form of phenomena, but as being a phase-field having physical reality. Hence, as the fact of the propagation of action and of the existence of rigid bodies leads us to found the affine relationship on the *metrical* character of the world which lies a grade lower, it immediately suggests itself to us, not only to identify the co-efficients of the quadratic groundform $g_{ik}dx_idx_k$ with the potentials of the gravitational field, but also to identify **the co-efficients of the linear groundform $\phi_i dx_i$ with the electromagnetic potentials.** The electromagnetic field and the electromagnetic forces are then derived from the metrical structure of the world or the *metrics*, as we may call it. No other truly essential actions of forces are, however, known to us besides those of gravitation and electromagnetic actions; for all the others statistical physics presents some reasonable argument which traces them back to the above two by the method of mean values. We thus arrive at the inference: **The world is a (3 + 1)-dimensional metrical manifold; all physical field-phenomena are expressions of the metrics of the world.** (Whereas the old view was that the four-dimensional metrical continuum is the scene of

* It must be recalled in this connection that the spatial direction-picture which a point-eye with a given world-line receives at every moment from a given region of the world, depends only on the ratios of the g_{ik}'s, inasmuch as this is true of the geodetic null-lines which are the determining factors in the propagation of light.

physical phenomena; the physical essentialities themselves are, however, things that exist "in" this world, and we must accept them in type and number in the form in which experience gives us cognition of them: nothing further is to be "comprehended" of them.) We shall use the phrase "state of the world-æther" as synonymous with the word "metrical structure," in order to call attention to the character of reality appertaining to metrical structure; but we must beware of letting this expression tempt us to form misleading pictures. In this terminology the fundamental theorem of infinitesimal geometry states that the guiding field, and hence also gravitation, is determined by the state of the æther. The antithesis of "physical state" and "gravitation" which was enunciated in § 28 and was expressed in very clear terms by the division of Hamilton's Function into two parts, is overcome in the new view, which is uniform and logical in itself. Descartes' dream of a purely geometrical physics seems to be attaining fulfilment in a manner of which he could certainly have had no presentiment. The quantities of intensity are sharply distinguished from those of magnitude.

The linear groundform $\phi_i dx_i$ is determined except for an additive total differential, but the tensor of distance-curvature

$$f_{ik} = \frac{\partial \phi_i}{\partial x_k} - \frac{\partial \phi_k}{\partial x_i}$$

which is derived from it, is free of arbitrariness. According to Maxwell's Theory the same result obtains for the electromagnetic potential. The electromagnetic field-tensor, which we denoted earlier by F_{ik}, is now to be identified with the distance-curvature f_{ik}. If our view of the nature of electricity is true, then the first system of Maxwell's equations

$$\frac{\partial f_{ik}}{\partial x_l} + \frac{\partial f_{kl}}{\partial x_i} + \frac{\partial f_{li}}{\partial x_k} = 0 \quad . \quad . \quad . \quad . \quad (67)$$

is an intrinsic law, the validity of which is wholly independent of whatever physical laws govern the series of values that the physical phase-quantities actually run through. In a four-dimensional metrical manifold the simplest integral invariant that exists at all is

$$\int \mathfrak{l}\, dx = \tfrac{1}{4}\int f_{ik}\mathfrak{f}^{ik} dx \quad . \quad . \quad . \quad . \quad (68)$$

and it is just this one, in the form of *Action*, on which Maxwell's Theory is founded! We have accordingly a good right to claim that the whole fund of experience which is crystallised in Maxwell's Theory weighs in favour of the world-metrical nature of electricity. And since it is impossible to construct an integral invariant at all of such a simple structure in manifolds of more or less than four dimensions the new point of view does not only lead to a deeper understanding of Maxwell's Theory but the fact that the world is four-dimensional, which has hitherto always been accepted as merely "accidental," becomes intelligible through it. In the linear groundform $\phi_i dx_i$ there is an arbitrary factor in the form of an additive total differential, but there is not a factor of proportionality; the quantity *Action* is a pure number. But this is only as it should be, if the theory is to be in agreement with that atomistic structure of the world which, according to the most recent results (Quantum Theory), carries the greatest weight.

The **statical case** occurs when the co-ordinate system and the calibration may be chosen so that the linear groundform becomes equal to ϕdx_0 and the quadratic groundform becomes equal to

$$f^2 dx_0^2 - d\sigma^2$$

whereby ϕ and f are not dependent on the time x_0, but only on the space-co-ordinates x_1, x_2, x_3, whilst $d\sigma^2$ is a definitely positive quadratic differential form in the three space-variables. This particular form of the groundform (if we disregard quite particular cases) remains unaffected by a transformation of co-ordinates and a re-calibration only if x_0 undergoes a linear transformation of its own, and if the space-co-ordinates are likewise transformed only among themselves, whilst the calibration ratio must be a constant. Hence, in the statical case, we have a three-dimensional Riemann space with the groundform $d\sigma^2$ and two scalar fields in it: the electrostatic potential ϕ, and the gravitational potential or the velocity of light f. The length-unit and the time-unit (centimetre, second) are to be chosen as arbitrary units; $d\sigma^2$ has dimensions cm^2, f has dimensions cm . sec^{-1}, and ϕ has sec^{-1}. Thus, as far as one may speak of a space at all in the general theory of relativity (namely, in the statical case), it appears as a **Riemann** space, and not as one of the more general type, in which the transference of distances is found to be non-integrable.

We have the case of the special theory of relativity again, if the co-ordinates and the calibration may be chosen so that

$$ds^2 = dx_0^2 - (dx_1^2 + dx_2^2 + dx_3^2).$$

If x_i, $\bar{x}_i$ denote two co-ordinate systems for which this normal form for ds^2 may be obtained, then the transition from x_i to $\bar{x}_i$ is a conformal transformation, that is, we find

$$dx_0^2 - (dx_1^2 + dx_2^2 + dx_3^2)$$

except for a factor of proportionality, is equal to

$$d\bar{x}_0^2 - (d\bar{x}_1^2 + d\bar{x}_2^2 + d\bar{x}_3^2).$$

The conformal transformations of the four-dimensional Minkowski world coincide with spherical transformations (*vide* note 34), that is, with those transformations which convert every "sphere" of the world again into a sphere. A sphere is represented by a linear homogeneous equation between the homogeneous "hexaspherical" co-ordinates

$$u_0 : u_1 : u_2 : u_3 : u_4 : u_5 = x_0 : x_1 : x_2 : x_3 : \frac{(xx) + 1}{2} : \frac{(xx) - 1}{2}$$

where $$(xx) = x_0^2 - (x_1^2 + x_2^2 + x_3^2).$$

They are bound by the condition

$$u_0^2 - u_1^2 - u_2^2 \quad u_3^2 - u_4^2 + u_5^2 = 0.$$

The spherical transformations therefore express themselves as those linear homogeneous transformations of the u_i's which leave this condition, as expressed in the equation, invariant. Maxwell's equations of the æther, in the form in which they hold in the special theory of relativity, are therefore invariant not only with respect to the 10-parameter group of the linear Lorentz transformations but also indeed with respect to the more comprehensive 15-parameter group of spherical transformations (*vide* note 35).

To test whether the new hypothesis about the nature of the electromagnetic field is able to account for phenomena, we must work out its implications. We choose as our initial physical law a Hamilton principle which states that the change in the *Action* $\int \mathbf{W} dx$ for every infinitely small variation of the metrical structure of the world that vanishes outside a finite region is zero. The *Action* is an invariant, and hence **W** is a scalar-density (in the true sense) which is derived from the metrical structure. Mie, Hilbert, and Einstein assumed the *Action* to be an invariant with respect to transformations of the co-ordinates. We have here to add the further limitation that it must also be invariant with respect to the process of re-calibration, in which ϕ_i, g_{ik} are replaced by

$$\phi_i - \frac{1}{\lambda}\frac{\partial \lambda}{\partial x_i} \text{ and } \lambda g_{ik}, \text{ respectively,} \qquad . \quad . \quad (69)$$

in which λ is an arbitrary positive function of position. We assume that **W** is an expression of the second order, that is, built up, on the one hand, of the g_{ik}'s and their derivatives of the first and second order, on the other hand, of the ϕ_i's and their derivatives of the first order. The simplest example is given by Maxwell's *density of action* **l**. But we shall here carry out a general investigation without binding ourselves to any particular form of **W** at the beginning. According to Klein's method, used in § 28 (and which will only now be applied with full effect), we shall here deduce certain mathematical identities, which are valid for every scalar-density $\mathbf{W}$ which has its origin in the metrical structure.

I. If we assign to the quantities ϕ_i, g_{ik}, which describe the metrical structure relative to a system of reference, infinitely small increments $\delta\phi_i$, δg_{ik}, and if $\mathfrak{X}$ denote a finite region of the world, then the effect of partial integration is to separate the integral of the corresponding change $\delta\mathbf{W}$ in $\mathbf{W}$ over the region $\mathfrak{X}$ into two parts: (*a*) a divergence integral and (*b*) an integral whose integrand is only a linear combination of $\delta\phi_i$ and δg_{ik}, thus

$$\int_{\mathfrak{X}} \delta\mathbf{W}dx = \int_{\mathfrak{X}} \frac{\partial(\delta\mathbf{v}^k)}{\partial x_k}dx + \int_{\mathfrak{X}} (\mathbf{w}^i\delta\phi_i + \tfrac{1}{2}\mathbf{W}^{ik}\delta g_{ik})dx \quad . \quad (70)$$

whereby $\mathbf{W}^{ki} = \mathbf{W}^{ik}$.

The $\mathbf{w}^i$'s are components of a contra-variant vector-density, but the $\mathbf{W}^k_i$'s are the components of a mixed tensor-density of the second order (in the true sense). The $\delta\mathbf{v}^k$'s are linear combinations of

$$\delta\phi_a, \quad \delta g_{a\beta} \text{ and } \delta g_{a\beta,\,i} \qquad \left[g_{a\beta,\,i} = \frac{\partial g_{a\beta}}{\partial x_i}\right].$$

We indicate this by the formula

$$\delta\mathbf{v}^k = (ka)\delta\phi_a + (ka\beta)\delta g_{a\beta} + (kia\beta)\delta g_{a\beta,\,i}.$$

The $\delta\mathbf{v}^k$'s are defined uniquely by equation (70) only if the normalising condition that the co-efficients $(kia\beta)$ be symmetrical in the indices k and i is added. In the normalisation the $\delta\mathbf{v}^k$'s are components of a vector-density (in the true sense), if the $\delta\phi_i$'s are regarded as the components of a co-variant vector of weight zero and the δg_{ik}'s as the components of a tensor of weight unity. (There is, of course, no objection to applying another normalisation in place of this one, provided that it is invariant in the same sense.)

First of all, we express that $\int_{\mathfrak{X}} \mathbf{W}dx$ is a calibration invariant, that is, that it does not alter when the calibration of the world is altered infinitesimally. If the calibration ratio between the altered and the original calibration is $\lambda = 1 + \pi$, π is an infinitesimal scalar-field which characterises the event and which may be assigned arbitrarily. As a result of this process, the fundamental quantities assume, according to (69), the following increments:

$$\delta g_{ik} = \pi g_{ik}, \quad \delta\phi_i = -\frac{\partial\pi}{\partial x_i} \quad . \quad . \quad . \quad (71)$$

If we substitute these values in $\delta\mathbf{v}^k$, let the following expressions result:

$$\mathbf{s}^k(\pi) = \pi \,.\, \mathbf{s}^k + \frac{\partial \pi}{\partial x_a} \,.\, \mathbf{h}^{ka} \quad . \quad . \quad . \quad . \quad (72)$$

They are the components of a vector-density which depends on the scalar-field π in a linear-differential manner. It further follows from this, that, since the $\frac{\partial \pi}{\partial x_a}$'s are the components of a co-variant vector-field which is derived from the scalar-field, $\mathbf{s}^k$ is a vector-density, and $\mathbf{h}^{ka}$ is a contra-variant tensor-density of the second order. The variation (70) of the integral of Action must vanish on account of its calibration invariance; that is, we have

$$\int_{\mathfrak{X}} \frac{\partial \mathbf{s}^k(\pi)}{\partial x_k} dx + \int_{\mathfrak{X}} \left(- \mathbf{w}^i \frac{\partial \pi}{\partial x_i} + \tfrac{1}{2} \mathbf{W}^i_i \pi \right) dx = 0.$$

If we transform the first term of the second integral by means of partial integration, we may write, instead of the preceding equation,

$$\int_{\mathfrak{X}} \frac{\partial (\mathbf{s}^k(\pi) - \pi \mathbf{w}^k)}{\partial x_k} dx + \int_{\mathfrak{X}} \pi \left(\frac{\partial \mathbf{w}^i}{\partial x_i} + \tfrac{1}{2} \mathbf{W}^i_i \right) dx = 0 \quad . \quad (73)$$

This immediately gives the identity

$$\frac{\partial \mathbf{w}^i}{\partial x_i} + \tfrac{1}{2} \mathbf{W}^i_i = 0 \quad . \quad . \quad . \quad . \quad . \quad (74)$$

in the manner familiar in the calculus of variations. If the function of position on the left were different from 0 at a point x_i, say positive, then it would be possible to mark off a neighbourhood $\mathfrak{X}$ of this point so small that this function would be positive at every point within $\mathfrak{X}$. If we choose this region for $\mathfrak{X}$ in (73), but choose for π a function which vanishes for points outside $\mathfrak{X}$ but is > 0 throughout $\mathfrak{X}$, then the first integral vanishes, but the second is found to be positive—which contradicts equation (73). Now that this has been ascertained, we see that (73) gives

$$\int_{\mathfrak{X}} \frac{\partial (\mathbf{s}^k(\pi) - \pi \mathbf{w}^k)}{\partial x_k} dx = 0.$$

For a given scalar-field π it holds for every finite region $\mathfrak{X}$, and consequently we must have

$$\frac{\partial (\mathbf{s}^k(\pi) - \pi \mathbf{w}^k)}{\partial x_k} = 0 \quad . \quad . \quad . \quad . \quad (75)$$

If we substitute (72) in this, and observe that, for a particular point, arbitrary values may be assigned to π, $\frac{\partial \pi}{\partial x}$, $\frac{\partial^2 \pi}{\partial x_i \partial x_k}$, then this single formula resolves into the identities:

$$\frac{\partial \mathbf{s}^k}{\partial x_k} = \frac{\partial \mathbf{w}^k}{\partial x_k}; \quad \mathbf{s}^i + \frac{\partial \mathbf{h}^{\alpha i}}{\partial x_\alpha} = \mathbf{w}^i; \quad \mathbf{h}^{\alpha\beta} + \mathbf{h}^{\beta\alpha} = 0 \; . \quad (75_{1,2,3})$$

According to the third identity, $\mathbf{h}^{ik}$ is a linear tensor-density of the second order. In view of the skew-symmetry of $\mathbf{h}$ the first is a result of the second, since

$$\frac{\partial^2 \mathbf{h}^{\alpha\beta}}{\partial x_\alpha \partial x_\beta} = 0.$$

II. We subject the world-continuum to an infinitesimal deformation, in which each point undergoes a displacement whose components are ξ^i; let the metrical structure accompany the deformation without being changed. Let δ signify the change occasioned by the deformation in a quantity, if we remain at the same space-time point, δ' the change in the same quantity if we share in the displacement of the space-time point. Then, by (20), (21′), (71)

$$\left.\begin{aligned} -\delta\phi_i &= \left(\phi_r \frac{\partial \xi^r}{\partial x_i} + \frac{\partial \phi_i}{\partial x_r}\xi^r\right) + \frac{\partial \pi}{\partial x_i} \\ -\delta g_{ik} &= \left(g_{ir}\frac{\partial \xi_r}{\partial x_k} + g_{kr}\frac{\partial \xi^r}{\partial x_i} + \frac{\partial g_{ik}}{\partial x_r}\xi^r\right) - \pi g_{ik} \end{aligned}\right\} \quad . \quad (76)$$

in which π denotes an infinitesimal scalar-field that has still been left arbitrary by our conventions. The invariance of the *Action* with respect to transformation of co-ordinates and change of calibration is expressed in the formula which relates to this variation:

$$\delta' \int_{*} \mathbf{W} dx = \int_{*} \left\{ \frac{\partial(\mathbf{W}\xi^k)}{\partial x_k} + \delta \mathbf{W} \right\} dx = 0 \quad . \quad . \quad (77)$$

If we wish to express the invariance with respect to the co-ordinates alone we must make $\pi = 0$; but the resulting formulæ of variation (76) have not then an invariant character. This convention, in fact, signifies that the deformation is to make the two groundforms vary in such a way that the measure l of a line-element remains unchanged, that is, $\delta' l = 0$. This equation does not, however, express the process of congruent transference of a distance, but indicates that

$$\delta' l = -l(\phi_i \delta' x_i) = -l(\phi_i \xi^i).$$

Accordingly, in (76) we must choose π not equal to zero but equal to $-(\phi_i \xi^i)$ if we are to arrive at invariant formulæ, namely,

$$\left.\begin{aligned} -\delta\phi_i &= f_{ir}\xi^r \\ -\delta g_{ik} &= \left(g_{ir}\frac{\partial\xi^r}{\partial x_k} + g_{kr}\frac{\partial\xi^r}{\partial x_i}\right) + \left(\frac{\partial g_{ik}}{\partial x_r} + g_{ik}\phi_r\right)\xi^r \end{aligned}\right\} \quad . \quad (78)$$

The change in the two groundforms which it represents is one that makes *the metrical structure appear carried along unchanged by the deformation and every line-element to be transferred congruently.* The invariant character is easily recognised analytically, too; particularly in the case of the second equation (78), if we introduce the mixed tensor

$$\frac{\partial\xi^i}{\partial x_k} + \Gamma^i_{kr}\xi^r = \xi^i_k.$$

The equation then becomes

$$-\delta g_{ik} = \xi_{ik} + \xi_{ki}.$$

Now that the calibration invariance has been applied in I, we may in the case of (76) restrict ourselves to the choice of π, which was discussed just above, and which we found to be alone possible from the point of view of invariance.

For the variation (78) let

$$\mathbf{W}\xi^k + \delta\mathbf{v}^k = \mathbf{S}^k(\xi).$$

$\mathbf{S}^k(\xi)$ is a vector-density which depends in a linear differential manner on the arbitrary vector-field ξ^i. We write in an explicit form

$$\mathbf{S}^k(\xi) = \mathbf{S}^k_i\xi^i + \overline{\mathbf{H}}^{k\alpha}_i\frac{\partial\xi^i}{\partial x_\alpha} + \tfrac{1}{2}\mathbf{H}^{k\alpha\beta}_i\frac{\partial^2\xi^i}{\partial x_\alpha\partial x_\beta}$$

(the last co-efficient is, of course, symmetrical in the indices α, β). The fact that $\mathbf{S}^k(\xi)$ is a vector-density dependent on the vector-field ξ^i expresses most simply and most fully the character of invariance possessed by the co-efficients which occur in the expression for $\mathbf{S}^k(\xi)$; in particular, it follows from this that the $\mathbf{S}^k_i$'s are not components of a mixed tensor-density of the second order: we call them the components of a "pseudo-tensor-density". If we insert in (77) the expressions (70) and (78), we get an integral, whose integrand is

$$\frac{\partial\mathbf{S}^k(\xi)}{\partial x_k} - \xi^i\left\{f_{ki}\mathbf{w}^k + \tfrac{1}{2}\left(\frac{\partial g_{\alpha\beta}}{\partial x_i} + g_{\alpha\beta}\phi_i\right)\mathbf{W}^{\alpha\beta}\right\}\mathbf{W}^k_i\frac{\partial\xi^i}{\partial x_k}.$$

On account of

$$\frac{\partial g_{\alpha\beta}}{\partial x_i} + g_{\alpha\beta}\phi_i = \Gamma_{\alpha,\beta i} + \Gamma_{\beta,\alpha i}$$

and of the symmetry of $\mathbf{W}^{\alpha\beta}$ we find

$$\tfrac{1}{2}\left(\frac{\partial g^{\alpha\beta}}{\partial x_i} + g_{\alpha\beta}\phi_i\right)\mathbf{W}^{\alpha\beta} = \Gamma_{\alpha,\beta i}\mathbf{W}^{\alpha\beta} = \Gamma^\alpha_{\beta i}\mathbf{W}^\beta_\alpha.$$

If we apply partial integration to the last member of the integrand, we get

$$\int_{*}\frac{\partial(\mathbf{S}^k(\xi) - \mathbf{W}^k_i\xi^i)}{\partial x_k}\,dx + \int_{*}[\;.\;.\;.\;]_i\xi^i dx = 0.$$

According to the method of inference used above we get from this the identities:

$$[\;.\;.\;.\;]_i, \text{ that is, } \left(\frac{\partial \mathbf{W}^k_i}{\partial x_k} - \Gamma^\alpha_\beta \mathbf{W}^\beta_\alpha\right) + f_{ik}\mathfrak{w}^k = 0 \quad . \quad (79)$$

and

$$\frac{\partial(\mathbf{S}^k(\xi) - \mathbf{W}^k_i\xi^i)}{\partial x_k} = 0 \; . \quad . \quad . \quad . \quad (80)$$

The latter resolves into the following four identities:

$$\left.\begin{array}{ll} \dfrac{\partial \mathbf{S}^k_i}{\partial x^k} = \dfrac{\partial \mathbf{W}^k_i}{\partial x_k}; & \mathbf{S}^k_i + \dfrac{\partial \overline{\mathbf{H}}_i^{\alpha k}}{\partial x_\alpha} = \mathbf{W}^k_i \\ (\overline{\mathbf{H}}^{\alpha\beta}_i + \overline{\mathbf{H}}^{\beta\alpha}_i) + \dfrac{\partial \mathbf{H}^{\gamma\alpha\beta}_i}{\partial x_\gamma} = 0; & \mathbf{H}^{\alpha\beta\gamma}_i + \mathbf{H}^{\beta\gamma\alpha}_i + \mathbf{H}^{\gamma\alpha\beta}_i = 0 \end{array}\right\} (80_{1,2,3,4})$$

If from $(_4)$ we replace in $(_3)$

$$\mathbf{H}^{\gamma\alpha\beta}_i \text{ by } - \mathbf{H}^{\alpha\beta\gamma}_i - \mathbf{H}^{\beta\alpha\gamma}_i$$

we get that

$$\overline{\mathbf{H}}^{\alpha\beta}_i - \frac{\partial \mathbf{H}^{\alpha\beta\gamma}_i}{\partial x_\gamma} = \mathbf{H}^{\alpha\beta}_i$$

is skew-symmetrical in the indices α, β. If we introduce $\mathbf{H}^{\alpha\beta}_i$ in place of $\overline{\mathbf{H}}^{\alpha\beta}_i$ we see that $(_3)$ and $(_4)$ are merely statements regarding symmetry, but $(_2)$ becomes

$$\mathbf{S}^k_i + \frac{\partial \mathbf{H}^{\alpha k}_i}{\partial x_\alpha} + \frac{\partial^2 \mathbf{H}^{\alpha\beta k}_i}{\partial x_\alpha \partial x_\beta} = \mathbf{W}^k_i \quad . \quad . \quad . \quad (81)$$

$(_1)$ follows from this because, on account of the conditions of symmetry

$$\frac{\partial^2 \mathbf{H}^{\alpha\beta}_i}{\partial x_\alpha \partial x_\beta} = 0, \text{ we get } \frac{\partial^3 \mathbf{H}^{\alpha\beta\gamma}_i}{\partial x_\alpha \partial x_\beta \partial x_\gamma} = 0$$

Example.—In the case of Maxwell's Action-density we have, as is immediately obvious

$$\delta \mathfrak{v}^k = \mathfrak{f}^{ik}\delta\phi_i.$$

Consequently

$$\mathfrak{s}^i = 0, \mathfrak{h}^{ik} = \mathfrak{f}^{ik}; \; \mathbf{S}^k_i = \mathfrak{l}\delta^k_i - f_{i\alpha}\mathfrak{f}^{k\alpha}, \text{ and the quantities } \mathbf{H} = 0.$$

Hence our identities lead to

$$\mathbf{w}^i = \frac{\partial \mathfrak{f}^{\alpha i}}{\partial x_\alpha} \quad \frac{\partial \mathbf{w}^i}{\partial x_i} = 0, \qquad \mathbf{W}^i_i = 0$$

$$\mathbf{W}^k_i = \mathbf{S}^k_i \quad \left(\frac{\partial \mathbf{S}^k_i}{\partial x_k} - \frac{1}{2}\frac{\partial g_{\alpha\beta}}{\partial x_i}\mathbf{S}^{\alpha\beta}\right) + f_{i\alpha}\frac{\partial \mathfrak{f}^{\beta\alpha}}{\partial x_\beta} = 0.$$

We arrived at the last two formulæ by calculation earlier, the former on page 230, the latter on page 167; the latter was found to express the desired connection between Maxwell's tensor-density $\mathbf{S}^k_i$ of the field-energy and the ponderomotive force.

Field Laws and Theorems of Conservation.—If, in (70), we take for δ an arbitrary variation which vanishes outside a finite region, and for $\mathfrak{X}$ we take the whole world or a region such that, outside it, $\delta = 0$, we get

$$\int \delta \mathbf{W} dx = \int (\mathbf{w}^i \delta\phi_i + \tfrac{1}{2}\mathbf{W}^{ik}\delta g_{ik})dx.$$

If $\int \mathbf{W} dx$ is the *Action*, we see from this that the following invariant laws are contained in Hamilton's Principle:

$$\mathbf{w}^i = 0 \qquad \mathbf{W}^k_i = 0.$$

Of these, we have to call the former the electromagnetic laws, the latter the gravitational laws. Between the left-hand sides of these equations there are five identities, which have been stated in (74) and (79). Thus there are among the field-equations five superfluous ones corresponding to the transition (dependent on five arbitrary functions) from one system of reference to another.

According to (75_2) the electromagnetic laws have the following form:

$$\frac{\partial \mathbf{h}^{ik}}{\partial x_k} = \mathbf{s}^i \qquad \text{[and (67)]} \qquad . \quad . \quad . \quad (82)$$

in full agreement with Maxwell's Theory; $\mathbf{s}^i$ is the density of the 4-current, and the linear tensor-density of the second order $\mathbf{h}^{ik}$ is the electromagnetic density of field. Without specialising the *Action* at all we can read off the whole structure of Maxwell's Theory from the calibration invariance alone. The particular form of Hamilton's function $\mathbf{W}$ affects only the formulæ which state that current and field-density are determined by the phase-quantities ϕ_i, g_{ik} of the æther. In the case of Maxwell's Theory in the restricted sense ($\mathbf{W} = \mathbf{l}$), which is valid only in empty space, we get $\mathbf{h}^{ik} = \mathfrak{f}^{ik}$, $\mathbf{s}^i = 0$, which is as it should be.

Just as the $\mathbf{s}^i$'s constitute the density of the 4-current, so the scheme of $\mathbf{S}^k_i$'s is to be interpreted as the pseudo-tensor-density of the energy. In the simplest case, $\mathbf{W} = 1$, this explanation becomes identical with that of Maxwell. According to (75_1) and (80_1) **the theorems of conservation**

$$\frac{\partial \mathfrak{s}^i}{\partial x_i} = 0, \qquad \frac{\partial \mathbf{S}_i^k}{\partial x_k} = 0$$

are generally valid; and, indeed, they follow in two ways from the field laws. For $\frac{\partial \mathfrak{s}^i}{\partial x_i}$ is not only identically equal to $\frac{\partial \mathfrak{w}}{\partial x_i}$, but also to $-\frac{1}{2}\mathbf{W}_i^i$, and $\frac{\partial \mathbf{S}_i^k}{\partial x_k}$ is not only identically equal to $\frac{\partial \mathbf{W}_i^k}{\partial x_k}$, but also to $\Gamma_{i\beta}^{\alpha}\mathbf{W}_{\alpha}^{\beta} - f_{ik}\mathfrak{w}^k$. The form of the gravitational equations is given by (81). The field laws and their accompanying laws of conservation may, by (75) and (80), be summarised conveniently in the two equations

$$\frac{\partial \mathfrak{s}^i(\pi)}{\partial x_i} = 0, \qquad \frac{\partial \mathbf{S}^i(\xi)}{\partial x_i} = 0.$$

Attention has already been directed above to the intimate connection between the laws of conservation of the energy-momentum and the co-ordinate-invariance. To these four laws there is to be added the law of conservation of electricity, and, corresponding to it, there must, logically, be a property of invariance which will introduce a fifth arbitrary function; the calibration-invariance here appears as such. Earlier we derived the law of conservation of energy-momentum from the co-ordinate-invariance only owing to the fact that Hamilton's function consists of two parts, the *action*-function of the gravitational field and that of the "physical phase"; each part had to be treated differently, and the component results had to be combined appropriately (§ 33). If those quantities, which are derived from $\mathbf{W}\xi^k + \delta\mathfrak{v}^k$ by taking the variation of the fundamental quantities from (76) for the case $\pi = 0$, instead of from (78), are distinguished by a prefixed asterisk, then, in consequence of the co-ordinate-invariance, the "theorems of conservation" $\frac{\partial {}^*\mathbf{S}_i^k}{\partial x_k} = 0$ are generally valid. But the ${}^*\mathbf{S}_i^k$'s are not the energy-momentum components of the two-fold action-function which have been used as a basis since § 28. For the gravitational component ($\mathbf{W} = \mathbf{G}$) we defined the energy by means of ${}^*\mathbf{S}_i^k$ (§ 33), but for the electromagnetic component ($\mathbf{W} = \mathbf{L}$, § 28) we introduced $\mathbf{W}_i^k$ as the energy components. This second component $\mathbf{L}$ contains only the g_{ik}'s themselves, not their derivatives; for a quantity of this kind we have, by (80_2), $\mathbf{W}_i^k = \mathbf{S}_i^k$. Hence (**if we use the transformations**

which the fundamental quantities undergo during an infinitesimal alteration of the calibration), we can adapt the two different definitions of energy to one another although we cannot reconcile them entirely. These discrepancies are removed only here since it is the new theory which first furnishes us with an explanation of the current $\mathbf{s}^i$, of the electromagnetic density of field $\mathbf{h}^{ik}$, and of the **energy** $\mathbf{S}_i^k$, which is no longer bound by the assumption that the *Action* is composed of two parts, of which the one does not contain the ϕ_i's and their derivatives, and the other does not contain the derivatives of the g^{ik}'s. The virtual deformation of the world-continuum which leads to the definition of $\mathbf{S}_i^k$ must, accordingly, carry along the metrical structure and the line-elements "unchanged" in **our** sense and not in that of **Einstein.** The laws of conservation of the $\mathbf{s}^i$'s and the $\mathbf{S}_i$'s are then likewise not bound by an assumption concerning the composition of the *Action*. Thus, after the total energy had been introduced in § 33, we have once again passed beyond the stand taken in § 28 to a point of view which gives a more compact survey of the whole. What is done by Einstein's theory of gravitation with respect to the equality of inertial and gravitational matter, namely, that it recognises their identity as necessary but not as a consequence of an undiscovered law of physical nature, is accomplished by the present theory with respect to the facts that find expression in the structure of Maxwell's equations and the laws of conservation. Just as is the case in § 33 in which we integrate over the cross-section of a canal of the system, so we find here that, as a result of the laws of conservation, if the $\mathbf{s}^i$'s and $\mathbf{S}_i^k$'s vanish outside the canal, the system has a constant charge e and a constant energy-momentum J. Both may be represented, by Maxwell's equations (82) and the gravitational equations (81), as the flux of a certain spatial field through a surface Ω that encloses the system. If we regard this representation as a definition, the integral theorems of conservation hold, even if the field has a real singularity within the canal of the system. To prove this, let us replace this field within the canal in any arbitrary way (preserving, of course, a continuous connection with the region outside it) by a regular field, and let us define the $\mathbf{s}^i$'s and the $\mathbf{S}_i^k$'s by the equations (82), (81) (in which the right-hand sides are to be replaced by zero) in terms of the quantities $\mathbf{h}$ and $\mathbf{H}$ belonging to the altered field. The integrals of these fictitious quantities $\mathbf{s}^0$ and $\mathbf{S}_i^0$, which are to be taken over the cross-section of the canal (the interior of Ω), are constant; on the other hand, they coincide with the fluxes mentioned above over the surface Ω, since on Ω the imagined field coincides with the real one.

§ 36. Application of the Simplest Principle of Action. The Fundamental Equations of Mechanics

We have now to show that if we uphold our new theory it is possible to make an assumption about **W** which, as far as the results that have been confirmed in experience are concerned, agrees with Einstein's Theory. The simplest assumption* for purposes of calculation (I do not insist that it is realised in nature) is:

$$\mathbf{W} = -\tfrac{1}{4}F^2\sqrt{g} + a\mathbf{l} \quad . \quad . \quad . \quad (83)$$

The quantity *Action* is thus to be composed of the volume, measured in terms of the radius of curvature of the world as unit of length (cf. (62), § 17) and of Maxwell's action of the electromagnetic field; the positive constant a is a pure number. It follows that

$$\delta\mathbf{W} = -\tfrac{1}{2}F\delta(F\sqrt{g}) + \tfrac{1}{4}F^2\delta\sqrt{g} + a\delta\mathbf{l}.$$

We assume that $-F$ is positive; the calibration may then be uniquely determined by the postulate $F = -1$; thus

$$\delta\mathbf{W} = \text{the variation of } \tfrac{1}{2}F\sqrt{g} + \tfrac{1}{4}\sqrt{g} + a\mathbf{l}.$$

If we use the formula (61), § 17 for F, and omit the divergence

$$\delta\frac{(\partial\sqrt{g}\phi^i)}{\partial x_i}$$

which vanishes when we integrate over the world, and if, by means of partial integration, we convert the world-integral of $\delta(\tfrac{1}{2}R\sqrt{g})$ into the integral of $\delta\mathbf{G}$ (§ 28), then our principle of action takes the form

$$\delta\int\mathbf{V}dx = 0, \text{ and we get } \mathbf{V} = \mathbf{G} + a\mathbf{l} + \tfrac{1}{4}\sqrt{g}\{1 - 3\,(\phi_i\phi^i)\} \quad (84)$$

This normalisation denotes that we are measuring with cosmic measuring rods. If, in addition, we choose the co-ordinates x_i so that points of the world whose co-ordinates differ by amounts of the order of magnitude 1, are separated by cosmic distances, then we may assume that the g_{ik}'s and the ϕ_i's are of the order of magnitude 1. (It is, of course, a fact that the potentials vary perceptibly by amounts that are extraordinarily small in comparison with cosmic distances.) By means of the substitution $x_i = \epsilon x'_i$ we introduce co-ordinates of the order of magnitude in general use (that is having dimensions comparable with those of the human body); ϵ is a very small constant. The g_{ik}'s do not change during this transformation,

* *Vide* note 36.

if we simultaneously perform the re-calibration which multiplies ds^2 by $\frac{1}{\epsilon^2}$. In the new system of reference we then have

$$g'_{ik} = g_{ik}, \qquad \phi'_i = \epsilon\phi_i; \qquad F' = -\epsilon^2.$$

$\frac{1}{\epsilon}$ is accordingly, in our ordinary measures, the radius of curvature of the world. If g_{ik}, ϕ_i retain their old significance, but if we take x_i to represent the co-ordinates previously denoted by x'_i, and if Γ^r_{ik} are the components of the affine relationship corresponding to these co-ordinates, then

$$\mathbf{V} = (\mathbf{G} + a\mathbf{l}) + \frac{\epsilon^2}{4}\sqrt{g}\{1 - 3(\phi_i\phi^i)\},$$

$$\Gamma^r_{ik} = \begin{Bmatrix} ik \\ r \end{Bmatrix} + \tfrac{1}{2}\epsilon^2(\delta^r_i\phi_k + \delta^r_k\phi_i - g_{ik}\phi^r).$$

Thus, by neglecting the exceedingly small cosmological terms, we arrive exactly at the classical Maxwell-Einstein theory of electricity and gravitation. To make the expression correspond exactly with that of § 34 we must set $\frac{\epsilon^2}{2} = \lambda$. Hence our theory necessarily gives us Einstein's cosmological term $\frac{1}{2}\lambda\sqrt{g}$. The uniform distribution of electrically neutral matter at rest over the whole of (spherical) space is thus a state of equilibrium which is compatible with our law. But, whereas in Einstein's Theory (cf. § 34) there must be a pre-established harmony between the universal physical constant λ that occurs in it, and the total mass of the earth (because each of these quantities in themselves already determine the curvature of the world), here (where λ **denotes** merely the curvature), we have that the mass present in the world **determines** the curvature. It seems to the author that just this is what makes Einstein's cosmology physically possible. In the case in which a physical field is present, Einstein's cosmological term must be supplemented by the further term $-\frac{3}{2}\lambda\sqrt{g}(\phi_i\phi^i)$; and in the components Γ^r_{ik} of the gravitational field, too, a cosmological term that is dependent on the electromagnetic potentials occurs. Our theory is founded on a definite unit of electricity; let it be e in ordinary electrostatic units. Since, in (84), if we use these units, $\frac{2\kappa}{c^2}$ occurs in place of a, we have

$$\frac{2e^2\kappa}{c^2} = \frac{a}{-F}, \quad \frac{e\sqrt{\kappa}}{c} = \frac{1}{\epsilon}\sqrt{\frac{a}{2}}:$$

our unit is that quantity of electricity whose gravitational radius is $\sqrt{\frac{a}{2}}$ times the radius of curvature of the world. It is, therefore, like the quantum of action 1, of cosmic dimensions. The cosmological factor which Einstein added to his theory later is part of ours from the very beginning.

Variation of the ϕ_i's gives us Maxwell's equations.

$$\frac{\partial \mathbf{f}^{ik}}{\partial x_k} = \mathbf{s}^i$$

and, in this case, we have simply

$$\mathbf{s}^i = -\frac{3\lambda}{a}\phi_i \sqrt{g}.$$

Just as according to Maxwell the æther is the seat of energy and mass so we obtain here an electric charge (plus current) diffused thinly throughout the world. Variatio of the g_{ik}'s gives the gravitational equations

$$\mathbf{R}_i^k - \frac{\mathbf{R} + \lambda\sqrt{g}}{2}\delta_i^k = a\mathbf{T}_i^k \quad . \quad . \quad . \quad . \quad . \quad . \quad (85)$$

where

$$\mathbf{T}_i^k = \{1 + \tfrac{1}{2}(\phi_r \mathbf{s}^r)\}\,\delta_i^k - f_{ir}\mathbf{f}^{kr} - \phi_i \mathbf{s}^k.$$

The conservation of electricity is expressed in the divergence equation

$$\frac{\partial(\sqrt{g}\phi^i)}{\partial x_i} = 0 \quad . \quad . \quad . \quad . \quad . \quad . \quad (86)$$

This follows, on the one hand, from Maxwell's equations, but must, on the other hand, be derivable from the gravitational equations according to our general results. We actually find, by contracting the latter equations with respect to ik, that

$$R + 2\lambda = \tfrac{3}{2}(\phi_i \phi^i)$$

and this in conjunction with $-F = 2\lambda$ again gives (86). We get for the pseudo-tensor-density of the energy-momentum, as is to be expected

$$\mathbf{S}_i^k = a\mathbf{T}_i^k + \left\{\mathbf{G} + \tfrac{1}{2}\lambda\sqrt{g}\delta_i^k - \tfrac{1}{2}\frac{\partial g_{\alpha\beta}}{\partial x_i}\mathbf{G}^{\alpha\beta,k}\right\}.$$

From the equation $\delta'\int \mathbf{V}dx = 0$ for a variation δ' which is produced by the displacement in the true sense [from formula (76) with $\xi^i =$ const., $\pi = 0$], we get

$$\frac{\partial(*\mathbf{S}_i^k \xi^i)}{\partial x_k} = 0 \quad . \quad . \quad . \quad . \quad . \quad . \quad (87)$$

where

$$^*\mathbf{S}_i^k = \mathbf{V}\delta_i^k - \tfrac{1}{2}\frac{\partial g_{\alpha\beta}}{\partial x_i}\mathbf{G}^{\alpha\beta,k} + \alpha\frac{\partial\phi}{\partial x_i}\mathbf{f}^{kr}.$$

To obtain the conservation theorems, we must, according to our earlier remarks, write Maxwell's equations in the form

$$\frac{\partial\left(\pi\mathbf{s}^i + \dfrac{\partial\pi}{\partial x_k}\mathbf{f}^{ik}\right)}{\partial x_i} = 0$$

then set $\pi = -(\phi_i\xi^i)$, and, after multiplying the resulting equation by α, add it to (87). We then get, in fact,

$$\frac{\partial(\mathbf{S}_i^k\xi^i)}{\partial x_k} = 0.$$

The following terms occur in $\mathbf{S}_i^k$: the Maxwell energy-density of the electromagnetic field

$$\mathbf{l}\delta_i^k - f_{ir}\mathbf{f}^{kr},$$

the gravitational energy

$$\mathbf{G}\delta_i^k - \tfrac{1}{2}\frac{\partial g_{\alpha\beta}}{\partial x_i}\mathbf{G}^{\alpha\beta,k}$$

and the supplementary cosmological terms

$$\tfrac{1}{2}(\lambda\sqrt{g} + \phi_r\mathbf{s}^r)\delta_i^k - \phi_i\mathbf{s}^k.$$

The statical world is by its own nature calibrated. The question arises whether $F = \text{const.}$ for this calibration. The answer is in the affirmative. For if we re-calibrate the statical world in accordance with the postulate $F = -1$ and distinguish the resulting quantities by a horizontal bar, we get

$$\bar{\phi}_i = -\frac{F_i}{F}, \quad \text{where we set } F_i = \frac{\partial F}{\partial x_i}\,(i = 1, 2, 3)$$

$$\bar{g}_{ik} = -Fg_{ik}, \text{ that is, } \bar{g}^{ik} = -\frac{g^{ik}}{F}, \quad \sqrt{\bar{g}} = F^2\sqrt{g}$$

and equation (86) gives

$$\sum_{i=1}^{3}\frac{\partial\mathbf{F}^i}{\partial x_i} = 0 \qquad (\mathbf{F}^i = \sqrt{g}F^i)$$

From this, however, it follows that $F = \text{const.}$

From the fact that a further electrical term becomes added to Einstein's cosmological term, the existence of a material particle becomes possible without a mass horizon becoming necessary. The particle is necessarily charged electrically. If, in order to determine the radially symmetrical solutions for the statical case, we again use the old terms of § 31, and take ϕ to mean the electrostatic potential, then the integral whose variation must vanish, is

$$\int \mathbf{V} r^2 dr = \int \left\{ w\Delta' - \frac{ar^2\phi'^2}{2\Delta} + \frac{\lambda r^2}{2}\left(\Delta - \frac{3h^2\phi^2}{2\Delta}\right)\right\} dr$$

(the accent denotes differentiation with respect to r). Variation of w, Δ, and ϕ, respectively, leads to the equations

$$\Delta\Delta' = \frac{3\lambda}{4} h^4\phi^2 r$$

$$w' = \frac{\lambda r^2}{2}\left(1 + \frac{3}{2}\frac{h^2\phi^2}{\Delta^2}\right) + \frac{a}{2}\frac{r^2\phi'^2}{\Delta^2}$$

$$\left(\frac{r^2\phi'}{\Delta}\right)' = \frac{3}{2a}\frac{h^2r^2\phi}{\Delta}.$$

As a result of the normalisations that have been performed, the spatial co-ordinate system is fixed except for a Euclidean rotation, and hence h^2 is uniquely determined. In f and ϕ, as a result of the free choice of the unit of time, a common constant factor remains arbitrary (a circumstance that may be used to reduce the order of the problem by 1). If the equator of the space is reached when $r = r_0$, then the quantities that occur as functions of $z = \sqrt{r_0^2 - r^2}$ must exhibit the following behaviour for $z = 0$: f and ϕ are regular, and $f \neq 0$; h^2 is infinite to the second order, Δ to the first order. The differential equations themselves show that the development of h^2z^2 according to powers of z begins with the term h_0^2, where

$$h_0^2 = \frac{2r_0^2}{\lambda r_0^2 - 2}$$

—this proves, incidentally, that λ must be positive (the curvature F negative) and that $r_0^2 > \frac{2}{\lambda}$—whereas for the initial values of f_0, ϕ, of f and ϕ we have

$$f_0^2 = \frac{3\lambda}{4} h_0^2\phi_0^2.$$

If diametral points are to be identified, ϕ must be an even function of z, and the solution is uniquely determined by the initial values for $z = 0$, which satisfy the given conditions (*vide* note 37). It cannot remain regular in the whole region $0 \leq r \leq r_0$, but must, if we let r decrease from r_0, have a singularity at least ultimately when $r = 0$. For otherwise it would follow, by multiplying the differential equation of ϕ by ϕ, and integrating from 0 to r_0, that

$$\int_0^{r_0} \frac{r^2}{\Delta}\left(\phi'^2 + \frac{3}{2a} h^2\phi^2\right) dr = 0.$$

Matter is accordingly a true singularity of the field. The fact that the phase-quantities vary appreciably in regions whose linear dimensions are very small in comparison with $\frac{1}{\sqrt{l}}$ may be explained, perhaps, by the circumstance that a value must be taken for r_0^2 which is enormously great in comparison with $\frac{1}{\lambda}$. The fact that all elementary particles of matter have the same charge and the same mass seems to be due to the circumstance that they are all embedded in the same world (of the same radius r_0); this agrees with the idea developed in § 32, according to which the charge and the mass are determined from infinity.

In conclusion, we shall set up the mechanical equations that govern the motion of a material particle. In actual fact we have not yet derived these equations in a form which is admissible from the point of view of the general theory of relativity; we shall now endeavour to make good this omission. We shall also take this opportunity of carrying out the intention stated in § 32, that is, to show that in general the inertial mass is the flux of the gravitational field through a surface which encloses the particle, even when the matter has to be regarded as a singularity which limits the field and lies, so to speak, outside it. In doing this we are, of course, debarred from using a substance which is in motion; the hypotheses corresponding to the latter idea, namely (§ 27):

$$dmds = \mu dx, \qquad \mathbf{T}_i^k = \mu u_i u^k$$

are quite impossible here, as they contradict the postulated properties of invariance. For, according to the former equation, μ is a scalar-density of weight $\frac{1}{2}$, and, according to the latter, one of weight 0, since $\mathbf{T}_i^k$ is a tensor-density in the true sense. And we see that these initial conditions are impossible in the new theory for the same reason as in Einstein's Theory, namely, because they lead to a false value for the mass, as was mentioned at the end of § 33. This is obviously intimately connected with the circumstance that the integral $\int dmds$ has now no meaning at all, and hence cannot be introduced as "substance-action of gravitation". We took the first step towards giving a real proof of the mechanical equations in § 33. There we considered the special case in which the body is completely isolated, and no external forces act on it.

From this we see at once that we must start from the laws of conservation

$$\frac{\partial \mathbf{S}_i^k}{\partial x_k} = 0 \quad . \quad . \quad . \quad . \quad . \quad (89)$$

which hold for the **total energy.** Let a volume Ω, whose dimensions are great compared with the actual essential nucleus of the particle, but small compared with those dimensions of the external field which alter appreciably, be marked off around the material particle. In the course of the motion Ω describes a canal in the world, in the interior of which the current filament of the material particle flows along. Let the co-ordinate system consisting of the "time-co-ordinate" $x_0 = t$ and the "space-co-ordinates" x_1, x_2, x_3, be such that the spaces $x_0 =$ const. intersect the canal (the cross-section is the volume Ω mentioned above). The integrals

$$\int_{\Omega} \mathbf{S}_i^0 dx_1 dx_2 dx_3 = J_i$$

which are to be taken in a space $x_0 =$ const. over Ω, and which are functions of the time alone, represent the energy ($i = 0$) and the momentum ($i = 1, 2, 3$) of the material particle. If we integrate the equation (89) in the space $x_0 =$ const. over Ω, the first member ($k = 0$) gives the time-derivative $\frac{dJ_i}{dt}$; the integral sum over the three last terms, however, becomes transformed by Gauss' Theorem into an integral K_i which is to be taken over the surface of Ω. In this way we arrive at the mechanical equations

$$\frac{dJ_i}{dt} = K_i \quad . \quad . \quad . \quad . \quad . \quad . \quad (90)$$

On the left side we have the components of the "inertial force," and on the right the components of the external "field-force". Not only the field-force but also the four-dimensional momentum J_i may be represented, in accordance with a remark at the end of § 35, as a flux through the surface of Ω. If the interior of the canal encloses a real singularity of the field the momentum must, indeed, be defined in the above manner, and then the device of the "fictitious field," used at the end of § 35, leads to the mechanical equations proved above. *It is of fundamental importance to notice that in them only such quantities are brought into relationship with one another as are determined by the course of the field outside the particle* (on the surface of Ω), *and have nothing to do with the singular states or phases in its interior.* The antithesis of kinetic and potential which receives expression in the fundamental law of mechanics does not, indeed, depend actually on the separation of energy-momentum into one part belonging to the external field and another belonging to the particle (as we pictured it in § 25), but rather on this juxtaposition, conditioned by the resolution into space and time, of the first and the three last members of the divergence equations which make up the laws of conservation, that is, on the circumstance that the singularity canals of the material particles have an infinite extension in only **one** dimension, but are very limited in **three** other dimensions. This stand was taken most definitely by Mie in the third part of his epoch-making *Foundations of a Theory of Matter*, which deals with "Force and Inertia" (*vide* note 38). Our next object is to work out the full consequences of this view for the principle of action adopted in this chapter.

To do this, it is necessary to ascertain exactly the meaning of the electromagnetic and the gravitational equations. If we discuss Maxwell's equations first, we may disregard gravitation entirely and take the point of view presented by the special theory of relativity. We should be reverting to the notion of substance if we were to interpret the Maxwell-Lorentz equation

$$\frac{\partial f^{ik}}{\partial x_k} = \rho u^i$$

so literally as to apply it to the volume-elements of an electron. Its true meaning is rather this: Outside the Ω-canal, the homogeneous equations

$$\frac{\partial f^{ik}}{\partial x_k} = 0 \text{ hold} \quad . \quad . \quad . \quad . \quad (91)$$

The only statical radially symmetrical solution $\bar{f}_{ik}$ of (91) is that derived from the potential $\frac{e}{r}$; it gives the flux e (and not 0, as it would be in the case of a solution of (91) which is free from singularities) of the electric field through an envelope Ω enclosing the particle. On account of the linearity of equations (91), these properties are not lost when an arbitrary solution f_{ik} of equations (91), free from singularities, is added to $\bar{f}_{ik}$; such a one is given by $f_{ik} =$ const. **The field which surrounds the moving electron must be of the type:** $f_{ik} + \bar{f}_{ik}$, if we introduce at the moment under consideration a co-ordinate system in which the electron is at rest. This assumption concerning the constitution of the field outside Ω is, of course, justified only when we are dealing with quasi-stationary motion, that is, when the world-line of the particle deviates by a sufficiently small amount from a straight line. The term ρu^i in Lorentz's equation is to express the general effect of the charge-singularities for a region that contains many electrons. But it is clear that this assumption comes into question only for **quasi-stationary motion.** Nothing at all can be asserted about what happens during rapid acceleration. The opinion which is so generally current among physicists nowadays, that, according to classical electrodynamics, a greatly accelerated particle emits radiation, seems to the author quite unfounded. It is justified only if Lorentz's equations are interpreted in the too literal fashion repudiated above, and if, also, it is assumed that the constitution of the electron is not modified by the acceleration. **Bohr's Theory of the Atom** has led to the idea that there are individual stationary orbits for the electrons circulating in the atom, and that they may move permanently in these orbits without emitting radiations; only when an electron jumps from one stationary orbit to another is the energy that is lost by the atom emitted as electromagnetic energy of vibration (*vide* note 39). If matter is to be regarded as a boundary-singularity of the field, our field-equations make assertions only about **the possible states of the field,** and **not about the conditioning of the states of the field by the matter.** This gap is filled by the **Quantum Theory** in a manner of which the underlying principle is not yet fully grasped. The above assumption about the singular component $\bar{f}$ of the field surrounding the particle is, in our opinion, true for a quasi-stationary electron. We may, of course, work out other assumptions. If, for example, the particle is a radiating atom, the $\bar{f}_{ik}$'s will have to be represented as the field of an oscillating Hertzian dipole. (This is a possible state of the field which is caused by matter in a manner which, according to Bohr, is quite different from that imagined by Hertz.)

As far as gravitation is concerned, we shall for the present adopt the point of view of the original Einstein Theory. In it the (homogeneous) gravitational equations have (according to § 31) a statical radially symmetrical solution, which depends **on a single constant m, the mass.** The flux of a gravitational field through a sufficiently great sphere described about the centre is not equal to 0, as it should be if the solution were free from singularities, but equal to m. We assume that this solution is characteristic of the moving particle in the following sense: We consider the values traversed by the g_{ik}'s outside the canal to be extended over the canal, by supposing the narrow deep furrow, which the path of the material particle cuts out in the metrical picture of the world, to be smoothed out, and by treating the stream-filament of the particle as a line in this smoothed-out metrical field. Let ds be the corresponding proper-time differential. For a point of the stream-filament we may introduce a ("normal") co-ordinate system such that, at that point,

$$ds^2 = dx_0^2 - (dx_1^2 + dx_2^2 + dx_3^2)$$

the derivatives $\frac{\partial g_{\alpha\beta}}{\partial x_i}$ vanish, and the direction of the stream-filament is given by

$$dx_0 : dx_1 : dx_2 : dx_3 = 1 : 0 : 0 : 0.$$

In terms of these co-ordinates the field is to be expressed by the above-mentioned statical solution (only, of course, in a certain neighbourhood of the world-point under consideration, from which the canal of the particle is to be cut out). If we regard the normal co-ordinates x_i as Cartesian co-ordinates in a four-dimensional Euclidean space, then the picture of the world-line of the particle becomes a definite curve in the Euclidean space. Our assumption is, of course, admissible again only if the motion is quasi-stationary, that is, if this picture-curve is only slightly curved at the point under consideration. (The transformation of the homogeneous gravitational equations into non-homogeneous ones, on the right side of which the tensor $\mu u_i u_k$ appears, takes account of the singularities, due to the presence of masses, by fusing them into a continuum; this assumption is legitimate only in the quasi-stationary case.)

To return to the derivation of the mechanical equations! We shall use, once and for all, the calibration normalised by F = const., and we shall neglect the cosmological terms outside the canal. The influence of the charge of the electron on the gravitational field is, as we know from § 32, to be neglected in comparison with the influence of the mass, provided the distance from the particle is sufficiently great. Consequently, if we base our calculations on the normal co-ordinate system, we may assume the gravitational field to be that mentioned above. The determination of the electromagnetic field is then, as in the gravitational case, a linear problem; it is to have the form $f_{ik} + \bar{f}_{ik}$ mentioned above (with f_{ik} = const. on the surface of Ω). But this assumption is compatible with the field-laws only if e = const. To prove this, we shall deduce from a fictitious field that fills the canal regularly and that links up with the really existing field outside, that

$$\frac{\partial \mathfrak{f}^{ik}}{\partial x_k} = \mathfrak{s}^i, \quad \int_\Omega \mathfrak{s}^0 dx_1 dx_2 dx_3 = e^*$$

in any arbitrary co-ordinate system; e^* is independent of the choice of the fictitious field, inasmuch as it may be represented as a field-flux through the surface of Ω. Since (if we neglect the cosmological terms) the $\mathfrak{s}^i$'s on this surface vanish, the equation of definition gives us, if $\frac{\partial \mathfrak{s}^i}{\partial x_i} = 0$ is integrated, $\frac{de^*}{dt} = 0$; moreover, the arguments set out in § 33 show that e^* is independent of the co-ordinate system chosen. If we use the normal co-ordinate system at one point, the representation of e^* as a field-flux shows that $e^* = e$.

Passing on from the charge to the momentum, we must notice at once that, with regard to the representation of the energy-momentum components by means of field-fluxes, we may not refer to the general theory of § 35, because, by applying the process of partial integration to arrive at (84), we sacrificed the co-ordinate invariance of our *Action*. Hence we must proceed as follows. With the help of the fictitious field which bridges the canal regularly, we define $a\mathbf{S}_i^k$ by means of

$$(\mathbf{R}_i^k - \tfrac{1}{2}\delta_i^k\mathbf{R}) + \left(\mathbf{G}\delta_i^k - \tfrac{1}{2}\frac{\partial g_{\alpha\beta}}{\partial x_i}\mathbf{G}^{\alpha\beta,k}\right).$$

The equation

$$\frac{\partial \mathbf{S}_i^k}{\partial x_k} = 0 \quad . \quad . \quad . \quad . \quad . \quad (92)$$

is an identity for it. By integrating (92) we get (90), whereby

$$J_i = \int_\Omega \mathbf{S}_i^0 dx_1 dx_2 dx_3.$$

K_i expresses itself as the field-flux through the surface Ω. In these expressions the fictitious field may be replaced by the real one, and, moreover, in accordance with the gravitational equations, we may replace

$$\frac{1}{a}(\mathbf{R}_i^k - \tfrac{1}{2}\delta_i^k\mathbf{R}) \text{ by } \mathbf{l}\delta_i^k - f_{ir}\mathbf{f}^{kr}.$$

If we use the normal co-ordinate system the part due to the gravitational energy drops out; for its components depend not only linearly but also quadratically on the (vanishing) derivatives $\frac{\partial g_{\alpha\beta}}{\partial x_i}$. We are, therefore, left with only the electromagnetic part, which is to be calculated along the lines of Maxwell. Since the components of Maxwell's energy-density depend quadratically on the field $f + \bar{f}$, each of them is composed of three terms in accordance with the formula

$$(f + \bar{f})^2 = f^2 + 2\bar{f}f + \bar{f}^2.$$

In the case of each, the first term contributes nothing, since the flux of a constant vector through a closed surface is 0. The last term is to be neglected since it contains the weak field $\bar{f}$ as a square; the middle term alone remains. But this gives us

$$K_i = ef_{0i}$$

Concerning the momentum-quantities we see (in the same way as in § 33, by using identities (92) and treating the cross-section of the stream-filament as infinitely small in comparison with the external field) (1) that, for co-ordinate transformations that are to be regarded as linear in the cross-section of the canal, the J_i's are the co-variant components of a vector which is independent of the co-ordinate system; and (2) that if we alter the fictitious field occupying the canal (in § 33 we were concerned, not with this, but with a charge of the co-ordinate system in the canal) the quantities J_i retain their values. In the normal co-ordinate system, however, for which the gravitational field that surrounds the particle has the form calculated in § 31, we find that, since the fictitious field may be chosen as a statical one, according to page 272 : $J_1 = J_2 = J_3 = 0$, and $J_0 =$ the flux of a spatial vector-density through the surface of Ω, and hence $= m$. On account of the property of co-variance possessed by J_i, we find that not only at the point of the canal under consideration, but also just before it and just after it

$$J_i = mu_i \qquad \left(u^i = \frac{dx_i}{ds}\right).$$

Hence the equations of motion of our particle expressed in the normal co-ordinate system are

$$\frac{d(mu_i)}{dt} = ef_{0i} \quad . \quad . \quad . \quad . \quad . \quad (93)$$

The 0th of these equations gives us : $\frac{dm}{dt} = 0$; thus the field equations require that the mass be constant. But in any arbitrary co-ordinate system we have :

$$\frac{d(mu_i)}{ds} - \tfrac{1}{2}\frac{\partial g_{\alpha\beta}}{\partial x_i} mu^\alpha u^\beta = e\,.f_{ki}u^k \quad . \quad . \quad (94)$$

For the relations (94) are invariant with respect to co-ordinate transformations, and agree with (93) in the case of the normal co-ordinate system. *Hence, according to the field-laws, a necessary condition for a singularity canal, which is to fit into the remaining part of the field, and in the immediate neighbourhood of which the field has the required structure, is that the quantities e and m that characterise the singularity at each point of the canal remain constant along the canal, but that the world-direction of the canal satisfy the equations*

$$\frac{du_i}{ds} - \tfrac{1}{2}\frac{\partial g_{\alpha\beta}}{\partial x^i} u^\alpha u^\beta = \frac{e}{m}\,.f_{ki}\,.\,u^k.$$

In the light of these considerations, it seems to the author that the opinion expressed in § 25 stating that mass and field-energy are identical is a premature inference, and the whole of Mie's view of matter assumes a fantastic, unreal complexion. It was, of course, a natural result of the special theory of relativity that we should come to this conclusion. It is only when we arrive at the general theory that we find it possible to represent the mass as a field-flux, and to ascribe to the world relationships such as obtain in Einstein's *Cylindrical World* (§ 34), when there are cut out of it canals of circular cross-section which stretch to infinity in both directions. This view of m states not only that inertial and gravitational masses are identical in nature, but also that mass as the **point of attack** of the metrical field is identical in nature with mass as the **generator** of the metrical field. That which is physically important in the statement that energy has inertia still persists in spite of this. For example, a radiating particle loses inertial mass of exactly the same amount as the electromagnetic energy that it emits. (In this example Einstein first recognised the intimate relationship between energy and inertia.) This may be proved simply and rigorously from our present point of view. Moreover, the new standpoint in no wise signifies a relapse to the old idea of substance, but it deprives of meaning the problem of the cohesive pressure that holds the charge of the electron together.

With about the same reasonableness as is possessed by Einstein's Theory we may conclude from our results that a **clock** in quasi-stationary motion indicates the proper time $\int ds$ which corresponds to the normalisation $F =$ const.* If during the motion of a clock (e.g. an atom) with infinitely small period, the world-distance traversed by it during a period were to be transferred congruently from period to period in the sense of our world-geometry, then two clocks which set out from the same world-point A with the same period, that is, which traverse congruent world-distances in A during their first period will have, in general, different periods when they meet at a later world-point B. The orbital motion of the electrons in the atom can, therefore, certainly not take place in the way described, independently of their previous

* The invariant quadratic form $F \cdot ds^2$ is very far from being distinguished from all other forms of the type $E \cdot ds^2$ (E being a scalar of weight -1) as is the ds^2 of Einstein's Theory, which does not contain the derivatives of the potentials at all. For this reason the inference made in our calculation of the **displacement towards the infra-red** (p. 246), that similar atoms radiate the same frequency measured in the proper time ds corresponding to the normalisation $F =$ const., is by no means as convincing as in the theory of Einstein: it loses its validity altogether if a principle of action other than that here discussed holds.

histories, since the atoms emit spectral lines of definite frequencies. Neither does a measuring rod at rest in a statical field undergo a congruent transference; for the measure $l = d\sigma^2$ of a measuring rod at rest does not alter, whereas for a congruent transference it would have to satisfy the equation $\frac{dl}{dt} = -l \cdot \phi$. What is the source of this discrepancy between the conception of congruent transference and the behaviour of measuring rods, clocks, and atoms? We may distinguish two modes of determining a quantity in nature, namely, that of **persistence** and that of **adjustment.** This difference is illustrated in the following example. We may prescribe to the axis of a rotating top any arbitrary direction in space; but once this arbitrary initial direction has been fixed the direction of the axis of the top when left to itself is determined from it for all time by a **tendency of persistence** which is active from one moment to another; at each instant the axis experiences an infinitesimal parallel displacement. Diametrically opposed to this is the case of a magnet needle in the magnetic field. Its direction is determined at every moment, independently of the state of the system at other moments, by the fact that the system, in virtue of its constitution, **adjusts** itself to the field in which it is embedded. There is no *a priori* ground for supposing a pure transference, following the tendency of persistence, to be integrable. But even if this be the case, as, for example, for rotations of the top in Euclidean space, nevertheless two tops which set out from the same point with axes in the same position, and which meet after the lapse of a great length of time, will manifest any arbitrary deviations in the positions of the axes, since they can never be fully removed from all influences. Thus although, for example, Maxwell's equations for the charge e of an electron make necessary the equation of conservation $\frac{de}{dt} = 0$, this does not explain why an electron itself after an arbitrarily long time still has the same charge, and why this charge is the same for all electrons. This circumstance shows that the charge is determined not by persistence but by adjustment: there can be only **one** state of equilibrium of negative electricity, to which the corpuscle adjusts itself afresh at every moment. The same reason enables us to draw the same conclusion for the spectral lines of the atoms, for what is common to atoms emitting equal frequencies is their constitution and not the equality of their frequencies at some moment when they were together far back in time. In the same way, obviously, the length of a measuring rod is determined by adjustment; for it would be impossible to give to **this** rod at **this** point of the field any length, say two or three times as great as the one that it now has, in the way that I can prescribe its direction arbitrarily. The world-curvature makes it theoretically possible to determine a length by adjustment. In consequence of its constitution the rod assumes a length which has such and such a value in relation to the radius of curvature of the world. (Perhaps the time of rotation of a top gives us an example of a time-length that is determined by persistence; if what we assumed above is true for direction then at each moment of the motion of the top the rotation vector would experience a parallel displacement.) We may briefly summarise as follows: The affine and metrical relationship is an *a priori* datum telling us how vectors and lengths alter, **if they happen to follow the tendency of persistence.** But to what extent this is the case in nature, and in what proportion persistence and adjustment modify one another, can be found only by starting from the physical laws that hold, i.e. from the principle of action.

The subject of the above discussion is the principle of action, compatible with the new axiom of calibration invariance, which most nearly approaches the Maxwell-Einstein theory. We have seen that it accounts equally well for all the phenomena which are explained by the latter theory and, indeed, that it has decided advantages so far as the deeper problems, such as the cosmological problems and that of matter are concerned. Nevertheless, I doubt whether the Hamiltonian function (83) corresponds to reality. We may certainly assume that $\mathbf{W}$ has the form $W\sqrt{g}$, in which W is an invariant of weight -2 formed in a perfectly rational manner from the components of curvature. Only **four** of these invariants may be set up, from which every other may be built up linearly by means of numerical co-efficients (*vide* note 40). One of these is Maxwell's:

$$l = \tfrac{1}{4} f_{ik} f^{ik} . \qquad . \quad . \quad . \quad . \quad . \quad (95)$$

another is the F^2 used just above. But curvature is by its nature a linear matrix-tensor of the second order: $\mathsf{F}_{ik} dx_i \delta x_k$. According to the same law by which (95), the square of the numerical value, is produced from the distance-curvature f_{ik} we may form

$$\tfrac{1}{4} \mathsf{F}_{ik} \mathsf{F}^{ik} \quad . \quad . \quad . \quad . \quad . \quad . \quad (96)$$

from the total curvature. The multiplication is in this case to be interpreted as a composition of matrices; (96) is therefore itself again a matrix. But its trace L is a scalar—of weight -2. The two quantities L and l seem to be invariant and of the kind sought, and they can be formed most naturally from the curvature; invariants of this natural and simple type, indeed, exist only in a four-dimensional world at all. It seems more probable that W is a linear combination of L and l. Maxwell's equations become then as above: (when the calibration has been normalised by F = const.) $\mathbf{s}^i$ = a constant multiple of $\sqrt{g}\phi^i$, and $\mathbf{h}^{ik} = \mathbf{f}^{ik}$. The gravitational laws in the statical case here, too, agree to a first approximation with Newton's laws. Calculations by Pauli (*vide* note 41) have indeed disclosed that the field determined in § 31 is not only a rigorous solution of Einstein's equations, but also of those favoured here, so that the amount by which the perihelion of Mercury's orbit advances and the amount of the deflection of light rays owing to the proximity of the sun at least do not conflict with these equations. But in the question of the mechanical equations and of the relationship holding between the results obtained by measuring-rods and clocks on the one hand and the quadratic form on the other, the connecting link with the old theory seems to be lost; here we may expect to meet with new results.

One serious objection may be raised against the theory in its present state: it does not account for the **inequality of positive and negative electricity** (*vide* note 42). There seem to be two ways out of this difficulty. Either we must introduce into the law of action a square root or some other irrationality; in the discussion on Mie's theory, it was mentioned how the desired inequality could be caused in this way, but it was also pointed out what obstacles lie in the way of such an irrational *Action*. Or, secondly, there is the following view which seems to the author to give a truer statement of reality. We have here occupied ourselves only with the **field** which satisfies certain generally invariant functional laws. It is quite a different matter to inquire into the **excitation** or **cause** of the field-phases that appear to be possible according to these laws; it directs our attention to the reality lying beyond the field. Thus in the æther there may exist convergent as well as divergent electromagnetic waves; but only the latter event can be brought about by an atom, situated at the centre, which emits energy owing to the jump of an electron from one orbit to another in accordance with Bohr's hypothesis. This example shows (what is immediately obvious from other considerations) that the idea of causation (in contradistinction to functional relation) is intimately connected with the **unique direction of progress characteristic of Time,** namely **Past → Future.** This oneness of sense in Time exists beyond doubt—it is, indeed, the most fundamental fact of our perception of Time—but *a priori* reasons exclude it from playing a part in physics of the field. But we saw above (§ 33) that the sign, too, of an isolated system is fully determined, as soon as a definite sense of flow, Past → Future, has been prescribed to the world-canal swept out by the system. This connects the inequality of positive and negative electricity with the inequality of Past and Future; but the roots of this problem are not in the field, but lie outside it. Examples of such regularities of structure that concern, not the field, but the causes of the field-phases are instanced: by the existence of cylindrically shaped boundaries of the field: by our assumptions above concerning the constitution of the field in their immediate neighbourhood: lastly, and above all, by the facts of the quantum theory. But the way in which these regularities have hitherto been formulated are, of course, merely provisional in character. Nevertheless, it seems that the **theory of statistics** plays a part in it which is fundamentally necessary. We must here state in unmistakable language that physics at its present stage can in no wise be regarded as lending support to the belief that there is a causality of physical nature which is founded on rigorously exact laws. The extended field, "æther," is merely the *transmitter* of effects and is, of itself, powerless; it plays a part that is in no wise different from that which space with its rigid Euclidean metrical structure plays, according to the old view; but now the rigid motionless character has become transformed into one which gently yields and adapts itself. But freedom of action in the world is no more restricted by the rigorous laws of field physics than it is by the validity of the laws of Euclidean geometry according to the usual view.

If Mie's view were correct, we could recognise the field as objective reality, and physics would no longer be far from the goal of giving so complete a grasp of the nature of the physical world, of matter, and of natural forces, that logical necessity would extract from this insight the unique laws that underlie the occurrence of physical events. For the present, however, we must reject these bold hopes. The laws of the metrical field deal less with reality itself than with the shadow-like extended medium that serves as a link between material things, and with the formal constitution of this medium that gives it the power of transmitting effects. **Statistical physics,** through the quantum theory, has already reached a deeper stratum of reality than is accessible to field physics; but the problem of matter is still wrapt in deepest gloom. But even if we recognise the limited range of field physics, we must gratefully acknowledge the insight to which it has helped us. Whoever looks back over the ground that has been traversed, leading from the Euclidean metrical structure to the mobile metrical field which depends on matter, and which includes the field phenomena of gravitation and electromagnetism; whoever endeavours to get a complete survey of what could be represented only successively and fitted into an articulate manifold, must be overwhelmed by a feeling of freedom won—the mind has cast off the fetters which have held it captive. He must feel transfused with the conviction that reason is not only a human, a too human, makeshift in the struggle for existence, but that, in spite of all disappointments and errors, it is yet able to follow the intelligence which has planned the world, and that the consciousness of each one of us is the centre at which the One Light and Life of Truth comprehends itself in Phenomena. Our ears have caught a few of the fundamental chords from that harmony of the spheres of which Pythagoras and Kepler once dreamed.

Note 28. (273). Cf. G. Nordström, On the mass of a material system according to the Theory of Einstein, Akad. v. Wetensch., Amsterdam, vol xx., No. 7 (Dec. 29th, 1917).

Note 29. (275). Hilbert (l.c.[8]), 2 Mitt.

Note 30. (276). Einstein, Sitzungsber. d. Preuss. Akad. d. Wissensch., 1917 6, p. 142.

Note 31. (280). Weyl, Physik. Zeitschr., Bd. 20 (1919), p. 31.

Note 32. (282). Cf. de Sitter's Mitteilungen im Versl. d. Akad. v. Wetensch. te Amsterdam, 1917, as also his series of concise articles: On Einstein's theory of gravitation and its astronomical consequences (Monthly Notices of the R. Astronom. Society); also F. Klein (l.c.[27]).

Note 33. (282). The theory contained in the two following articles were developed by Weyl in the Note "Gravitation und Elektrizität," Sitzungsber. d. Preuss. Akad. d. Wissensch., 1918, p. 465. Cf. also Weyl, Eine neue Erweiterung der Relativitätstheorie, Ann. d. Physik, Bd. 59 (1919). A similar tendency is displayed (although obscure to the present author in essential points) in E. Reichenbächer (Grundzüge zu einer Theorie der Elektrizität und Gravitation, Ann. d. Physik, Bd. 52 [1917], p. 135; also Ann. d. Physik, Bd. 63 [1920], pp. 93-144). Concerning other attempts to derive Electricity and Gravitation from a common root cf. the articles of Abraham quoted in Note 4; also G. Nordström, Physik. Zeitschr., 15 (1914), p. 504; E. Wiechert, Die Gravitation als elektrodynamische Erscheinung, Ann. d. Physik, Bd. 63 (1920), p. 301.

Note 34. (286). This theorem was proved by Liouville: Note IV in the appendix to G. Monge, Application de l'analyse à la géométrie (1850), p. 609.

Note 35. (286). This fact, which here appears as a self-evident result, had been previously noted: E. Cunningham, Proc. of the London Mathem. Society (2), vol. viii. (1910), pp. 77-98; H. Bateman, *idem*, pp. 223-64.

Note 36. (295). Cf. also W. Pauli, Zur Theorie der Gravitation und der Elektrizität von H. Weyl, Physik. Zeitschr., Bd. 20 (1919), pp. 457-67. Einstein arrived at partly similar results by means of a further modification of his gravitational equations in his essay: Spielen Gravitationsfelder im Aufbau der materiellen Elementarteilchen eine wesentliche Rolle? Sitzungsber. d. Preuss. Akad. d. Wissensch., 1919, pp. 349-56.

Note 37. (299). Concerning such existence theorems at a point of singularity, *vide* Picard, Traité d'Analyse, t. 3, p. 21.

Note 38. (302). Ann. d. Physik, Bd. 39 (1913).

Note 39. (303). As described in the book by Sommerfeld, Atombau and Spektrallinien, Vieweg, 1919 and 1921.

Note 40. (309). This was proved by R. Weitzenböck in a letter to the present author; his investigation will appear soon in the Sitzungsber. d. Akad. d. Wissensch. in Wien.

Note 41. (310). W. Pauli, Merkur-Perihelbewegung und Strahlenablenkung in Weyl's Gravitationstheorie, Verhandl. d. Deutschen physik. Ges., Bd. 21 (1919), p. 742.

Note 42. (310). Pauli (l.c.[36]).

Quantenmechanische Deutung der Theorie von Weyl[1]).

Von F. London in Stuttgart.

(Eingegangen am 25. Februar 1927.)

Kapitel I. Die Theorie von Weyl.

Die Idee einer „reinen Nahgeometrie", zuerst von Riemann konzipiert, hat bekanntlich kürzlich durch Weyl eine außerordentlich schöne und einfache Vervollständigung erfahren. Man kann den Riemannschen Raumbegriff betrachten als die Aufhebung des Vorurteils, daß die Krümmungsverhältnisse an einer Stelle des Raumes verbindlich sein müßten für die Krümmung an allen anderen. Um dieser Aussage Riemanns einen Sinn zu geben, war zunächst die Annahme notwendig, daß der Maßstab, welcher an jeder Stelle zur Bestimmung der Koeffizienten g_{ik} der metrischen Fundamentalform

$$ds^2 = g_{ik}\, dx^i\, dx^k$$

zur Anwendung gelangt, ein „starrer" Maßstab sei.

Demgegenüber macht Weyl mit Recht geltend, daß die Annahme eines solchen starren Maßstabes einer radikalen Nahgeometrie zuwider sei, daß nur die Verhältnisse der g_{ik} an einer Stelle, nicht ihre Absolutbeträge, sinngemäß festgelegt werden können, und dementsprechend setzt er für die Änderung dl eines Eichmaßstabes von der Länge l bei einer infinitesimalen Verschiebung dx^i an:

$$dl = l\varphi_i\, dx^i, \tag{1}$$

wobei die Proportionalitätsfaktoren φ_i Funktionen des Ortes sind, Charakteristika der Maßverhältnisse des Raumes — ähnlich den g_{ik}. Oder, wenn man (1) integriert:

$$l = l_0\, e^{\int \varphi_i\, dx^i} \tag{2}$$

[1]) Vorgetragen teilweise auf der Tagung des Gauvereins Württemberg der D. Phys. Ges. Stuttgart, am 18. Dezember 1926; vgl. auch einen vorläufigen zusammenfassenden Bericht in Naturwiss. 15, 187, 1927.

($l_0 = l$ am Anfang der Verschiebung). Das Eichmaß ist im allgemeinen vom Wege abhängig (nicht integrabel), es sei denn, daß die Größen

$$f_{ik} = \frac{\partial \varphi_i}{\partial x^k} - \frac{\partial \varphi_k}{\partial x^i} \tag{3}$$

verschwinden. Über diese Größen f_{ik} kann man laut ihrer Definition (3) die Identität aussprechen (die Dimensionenzahl der Mannigfaltigkeit sei 4):

$$\frac{\partial f_{ik}}{\partial x^l} + \frac{\partial f_{kl}}{\partial x^i} + \frac{\partial f_{li}}{\partial x^k} = 0 \qquad i \neq k \neq l, \quad i, k, l = 1, 2, 3, 4. \tag{4}$$

Die formale Übereinstimmung dieser vier Gleichungen mit dem einen System der Maxwellschen Gleichungen

$$\operatorname{rot} \mathfrak{E} + \frac{1}{c} \dot{\mathfrak{H}} = 0,$$

$$\operatorname{div} \mathfrak{H} = 0,$$

sowie einige weitere formale Analogien haben Weyl zu dem Schluß geführt, die φ_i seien bis auf einen konstanten Proportionalitätsfaktor zu identifizieren mit den Komponenten Φ_i des elektromagnetischen Viererpotentials, die f_{ik} entsprechend mit den elektromagnetischen Feldstärken $\mathfrak{E}$, $\mathfrak{H}$. In folgerichtiger Ergänzung der geometrischen Deutung der Gravitation durch die variablen Krümmungen des Riemannschen Raumes dachte sich Weyl den noch übrigbleibenden Teil physikalischer Wirkungen, das elektromagnetische Feld, ebenfalls als eine Eigenschaft der Maßverhältnisse des Raumes, charakterisiert durch die Variabilität des Eichmaßes. Es ist also zu schreiben:

$$l = l_0 \, e^{\alpha \int \Phi_i \, dx_i} \qquad (\alpha = \text{Proportionalitätsfaktor}). \tag{2a}$$

Man wird die ungeheure Kühnheit bewundern, mit welcher Weyl allein auf Grund dieser ganz formalen Zuordnung seine Lehre von der eichgeometrischen Deutung des Elektromagnetismus aufgespürt hat: In der Gravitationstheorie war es eine physikalische Tatsache, das Prinzip der Äquivalenz zwischen träger und schwerer Masse, welche Einstein zu seiner geometrischen Deutung anregte. In der Theorie der Elektrizität dagegen war eine solche Tatsache nicht bekannt: Es bestand keine Veranlassung, an einen universellen Einfluß des elektromagnetischen Feldes auf die sogenannten starren Maßstäbe (bzw. Uhren) zu denken. Ganz im Gegenteil, die Atomuhren z. B. repräsentieren Maßstäbe, deren Unabhängigkeit von der Vorgeschichte durch die Schärfe der Spektrallinien belegt ist, im Widerspruch zu dem nicht integrablen

Maße (2a), welches Weyl im magnetischen Felde annimmt. Es bedurfte wohl einer ungewöhnlich klaren metaphysischen Überzeugung, die Weyl solchen elementarsten Erfahrungen zum Trotz nicht von dem Gedanken abgehen ließ, daß die Natur von diesen schönen ihr gebotenen geometrischen Möglichkeiten Gebrauch machen müsse. Er hielt an seiner Auffassung fest und entzog die eben geschilderten Widersprüche der Diskussion durch eine etwas dunkle Umdeutung des Begriffs „realer Maßangabe", womit nun allerdings seiner Theorie ihr so prägnanter physikalischer Sinn genommen war und sie dadurch sehr an Überzeugungskraft verlor.

Auf diese abstrakte Ausgestaltung der Theorie brauche ich nicht einzugehen. Ich werde vielmehr zeigen, daß gerade der prägnanten ursprünglichen Fassung der Weylschen Theorie eine noch viel größere Spannkraft innewohnt, als ihr Urheber bereits wirksam gemacht hat, daß man nämlich in ihr nichts geringeres als einen folgerichtigen Weg zur Undulationsmechanik zu erblicken hat, unter deren Gesichtspunkten sie erst eine unmittelbar verständliche physikalische Bedeutung gewinnt.

Kapitel II. Die Undulationsmechanik von de Broglie und die Theorie von Weyl.

Als „Theorie von de Broglie" bezeichne ich jene noch unvollkommene Vorstufe der Undulationsmechanik, in welcher die Wellenfunktion der Bewegung eines Elektrons (auf welche wir uns hier beschränken)

$$\psi = e^{\frac{2\pi i}{h} W(x_i)} \qquad i = 1, 2, 3, 4 \tag{5}$$

aus einer vollständigen Lösung W der Hamilton-Jacobischen partiellen Differentialgleichung

$$\left(\frac{\partial W}{\partial x^i} - \frac{e}{c}\Phi_i\right)\left(\frac{\partial W}{\partial x_i} - \frac{e}{c}\Phi^i\right) = -m_0^2 c^2 \tag{6}$$

hervorgeht, wobei die Integrationskonstanten in bekannter Weise so zu bestimmen sind, daß ψ eine eindeutige Funktion des Raumes, d. h. W additivperiodisch wird, mit einem ganzzahligen Multiplum der Planckschen Konstanten als Periode.

Wenn man Ernst macht mit der radikalen Kontinuumsauffassung der Materie, mit der Auflösung des diskontinuierlich abgegrenzten Elektrons in eine stetig in Raum und Zeit veränderliche Feldgröße, wie es durch diese de Brogliesche und konsequenter durch die später zu betrachtende Schrödingersche Theorie nahegelegt[1]) wird, so gelangt man in eine außerordentliche prinzipielle Schwierigkeit, wenn man untersucht, welchen Sinn man überhaupt metrischen Aussagen innerhalb des Undulationskontinuums beizulegen hat. Denn in diesem schwingenden und fluktuierenden unendlich ausgebreiteten Medium, welches an die Stelle des abgegrenzten Elektrons getreten ist, findet man keine unveränderlichen Diskontinuitäten, keine starren Körper, welche als reproduzierbare Maßstäbe die Festlegung einer Maßbestimmung gestatten könnten.

Ich vertrete durchaus nicht die Auffassung, daß, um von Geometrie im atomaren Gebiete zu reden, eine ausführbare Meßvorschrift angegeben werden müsse; von einer solchen kann ja auch in der Elektronentheorie nicht die Rede sein. Aber wenn man irgend einen definierten Sinn mit einer metrischen Angabe verbinden will, scheint mir, ist das mindeste, was man verlangen kann: die Angabe irgend eines realen Gegenstandes (als „Prototyp"), auf welchen die metrische Aussage bereits bezogen ist: Eines Elektronendurchmessers oder -abstands usw., wenngleich eine solche Aussage noch in einem sehr problematischen Zusammenhang zu einer ausführbaren Messung stehen mag.

Aber ein solcher realer Gegenstand ist in dem Undulationskontinuum nicht vorhanden. Der Satz der Identität ist in dem πάντα ῥεῖ entstehender und zerfließender Wellen nicht anzuwenden, kein Merkmal im Kontinuum festzuhalten, welches geeignet wäre, ein reproduzierbares Maß zu bilden. Die prinzipielle Lage, in die man hier versetzt ist, wäre völlig hoffnungslos, hätte nicht Weyl in seiner Verallgemeinerung des Riemannschen Raumbegriffs bereits einen Raumtypus geschaffen, in welchem gerade die Nichtreproduzierbarkeit der Eicheinheit als konsequentes Postulat einer radikalen Nahgeometrie vorgesehen ist. War bisher diese Theorie im Weltbild der diskontinuierlichen Elektronentheorie eine überflüssige Belastung, da man ja gerade in den Elektronen reproduzierbare Maßgrößen zu besitzen glaubte, so hat sich jetzt die Sachlage von Grund auf geändert. Man ist geradezu gezwungen, sich

[1]) Es sprechen bekanntlich wichtige Gründe dafür, auf welche vor allem von Born und seinen Mitarbeitern hingewiesen wurde, daß der ganze Undulationsformalismus statistisch umzudeuten ist. Insofern die Ladungsdichte als eine statistische Gewichtsfunktion umgedeutet wird, ist es unschwer einzusehen, daß dieselbe Unbestimmtheit hinsichtlich der Anwendbarkeit des Satzes der Identität, auf die wir hier hinweisen, sich hinüber übersetzt. Aber da jene Auffassung zunächst jede Interpretation in Raum und Zeit ablehnt, hat für sie die Beziehung zur Weylschen Raumlehre geringes Interesse.

auf den allgemeinen Weylschen Raumbegriff zurückzuziehen und zu versuchen, ihn auf das Schrödingersche Kontinuum anzuwenden. Da enthüllt sich nun ein einfacher Zusammenhang.

§ 1. Nehmen wir einmal an, wir besäßen bereits einen Maßstab l der sich nach der Weylschen Vorschrift (2 a) verändert, und führen ihn im ψ-Felde herum. Und zwar werde er mit der Strömungsgeschwindigkeit der Materie, der Gruppenvierergeschwindigkeit

$$u^i = \frac{dx^i}{d\tau} = \frac{1}{m_0}\left(\frac{\partial W}{\partial x_i} - \frac{e}{c}\Phi^i\right) \tag{7}$$

geführt.

Ich behaupte, mit dieser naheliegenden Vorschrift über den Weg wird Weyls Skalar l numerisch identisch mit dem de Broglieschen Feldskalar ψ. Hierzu sind noch zwei Präzisierungen zu treffen:

In dem Weylschen Eichmaß war noch ein Faktor α unbestimmt gelassen; für diesen mache ich die Hypothese, er sei gleich $\frac{2\pi i e}{hc}$. Also

$$l = l_0 e^{\frac{2\pi i}{h}\int \frac{e}{c}\varphi_i\, dx^i} \tag{2a}$$

Schließlich noch: ich benutze nicht genau das ψ aus Gleichung (5), sondern das mit dem Faktor $e^{\frac{2\pi i}{h} m_0 c^2 \tau}$ versehene fünfdimensionale ψ, wie es den Vorschlägen von Klein, Fock und Kudar entspricht, wobei unter τ die Eigenzeit[1]) zu verstehen ist. Es sei also jetzt

$$\psi = e^{\frac{2\pi i}{h}(W + m_0 c^2 \tau)} \tag{5a}$$

oder

$$= e^{\frac{2\pi i}{h}\left\{\int \frac{\partial W}{\partial x^i} dx^i + m_0 c^2 \tau\right\}}.$$

Diese Größe ψ ist zu vergleichen mit dem entlang der Strömung des Kontinuums geführten Weylschen Eichmaße (2 a). Man erhält:

$$\frac{\psi}{l} = \frac{1}{l_0} e^{\frac{2\pi i}{h}\left\{\int\left(\frac{\partial W}{\partial x^i} - \frac{e}{c}\varphi^i\right) dx^i + m_0 c^2 \tau\right\}},$$

hier sind die dx^i gemäß der durch (7) angegebenen Strömung zu führen:

$$= \frac{1}{l_0}\cdot e^{\frac{2\pi i}{h}\left\{\int\left(\frac{\partial W}{\partial x^i} - \frac{e}{c}\varphi_i\right)\left(\frac{\partial W}{\partial x^i} - \frac{e}{c}\varphi^i\right)\frac{d\tau}{m_0} + m_0 c^2 \tau\right\}}.$$

[1]) Diese Auffassung von τ, die auf Kudar, Ann. d. Phys. **81**, 632, 1926, zurückgeht, steht durchaus in Übereinstimmung mit der kürzlich diskutierten Deutung als Winkelkoordinate der Eigenrotationsbewegung des Elektrons (Naturwissenschaften **15**, 15, 1927). Denn dieser Drehwinkel ist als eine vom Elektron mitgeführte Uhr anzusehen. Er transformiert sich wie die Eigenzeit.

Infolge der Hamilton-Jacobischen Differentialgleichung (6) ist der Integrand $= -m_0 c^2$, man erhält:

$$\frac{\psi}{l} = \frac{1}{l_0} \cdot e^{\frac{2\pi i}{h} \cdot \text{const}} = \text{const.} \tag{8}$$

Der physikalische Gegenstand ist gefunden, der sich so verhält wie das Weylsche Maß: die komplexe Amplitude der de Broglieschen Welle; sie also erfährt im elektromagnetischen Felde genau den Einfluß, welchen Weyl für sein Eichmaß postuliert hat und dem er — als ein leerlaufendes Glied der damaligen Physik — eine metaphysikalische Existenz zuweisen mußte. Sie also ist sozusagen das Prototyp des Weylschen Maßes. Und ähnlich wie es in der Gravitationstheorie in unserem Belieben steht, von abgelenkten Lichtstrahlen und Massen oder aber von ihrer geodätischen Bewegung in einem Riemannschen Raum zu reden, so gibt uns (8) die Möglichkeit, den de Broglieschen Schwingungsvorgang der Materie und seine Beeinflussung durch die elektrischen Potentiale geometrisch zu deuten durch einen homogen mit Materie ausgefüllten Weylschen Raum, dessen metrischer Zusammenhang jedoch nicht integrabel ist.

Bei fehlendem elektromagnetischen Felde soll nach (2 a) das Eichmaß eine Konstante sein. Man müßte also auch einen konstanten Wert der de Broglieschen Wellenfunktion erhalten, wenn man sie mit der zugehörigen Strom-, d. h. Gruppengeschwindigkeit (v stets $< c$) verfolgt. Das scheint ein Widerspruch zu den grundlegendsten Ergebnissen de Broglies zu sein, nach welchen die Phasen seiner Wellen mit einer sehr viel größeren Phasengeschwindigkeit $\left(u = \frac{c^2}{v}\right)$ fortschreiten. Aber das ist hier nicht zutreffend, denn oben wurde nicht genau das de Brogliesche, sondern das fünfdimensional erweiterte ψ verwandt, welches dispersionsfrei ist und demgemäß fällt hier die Unterscheidung zwischen Gruppen- und Phasengeschwindigkeit fort. Man überzeugt sich auch leicht unmittelbar, daß die ebene Welle

$$\psi = e^{-\frac{2\pi i}{h}\left(\frac{m_0 c^2}{\sqrt{1-\beta^2}} t - \frac{m_0 v}{\sqrt{1-\beta^2}} x - m_0 c^2 \tau\right)} \qquad \left(\beta = \frac{v}{c}\right)$$

in der Tat beim Verfolgen mit der Geschwindigkeit v konstante Phase zeigt.

Ein weiterer Einwand, daß wir hier ψ, eine Dichte, mit einer Länge l vergleichen, scheint mir ebenfalls keine Schwierigkeit zu bieten.

Man müßte ψ von vornherein mit l^{-3} vergleichen, was nur eine Änderung in der Wahl des unbestimmten Faktors α bedeuten würde. Naturgemäßer wäre es wohl, aus dem hier aufgedeckten Zusammenhang zu entnehmen, daß der Weylschen Eichgröße l von vornherein dieselbe Dimension beizulegen ist, wie dem de Broglieschen ψ. Innerhalb der Weylschen Theorie konnte eine solche Aussage nicht getroffen werden, da in ihr nichts über die „Natur" von l bekannt war.

Eine ernstlichere Schwierigkeit scheint die komplexe Form der Streckenübertragung dem Verständnis aufzugeben. Es ist hierbei durchaus nicht zulässig, sich etwa auf den Realteil zu beschränken. Man hat hierin ein Gegenstück dafür zu sehen, daß die Wellenfunktion ψ selbst wesentlich komplex aufzufassen ist, besser gesagt, eine Zusammenfassung von zwei physikalischen Zustandsgrößen, nämlich $\psi\bar{\psi}$ und dem Realteil von $\frac{h}{2\pi i}\ln\psi$, darstellt. In diesem Sinne ist es auch zu verstehen, daß im Variationsproblem der Wellenmechanik ψ und $\bar{\psi}$ unabhängig voneinander zu variieren sind. Was es aber nun bedeuten soll, daß jede Strecke als eine komplexe Größe aufzufassen ist, und daß sich die ganze Weylsche Variabilität des Streckenmaßes als eine Änderung einzig der Phase unter Beibehaltung des Absolutbetrages herausstellt, das möchte ich noch nicht zur Diskussion bringen.

§ 2. Aber noch besteht der Einwand, auf den wir oben hinwiesen, daß die Erfahrung gegen die Nichtintegrabilität des Eichmaßes spricht. Man sieht jetzt bereits voraus, wie sich diese Schwierigkeit lösen muß: Die Quantentheorie erlaubt der Materie nur eine diskrete Reihe von Bewegungszuständen, und man vermutet, daß diese ausgezeichneten Bewegungen das Eichmaß nur derartig zu transportieren gestatten, daß die Phase bei Rückkehr an den Ausgangspunkt gerade eine ganzzahlige Anzahl von Umläufen durchgemacht hat, so daß trotz der Nichtintegrabilität der Streckenübertragung das Eichmaß an jeder Stelle stets in eindeutiger Weise realisiert wird. In der Tat erinnert man sich an die Resonanzeigenschaft der de Broglieschen Wellen, dieselbe, durch welche die alte Sommerfeld-Epsteinsche Quantenbedingung von de Broglie zuerst so folgenreich umgedeutet wurde. Diese ist allerdings an die Phasengeschwindigkeit geknüpft; aber infolge der fünfdimensionalen Erweiterung der Wellenfunktion ist der Schwingungsvorgang dispersionsfrei, und unsere Stromgeschwindigkeit wird infolgedessen identisch mit der Phasengeschwindigkeit. Hierdurch und infolge der Identität der Wellenfunktion ψ mit dem Weylschen Maße erscheint es also bereits erwiesen[1]), daß auch das Weylsche Maß, wenn ich es nur entlang der quantentheoretisch möglichen Materieströmung führe, an der Resonanz der de Broglieschen Wellen teil hat und trotz der Nichtintegrabilität des Differentialausdrucks (2 a) im elektromagnetischen Felde dennoch zu einer eindeutigen Maßbestimmung an jeder Stelle führt. Hätte man die Eindeutigkeit des Maßbegriffes als eine allgemein anerkannte Erfahrungstatsache der Weylschen Theorie axiomatisch angeschlossen, so wäre man folgerichtig auf das System der diskreten Bewegungszustände der „klassischen" Quantentheorie und ihre de Broglieschen Wellen geführt worden.

Ich möchte diesen Gegenstand nicht verlassen, ohne darauf aufmerksam zu machen, daß diese Resonanzeigenschaft des Weylschen Streckenmaßes, die uns hier als charakteristischer Satz der Undulationsmechanik entgegentritt, von Schrödinger[2]) bereits 1922 als eine „bemerkenswerte Eigenschaft der Quantenbahnen" vermutet und an einer Anzahl von Beispielen demonstriert worden ist, ohne daß sie damals in ihrer Bedeutung erkannt wurde. Es wurde auch die Möglichkeit von $\alpha = 2\pi i \cdot \frac{e}{hc}$ ins Auge gefaßt, aber ihr nicht der Vorzug vor einer anderen Wahl von α zuerkannt. Schon damals also hatte Schrödinger die charakteristischen wellenmechanischen Periodizitäten in der Hand, welchen er später unter so ganz anderen Gesichtspunkten wieder begegnen sollte.

Es ist deshalb vielleicht nicht überflüssig, wenn ich diese Schrödingersche Vermutung auch unabhängig von den wellenmechanischen Zusammenhängen, wie sie ursprünglich gemeint war, als einen Satz der „klassischen" Quantentheorie beweise. Es ist also behauptet: Der Streckenexponent des Weylschen Maßes, geführt über eine räumlich geschlossene Quantenbahn, ist ein ganzzahliges Multiplum der Planckschen Konstanten:

$$\oint \frac{e}{c} \Phi_i \, dx^i = n h. \tag{9}$$

Um das zu beweisen, benutzt man die bereits in § 1 verwendete Relation:

$$\int \left(\frac{\partial W}{\partial x^i} - \frac{e}{c} \Phi_i \right) dx^i = - \int m_0 c^2 \, d\tau = - \int m_0 c^2 \sqrt{1 - \left(\frac{v}{c}\right)^2} \, dt.$$

[1]) Diese Schlußweise ist nicht präzise, sie wird sogleich richtiggestellt werden.

[2]) E. Schrödinger, ZS. f. Phys. 12, 13, 1922.

Infolge der Quantenbedingungen

$$\sum_1^3 {}_i \oint \frac{\partial W}{\partial x^i} d x^i = n h$$

erhält man hieraus:

$$\oint \left(\frac{\partial W}{\partial x_4} d x_4 - \frac{e}{c} \Phi_i d x^i \right) = - n h - \oint m_0 c^2 \sqrt{1 - \left(\frac{v}{c} \right)^2} d t.$$

Vorausgesetzt, daß ein Energieintegral existiert, ist

$$\frac{\partial W}{\partial x_4} d x_4 = - (E_{\text{kin}} + E_{\text{pot}}) d t,$$

also:

$$- \oint \frac{e}{c} \Phi_i d x^i = - n h + \oint \left(- m_0 c^2 \sqrt{1 - \left(\frac{v}{c} \right)^2} + E_{\text{kin}} + E_{\text{pot}} \right) d t.$$

Hier verschwinden die Integrale auf der rechten Seite infolge der relativistischen Verallgemeinerung des Virialsatzes [1]) unter der Voraussetzung, daß das Potential homogen vom Grade -1 in den x^i ist, woraus die Behauptung (9) unmittelbar folgt.

Man sieht aus dieser Ableitung, daß nur unter zwei Voraussetzungen der Eindeutigkeitsbeweis des Weylschen Eichmaßes gelingt. Diese Voraussetzungen (insbesondere die erste) sind offenbar sehr wesentlich und sie werden sich sicher nicht völlig umgehen lassen. Sie garantieren gewisse stationäre Verhältnisse im Raume, die es überhaupt erst

---

[1]) Mir ist ein Beweis der relativistischen Verallgemeinerung des Virialsatzes in der Literatur nicht bekannt, deshalb will ich ihn hier mitteilen. Es ist

$$\oint \left(- m_0 c^2 \sqrt{1 - \left(\frac{v}{c} \right)^2} + \frac{m_0 c^2}{\sqrt{1 - \left(\frac{v}{c} \right)^2}} + E_{\text{pot}} \right) d t = \oint \left(\frac{m_0 v^2}{\sqrt{1 - \left(\frac{v}{c} \right)^3}} + E_{\text{pot}} \right) d t$$

$$= \oint \left(\sum_1^3 {}_i p_i \frac{d x^i}{d t} + E_{\text{pot}} \right) d t.$$

Hieraus durch Produktintegration unter Beachtung der Periodizitätsbedingung:

$$= \oint \left(- \sum_1^3 x^i \frac{d p_i}{d t} + E_{\text{pot}} \right) d t,$$

$\frac{d p_i}{d t}$ ist infolge der Bewegungsgleichungen $= - \frac{\partial E_{\text{pot}}}{\partial x^i}$, man erhält also

$$= \oint \left(\sum_1^3 x^i \frac{\partial E_{\text{pot}}}{\partial x^i} + E_{\text{pot}} \right) d t.$$

Hier verschwindet der Integrand infolge des Eulerschen Satzes über homogene Funktionen.

gestatten, von räumlich geschlossenen Bahnen in der Minkowskischen Welt zu reden, eine Aussage, welche im allgemeinen vom Bezugssystem durchaus abhängig ist. Man wird deshalb diese Voraussetzungen als Bedingungen der Möglichkeit für die Anwendung des Satzes der Identität auf den Raum zu bezeichnen haben.

Meist werden die Bahnkurven nicht exakt periodisch, sondern nur quasiperiodisch sein. Dann kann man unter geeigneten Stetigkeitsvoraussetzungen beweisen, daß bei hinreichend guter Annäherung an die Ausgangspunkte das Weylsche Maß bis auf einen vorgegebenen beliebig kleinen Betrag mit seinem ursprünglichen Wert übereinstimmt. Mehr braucht man auch nicht zu verlangen.

Daß hierbei stets der Transport der Eichstrecke mit der Geschwindigkeit (7) der Materie zu erfolgen hat, erscheint außerordentlich befriedigend; denn ein Transport mit anderer Geschwindigkeit wäre quantentheoretisch (bzw. mechanisch) garnicht möglich. Eine nähere Rechtfertigung dieser Zusammenhänge und ihren Einbau in eine erkenntnistheoretisch begründete Theorie des Maßes möchte ich jedoch noch verschieben, da hierzu noch wesentlich andere Gesichtspunkte namhaft gemacht werden müssen. Wenn wir auch gesehen haben, wie die Weylschen Ideen eine nicht vorauszuahnende Verkörperung in den gegenwärtigen physikalischen Anschauungen gefunden haben, so glaube ich doch nicht, daß man sich mit dem Gewonnenen bereits zufrieden geben kann. Ich habe die Kontinuumsauffassung der Quantenmechanik hier mit einer Einseitigkeit in den Vordergrund gestellt, welche nicht meiner Überzeugung entspricht. Immerhin schien es mir wünschenswert, zunächst diesen Gedanken mit einiger Konsequenz bis zu Ende zu verfolgen. In diesem Sinne sind die Ausführungen des folgenden Kapitels durchaus als Provisorium zu betrachten. Ich hoffe, auf den ganzen Zusammenhang unter allgemeineren physikalischen Gesichtspunkten demnächst zurückzukommen.

Kapitel III. Quantenmechanische Umdeutung der Theorie von Weyl.

Die Untersuchungen des vorigen Kapitels erstreckten sich ausdrücklich auf die als „de Brogliesche Theorie" gekennzeichnete Vorstufe der Quantenmechanik. Sie werden daher falsch, wenn man sie unmittelbar auf die Schrödingersche Theorie übertragen wollte — wenigstens in dem Gebiete, wo beide Theorien auseinandergehen. Man kann aber jedenfalls bereits sagen, daß unsere Resultate asymptotisch richtig bleiben müssen in der Grenze großer Quantenzahlen, da beide Theorien dort ineinander übergehen.

Man kann den Fortschritt zur Schrödingerschen Form der Wellenmechanik dahin charakterisieren, daß sie der Tatsache der „Eingemeindung" der Trajektorien der klassischen Mechanik, denen de Broglie zunächst nur äußerlich durch (5) eine Welle aufgeprägt hatte, zu einem zusammenhängenden Wellenkontinuum Rechnung trägt. In der geometrischen Optik ist die Betrachtung der einzelnen losgelösten Trajektorien und die der Wellenfronten physikalisch äquivalent. In der Wellenoptik dagegen erfährt ein einzelner Wellenstrahl, wenn er einer Front von Strahlen einverleibt wird, einen gewissen Einfluß durch seine Nachbarn. Daß dieser Einfluß zum Ausdruck kommt, ist die charakteristische Aussage der Schrödingerschen Theorie, wenn sie die Wellenfunktion ψ anstatt durch eine Jacobische Differentialgleichung (6) durch eine Wellengleichung beschreibt. Bei Zerlegung in imaginären und reellen Bestandteil lautet die Schrödingersche Wellengleichung für $\psi = |\psi| e^{\frac{2\pi i}{h} W}$ (W reell):

$$\left.\begin{aligned} \left(\frac{h}{2\pi i}\right)^2 \frac{\Box|\psi|}{|\psi|} + \left(\frac{\partial W}{\partial x^i} - \frac{e}{c}\Phi_i\right)\left(\frac{\partial W}{\partial x_i} - \frac{e}{c}\Phi^i\right) + m^2 c^2 = 0, \\ \frac{\partial}{\partial x_k}\left\{|\psi|^2 \frac{e}{m}\left(\frac{\partial W}{\partial x_k} - \frac{e}{c}\Phi_k\right)\right\} = 0. \end{aligned}\right\} \quad (10)$$

In dieser Darstellung erkennt man den Gegensatz zur de Broglieschen Theorie in dem Auftreten des Gliedes $\frac{\Box|\psi|}{|\psi|}$. Zugleich wird hier auch sichtbar, daß es sich um ein Problem mit zwei unbekannten reellen Funktionen handelt. Die zweite Gleichung ist die Kontinuitätsgleichung des Stromes, dessen vier Komponenten durch die geschweiften Klammern eingefaßt werden.

Es ist keine Frage, daß wir gegenwärtig der Schrödingerschen Theorie ihrer Idee nach und wegen ihrer besseren Übereinstimmung mit der Erfahrung unbedingt vor der de Broglieschen den Vorzug zu geben haben. In ihrer Diskrepanz mit der Weylschen Theorie haben wir gewiß keinen Mangel der Schrödingerschen Theorie zu sehen.

Wenn man beachtet, daß sich die Abweichungen charakteristisch bei kleinen Quantenzahlen einstellen, so kann kein Zweifel sein, worauf die Schwierigkeit zurückzuführen sein wird: Die Weylsche Theorie ist ihrer ganzen Kompetenz nach sozusagen auf die klassische Mechanik und somit auch auf die ihr zugeordnete de Brogliesche Theorie zugeschnitten. Es ist demzufolge von ihr gar nicht zu erwarten oder zu verlangen, daß sie auf die Schrödingersche Theorie bereits paßt. Die Aufgabe muß vielmehr sein, an der jetzt veralteten Weylschen Theorie den entsprechenden Schritt zu vollziehen, welcher von de Broglie zu Schrödinger führt, sie muß ihrerseits entsprechend der quantenmechanischen Korrektur der klassischen Gesetze modifiziert werden.

Man kann voraussehen, in welcher Richtung die Korrektur des Weylschen Maßes geschehen wird. Bisher war angenommen, daß die vier Potentiale Φ_i, welche eine vollständige Beschreibung des elektromagnetischen Feldes liefern, einzig für die Streckenverschiebung maßgebend seien (2a). Jetzt hat sich die Sachlage insofern geändert, als zu den vier Zustandsgrößen des Feldes Φ_i als fünfte das Schrödingersche ψ getreten ist, welches in vieler Hinsicht — vor allem in der Darstellung durch ein Variationsproblem [1]) — symmetrisch den Feldgrößen Φ_i gegenübersteht. Die Materie, in der elektronentheoretischen Auffassung hinter undurchdringliche Grenzflächen aus dem Felde verbannt oder in die Singularitäten desselben verwiesen, ist jetzt über den ganzen Raum ausgebreitet, und während man in der Weylschen Theorie sich mit Recht einen Maßstab im „leeren" Raum nur von den dort herrschenden elektromagnetischen Potentialen beeinflußt dachte, wird jetzt dem Umstand Rechnung zu tragen sein, daß die alte Trennung zwischen der „undurchdringlichen" Materie und dem κενὸν aufgehoben ist und man sich stets sozusagen im Innern der alles durchdringenden [2]) neuen Substanz $|\psi|$ befindet.

Es ist also zu erwarten, daß außer den äußeren elektromagnetischen Feldgrößen noch eine innere, die allein von $|\psi|$ abhängt, zu berücksichtigen sein wird. Madelung [3]) hat das „Potential" dieser inneren Wirkung des ψ-Feldes auf sich selbst angegeben. Ich möchte als relativistische Verallgemeinerung desselben vorschlagen:

$$e\Phi_5 = m_0 c^2 \left(1 - \sqrt{1 + \left(\frac{h}{2\pi i}\right)^2 \frac{\Box |\psi|}{m_0^2 c^2 |\psi|}}\right). \tag{11}$$

---

[1]) E. Schrödinger, Ann. d. Phys. **82**, 265, 1927.

[2]) Denn ψ genügt einer linearen Differentialgleichung. Superpositionsprinzip! Dennoch scheint die Eigenschaft der Undurchdringlichkeit in Form des Pauliverbots ihren quantenmechanischen Ausdruck zu finden. (P. Ehrenfest, Naturwissenschaften **15**, 161, 1927.)

[3]) E. Madelung, ZS. f. Phys. **40**, 322, 1926.

Das Wort „Potential" ist mit Vorsicht zu gebrauchen. Φ_5 entspricht nicht etwa dem „skalaren" Potential Φ_4, welches relativistisch als zeitliche Komponente eines Vierervektors figuriert, sondern ist auch relativistisch ein invarianter Skalar. Dementsprechend kann Φ_5 auch nicht die Streckenänderung längs einer bestimmten Weltrichtung regieren. Wenn man überhaupt einen Einfluß auf das Eichmaß annehmen will, so kann er nur vom Betrage der vierdimensionalen Streckenverschiebung abhängen, nicht von ihrer Richtung. Führt man dementsprechend durch das Weltlinienelement $dx_5 = c\,d\tau$ (τ = Eigenzeit) eine fünfte Koordinate ein, welche nicht unabhängig von den übrigen dx_i ist, sondern sich ihnen durch die Bedingung

$$dx_1^2 + dx_2^2 + dx_3^2 + dx_4^2 + dx_5^2 = 0 \tag{12}$$

gleichberechtigt zugesellt [1]), so wird man vermuten, daß

$$l = l_0 e^{\frac{2\pi i}{h}\int \sum_1^5 \frac{e}{c}\Phi_i\,dx^i} \tag{13}$$

die quantenmechanische Verallgemeinerung des Weylschen Streckenmaßes darstellt.

Um die Identität von (13) mit der Schrödingerschen Wellenfunktion nachzuweisen, müssen wir zunächst angeben, längs welchen Weges das verallgemeinerte Streckenmaß (13) zu führen ist. Man wird wieder den Transport mit der Strömungsgeschwindigkeit der Materie vorschreiben wollen. Hierbei ist aber zu beachten, daß jetzt die Komponenten u^i der Vierergeschwindigkeit nicht durch (7) gegeben werden, obwohl die Darstellung des Stromes in der zweiten Gleichung (10) die Abtrennung des Faktors $e\psi\bar{\psi}$ als Ruhladungsdichte nahe legt. Die derart abgetrennten Geschwindigkeitskomponenten würden nämlich wegen (10_1) nicht die Identität der Vierergeschwindigkeit [2])

$$u_k u^k = \frac{dx_k}{d\tau}\frac{dx^k}{d\tau} = -c^2 \tag{12'}$$

[1]) Das Auftreten dieser fünfdimensionalen quadratischen Form ist im Sinne der Weylschen Forderung der Eichinvarianz ganz konsequent. Das Weltlinienelement $d\tau$ bzw. dx_5 ist zwar eine relativistische Invariante, aber keine Eichinvariante (Übergang zu einer anderen Eicheinheit ändert $d\tau$), wohl aber ist das Verschwinden der quadratischen Form (12) eichinvariant. — Offenbar sind in diesem Sinne die fünfdimensionalen Ansätze von Kaluza zu verstehen.

[2]) Wenn nichts anderes angegeben, sind im folgenden die Summationen über gleiche Indizes stets von 1 bis 4 wie bisher zu verstehen.

erfüllen. Es ist vielmehr zu schreiben

$$\frac{d x_k}{d x_5} \equiv \frac{u_k}{c} = \frac{\psi \bar{\psi}}{\varrho} \cdot \frac{e}{m_0 c} \left(\frac{\partial W}{dx} - \frac{e}{c} \Phi_k \right), \tag{7a}$$

wobei der Faktor

$$\varrho = e \psi \bar{\psi} \sqrt{1 + \left(\frac{h}{2 \pi i} \right)^2 \frac{\Box |\psi|}{m_0{}^2 c^2 |\psi|}} = e \psi \bar{\psi} \left(1 - \frac{e}{m_0 c^2} \Phi_5 \right) \tag{14}$$

als „Ruhladungsdichte“ abgetrennt ist.

In dieser Bezeichnung erhält man

$$e \Phi_5 = m_0 c^2 \left(1 - \frac{\varrho}{e \psi \bar{\psi}} \right) \tag{11a}$$

und die erste Schrödingersche Gleichung lautet in fünfdimensionaler Fassung [1]):

$$\sum_1^5 {}_i \left(\frac{\partial W}{d x_i} - \frac{e}{c} \Phi^i \right) \left(\frac{\partial W}{d x^i} - \frac{e}{c} \Phi_i \right) = 0. \tag{10a}$$

Wir vergleichen jetzt die Strecke l (13) entlang der Strömung (7 a) mit dem Schrödingerschen Skalar ψ. Man erhält für ψ / l

$$\frac{\psi}{l} = \frac{|\psi|}{l_0} e^{\frac{2 \pi i}{h} \int \sum_1^5 \left(\frac{\partial W}{\partial x_i} - \frac{e}{c} \Phi^i \right) d x_i},$$

(7 a) ergibt:

$$= \frac{|\psi|}{l_0} e^{\frac{2 \pi i}{h} \int \sum_1^4 \frac{\psi \bar{\psi}}{\varrho} \frac{e}{mc} \left(\frac{\partial W}{\partial x_i} - \frac{e}{c} \Phi^i \right) \left(\frac{\partial W}{\partial x^i} - \frac{e}{c} \Phi_i \right) d x_5 + \left(\frac{\partial W}{\partial x_5} - \frac{e}{c} \Phi_5 \right) d x_5},$$

(11 a) ergibt:

$$= \frac{|\psi|}{l_0} e^{\frac{2 \pi i}{h} \int \frac{\psi \bar{\psi}}{\varrho} \frac{e}{mc} \sum_1^5 \left(\frac{\partial W}{\partial x_i} - \frac{e}{c} \Phi^i \right) \left(\frac{\partial W}{\partial x^i} - \frac{e}{c} \Phi_i \right) \cdot d x_5}$$

$$= \frac{|\psi|}{l_0}.$$

Letzteres wegen (10 a). Man erhält also zunächst nicht $\psi / l =$ konst, sondern

$$\frac{\psi}{l} = \frac{|\psi|}{l_0}, \tag{8a}$$

---

[1]) Hierbei ist zu beachten, daß Φ_5 seinerseits noch selbst eine erst zu bestimmende Unbekannte ist. Bekanntlich ist es ein noch unverstandenes Wunder, warum das gleiche nicht für die Potentiale Φ_1, Φ_2, Φ_3, Φ_4 gilt, wie man erwarten müßte. (E. Schrödinger, Ann. d. Phys. **82**, 265, 1927.) $\frac{\partial W}{\partial x_5}$ ist $= m_0 c$ [vgl. (5 a)].

welches eine eindeutige Ortsfunktion ist[1]). Aber die Potentiale Φ_k sind nur bis auf einen additiven Gradienten physikalisch festgelegt; führe ich statt ihrer

$$\Phi_k^* = \Phi_k - \frac{hc}{2\pi i e}\frac{\partial}{\partial x^k}\ln|\psi|$$

als Potentiale ein, was die elektromagnetischen Feldstärken unberührt läßt, so folgt $\psi/l =$ konst.

Die auf der Resonanz der Wellen beruhende Eindeutigkeit des mit der Strömung mitgeführten Eichmaßes überträgt sich natürlich jetzt ohne weiteres aus der de Broglieschen auf die Schrödingersche Theorie, so daß wir den Überlegungen des 2. Kapitels hier nichts hinzuzufügen haben.

Stuttgart, Physik. Inst. d. techn. Hochschule, 27. Februar 1927.

---

[1]) Man kann diese Beweisführung im Sinne der fünfdimensionalen Geometrie sinngemäßer folgendermaßen aussprechen:

$$\left(\frac{\partial W}{\partial x^i} - \frac{e}{c}\Phi_i\right) \text{ ist parallel dem Fünferstrom } j_i = \frac{e}{m}\psi\bar{\psi}\left(\frac{\partial W}{\partial x^i} - \frac{e}{c}\Phi_i\right),$$

dx^i soll parallel dem Fünferstrom j^i gewählt werden.

Der Fünferstrom ist orthogonal auf sich selbst $\left(\sum_1^5 j_i j^i = 0\right)$; also ist j_i auch orthogonal auf dx^i und also $\sum_1^5 \left(\frac{\partial W}{\partial x^i} - \frac{e}{c}\Phi_i\right) dx^i = 0$.

Ich verdanke diese schöne Formulierung einer Mitteilung von Herrn A. Landé. Hierbei ist die 5. Komponente des Fünferstroms $j_5 = \varrho c$.

1

Quantum mechanical interpretation of Weyl's theory[1]

By **F. London** in Stuttgart

(Received on 25 February 1927.)

Chapter I. Weyl's Theory

As is generally known, recently the idea of a "purely local geometry", first conceived by Riemann, has been completed by Weyl in a remarkably beautiful and simple way. Riemann's notion of space can be viewed as the lifting of the prejudice that the curvature conditions at one point of space control the curvature at all others. In order to make sense of this statement by Riemann, it had to be assumed that the scale which is used at each point to determine the coefficients g_{ik} of the metric fundamental form,

$$ds^2 = g_{ik} dx^i dx^k,$$

is a "rigid" scale.

In contrast to this Weyl claims - and rightly so - that the assumption of such a rigid scale is against a purely local geometry, and that one can only determine sensibly the ratios of the g_{ik} at a point, but not their absolute values. Thus, he makes the ansatz for the change dl of a length scale l under an infinitesimal shift dx^i:

$$dl = l\varphi_i dx^i, \tag{1}$$

where the proportionality factors φ_i are functions of the position in space, *i.e.* characteristics of the scaling relations of space - similar to the g_{ij}. Or, if one integrates (1):

$$l = l_0 e^{\int \varphi_i dx^i} \tag{2}$$

($l_0 = l$ at the beginning of the shift). In general, the gauge scale is path-dependent (non-integrable), unless the quantities

$$f_{ik} = \frac{\partial \varphi_i}{\partial x^k} - \frac{\partial \varphi_k}{\partial x^i} \tag{3}$$

[1] Presented partly at the meeting of the Gauverein Württemberg of the German Physics Society Stuttgart, on 18 December 1926; cf. also a preliminary summary in *Naturwissen.* **15**, 187, 1927

2

vanish. According to their definition (3) the quantities f_{ik} obey the identity (assume that the manifold is of dimension four):

$$\frac{\partial f_ik}{\partial x^l} + \frac{\partial f_{kl}}{\partial x^i} + \frac{\partial f_{li}}{\partial x^k} = 0, \ i \neq k \neq l, \ \ i, k, l = 1, 2, 3, 4 \tag{4}$$

The formal match of these four equations with the system of Maxwell's equations

$$\begin{aligned} \text{rot } \mathcal{E} + \frac{1}{c}\dot{\mathcal{H}} &= 0, \\ \text{div } \mathcal{H} &= 0, \end{aligned}$$

and also other formal analogies have led Weyl to the conclusion that the φ_i should be identified, up to a constant proportionality factor, with the components Φ_i of the electromagnetic four-potential, and accordingly the f_{ik} with the electromagnetic field strengths $\mathcal{E}, \mathcal{H}$. By logically extending the geometric interpretation of gravitation as variable curvature in Riemannian space, Weyl imagined the remaining type of physical forces, the electromagnetic field, also as a property of the scaling relations of space, characterized by the variability of the gauge scale. Thus, one has to write:

$$l = l_0 e^{\alpha \int \phi_i dx_i} \quad (\alpha = \text{ proportionality factor}).$$

One must admire the tremendous courage with which Weyl tracked down his theory of the gauge-geometric interpretation of electromagnetism solely on the grounds of this completely formal relation: in the case of the theory of gravity it was a physical fact, namely, the principle of equivalence of inertial and gravitational mass, which prompted Einstein to his geometric interpretation. In the theory of electricity, however, no such fact was known: there was no reason to think of a universal influence of the electromagnetic field on the so-called rigid scales (or clocks). On the contrary, for example atomic clocks do represent scales, whose independence of the history has been demonstrated by the sharpness of the spectral lines, contradicting to the non-integrable measure (2a), which Weyl assumes in a magnetic field. Apparently, a remarkably clear metaphysical conviction was needed, that caused Weyl - in defiance of such elementary experiences - not to deviate from the thought that Nature must use the geometric possibilities given to her. He held on to his belief and avoided a discussion of the aforementioned contradictions by a somewhat obscure re-interpretation of the notion "real measurement". This, however, removed the theory's striking physical meaning, and it became far less convincing.

I do not need to dwell on this abstract development of the theory. I will rather show that exactly this concise original version of Weyl's theory contains even more possibilities than has been used by its author, namely that one has to view it as not less than the logical path to wave mechanics, and only in the light of this does it gain a directly comprehensible physical meaning.

3

Chapter II: De Broglie's wave mechanics and Weyl's theory

I will call the still imperfect precursor of wave mechanics "de Broglie's theory". There the wave function describing the motion of *one* electron (to which we will restrict ourselves here)

$$\psi = e^{\frac{2\pi i}{h} W(x_i)} \quad i = 1, 2, 3, 4 \tag{5}$$

is determined by a complete solution W of the Hamilton-Jacobi partial differential equation

$$\left(\frac{\partial W}{\partial x^i} - \frac{e}{c}\Phi_i\right)\left(\frac{\partial W}{\partial x_i} - \frac{e}{c}\Phi^i\right) = -m_0^2 c^2 \tag{6}$$

where the integration constants are to be determined in a known fashion such that ψ is a single-valued function of space, *i.e.* W is additively periodic, where the period is an integer multiple of Planck's constant.

If one takes the radical continuum hypothesis seriously, and also the dissolution of the discrete electron into a field variable that depends continuously on space and time, as it is suggested by de Broglie's theory and more consistently by Schrödinger's theory[2] to be discussed below, then one is faced with an extraordinary, fundamental difficulty if one analyzes which meaning should be attached to metric statements within the wave continuum. For in this oscillating and fluctuating, infinitely extended medium, which took the place of the discrete electron, there are no invariable discontinuities, no rigid bodies which, as reproducible scales, could allow a determination of a measure.

I by no means take the view that a practicable measurement must be given in order to speak of the geometry in the atomic region; after all one cannot say that such a prescription exists in the electron theory. But if one wants to assign a defined meaning to a metric specification, it seems to me that the least one can require is the specification of some real object (as a "prototype"), which the metric statement is already related to: an electron diameter or distance, *etc.*, although such a statement might still have a problematic relation to a feasible measurement.

Yet such a real object is not present in the wave continuum. The theorem of identity is not to be applied in the πάντα ῥεῖ of forming and dissipating waves, in the continuum no attribute is to be fixed that would qualify as a reproducible measure. The basic situation one finds oneself in would be completely hopeless, had Weyl not already in his generalisation of Riemann's notion of space created a type of space in which exactly the impossibilty to reproduce the gauge unit is provided as a consistent postulate of a radical local geometry. While until now this theory was an unnecessary burden in the world view of the discontinuous

[2] It is known that there are important reasons, which were pointed out primarily by Born and his collaborators, that the whole wave formalism should be re-interpreted statistically. Insofar as the charge density is re-interpreted as a statistical weight function it is not difficult to see that the same uncertainty with respect to the applicability of the identity theorem, which we point out here, reappears. However, since this point of view initially rejects any interpretation in space and time, it has little interest in the relation to Weyl's theory of space.

electron theory, which was believed to provide exactly these reproducible scales, the situation has fundamentally changed now. One is almost forced to withdraw to Weyl's general notion of space and to try to apply it to Schrödinger's continuum. Then a simple relationship is revealed.

§1. Let us assume we already had a scale l which changes according to Weyl's prescription (2a), and we move it around in the ψ field. Namely, it is moved with the flow velocity of matter, the group velocity

$$u^i = \frac{dx^i}{d\tau} = \frac{1}{m_0}\left(\frac{\partial W}{\partial x_i} - \frac{e}{c}\Phi^i\right). \tag{7}$$

I claim that with this quite natural prescription for the path Weyl's scalar l becomes numerically identical with de Broglie's field scalar ψ. For this, two issues have to be made more precise:

In Weyl's measure the factor α was left undetermined; for this I propose the hypothesis that it be equal to $\frac{2\pi i e}{hc}$. Thus,

$$l = l_0 e^{\frac{2\pi i}{h}\int \frac{e}{c}\phi_i dx^i}. \tag{2a}$$

Also: I do not use exactly the ψ of equation (5), but the five-dimensional ψ, multiplied by a factor $e^{\frac{2\pi i}{h}m_0c^2\tau}$, corresponding to the proposals by Klein, Fock and Kudar, where τ is to be understood as the proper time.[3] Thus let now

$$\psi = e^{\frac{2\pi i}{h}(W + m_0c^2\tau)} \tag{5a},$$

or

$$= e^{\frac{2\pi i}{h}\left\{\int \frac{\partial W}{\partial x^i}dx^i + m_0c^2\tau\right\}}.$$

This quantity has to be compared with Weyl's gauge scale (2a) which is carried along the flow of the continuum. One gets

$$\frac{\psi}{l} = \frac{1}{l_0}e^{\frac{2\pi i}{h}\left\{\int\left(\frac{\partial W}{\partial x^i} - \frac{e}{c}\Phi_i\right)dx^i + m_0c^2\tau\right\}},$$

here the displacements dx^i are to be given by the flow (7):

$$= \frac{1}{l_0}.e^{\frac{2\pi i}{h}\left\{\int\left(\frac{\partial W}{\partial x^i} - \frac{e}{c}\Psi_i\right)\left(\frac{\partial W}{\partial x_i} - \frac{e}{c}\Psi^i\right)\frac{d\tau}{m_0} + m_0c^2\tau\right\}}.$$

As a consequence of the Hamilton-Jacobi differential equation (6) the integrand is $= -m_0c^2$, and one gets:

$$\frac{\psi}{l} = \frac{1}{l_0}.e^{\frac{2\pi i}{h}.\mathrm{const}} = \mathrm{const} \tag{8}$$

[3]This sense of τ, which goes back to Kudar, *Ann. d. Phys.* **81**, 632, 1926, is indeed in agreement with the recently discussed interpretation as an angular coordinate of the self-rotational motion of the electron (*Naturwissenschaften*, **15**, 15, 1927). For this rotation angle is to be viewed as a clock carried by the electron. It transforms like the proper time.

5

The physical object, which behaves as Weyl's measure, is found: the complex amplitude of the de Broglie wave; it is the wave that experiences exactly the influence that Weyl has postulated for his gauge scale and to which he - as a redundant element of the physics at that time - had to assign a metaphysical existence. It is so to speak the prototype of Weyl's scale. And as in the theory of gravity we are free to speak either of deflected light rays and masses, or of their geodesic motion in a Riemann space, so does (8) allow us to interpret the oscillation of matter and the influence exerted upon it by the electric potentials geometrically as a homogeneous Weyl space filled with matter, whose metric connection, however, is not integrable.

According to (2a), in the absence of an electromagnetic field the gauge scale should be a constant. Therefore, one should also get a constant value for de Broglie's wave function if one follows it with the associated flow velocity, *i.e.* group velocity (v always $< c$). This seems to be in contradiction with the fundamental results of de Broglie, according to which the phases of his waves propagate with a much higher phase velocity ($u = c^2/v$). But this is not the case here, since not exactly de Broglie's, but the five-dimensional extended ψ was used above, and this is dispersion-free, with no distinction between group and phase velocity. One can easily convince oneself directly that the plane wave

$$\psi = e^{-\frac{2\pi i}{h}\left(\frac{m_0 c^2}{\sqrt{1-\beta^2}}t - \frac{m_0 v}{\sqrt{1-\beta^2}}x - m_o c^2 \tau\right)} \qquad \left(\beta = \frac{v}{c}\right)$$

indeed has a constant phase along a trajectory with velocity v.

Another objection, namely that here we compare ψ, a density, with a length l does not seem to present any difficulty to me. One should compare ψ with l^{-3} from the outset, which would only imply a change in the choice of the undetermined factor α. It would probably be more natural to learn from the connection revealed here that the same dimension has to be assigned to Weyl's gauge scale l as to de Broglie's ψ right from the outset. Within Weyl's theory one could not make such a statement, because there one did not know anything about the "nature" of l.

A more serious difficulty for the understanding seems to be the complex form of the path connection. It is indeed not possible to restrict oneself to the real part. One has to view this as the counterpart of the fact that the wave function ψ itself is to be understood as intrinsically complex, or rather, it represents a collection of two physical quantities, namely $\psi\bar{\psi}$ and the real part of $\frac{h}{2\pi i}\ln\psi$. That in the variational problem of wave mechanics ψ and $\bar{\psi}$ are to be varied independently is also to be understood in this way. Yet what it means that any spatial interval has to be viewed as a complex quantity, and that the whole of Weyl's variability of the distance measure turns out to be a change solely of the phase at constant modulus, I do not want to discuss as yet.

§ 2. But there is still the objection we pointed out above, that experience speaks against the non-integrability of the gauge scale. One can already see how this difficulty needs to be resolved: quantum theory allows matter to have only a discrete series of states of motion, and one suspects that these distinguished motions allow for the measure to be transported only in such a way that on return to the starting point the phase has completed an integer number of rotations, so that in spite of the non-integrability of the connection at every point the gauge scale is unambiguous. Indeed, one is reminded of the resonance property of de Broglie waves, that by which the old Sommerfeld-Epstein quantization condition was first re-interpreted so influentially by de Broglie. This, however, is linked with the phase velocity; yet, as a consequence of the five-dimensional extension of the wave function the oscillation is dispersion-free and therefore the flow velocity becomes identical to the phase velocity. Because of this and the identity of the wave function ψ with Weyl's measure, it seems already proven[4] that also the Weyl measure, if I move it only along a matter flow that is quantum theoretically possible, shares the resonance of de Broglie waves, and despite the non-integrability of the differential expression (2a) it leads to an unambiguous assignment of a measure at every point. If the uniqueness property had been added axiomatically as a generally accepted empirical fact, then one would have logically been led to the system of discrete states of the "classical" quantum theory and its de Broglie waves.

I don't want to leave this matter without pointing out that the resonance property of Weyl's measure for distances, which we encounter here as a characteristic feature of wave mechanics, was conjectured by Schrödinger[5] already in 1922 as a "remarkable property of quantum orbits", and demonstrated in a number of examples, without its relevance being recognized. The possibility of $\alpha = 2\pi i \frac{e}{hc}$ was also considered, but its advantages over other choices was not acknowledged. Already at that time, Schrödinger had appreciated the characteristic wave mechanical periodicities, which he would meet again later from quite different points of view.

It is therefore probably not unnecessary for me to prove this conjecture of Schrödinger's as it was originally intended, as a theorem in "classical" wave mechanics, independently of the wave mechanical context. The claim is: the exponent of the Weyl measure, integrated along a spatially closed quantum orbit is an integer multiple of Planck's constant:

$$\oint \frac{e}{c}\Phi_i dx^i = nh. \tag{9}$$

In order to prove this, one employs the relation already used in §1:

$$\int \left(\frac{\partial W}{\partial x^i} - \frac{e}{c}\Phi_i\right) dx^i = -\int m_o c^2 d\tau = -\int m_0 c^2 \sqrt{1 - \left(\frac{v}{c}\right)^2} dt.$$

As a consequence of the quantum conditions

$$\sum_1^3 \oint \frac{\partial W}{\partial x^i} dx^i = nh$$

[4]This method of inference is not precise, it will be corrected immediately.

[5]E. Schrödinger, ZS. f. Phys. **12**, 13, 1922.

one obtains from this

$$-\oint\left(\frac{\partial W}{\partial x_4}dx_4-\frac{e}{c}\Phi_i dx^i\right)=-nh=\oint m_o c^2\sqrt{1-\left(\frac{v}{c}\right)^2}\,dt.$$

Assuming that an energy integral exists,

$$\frac{\partial W}{\partial x_4}dx_4=-(E_{\mathrm{kin}}+E_{\mathrm{pot}})dt,$$

thus:

$$\oint\frac{e}{c}\Phi_i dx^i=-nh+\oint\left(-m_0c^2\sqrt{1-\left(\frac{v}{c}\right)^2}+E_{\mathrm{kin}}+E_{\mathrm{pot}}\right)dt.$$

Here the integrals on the RHS vanish due to the relativistic generalization of the virial theorem[6] if it assumed that the potential is homogeneous of degree -1 in the x^i, from which the claim (9) follows immediately.

One can see from this derivation, that the proof of the uniqueness of the Weyl measure succeeds only under two assumptions. Apparently, these assumptions (in particular the first) are rather essential, and it is not possible to avoid them completely. They guarantee certain stationary configurations in space; only these allow us to speak of spatially closed orbits in the Minkowskian world in the first place, a statement, which in general depends on the frame of reference. Thus, one is forced to call these assumptions conditions for the possibility of the application of the identity theorem to the space.

Usually the orbits will not be exactly periodic, but only quasi-periodic. Then one can prove under appropriate continuity assumptions that for sufficiently good

[6] No proof of the relativistic generalisation of the virial theorem in the literature is known to me, so I would like to communicate it here. It is

$$\oint\left(-m_0c^2\sqrt{1-\left(\frac{v}{c}\right)^2}+\frac{m_0c^2}{\sqrt{1-\left(\frac{v}{c}\right)^2}}+E_{\mathrm{pot}}\right)dt=\oint\left(\frac{m_0v^2}{\sqrt{1-\left(\frac{v}{c}\right)^2}}+E_{\mathrm{pot}}\right)dt$$

$$=\oint\left(\sum_{i=1}^{3}p_i\frac{dx^i}{dt}+E_{\mathrm{pot}}\right)dt$$

From this by integration by parts and with the periodicity condition in mind:

$$\oint\left(-\sum_{1}^{3}x^i\frac{dp_i}{dt}+E_{\mathrm{pot}}\right)dt$$

Because of the equations of motion $\frac{dp_i}{dt}-\frac{\partial E_{pot}}{\partial x^i}$, so one obtains

$$=\oint\left(\sum_{1}^{3}x^i\frac{\partial E_{\mathrm{pot}}}{\partial x^i}+E_{\mathrm{pot}}\right)dt$$

Here the integrand vanishes as a consequence of Euler's theorem on homogeneous functions.

approximations to the initial starting points the Weyl measure agrees with its original value up to a given, arbitrarily small amount. More than this is not needed anyway.

It seems extraordinarily satisfying that the transport of the gauge distance always has to happen at the velocity (7) of matter; for a transport with a different velocity would quantum theoretically (or mechanically) not be possible. However, I would like to postpone a closer justification of these connections and their implementation into an epistemologically founded theory of the measure, since for this other quite different issues need to be addressed. Although we have seen how Weyl's ideas have been incorporated into current physical conceptions, which could not have been predicted, I do not believe that one can already be satisfied with what has been gained. I have emphasized the continuum interpretation of quantum mechanics in a one-sided way, which does not correspond to own my belief. Nevertheless, it seemed desirable to me first to follow up this thought carefully to the end. In this sense the remarks in the next chapter have to be regarded as being provisional. I hope to return to the whole context from a more general physical point of view in the near future.

Chapter III. Quantum mechanical re-interpretation of Weyl's theory

The analysis of the previous chapter was explicitly restricted to the precursor of quantum mechanics, which I termed the "de Broglie's theory". It would thus be wrong, if one were to apply it directly to Schrödinger's theory - at least where the two theories diverge. However, one can at least say that our results have to be asymptotically correct in the limit of large quantum number, because there both theories coincide.

One could characterize the progress towards Schrödinger's form of wave mechanics by saying that this form takes into account the "incorporation" of the trajectories of classical mechanics, which de Broglie had imposed upon a wave only superficially by (5), into a connected wave continuum. In geometric optics the consideration of the individual decoupled wave is physically equivalent to the consideration of the wave fronts. In wave optics, however, a single ray experiences the influence of its neighbours when it is merged into a wave front. The expression of this influence, is the characteristic statement of Schrödinger's theory, when it describes the wave function ψ by a wave equation rather than the Jacobi differential equation (6). After decomposition into imaginary and real parts Schrödinger's wave equation for $\psi = |\psi| e^{\frac{2\pi i}{h} W}$ (W real) reads:

$$\left.\begin{aligned} \left(\frac{h}{2\pi i}\right)^2 \frac{\Box|\psi|}{|\psi|} + \left(\frac{\partial W}{\partial x^i} - \frac{e}{c}\Psi_i\right)\left(\frac{\partial W}{\partial x_i} - \frac{e}{c}\Psi^i\right) + m^2c^2 = 0, \\ \frac{\partial}{\partial x_k}\left\{|\psi|^2 \frac{e}{m}\left(\frac{\partial W}{\partial x_k} - \frac{e}{c}\Psi_k\right)\right\} = 0. \end{aligned}\right\} \tag{10}$$

In this representation one recognizes the difference to the Broglie theory: the appearance of the term $\frac{\Box|\psi|}{|\psi|}$. At the same time it becomes clear that this is a problem with two unknown real functions. The second equation is the continuity equation for the current, whose four components are given in the curly brackets.

9

There is no question that we currently really have to prefer Schrödinger's theory to de Broglie's due to its basic principle and because of its better match with experience. Surely, we do not have to view the discrepancy with respect to Weyl's theory as a flaw of Schrödinger's theory.

If one takes into account that the deviations appear typically at small quantum numbers, then there can be no doubt about what the difficulty reduces to: Weyl's theory is in all its competence so to speak tailored to classical mechanics and therefore also to de Broglie's theory associated to it. Thus, it cannot be expected or required of it that it is already compatible with Schrödinger's theory. The task should rather be to change the now outdated Weyl theory analagously to the step from de Broglie to Schrödinger; Weyl's theory must be modified according to the quantum mechanical corrections of the classical laws.

One can foresee in which direction the correction of Weyl's measure will happen. Up until now it was assumed that only the four potentials Φ_i, which provide a complete description of the electromagnetic field, are responsible for the shift in distance (2a). Now the state of affairs has changed insofar as in addition to the four state variables of the field Φ_i there is also the fifth, Schödinger's ψ, which in many respects - in particular in the representation by a variational problem[7] - is on a footing with the field variables Φ_i. Matter, which in the electron-theoretic approach is banished from the field behind impenetrable boundaries, or removed into its singularities, is now distributed across the whole space; and while in Weyl's theory one thought of a measure in "empty" space being influenced by the electromagnetic potentials, and rightly so, now it has to be taken into account that the old distinction between the "impenetrable" matter and the χενὸν does not exists anymore and that one is always so to speak in the interior of the new substance $|\psi|$ that can penetrate[8] everything.

Thus, it can be expected that in addition to the external electro-magnetic field variables, one also needs to take into account an internal one, which only depends on $|\psi|$. Madelung[9] has given the "potential" for this internal action of the ψ-field. I would like to propose as a relativistic generalisation of it:

$$\Phi_5 = m_o c^2 \left(1 - \sqrt{1 + \left(\frac{h}{2\pi i}\right)^2 \frac{\Box|\psi|}{m_0^2 c^2 |\psi|}}\right). \qquad (11)$$

One needs to be careful in using the word "potential". Φ_5 does not correspond to the "scalar" potential Φ_4, which relativistically plays the role of the time component of the four-vector, but it is a relativistic scalar: accordingly, Φ_5 cannot determine the change of the distance along a certain world direction. If one wants to assume any influence on the displacement at all, it can only depend on the modulus of the four-dimensional displacement, not on its direction. If a fifth coordinate is introduced accordingly by the world line element $dx_5 = cd\tau$

[7]E. Schrödinger, Ann. d. Phys. **82**, 265, 1927

[8]For ψ obeys a linear differential equation. Superposition principle! Nevertheless, the property of impenetrability seems to find its quantum mechanical expression in the form of Pauli's exclusion principle.

[9]E. Madelung, ZS. f. Phys., **40**, 322, 1926.

(τ = proper time), which is not independent of the other dx_i, but is on an equal footing with them[10], by the relation

$$dx_1^2 + dx_2^2 + dx_3^2 + dx_4^2 + dx_5^2 = 0 \tag{12}$$

then one will conjecture that

$$l = l_0 \exp\left[\frac{2\pi i}{h}\int\sum_1^5 \frac{e}{c}\Psi_i dx^i\right] \tag{13}$$

is the quantum mechanical generalisation of Weyl's measure for distances.

In order to prove the identity of (13) with Schrödinger's wave function we first have to specify along which path the generalised distance measure has to be transported. Again, one would like to prescribe the transport with the flow velocity of matter. However, it has to be noted, that the components u^i of the four-velocity are not given by (7), even though the representation of the current in the second equation (10) suggests the separation of the factor $e\psi\overline{\psi}$ as charge density at rest. That is to say, the velocity components thus defined would, due to (10_1) not obey the the four-velocity identity [11]

$$u_k u^k = \frac{dx_k}{d\tau}\frac{dx^k}{d\tau} = -c^2 \tag{12'}$$

One should rather write

$$\frac{dx_k}{dx_5} \equiv \frac{u_k}{c} = \frac{\psi\overline{\psi}}{\rho}.\frac{e}{m_0 c}\left(\frac{\partial W}{\partial x^k} - \frac{e}{c}\psi_k\right), \tag{7a}$$

where the factor

$$\rho = e\psi\overline{\psi}\sqrt{1 + \left(\frac{h}{2\pi i}\right)\frac{\Box|\psi|}{m_0^2 c^2|\psi|}} = e\psi\overline{\psi}\left(1 - \frac{e}{m_0 c^2}\phi_5\right) \tag{14}$$

is separated as "charge density at rest".

In this notation one obtains

$$e\Phi_5 = m_0 c^2\left(1 - \frac{\rho}{e\psi\overline{\psi}}\right) \tag{11a}$$

and the first Schrödinger equation in five-dimensional form[12] reads

$$\sum_1^5\left(\frac{\partial W}{\partial x_i} - \frac{e}{c}\Psi^i\right)\left(\frac{\partial W}{\partial x^i} - \frac{e}{c}\Phi_i\right) = 0 \tag{10a}$$

[10] The appearance of this five-dimensional quadratic form is entirely consistent with Weyl's claim for gauge invariance. Although the world line element $d\tau$, or dx_5, is a relativistic invariant, it is not a gauge invariant (transition to a different gauge unit changes $d\tau$), but the vanishing of the quadratic form (12) is gauge invariant. Apparently the five-dimensional ansätze by Kaluza have to be understood in this sense.

[11] Unless stated otherwise in the following the summation over same indices are always understood to run from 1 to 4.

[12] Here it has to be noted that Φ_5 itself is a variable, still to be determined. As known, it is a miracle still not understood why this is not true for the potentials $\Phi_1, \Phi_2, \Phi_3, \Phi_4$, in contrast to what one would expect. (E. Schrödinger, Ann. d. Phys. **82**, 265, 1927.) $\frac{\partial W}{\partial x_5}$ is $= m_0 c$ [cf. (5a)]

We now compare the distance l (13) along the flow (7a) with the Schrödinger scalar ψ. For ψ/l one obtains

$$\frac{\psi}{l} = \frac{|\psi|}{l} \exp\left[\frac{2\pi i}{h}\int\sum_1^5\left(\frac{\partial W}{\partial x_i} - \frac{e}{c}\Phi^i\right)dx_i\right]$$

(7a) gives

$$= \frac{|\psi|}{l}\exp\left[\frac{2\pi i}{h}\int\sum_1^4\frac{\psi\bar\psi}{\rho}\frac{e}{mc}\left(\frac{\partial W}{\partial x_i} - \frac{e}{c}\Psi^i\right)\left(\frac{\partial W}{\partial x^i} - \frac{e}{c}\Psi_i\right)dx_5 + \left(\frac{\partial W}{\partial d_5} - \frac{e}{c}\Psi_5\right)dx_5\right],$$

(11a) gives

$$= \frac{|\psi|}{l_0}\exp\left[\frac{2\pi i}{h}\int\frac{\psi\bar\psi}{\rho}\frac{e}{mc}\sum_1^5\left(\frac{\partial W}{\partial x_i} - \frac{e}{c}\Psi^i\right)\left(\frac{\partial W}{\partial x^i} - \frac{e}{c}\Psi_i\right).dx_5\right]$$

$$= \frac{|\psi|}{l_0}.$$

The last is due to (10a). Thus, initially one does not obtain $\psi/l = \text{const}$, but

$$\frac{\psi}{l} = \frac{|\psi|}{l_0}, \tag{8a}$$

which is an unambiguous function depending on the position.[13] However, the potentials Φ_i are physically determined only up to an additive gradient; if I introduce instead

$$\Phi_k^* = \Phi_k - \frac{hc}{2\pi i e}\frac{\partial}{\partial x^k}\ln|\psi|$$

as potentials, which leaves the electro-magnetic field strengths unaltered, then it follows $\psi/l = \text{const}$.

The single-valuedness of the measure that is carried with the current and which is based on the resonance of the waves, is now carried over naturally from the de Broglie's to Schrödinger's theory, so that we do not have to add anything to the considerations of chapter II.

Stuttgart, Dept. Phys. Techn. Univ., 27 February 1927.

[13]One can formulate the proof in the sense of five-dimensional geometry more accurately as follows:
$\left(\frac{\partial W}{\partial x^i} - \frac{e}{c}\Phi_i\right)$ is parallel to the five-current $j_i = \frac{e}{m}\psi\bar\psi\left(\frac{\partial W}{\partial x^i} - \frac{e}{c}\Phi_i\right)$, dx^i is to be chosen parallel to the five-current. The five-current is orthogonal to itself $\left(\sum_1^5 j_i j^i = 0\right)$; Thus, j_i is also orthogonal to dx^i and thus $\sum_1^5\left(\frac{\partial W}{\partial x^i} - \frac{e}{c}\Phi_i\right)dx^i = 0$.
I owe this nice formulation to a communication by Mr A. Landé. Here the 5th component of the five-current is $j_5 = \varrho c$.

Quantentheorie und fünfdimensionale Relativitätstheorie.

Von Oskar Klein in Kopenhagen.

(Eingegangen am 28. April 1926.)

Auf den folgenden Seiten möchte ich auf einen einfachen Zusammenhang hinweisen zwischen der von Kaluza[1]) vorgeschlagenen Theorie für den Zusammenhang zwischen Elektromagnetismus und Gravitation einerseits und der von de Broglie[2]) und Schrödinger[3]) angegebenen Methode zur Behandlung der Quantenprobleme andererseits. Die Theorie von Kaluza geht darauf hinaus, die zehn Einsteinschen Gravitationspotentiale g_{ik} und die vier elektromagnetischen Potentiale φ_i in Zusammenhang zu bringen mit den Koeffizienten γ_{ik} eines Linienelementes von einem Riemannschen Raum, der außer den vier gewöhnlichen Dimensionen noch eine fünfte Dimension enthält. Die Bewegungsgleichungen der elektrischen Teilchen nehmen hierbei auch in elektromagnetischen Feldern die Gestalt von Gleichungen geodätischer Linien an. Wenn dieselben als Strahlengleichungen gedeutet werden, indem die Materie als eine Art Wellenausbreitung betrachtet wird, kommt man fast von selbst zu einer partiellen Differentialgleichung zweiter Ordnung, die als eine Verallgemeinerung der gewöhnlichen Wellengleichung angesehen werden kann. Werden nun solche Lösungen dieser Gleichung betrachtet, bei denen die fünfte Dimension rein harmonisch auftritt mit einer bestimmten mit der Planckschen Konstante zusammenhängenden Periode, so kommt man eben zu den obenerwähnten quantentheoretischen Methoden.

§ 1. Fünfdimensionale Relativitätstheorie. Ich fange damit an, eine kurze Darstellung von der fünfdimensionalen Relativitätstheorie zu geben, die sich nahe an die Theorie von Kaluza anschließt, aber in einigen Punkten von derselben abweicht.

Betrachten wir ein fünfdimensionales Riemannsches Linienelement, für welches wir einen vom Koordinatensystem unabhängigen Sinn postulieren. Wir schreiben dasselbe:

$$d\sigma = \sqrt{\Sigma \gamma_{ik}\, dx^i\, dx^k}, \tag{1}$$

wo das Zeichen Σ, wie überall im folgenden, eine Summation über die doppelt vorkommenden Indizes von 0 bis 4 angibt. Hierbei bezeichnen $x^0 \ldots x^4$ die fünf Koordinaten des Raumes. Die 15 Größen γ_{ik} sind die kovarianten Komponenten eines fünfdimensionalen symmetrischen Tensors. Um von denselben zu den Größen g_{ik} und φ_i der gewöhnlichen Relativitätstheorie zu kommen, müssen wir gewisse spezielle Annahmen machen. Erstens müssen vier der Koordinaten, sagen wir x^1, x^2, x^3, x^4, stets den gewöhnlichen Zeitraum charakterisieren. Zweitens dürfen die Größen

[1]) Th. Kaluza, Sitzungsber. d. Berl. Akad. 1921, S. 966.
[2]) L. de Broglie, Ann. d. Phys. (10) **3**, 22, 1925. Thèses, Paris 1924.
[3]) E. Schrödinger, Ann. d. Phys. **79**, 361 und 489, 1926.

γ_{ik} nicht von der fünften Koordinate x^0 abhängen. Hieraus folgt, daß die erlaubten Koordinatentransformationen sich auf die folgende Gruppe beschränken[1]):

$$\left.\begin{aligned} x^0 &= x^{0\prime} + \psi_0(x^{1\prime}, x^{2\prime}, x^{3\prime}, x^{4\prime}), \\ x^i &= \psi_i(x^{1\prime}, x^{2\prime}, x^{3\prime}, x^{4\prime}) \qquad (i = 1, 2, 3, 4). \end{aligned}\right\} \tag{2}$$

Eigentlich hätten wir in der ersten Gleichung Konstante mal $x^{0\prime}$ anstatt $x^{0\prime}$ schreiben sollen. Die Beschränkung auf den Wert Eins der Konstante ist ja aber ganz unwesentlich.

Wie man leicht zeigt, bleibt γ_{00} bei den Transformationen (2) invariant. Die Annahme $\gamma_{00} =$ Const. ist deshalb zulässig. Die Vermutung liegt nahe, daß nur die Verhältnisse der γ_{ik} einen physikalischen Sinn haben. Dann ist diese Annahme nur eine immer mögliche Konvention. Indem wir die Maßeinheit von x^0 vorläufig unbestimmt lassen, setzen wir:

$$\gamma_{00} = \alpha. \tag{3}$$

Man zeigt ferner, daß die folgenden Differentialgrößen bei den Transformationen (2) invariant bleiben, nämlich[1]):

$$d\vartheta = dx^0 + \frac{\gamma_{0i}}{\gamma_{00}}\, dx^i, \tag{4}$$

$$ds^2 = \left(\gamma_{ik} - \frac{\gamma_{0i}\gamma_{0k}}{\gamma_{00}}\right) dx^i\, dx^k. \tag{5}$$

In diesen Ausdrücken soll über die doppelt vorkommenden Indizes von 1 bis 4 summiert werden. Bei solchen Summen wollen wir, wie üblich, das Summenzeichen fortlassen. Die Größen $d\vartheta$ und ds hängen in der folgenden Weise mit dem Linienelement $d\sigma$ zusammen:

$$d\sigma^2 = \alpha\, d\vartheta^2 + ds^2. \tag{6}$$

Auf Grund der Invarianz von $d\vartheta$ und γ_{00} folgt nun, daß die vier γ_{0i} $(i \neq 0)$, wenn x^0 festgehalten wird, sich wie die kovarianten Komponenten eines gewöhnlichen Vierervektors transformieren. Wenn x^0 mittransformiert wird, tritt noch der Gradient eines Skalars additiv hinzu. Dies bedeutet, daß die Größen:

$$\frac{\partial \gamma_{0i}}{\partial x^k} - \frac{\partial \gamma_{0k}}{\partial x^i}$$

[1]) Vgl. H. A. Kramers, Proc. Amsterdam 23, Nr. 7, 1922, wo eine an die nun folgenden Betrachtungen erinnernde Überlegung mit einem einfachen Beweis für die Invarianz von $d\vartheta$ und ds^2 gegeben ist.

sich wie die kovarianten Komponenten F_{ik} des elektromagnetischen Feldtensors transformieren. Die Größen γ_{0i} verhalten sich also vom invariantentheoretischen Gesichtspunkt wie die elektromagnetischen Potentiale φ_i. Wir nehmen deshalb an:

$$d\vartheta = dx^0 + \beta\varphi_i dx^i, \tag{7}$$

d. h.

$$\gamma_{0i} = \alpha\beta\varphi_i \qquad (i = 1, 2, 3, 4), \tag{8}$$

wo β eine Konstante bedeutet, und wo die φ_i so definiert sind, daß in rechtwinkligen Galileischen Koordinaten gilt:

$$\left.\begin{aligned}(\varphi_x, \varphi_y, \varphi_z) &= A,\\ \varphi_z &= -cV,\end{aligned}\right\} \tag{9}$$

wo A das gewöhnliche Vektorpotential, V das gewöhnliche skalare Potential und c die Lichtgeschwindigkeit bezeichnen.

Die Differentialform ds wollen wir mit dem Linienelement der gewöhnlichen Relativitätstheorie identifizieren. Wir setzen also

$$\gamma_{ik} = g_{ik} + \alpha\beta^2\varphi_i\varphi_k, \tag{10}$$

wobei wir die g_{ik} so wählen wollen, daß in rechtwinkligen Galileischen Koordinaten gilt:

$$ds^2 = dx^2 + dy^2 + dz^2 - c^2 dt^2. \tag{11}$$

Hiermit sind die Größen γ_{ik} auf bekannte Größen zurückgeführt. Das Problem ist nun, solche Feldgleichungen für die Größen γ_{ik} aufzustellen, daß sich für die g_{ik} und φ_i in genügender Annäherung die Feldgleichungen der gewöhnlichen Relativitätstheorie ergeben. Auf dieses schwierige Problem wollen wir hier nicht näher eingehen, sondern wir wollen nur zeigen, daß die gewöhnlichen Feldgleichungen von dem Gesichtspunkt der fünfdimensionalen Geometrie sich einfach zusammenfassen lassen. Wir bilden die Invariante:

$$P = \sum \gamma^{ik}\left[\frac{\partial\begin{Bmatrix} i\mu \\ \mu \end{Bmatrix}}{\partial x^k} - \frac{\partial\begin{Bmatrix} ik \\ \mu \end{Bmatrix}}{\partial x^\mu} + \begin{Bmatrix} i\mu \\ \nu \end{Bmatrix}\begin{Bmatrix} k\nu \\ \mu \end{Bmatrix} - \begin{Bmatrix} ik \\ \mu \end{Bmatrix}\begin{Bmatrix} \mu\nu \\ \nu \end{Bmatrix}\right], \tag{12}$$

wo γ^{ik} die kontravarianten Komponenten des fünfdimensionalen metrischen Fundamentaltensors sind und wo $\begin{Bmatrix} rs \\ i \end{Bmatrix}$ die Christoffelschen Dreiindizessymbole bezeichnen, also:

$$\begin{Bmatrix} rs \\ i \end{Bmatrix} = \frac{1}{2}\sum \gamma^{i\mu}\left(\frac{\partial\gamma_{\mu r}}{\partial x^s} + \frac{\partial\gamma_{\mu s}}{\partial x^r} - \frac{\partial\gamma_{rs}}{\partial x^\mu}\right). \tag{13}$$

In dem Ausdruck von P denken wir uns, daß alle Größen von x^0 unabhängig sind, und daß $\gamma_{00} = \alpha$ ist.

Betrachten wir nun das über ein geschlossenes Gebiet des fünfdimensionalen Raumes ausgeführte Integral:

$$J = \int P \sqrt{-\gamma}\, dx^0\, dx^1\, dx^2\, dx^3\, dx^4, \tag{14}$$

wo γ die Determinante der γ_{ik} bedeutet.

Wir bilden δJ durch Variieren der Größen γ_{ik} und $\frac{\partial \gamma_{ik}}{\partial x^l}$, wobei deren Randwerte nicht verändert werden sollen. Hierbei soll α als eine Konstante betrachtet werden. Das Variationsprinzip:

$$\delta J = 0 \tag{15}$$

führt dann zu den folgenden Gleichungen:

$$R^{ik} - \frac{1}{2} g^{ik} R + \frac{\alpha \beta^2}{2} S^{ik} = 0 \quad (i, k = 1, 2, 3, 4) \tag{16a}$$

und

$$\frac{\partial \sqrt{-g}\, F^{i\mu}}{\partial x^\mu} = 0 \quad (i = 1, 2, 3, 4), \tag{16b}$$

wo R die Einsteinsche Krümmungsinvariante, R^{ik} die kontravarianten Komponenten des Einsteinschen Krümmungstensors, g^{ik} die kontravarianten Komponenten des Einsteinschen Fundamentaltensors, S^{ik} die kontravarianten Komponenten des elektromagnetischen Energie-Impulstensors, g die Determinante der g_{ik} und schließlich $F^{i\mu}$ die kontravarianten Komponenten des elektromagnetischen Feldtensors bedeuten. Setzen wir

$$\frac{\alpha \beta^2}{2} = \varkappa, \tag{17}$$

wo $\varkappa$ die von Einstein gebrauchte Gravitationskonstante bedeutet, so sehen wir, daß die Gleichungen (16a) in der Tat mit den Gleichungen der Relativitätstheorie für das Gravitationsfeld und (16b) mit den generalisierten Maxwellschen Gleichungen der Relativitätstheorie identisch sind für einen materiefreien Feldpunkt[1]).

Wenn wir uns auf die in der Elektronentheorie und der Relativitätstheorie übliche schematische Behandlungsweise der Materie beschränken, können wir die gewöhnlichen Gleichungen für den nicht materiefreien Fall in ähnlicher Weise erhalten. Wir ersetzen P in (14) durch

$$P + \varkappa \Sigma \gamma_{ik} \Theta^{ik}.$$

Um die Θ^{ik} zu definieren, wollen wir erst den auf ein Elektron oder einen Wasserstoffkern bezüglichen Tensor:

$$\vartheta^{ik} = \frac{dx^i}{dl} \frac{dx^k}{dl} \tag{18}$$

[1]) Siehe z. B. W. Pauli, Relativitätstheorie, S. 719 und 724.

betrachten, wo dx^i die Lageänderungen des Teilchens bezeichnen, und dl ein gewisses invariantes Differential bedeutet. Die Θ^{ik} sollen gleich der auf die Volumeneinheit bezogenen Summe der ϑ^{ik} für die verschiedenen Teilchen sein. Wir kommen dann wieder zu Gleichungen vom gewöhnlichen Typus, die mit den gewöhnlichen Feldgleichungen identisch werden, wenn wir setzen:

$$v_0 \frac{d\tau}{dl} = \pm \frac{e}{\beta c}, \tag{19}$$

$$\frac{d\tau}{dl} = \begin{cases} \sqrt{M} \\ \sqrt{m} \end{cases}, \tag{20}$$

wo allgemein

$$v_i = \Sigma \gamma_{i\mu} \frac{dx^\mu}{dl} \tag{21}$$

die kovarianten Komponenten des fünfdimensionalen Geschwindigkeitsvektors v^i sind, wo

$$v^i = \frac{dx^i}{dl}. \tag{22}$$

Ferner bedeuten e das elektrische Elementarquantum, M und m die Massen von Wasserstoffkern bzw. Elektron. Dabei gilt das obere Wertsystem für den Kern, das untere für das Elektron. Weiter ist

$$d\tau = \frac{1}{c} \sqrt{-ds^2}$$

das Differential der Eigenzeit.

Aus den Feldgleichungen folgen natürlich auf gewöhnliche Weise die Bewegungsgleichungen für materielle Teilchen und die Kontinuitätsgleichung. Die Rechnungen, die dazu führen, können von unserem Standpunkt aus einfach zusammengefaßt werden. Wie man leicht sieht, sind nämlich unsere Feldgleichungen mit den folgenden 14 Gleichungen äquivalent:

$$P^{ik} - \tfrac{1}{2}\gamma^{ik} P + \varkappa \Theta^{ik} = 0 \tag{23}$$

($i, k = 0, 1, 2, 3, 4$, aber nicht beide Null), wo die P^{ik} die kontravarianten Komponenten des verjüngten fünfdimensionalen Krümmungstensors sind (den R^{ik} entsprechend). Die in Frage stehenden Gleichungen folgen nun durch Divergenzbildung von (23). Hieraus folgt, daß sich die elektrischen Teilchen auf fünfdimensionalen geodätischen Linien bewegen, die den Bedingungen (19) und (20) genügen [1]). Wie man

[1]) Die speziellen Werte von $\frac{d\tau}{dl}$ sind natürlich in diesem Zusammenhang ohne Bedeutung. Wesentlich ist hier nur $\frac{d\tau}{dl} = \text{const.}$

sofort sieht, sind diese Bedingungen eben deshalb mit den Gleichungen der geodätischen Linien verträglich, weil x^0 in den γ_{ik} nicht vorkommt.

Es muß hier daran erinnert werden, daß wohl keine genügenden Gründe für die exakte Gültigkeit der Einsteinschen Feldgleichungen vorliegen. Immerhin möchte es nicht ohne Interesse sein, daß sich sämtliche 14 Feldgleichungen in so einfacher Weise vom Standpunkt der Theorie von Kaluza zusammenfassen lassen.

§ 2. Die Wellengleichung der Quantentheorie. Wir gehen nun dazu über, die Theorie der stationären Zustände und die damit zusammenhängenden charakteristischen Abweichungen von der Mechanik, die in der neueren Quantentheorie zum Vorschein kommen, in Beziehung zu der fünfdimensionalen Relativitätstheorie zu bringen. Betrachten wir zu diesem Zweck die folgende Differentialgleichung, die sich auf unseren fünfdimensionalen Raum beziehen soll und als eine einfache Verallgemeinerung der Wellengleichung betrachtet werden kann:

$$\sum a^{ik}\left(\frac{\partial^2 U}{\partial x^i \partial x^k} - \sum \begin{Bmatrix} i\,k \\ r \end{Bmatrix} \frac{\partial U}{\partial x^r}\right) = 0. \tag{24}$$

Hier bedeuten die a^{ik} die kontravarianten Komponenten eines fünfdimensionalen symmetrischen Tensors, die gewisse Funktionen der Koordinaten sein sollen. Die Gleichung (24) besteht unabhängig vom Koordinatensystem.

Betrachten wir erst eine durch (24) bestimmte Wellenausbreitung, die dem Grenzfall der geometrischen Optik entspricht. Wir kommen dazu, wenn wir setzen:

$$u = A e^{i\omega\Phi} \tag{25}$$

und ω als so groß annehmen, daß in (24) nur die mit ω^2 proportionalen Glieder berücksichtigt zu werden brauchen. Wir bekommen dann:

$$\sum a^{ik} \frac{\partial \Phi}{\partial x^i} \frac{\partial \Phi}{\partial x^k} = 0, \tag{26}$$

eine Gleichung, die der Hamilton-Jacobischen partiellen Differentialgleichung der Mechanik entspricht. Setzen wir

$$p_i = \frac{\partial \Phi}{\partial x^i}, \tag{27}$$

so können die Differentialgleichungen der Strahlen bekanntlich in der folgenden Hamiltonschen Form geschrieben werden:

$$\frac{d p_i}{-\dfrac{\partial H}{\partial x^i}} = \frac{d x^i}{\dfrac{\partial H}{\partial p_i}} = d\lambda, \tag{28}$$

wo

$$H = \frac{1}{2} \sum a^{ik} p_i p_k. \tag{29}$$

Aus (26) folgt noch

$$H = 0. \tag{30}$$

Eine andere Darstellung dieser Gleichungen, die der Lagrangeschen Form entspricht, ergibt sich durch den Umstand, daß die Strahlen als geodätische Nullinien der Differentialform:

$$\sum a_{ik}\, d x^i\, d x^k$$

betrachtet werden können, wo die a_{ik} die zu den a^{ik} reziproken Größen bedeuten, also

$$\sum a_{i\mu}\, a^{k\mu} = \delta_i^k = \begin{cases} 1, & i = k \\ 0, & i \neq k \end{cases}. \tag{31}$$

Setzen wir nun

$$\sum a_{ik}\, d x^i\, d x^k = \mu\, (d\vartheta)^2 + d s^2, \tag{32}$$

so können wir durch passende Wahl der Konstante μ erreichen, daß unsere Strahlengleichungen mit den Bewegungsgleichungen elektrischer Teilchen identisch werden. Setzen wir, um dies einzusehen:

$$L = \frac{1}{2} \mu \left(\frac{d\vartheta}{d\lambda}\right)^2 + \frac{1}{2} \left(\frac{d s}{d\lambda}\right)^2, \tag{33}$$

so folgt

$$p_0 = \frac{\partial L}{\partial \frac{d x^0}{d\lambda}} = \mu \frac{d\vartheta}{d\lambda} \tag{34}$$

und

$$p_i = \frac{\partial L}{\partial \frac{d x_i}{d\lambda}} = u_i \frac{d\tau}{d\lambda} + \beta\, p_0\, \varphi_i \quad (i = 1, 2, 3, 4), \tag{35}$$

wo $u_1 \ldots u_4$ die kovarianten Komponenten des gewöhnlichen Geschwindigkeitsvektors bedeuten.

Die Strahlengleichungen lauten nun:

$$\frac{d p_0}{d\lambda} = 0, \tag{36 a}$$

$$\frac{d p_i}{d\lambda} = \frac{1}{2} \frac{\partial g_{\mu\nu}}{\partial x^i} \frac{d x^\mu}{d^\lambda} \frac{d x^\nu}{d\lambda} + \beta p_0 \frac{\partial \varphi_\mu}{\partial x^i} \frac{d x^\mu}{d\lambda} \quad (i = 1, 2, 3, 4). \tag{36}$$

Aus

$$\mu\, d\vartheta^2 + d s^2 = \mu\, d\vartheta^2 - c^2\, d\tau^2 = 0$$

ergibt sich

$$\mu \frac{d\vartheta}{d\tau} = c \sqrt{\mu}. \tag{37}$$

Da nach (34) und (36 a) $\frac{d\vartheta}{d\lambda}$ und also auch $\frac{d\tau}{d\lambda}$ konstant ist, können wir λ so wählen, daß

$$\frac{d\tau}{d\lambda} = \begin{cases} M \text{ für den Wasserstoffkern,} \\ m \text{ für das Elektron.} \end{cases} \tag{38}$$

Ferner müssen wir, um zu den gewöhnlichen Bewegungsgleichungen zu gelangen, annehmen:

$$\beta p_0 = \begin{cases} +\frac{e}{c} \text{ für den Wasserstoffkern,} \\ -\frac{e}{c} \text{ für das Elektron.} \end{cases} \tag{39}$$

Aus (37) ergibt sich dann:

$$\mu = \begin{cases} \frac{e^2}{\beta^2 M^2 c^4} \text{ für den Wasserstoffkern,} \\ \frac{e^2}{\beta^2 m^2 c^4} \text{ für das Elektron.} \end{cases} \tag{40}$$

Die Gleichungen (35), (36) stimmen dann mit den gewöhnlichen Bewegungsgleichungen elektrischer Teilchen in Gravitationsfeldern und elektromagnetischen Feldern vollständig überein. Insbesondere sind die nach (35) definierten Größen p_i identisch mit den auf gewöhnliche Weise definierten generalisierten Momenten, was für die folgenden Überlegungen wichtig ist. Da wir β noch beliebig wählen können, wollen wir setzen:

$$\beta = \frac{e}{c}. \tag{41}$$

Es ergibt sich dann einfach:

$$p_0 = \begin{cases} +1 \text{ für den Wasserstoffkern,} \\ -1 \text{ für das Elektron,} \end{cases} \tag{39a}$$

und

$$\mu = \begin{cases} \frac{1}{M^2 c^2} \text{ für den Wasserstoffkern,} \\ \frac{1}{m^2 c^2} \text{ für das Elektron.} \end{cases} \tag{40a}$$

Wie man sieht, müssen wir in (37) für die Quadratwurzel das positive Zeichen im Falle des Kerns und das negative Zeichen im Falle des Elektrons wählen. Dies ist ja wenig befriedigend. Die Tatsache aber, daß man bei einem einzigen Werte von μ zwei verschiedene Klassen von Strahlen erhält, die sich gewissermaßen wie die positiven und negativen elektrischen Teilchen zueinander verhalten, könnte als ein Hinweis darauf angesehen werden, daß es vielleicht möglich ist, die Wellengleichung so abzuändern, daß sich die Bewegungsgleichungen beider Arten von Teilchen aus einem einzigen Wertsystem der Koeffizienten ergeben. Auf diese Frage wollen wir jetzt nicht weiter eingehen, sondern wir wollen dazu übergehen, die aus (32) folgende Wellengleichung im Falle des Elektrons etwas näher zu betrachten.

Da für das Elektron $p_0 = -1$ angenommen wurde, müssen wir nach (27) setzen:

$$\Phi = -x^0 + S(x^1, x^2, x^3, x^4). \tag{42}$$

Die Theorie von de Broglie ergibt sich nun, wenn wir die mit der Wellengleichung verträglichen, einem bestimmten Wert von ω entsprechenden stehenden Schwingungen aufsuchen, und dabei annehmen, daß die Wellenausbreitung nach den Gesetzen der geometrischen Optik vor sich geht. Dazu bedürfen wir des wohlbekannten Satzes von der Erhaltung der Phase, der sich sofort aus (28 und (30) ergibt. Es folgt nämlich:

$$\frac{d\Phi}{d\lambda} = \sum \frac{\partial \Phi}{\partial x^i}\frac{d x^i}{d\lambda} = \sum p_i \frac{\partial H}{\partial p_i} = 2H = 0. \tag{43}$$

Die Phase wird also von der Welle mitgeführt. Betrachten wir nun den einfachen Fall, wo sich Φ in zwei Teile spalten läßt, von denen der eine Teil nur von einer einzigen Koordinate, sagen wir x, abhängt, die mit der Zeit periodisch hin und her schwingt. Es wird dann eine stehende Schwingung möglich sein, die dadurch charakterisiert wird, daß eine in einem gewissen Augenblick durch (25) dargestellte harmonische Welle nach einer Periode von x mit derjenigen Welle in Phase zusammentrifft, die sich aus derselben Lösung (25) durch Einsetzen der neuen Werte von x^0, x^2, x^3, x^4 ergibt. Wegen der Erhaltung der Phase ist die Bedingung dafür einfach:

$$\omega \oint p\, dx = n \,.\, 2\pi, \tag{44}$$

wo n eine ganze Zahl bedeutet. Setzen wir:

$$\omega = \frac{2\pi}{h}, \tag{45}$$

wo h die Plancksche Konstante bedeutet, so ergibt sich also die gewöhnliche Quantenbedingung für eine separierbare Koordinate. Ähnliches gilt natürlich für ein beliebiges Periodizitätssystem. Die gewöhnliche Quantentheorie der Periodizitätssysteme entspricht also vollständig der Behandlung der Interferenzerscheinungen mittels der Annahme, daß sich die Wellen nach den Gesetzen der geometrischen Optik ausbreiten. Es mag noch hervorgehoben werden, daß wegen (42) die Beziehungen (44), (45) bei den Koordinatentransformationen (2) invariant bleiben.

Betrachten wir nun auch die Gleichung (24) in dem Falle, wo ω nicht so groß ist, daß wir nur die in ω quadratischen Glieder zu berücksichtigen brauchen. Wir beschränken uns dabei auf den einfachen Fall eines elektrostatischen Feldes. Dann haben wir in kartesischen Koordinaten:

$$\left.\begin{aligned} d\vartheta &= dx^0 - eV dt, \\ ds^2 &= dx^2 + dy^2 + dz^2 - c^2 dt^2. \end{aligned}\right\} \tag{46}$$

Also ergibt sich:

$$H = \frac{1}{2}(p_x^2 + p_y^2 + p_z^2) - \frac{1}{2c^2}(p_t + eVp_0)^2 + \frac{m^2c^2}{2}p_0^2. \tag{47}$$

In der Gleichung (24) können wir nun die mit $\left\{ {ik \atop r} \right\}$ proportionalen Größen vernachlässigen, denn die Dreiindizessymbole sind in diesem Falle nach (17) kleine mit der Gravitationskonstante $\varkappa$ proportionale Größen. Wir bekommen also[1]):

$$\varDelta U - \frac{1}{c^2}\frac{\partial^2 U}{\partial t^2} - \frac{2eV}{c^2}\frac{\partial^2 U}{\partial t \partial x^0} + \left(m^2c^2 - \frac{e^2V^2}{c^2}\right)\frac{\partial^2 U}{\partial x^{0\,2}} = 0. \tag{48}$$

Für U können wir, da V nur von x, y, z abhängt, in Übereinstimmung mit (42) und (45) ansetzen:

$$U = e^{-2\pi i\left(\frac{x^0}{h} - \nu t\right)}\psi(x, y, z). \tag{49}$$

Dies in (48) eingeführt, ergibt:

$$\varDelta\psi + \frac{4\pi^2}{c^2h^2}[(h\nu - eV)^2 - m^2c^4]\psi = 0. \tag{50}$$

Setzen wir noch:

$$h\nu = mc^2 + E, \tag{51}$$

so bekommen wir die von Schrödinger[2]) gegebene Gleichung, deren stehende Schwingungen bekanntlich Werten von E entsprechen, die mit

[1]) Außer durch das Auftreten von x^0, das ja für die Anwendungen unwesentlich ist, unterscheidet sich diese Gleichung von der Schrödingerschen Gleichung durch die Art, in welcher in (48) die Zeit auftritt. Als eine Stütze für diese Form der Quantengleichung kann angeführt werden, daß dieselbe in dem Fall, wo V harmonisch von der Zeit abhängt, wie man durch eine einfache Störungsrechnung zeigen kann, Lösungen besitzt, die sich in ähnlicher Weise zu der Dispersionstheorie von Kramers verhalten wie die Schrödingerschen Lösungen zu der Quantentheorie der Spektrallinien. Diese Bemerkung verdanke ich Dr. W. Heisenberg.

[2]) Schrödinger, l. c.

den aus der Heisenbergschen Quantentheorie berechneten Energiewerten identisch sind. Wie man sieht, ist E in dem Grenzfall der geometrischen Optik gleich der auf gewöhnliche Weise definierten mechanischen Energie. Die Frequenzbedingung besagt, wie Schrödinger hervorgehoben hat, nach (51), daß die zu dem System gehörenden Lichtfrequenzen den aus den verschiedenen Werten der Frequenz ν gebildeten Differenzen gleich sind.

§ 3. Schlußbemerkungen. Wie die Arbeiten von de Broglie sind obenstehende Überlegungen aus dem Bestreben entstanden, die Analogie zwischen Mechanik und Optik, die in der Hamiltonschen Methode zum Vorschein kommt, für ein tieferes Verständnis der Quantenerscheinungen auszunutzen. Daß dieser Analogie ein reeller physikalischer Sinn zukommt, scheint ja die Ähnlichkeit der Bedingungen für die stationären Zustände von Atomsystemen mit den Interferenzerscheinungen der Optik anzudeuten. Nun stehen bekanntlich Begriffe wie Punktladung und materieller Punkt schon der klassischen Feldphysik fremd gegenüber. Auch wurde ja öfters die Hypothese ausgesprochen, daß die materiellen Teilchen als spezielle Lösungen der Feldgleichungen aufzufassen sind, welche das Gravitationsfeld und das elektromagnetische Feld bestimmen. Es liegt nahe, die genannte Analogie zu dieser Vorstellung in Beziehung zu bringen. Denn nach dieser Hypothese ist es ja nicht so befremdend, daß die Bewegung der materiellen Teilchen Ähnlichkeiten aufweist mit der Ausbreitung von Wellen. Die in Rede stehende Analogie ist jedoch unvollständig, solange man eine Wellenausbreitung in einem Raum von nur vier Dimensionen betrachtet. Dies kommt schon in der variablen Geschwindigkeit der materiellen Teilchen zum Vorschein. Denkt man sich aber die beobachtete Bewegung als eine Art Projektion auf den Zeitraum von einer Wellenausbreitung, die in einem Raum von fünf Dimensionen stattfindet, so läßt sich, wie wir sahen, die Analogie vollständig machen. Mathematisch ausgedrückt heißt dies, daß die Hamilton-Jacobische Gleichung nicht als Charakteristikengleichung einer vierdimensionalen, wohl aber einer fünfdimensionalen Wellengleichung aufgefaßt werden kann. In dieser Weise wird man zu der Theorie von Kaluza geführt.

Obwohl die Einführung einer fünften Dimension in unsere physikalischen Betrachtungen von vornherein befremdend sein mag, wird eine radikale Modifikation der den Feldgleichungen zugrunde gelegten Geometrie doch wieder in ganz anderer Weise durch die Quantentheorie nahegelegt. Denn es ist bekanntlich immer weniger wahrscheinlich geworden, daß die Quantenerscheinungen eine einheitliche raumzeitliche Beschreibung zulassen, wogegen die Möglichkeit, diese Erscheinungen durch ein System von fünfdimensionalen Feldgleichungen darzustellen, wohl nicht von vornherein auszuschließen ist [1]). Ob hinter diesen Andeutungen von Möglichkeiten etwas Wirkliches besteht, muß natürlich die Zukunft entscheiden. Jedenfalls muß betont werden, daß die in dieser Note versuchte Behandlungsweise, sowohl was die Feldgleichungen als auch die Theorie der stationären Zustände betrifft, als ganz provisorisch zu betrachten ist. Dies kommt wohl besonders in der auf S. 898 erwähnten schematischen Behandlungsweise der Materie zum Vorschein, sowie in dem Umstand, daß die zwei Arten von elektrischen Teilchen durch verschiedene Gleichungen vom Schrödingerschen Typus behandelt werden. Auch wird die Frage ganz offen gelassen, ob man sich bei der Beschreibung der physikalischen Vorgänge mit den 14 Potentialen begnügen kann, oder ob die Schrödingersche Methode die Einführung einer neuen Zustandsgröße bedeutet.

Mit den in dieser Note mitgeteilten Überlegungen habe ich mich sowohl in dem Physikalischen Institut der University of Michigan, Ann Arbor, wie in dem hiesigen Institut für theoretische Physik beschäftigt. Ich möchte auch an dieser Stelle Prof. H. M. Randall und Prof. N. Bohr meinen wärmsten Dank aussprechen.

---

[1]) Bemerkungen dieser Art, die Prof. Bohr bei mehreren Gelegenheiten gemacht hat, haben einen entschiedenen Einfluß auf das Entstehen der vorliegenden Note gehabt.

Quantum Theory and five-dimensional Theory of Relativity

By Oskar Klein in Copenhagen.

Received on 28.April 1926. Zeitschrift für Physik. Bd. XXXVII, p. 895-906.

Abstract

On the following pages I would like to point out a simple connection between the theory suggested by Kaluza[1] on the connection between electromagnetism and gravity and the method suggested by de Broglie[2] and Schrödinger[3] for the treatment of problems in quantum theory. The theory of Kaluza essentially combines the ten Einstein gravitational potentials g_{ik} and the four electromagnetic potentials ϕ_i as the coefficients γ_{ik} of a line element of a Riemannian space, which has as well as the usual four dimensions an additional fifth dimension. The equations of motion of electrically charged particles then take the form of geodesic equations. If one interprets these as radiation equations by considering matter as a kind of wave propagation, one is lead almost automatically to a second order partial differential equation, which can be regarded as a generalization of the usual wave equation. If one then considers solutions to these equations which depend harmonically on the fifth coordinate, with a period depending on Planck's constant, one obtains the above mentioned quantum theoretic methods.

1 Five-dimensional Theory of Relativity

I begin with a brief account on the five-dimensional theory of relativity, which corresponds closely to the theory of Kaluza, but differs from this in several points.

Let us consider a five-dimensional Riemannian line element, for which we postulate a coordinate independent meaning. We write this as:

$$d\sigma = \sqrt{\sum \gamma_{ik} dx^i dx^k}, \tag{1}$$

where here, and in the following, the symbol $\sum$ denotes summation from 0 to 4 over repeated indices. Here $x^0 \cdots x^4$ denote the five coordinates of the space. The 15 quantities γ_{ik} are the covariant components of a five-dimensional symmetric tensor. To get from these to the quantities g_{ik} and ϕ_i of the ordinary theory of relativity, we need to make some special assumptions. First of all, four of the coordinates, say x^1, x^2, x^3, x^4, have to characterize the usual spacetime. Secondly, the quantities γ_{ik} need to be independent of the fifth coordinate x^0. From this follows that the allowed coordinate transformations are restricted to the following group:

$$\left.\begin{aligned} x^0 &= x^{0'} + \psi_0(x^{1'}, x^{2'}, x^{3'}, x^{4'}), \\ x^i &= \psi_i(x^{1'}, x^{2'}, x^{3'}, x^{4'}) \quad (i = 1, 2, 3, 4). \end{aligned}\right\} \tag{2}$$

[1]Th. Kaluza, Sitzungsber. d. Berl. Akad. 1921, p. 966.
[2]L. de Broglie, Ann. d. Phys. (10) **3**,22, 1925. Thèses, Paris 1924.
[3]E. Schödinger, Ann. d. Phys. **79**, 361 and 489, 1926.

In fact in the first equation we should have written constant times $x^{0'}$ instead of $x^{0'}$ But the restriction to the value 1 of this constant is immaterial.

One can easily prove, that γ_{00} is invariant under the transformation (2). Therefore the assumption γ_{00} =const. is allowed. It is natural to presume that only the ratios of γ_{ik} have a physical meaning. Then the above assumption is simply a choice of convention, which it is always possible to make. For now we leave the unit of x^0 undetermined and we set:

$$\gamma_{00} = \alpha. \tag{3}$$

Further one can show that the following differential quantities remain invariant under the transformation (2), namely[4]:

$$d\theta = dx^0 + \frac{\gamma_{0i}}{\gamma_{00}} dx^i, \tag{4}$$

$$ds^2 = \left(\gamma_{ik} - \frac{\gamma_{0i}\gamma_{0k}}{\gamma_{00}}\right) dx^i dx^k. \tag{5}$$

In these expression the summation is carried out over the indices 1 to 4. For such sums, we shall, as usual, leave out the summation sign. The quantities $d\theta$ and ds are related to the line element $d\sigma$ in the following way:

$$d\sigma^2 = \alpha d\theta^2 + ds^2. \tag{6}$$

It follows from the invariance of $d\theta$ and γ_{00} that the four γ_{0i} $(i \neq 0)$, when x^0 is fixed, transform like the components of an ordinary four-vector. When x^0 is transformed as well, a gradient of a scalar needs to be added. This means, that the quantities:

$$\frac{\partial \gamma_{0i}}{\partial x^k} - \frac{\partial \gamma_{0k}}{\partial x^i}$$

transform like the covariant components F_{ik} of the electromagnetic field tensor. Hence, the quantities γ_{0i} behave from the point of view of invariance theory like the electromagnetic potentials ϕ_i. We therefore assume:

$$d\theta = dx^0 + \beta \phi_i dx^i, \tag{7}$$

i.e.

$$\gamma_{0i} = \alpha\beta\phi_i \quad (i = 1, 2, 3, 4), \tag{8}$$

where β is a constant, and the ϕ_i are defined in such a way that in orthogonal Galilean coordinates the following holds:

$$\left.\begin{aligned} (\phi_x, \phi_y, \phi_z) &= A, \\ \phi_0 &= -cV, \end{aligned}\right\} \tag{9}$$

where A is the usual vector potential, V the usual scalar potential and c the speed of light.

We would like to identify the differential form ds with the line element of ordinary relativity. Therefore we set:

$$\gamma_{ik} = g_{ik} + \alpha\beta^2\phi_i\phi_k, \tag{10}$$

where we have chosen g_{ik} in such a way that in orthogonal Galilean coordinates:

$$ds^2 = dx^2 + dy^2 + dz^2 - c^2dt^2. \tag{11}$$

Thus the γ_{ik} have been expressed in terms of known quantities. The problem is now to find field equations for γ_{ik} such that these reduce to the equations for g_{ik} and ϕ_i of

[4] Cf. H. A. Kramers, Proc. Amsterdam **23**, Nr. 7, 1922, where a thought similar to the considerations below is presented, with a simple proof of the invariance of $d\theta$ and ds^2

ordinary relativity in a sufficiently accurate approximation. We shall not investigate this difficult problem further. We only want to show, that from the point of view of the five-dimensional theory, the ordinary field equations can be combined in a simple manner. We form the invariants:

$$P = \sum \gamma^{ik} \left[\frac{\partial \left\{ {i \mu \atop \mu} \right\}}{\partial x^k} - \frac{\partial \left\{ {i k \atop \mu} \right\}}{\partial x^\mu} + \left\{ {i \mu \atop \nu} \right\} \left\{ {i \nu \atop \mu} \right\} - \left\{ {i k \atop \mu} \right\} \left\{ {\mu \nu \atop \nu} \right\} \right] \tag{12}$$

where γ^{ik} are the contravariant components of the five-dimensional metric tensor and $\left\{ {r s \atop i} \right\}$ are the Christoffel symbols, that is:

$$\left\{ {r s \atop i} \right\} = \frac{1}{2} \sum \gamma^{i\mu} \left(\frac{\partial \gamma_{\mu r}}{\partial x^s} + \frac{\partial \gamma_{\mu s}}{\partial x^r} - \frac{\partial \gamma_{rs}}{\partial x^\mu} \right). \tag{13}$$

In the expression for P we regard all the quantities as being independent of x^0, and $\gamma_{00} = \alpha$. Let us now consider the following integral, which is evaluated over a closed region of the five-dimensional space:

$$J = \int P \sqrt{-\gamma} dx^0 dx^1 dx^2 dx^3 dx^4, \tag{14}$$

where γ denotes the determinant of γ_{ik}.

We then compute δJ by varying γ_{ik} and $\frac{\partial \gamma_{ik}}{\partial x^l}$, leaving the boundary values fixed. The variation principle:

$$\delta J = 0$$

implies the following equations:

$$R^{ik} - \frac{1}{2} g^{ik} R + \frac{\alpha \beta^2}{2} S^{ik} = 0 \quad (i, k = 1, 2, 3, 4) \tag{16a}$$

and

$$\frac{\partial \sqrt{-g} F^{i\mu}}{\partial x^\mu} = 0 \quad (i = 1, 2, 3, 4), \tag{16b}$$

where R is Einstein's curvature invariant, R^{ik} the contravariant components of the Einstein curvature tensor, g^{ik} the contravariant components of the metric tensor, S^{ik} the contravariant components of the electromagnetic energy-momentum tensor, g the determinant of g_{ik} and finally $F^{i\mu}$ are the contravariant components of the electromagnetic field tensor. If we set:

$$\frac{\alpha \beta^2}{2} = \kappa, \tag{17}$$

where κ is the gravitational constant, as used by Einstein, we see that indeed equation (16a) is identical to the equations of relativity for the gravitational field and (16b) is identical to the covariant form of Maxwell's equations in the absence of matter. [5].

If we restrict ourselves to the schematic treatment of matter, as it is common in the relativistic theory of electrons, we can obtain in a similar way the usual equations when matter is present. We replace P in (14) by

$$P + \kappa \sum \gamma_{ik} \Theta^{ik}.$$

To define Θ^{ik}, we shall first consider the tensor for one electron or one hydrogen nucleus:

$$\theta^{ik} = \frac{dx^i}{dl} \frac{dx^k}{dl} \tag{18}$$

[5]See e.g. W.Pauli, Relativitätstheorie, p. 719 and 724

Here dx^i is the change in position of the particle and dl denotes a certain invariant differential. The Θ^{ik} have to be equal to the sum of the θ^{ik} for the various particles, in a unit volume. Again, we obtain equations of the ordinary type, which reduce to the ordinary field equations, if we set:

$$v_0 \frac{d\tau}{dl} = \pm \frac{e}{\beta c}, \tag{19}$$

$$\frac{d\tau}{dl} = \begin{cases} \sqrt{M} \\ \sqrt{m} \end{cases}, \tag{20}$$

where generally

$$v_i = \sum \gamma_{i\mu} \frac{dx^\mu}{dl} \tag{21}$$

is the covariant component of the five-velocity vector

$$v^i = \frac{dx^i}{dl}. \tag{22}$$

Also, e is the fundamental electric quantum, M and m are the masses of the hydrogen nucleus and electron, respectively. Here, the upper value denotes the term for the nucleus and the lower the one for the electron. Further

$$d\tau = \frac{1}{c}\sqrt{-ds^2}$$

is the differential of the proper time.

The field equations imply in the usual way the matter equations of motion and the continuity equations. From our point of view we can now simply summarize the calculations, which lead to these. It can easily be seen that our field equations are equivalent to the following 14 equations:

$$P^{ik} - \frac{1}{2}\gamma^{ik} P + \kappa\Theta^{ik} = 0 \tag{23}$$

($i, k = 0, 1, 2, 3, 4$, but not both zero), where P^{ik} are the contravariant components of the contracted five-dimensional curvature tensor (corresponding to the R^{ik}). The equations in question can now be obtained by taking the divergence in (23). It follows, that electrically charged particles move along five-dimensional geodesic lines, which obey the conditions (19) and (20)[6]. It is immediate, that these conditions are compatible with the geodesic equations, because the γ_{ik} are independent of x^0.

It is necessary to remind the reader that there is not sufficient evidence to show that Einstein's field equations hold exactly. It is certainly of some interest that, from the point of view of Kaluza's theory, these 14 field equations can be summarized in such simple fashion.

2 The Wave equation of Quantum Theory

We will now give the theory of stationary states, and thus connect the characteristic deviations from mechanics, which appear in modern quantum theory, with five-dimensional relativity. For this purpose, let us consider the following differential equation, which is formulated for our five-dimensional space and which can be considered as a simple generalization of the wave equation:

$$\sum a^{ik}\left(\frac{\partial^2 U}{\partial x^i \partial x^k} - \sum \begin{Bmatrix} i\,k \\ r \end{Bmatrix} \frac{\partial U}{\partial x^r}\right) = 0. \tag{24}$$

[6]Obviously, the particular values of $\frac{d\tau}{dl}$ are irrelevant in this context. Of importance is here only $\frac{d\tau}{dl}$ =const.

Here, a^{ik} are the contravariant components of a five-dimensional symmetric tensor, which are functions of the coordinates. Equation (24) holds independently of the choice of coordinate system.

Let us begin by considering the wave evolution given by (24), in the special limit of geometric optics. We obtain this by setting:

$$u = Ae^{i\omega\Phi} \tag{25}$$

and taking ω so large that we have to consider only terms proportional to ω^2 in (24). We then obtain:

$$\sum a^{ik}\frac{\partial\Psi}{\partial x^i}\frac{\partial\Psi}{\partial x^k} = 0, \tag{26}$$

This equation corresponds to the Hamilton-Jacobi partial differential equation of mechanics. If we set:

$$p_i = \frac{\partial\Psi}{\partial x^i}, \tag{27}$$

we can write the differential equations for the rays in the following Hamiltonian form:

$$\frac{dp_i}{\frac{-\partial H}{\partial x^i}} = \frac{dx^i}{\frac{\partial H}{\partial p_i}} = d\lambda, \tag{28}$$

where

$$H = \frac{1}{2}\sum a^{ik}p_i p_k. \tag{29}$$

Further, (26) implies

$$H = 0. \tag{30}$$

A different, Lagrangian, form of this equation can be obtained using the fact that the rays can be viewed as geodesic null lines of the differential form:

$$\sum a_{ik}dx^i dx^k$$

where a_{ik} are the dual quantities to a^{ik}, that is

$$\sum a_{i\mu}a^{k\mu} = \delta_i^k = \begin{cases} 1, & (i = k), \\ 0, & (i \neq k) \end{cases} \tag{31}$$

Setting

$$\sum a_{ik}dx^i dx^k = \mu(d\theta)^2 + ds^2, \tag{32}$$

and suitably choosing the constant μ, we can make our ray equation coincide with the equations of motion of electrically charged particles. To see this, we set:

$$L = \frac{1}{2}\mu\left(\frac{d\theta}{d\lambda}\right)^2 + \frac{1}{2}\left(\frac{ds}{d\lambda}\right)^2, \tag{33}$$

which implies

$$p_0 = \frac{\partial L}{\partial\frac{dx^0}{d\lambda}} = \mu\frac{d\theta}{d\lambda} \tag{34}$$

and

$$p_i = \frac{\partial L}{\partial\frac{dx_i}{d\lambda}} = u_i\frac{d\tau}{d\lambda} + \beta p_0\phi_i \quad (i = 1,2,3,4). \tag{35}$$

where $u_1 \cdots u_4$ are the usual covariant components of the velocity vector.

The ray equations then become:

$$\frac{dp_0}{d\lambda} = 0, \tag{36a}$$

$$\frac{dp_i}{d\lambda} = \frac{1}{2}\frac{\partial g_{\mu\nu}}{\partial x^i}\frac{dx^\mu}{d\lambda}\frac{dx^\nu}{d\lambda} + \beta p_0 \frac{\partial \phi_\mu}{\partial x^i}\frac{dx^\mu}{d\lambda} \quad (i = 1, 2, 3, 4) \tag{36}$$

From

$$\mu d\theta^2 + ds^2 = \mu d\theta^2 - c^2 d\tau^2 = 0$$

it follows that

$$\mu \frac{d\theta}{d\tau} = c\sqrt{\mu}. \tag{37}$$

From (34) and (36a), $\frac{d\theta}{d\lambda}$ and so also $\frac{d\tau}{d\lambda}$ are constant, so we can choose λ such that:

$$\frac{d\tau}{d\lambda} = \begin{cases} M \text{ for the hydrogen nucleus,} \\ m \text{ for the electron.} \end{cases} \tag{38}$$

To obtain the usual equations of motion, we further have to assume that:

$$\beta p_0 = \begin{cases} \frac{e}{c} \text{ for the hydrogen nucleus,} \\ -\frac{e}{c} \text{ for the electron.} \end{cases} \tag{39}$$

It follows from (37) that:

$$\mu = \begin{cases} \frac{e^2}{\beta^2 M^2 c^4} \text{ for the hydrogen nucleus,} \\ \frac{e^2}{\beta^2 m^2 c^4} \text{ for the electron.} \end{cases} \tag{40}$$

The equations (35), (36) then exactly coincide with the usual equations of motion of electrically charged particles in gravitational and electromagnetic fields. In particular, the quantities p_i defined in (35) are identical with the generalized momenta defined in the usual way. This is important for the following reasons. We can still choose β, so we shall set:

$$\beta = \frac{e}{c}, \tag{41}$$

It then follows that:

$$p_0 = \begin{cases} +1 \text{ for the hydrogen nucleus,} \\ -1 \text{ for the electron,} \end{cases} \tag{39a}$$

and

$$\mu = \begin{cases} \frac{1}{M^2c^2} M \quad \text{for the hydrogen nucleus,} \\ \frac{1}{m^2c^2} m \quad \text{for the electron.} \end{cases} \tag{40a}$$

For the square-root in (37) we need to choose the positive sign in the case of the nucleus and the negative sign for the electron. This is not very satisfactory. Nevertheless, one could interpret the fact that two classes of solutions for rays appear for one value of μ, which in some sense relate to each other like positively and negatively charged particles, as a hint that it might be possible to alter the wave equation in such a way that the equations of motion of both types of particles result from one assignment of values to the coefficients. We will not investigate this question further. We turn to a more detailed analysis of the wave equation for the electron, as obtained from (32).

Since we assumed $p_0 = -1$ for the electron, (27) implies that we have to set:

$$\Phi = -x^0 + S(x^1, x^2, x^3, x^4). \tag{42}$$

The theory of de Broglie is obtained by finding the standing waves, which are consistent with the wave equation and correspond to a given value of ω, thus assuming that the wave evolution proceeds according to geometric optics. We need to make use of the well-known theorem of conservation of phase, which can be obtained directly from (28) and (30). Indeed, it follows that:

$$\frac{d\Psi}{d\lambda} = \sum \frac{\partial \Psi}{\partial x^i}\frac{dx^i}{d\lambda} = \sum p_i \frac{\partial H}{\partial p_i} = 2H = 0. \tag{43}$$

The phase is therefore transported with the wave. We shall now consider the case when Φ splits up into two parts. One of these parts will depend only on one coordinate, say x, oscillating periodically in time. This will allow for a standing wave, characterized by the property that, after a period of x, the harmonic wave, represented at a particular moment by (25) comes into phase with the wave which is obtained from the same solution (25) but with the new values for x^0, x^2, x^3, x^4 inserted. Because of the conservation of phase, the condition for this is simply:

$$\omega \oint p\, dx = n.2\pi, \tag{44}$$

where n is an integer. Now setting:

$$\omega = \frac{2\pi}{h}, \tag{45}$$

where h is the Planck constant, we obtain the usual quantum condition for one separable coordinate. A similar analysis obviously holds for arbitrary periodic systems. We can say therefore that, assuming that the waves evolve according to the laws of geometric optics, the usual quantum theory of periodic systems corresponds just to the analysis of interference. It may be worth noting, that the relations (44), (45) remain invariant under the coordinate transformation (2) due to (42). Let us now consider (24) in the case where ω is not so large as to allow consideration of the terms in ω squared only. We shall restrict ourselves to the simple case of an electrostatic field. In cartesian coordinates we then obtain:

$$\left.\begin{aligned} d\theta &= dx^0 - eV dt, \\ ds^2 &= dx^2 + dy^2 + dz^2 - c^2 dt^2. \end{aligned}\right\} \tag{46}$$

Hence it follows that:

$$H = \frac{1}{2}(p_x^2 + p_y^2 + p_z^2) - \frac{1}{2c^2}(P_t + eV p_o)^2 + \frac{m^2c^2}{2} p_0^2. \tag{47}$$

We can now neglect the quantities proportional to $\left\{ {i\,k \atop r} \right\}$ in equation (24), because in this case, by (17), these triple indexed quantities are small quantities, proportional to the gravitational constant κ. We therefore obtain[7]

$$\nabla U - \frac{1}{c^2}\frac{\partial^2 U}{\partial t^2} - \frac{2eV}{c^2}\frac{\partial^2 U}{\partial t \partial x^0} + \left(m^2c^2 - \frac{e^2V^2}{c^2}\right)\frac{\partial^2 U}{\partial x^{0\,2}} = 0. \tag{48}$$

Since V depends only on x, y, z, we can make the following ansatz for U, in accordance with (42) and (45):

$$U = e^{-2\pi i \left(\frac{x^0}{h} - \nu t\right)} \psi(x, y, z). \tag{49}$$

Inserting this into (48) results in:

$$\nabla \psi + \frac{4\ \pi^2}{c^2\ h^2}[(h\nu - eV)^2 - m^2c^4]\psi = 0. \tag{50}$$

Moreover setting:

$$h\nu = mc^2 + E, \tag{51}$$

[7]This equation differs from the Schrödinger equation, not only in the appearance of x^0, which for applications is irrelevant anyway, but also in the way in which time enters equation (48). To support this sort of quantum equation we note, that in the case when V depends harmonically on time, it has solutions which behave in a similar way to Kramer's dispersion theory as Schrödinger's solutions behave with respect to the quantum theory of spectral lines. This can be easily seen by perturbation theory. I owe this remark to Dr. W. Heisenberg.

we obtain Schrödinger's[8] equation, of which the standing wave solutions are known to give values for E identical to the values of the energy calculated from Heisenberg's quantum theory. In the limit of geometric optics, one sees that E corresponds to the mechanical energy defined in the usual way. Schödinger has emphasized, that the condition on the frequency means, by (51), that the frequencies of light corresponding to the system are equal to the difference of the values for ν.

3 Final Remarks

The above considerations arose, just like the works of de Broglie, from the endeavour to use the relation between mechanics and optics, appearing in the Hamiltonian method, in order to obtain a deeper understanding of quantum phenomena. It seems, that the analogy between the conditions on stationary states in atomic systems and the ones in optical interference phenomena, give a real physical meaning to this analogy. Already in the classical physics of fields the notions of point charge and of point matter are somewhat alien. Also, occasionally the hypothesis has been put forward that matter particles should be considered as special solutions to the field equations, which govern gravitational and electromagnetic fields. Relating this idea with the above analogy seems to suggest itself, because from this hypothesis it is not at all strange that the motion of matter particles has similarities to the evolution of waves. However, this analogy remains incomplete if one restricts oneself to wave evolution in a space of only four dimensions. This reveals itself already in the variability of the particle velocity. But, as we have seen, if one thinks of the observed motion of the particle as a kind of projection of a wave evolution in a space of five dimensions onto spacetime, then the analogy is complete. In mathematical terms this means, that the Hamilton-Jacobi equation cannot be considered as the characteristic equation of a four-, but only a five-dimensional wave equation. In this way one is lead to the theory of Kaluza.

Although it might initially appear alien to include a fifth dimension, this radical change in the geometry underlying the field equations is suggested naturally from an entirely different point of view by quantum theory. As is know, it seems more and more unlikely that quantum phenomena allow a unified spacetime description. But the possibility of describing these phenomena in terms of a system of five-dimensional field equations does not seem to be excluded from the start[9]. Whether there is something real behind these possibilities one will have to investigate in future. We would like to emphasize that the treatment presented in this note of the field equations as well as the theory of stationary states has to be regarded as entirely provisional. This is apparent in particular from the schematic treatment of matter in the paragraph containing (18), and the fact that two types of electrically charged particles are described by two different equations of Schrödinger-type. We also leave the question unanswered whether in describing physical processes it is sufficient to consider the 14 potentials, or whether Schrödinger's method means an introduction of further parameters.

I worked on the ideas presented in this note while at the Institute of Physics of the University of Michigan, Ann Arbor, as well as at the Institute of Theoretical Physics here. I would like to express my warmest gratitude towards Prof. H. M. Randall and Prof. N. Bohr.

[8] Schrödinger, l.c.

[9] Remarks of this sort, which Prof. Bohr has made on several occasions, had a crucial influence on the development of these notes.

Über die invariante Form der Wellen- und der Bewegungsgleichungen für einen geladenen Massenpunkt[1]).

Von V. Fock in Leningrad.

(Eingegangen am 30. Juli 1926.)

Die Schrödingersche Wellengleichung wird als invariante Laplacesche Gleichung und die Bewegungsgleichungen als diejenigen einer geodätischen Linie im fünfdimensionalen Raum geschrieben. Der überzählige fünfte Koordinatenparameter steht in enger Beziehung zu der linearen Differentialform der elektromagnetischen Potentiale.

In seiner noch nicht veröffentlichten Arbeit bedient sich H. Mandel[2]) des Begriffes des fünfdimensionalen Raumes, um die Gravitation und das elektromagnetische Feld von einem einheitlichen Standpunkte aus zu betrachten. Die Einführung eines fünften Koordinatenparameters scheint uns zur Aufstellung der Schrödingerschen Wellengleichung sowie der mechanischen Gleichungen in invarianter Form gut geeignet.

1. Spezielle Relativitätstheorie.

Die Lagrangesche Funktion für die Bewegung eines geladenen Massenpunktes ist, in leicht verständlicher Bezeichnungsweise,

$$L = -m c^2 \sqrt{1 - \frac{\mathfrak{v}^2}{c^2}} + \frac{e}{c}\, \mathfrak{A} \cdot \mathfrak{v} - e\varphi, \tag{1}$$

und die entsprechende Hamilton-Jacobische Gleichung (H. P.) lautet:

$$(\operatorname{grad} W)^2 - \frac{1}{c^2}\left(\frac{\partial W}{\partial t}\right)^2 - \frac{2e}{c}\left(\mathfrak{A} \cdot \operatorname{grad} W + \frac{\varphi}{c}\frac{\partial W}{\partial t}\right) + m^2 c^2 + \frac{e^2}{c^2}(\mathfrak{A}^2 - \varphi^2) = 0. \tag{2}$$

---

[1]) Die Idee dieser Arbeit ist in einem Gespräche mit Prof. V. Fréedericksz entstanden, dem ich auch manche wertvollen Ratschläge verdanke.

Anmerkung bei der Korrektur. Während diese Notiz im Druck war, ist die schöne Arbeit von Oskar Klein (ZS. f. Phys. **37**, 895, 1926) in Leningrad eingetroffen, in welcher der Verfasser zu Resultaten gelangt, die prinzipiell mit denen dieser Notiz identisch sind. Wegen der Wichtigkeit der Resultate dürfte aber vielleicht ihre Ableitung auf einem anderen Wege (Verallgemeinerung eines in meiner früheren Arbeit gebrauchten Ansatzes) von Interesse sein.

[2]) Der Verfasser hat mir liebenswürdigerweise die Möglichkeit gegeben, seine Arbeit im Manuskript zu lesen.

Analog dem in unserer früheren Arbeit[1]) gebrauchten Ansatze setzen wir hier

$$\operatorname{grad} W = \frac{\operatorname{grad} \psi}{\dfrac{\partial \psi}{\partial p}}; \quad \frac{\partial W}{\partial t} = \frac{\dfrac{\partial \psi}{\partial t}}{\dfrac{\partial \psi}{\partial p}}, \tag{3}$$

wo p einen neuen Parameter von der Dimension des Wirkungsquantums bezeichnet. Nach Multiplikation mit $\left(\frac{\partial \psi}{\partial p}\right)^2$ erhalten wir eine quadratische Form

$$Q = (\operatorname{grad} \psi)^2 - \frac{1}{c^2}\left(\frac{\partial \psi}{\partial t}\right)^2 - \frac{2e}{c}\frac{\partial \psi}{\partial p}\left(\mathfrak{A} \cdot \operatorname{grad} \psi + \frac{\varphi}{c}\frac{\partial \psi}{\partial t}\right) + \left[m^2 c^2 + \frac{e^2}{c^2}(\mathfrak{A}^2 - \varphi^2)\right]\left(\frac{\partial \psi}{\partial p}\right)^2. \tag{4}$$

Wir bemerken, daß die Koeffizienten der nullten, ersten und zweiten Potenz von $\frac{\partial \psi}{\partial p}$ vierdimensionale Invarianten sind. Ferner bleibt die Form Q invariant, wenn man

$$\left.\begin{aligned} \mathfrak{A} &= \mathfrak{A}_1 + \operatorname{grad} f, \\ \varphi &= \varphi_1 - \frac{1}{c}\frac{\partial f}{\partial t}, \\ p &= p_1 - \frac{e}{c} f \end{aligned}\right\} \tag{5}$$

setzt, wo f eine willkürliche Funktion der Koordinaten und der Zeit bezeichnet. Die letztere Transformation läßt auch die lineare Differentialform

$$d'\Omega = \frac{e}{mc^2}(\mathfrak{A}_x dx + \mathfrak{A}_y dy + \mathfrak{A}_z dz) - \frac{e}{mc}\varphi\, dt + \frac{1}{mc} dp \tag{6}$$

invariant[2]).

Wir wollen nun die Form Q als das Quadrat des Gradienten der Funktion ψ im fünfdimensionalen Raum (R_5) auffassen, und suchen das entsprechende Linienelement. Man findet leicht

$$ds^2 = dx^2 + dy^2 + dz^2 - c^2 dt^2 + (d'\Omega)^2. \tag{7}$$

[1]) V. Fock, Zur Schrödingerschen Wellenmechanik, ZS. f. Phys. 38, 242, 1926.

[2]) Das Zeichen d' soll andeuten, daß $d'\Omega$ kein vollständiges Differential ist.

Die Laplacesche Gleichung in R_5 lautet:

$$\Delta\psi - \frac{1}{c^2}\frac{\partial^2\psi}{\partial t^2} - \frac{2e}{c}\left(\mathfrak{A}\,.\,\mathrm{grad}\,\frac{\partial\psi}{\partial p} + \frac{\varphi}{c}\frac{\partial^2\psi}{\partial t\,\partial p}\right)$$

$$-\frac{e}{c}\frac{\partial\psi}{\partial p}\left(\mathrm{div}\,\mathfrak{A} + \frac{1}{c}\frac{\partial\varphi}{\partial t}\right) + \left[m^2c^2 + \frac{e^2}{c^2}(\mathfrak{A}^2 - \varphi^2)\right]\frac{\partial^2\psi}{\partial p^2} = 0. \quad (8)$$

Sie bleibt, ebenso wie (7) und (4), bei Lorentztransformationen und bei Transformationen (5) invariant.

Da die Koeffizienten der Gleichung (8) den Parameter p nicht enthalten, können wir die Abhängigkeit der Funktion ψ von p in Form eines Exponentialfaktors ansetzen, und zwar müssen wir, um Übereinstimmung mit der Erfahrung zu erhalten

$$\psi = \psi_0\, e^{2\pi i\frac{p}{h}} \quad (9)$$

setzen[1]). Die Gleichung für ψ_0 ist gegenüber Lorentztransformationen, nicht aber gegenüber Transformationen (5), invariant. Die Bedeutung des überzähligen Koordinatenparameters p scheint nämlich gerade darin zu liegen, daß er die Invarianz der Gleichungen in bezug auf die Addition eines beliebigen Gradienten zum Viererpotential bewirkt.

Es sei hier bemerkt, daß die Koeffizienten der Gleichung für ψ_0 im allgemeinen komplex sind.

Nimmt man ferner an, daß diese Koeffizienten von t nicht abhängen, und setzt man

$$\psi_0 = e^{-\frac{2\pi i}{h}(E+mc^2)t}\,\psi_1, \quad (10)$$

so erhält man für ψ_1 eine zeitfreie Gleichung, welche mit der in unserer früheren Arbeit aufgestellten Verallgemeinerung der Schrödingerschen Wellengleichung identisch ist. Diejenigen Werte von E, für welche eine Funktion ψ_1 existiert, die gewissen Endlichkeits- und Stetigkeitsforderungen genügt, sind dann die Bohrschen Energieniveaus. Aus den soeben angeführten Betrachtungen folgt, daß die Addition eines Gradienten zum Viererpotential keinen Einfluß auf die Energieniveaus ausüben kann. Die beiden mit den Vektorpotentialen $\mathfrak{A}$ und $\overline{\mathfrak{A}} = \mathfrak{A} - \mathrm{grad}\,f$ erhaltenen Funktionen ψ_1 und $\overline{\psi}_1$ würden sich nämlich nur um einen Faktor $e^{\frac{2\pi i e}{ch}f}$ vom absoluten Betrage 1 unterscheiden und folglich (bei sehr allgemeinen Voraussetzungen über die Funktion f) die gleichen Stetigkeitseigenschaften haben.

---

[1]) Das Auftreten des mit der Linearform verbundenen Parameters p in einer Exponentialfunktion könnte vielleicht mit einigen von E. Schrödinger (ZS. f. Phys. **12**, 13, 1923) bemerkten Beziehungen im Zusammenhange stehen.

2. Allgemeine Relativitätstheorie.

A. Wellengleichung. Für das Linienelement im fünfdimensionalen Raum setzen wir an:

$$\left.\begin{aligned} ds^2 &= \sum_{i,k=1}^{5} \gamma_{ik}\, dx_i\, dx_k \\ &= \sum_{i,k=1}^{4} g_{ik}\, dx_i\, dx_k + \frac{e^2}{m^2}\left(\sum_{i=1}^{5} q_i\, dx_i\right)^2. \end{aligned}\right\} \tag{11}$$

Hier sind die Größen g_{ik} die Komponenten des Einsteinschen Fundamentaltensors, die Größen q_i $(i = 1, 2, 3, 4)$ die durch c^2 dividierten Komponenten des Viererpotentials, also

$$\sum_{i=1}^{4} q_i\, dx_i = \frac{1}{c^2}(\mathfrak{A}_x\, dx + \mathfrak{A}_y\, dy + \mathfrak{A}_z\, dz - \varphi\, c\, dt), \tag{12}$$

die Größe q_5 eine Konstante und x_5 der überzählige Koordinatenparameter. Alle Koeffizienten sind reell und von x_5 unabhängig.

Die Größen g_{ik} und q_i hängen nur vom Felde, nicht aber von der Beschaffenheit des Massenpunktes ab; die letztere wird durch den Faktor $\frac{e^2}{m^2}$ repräsentiert. Zur Abkürzung wollen wir aber die von $\frac{e}{m}$ abhängigen Größen

$$\frac{e}{m}\, q_i = a_i \quad (i = 1, 2, 3, 4, 5) \tag{13}$$

einführen und folgende Verabredung treffen: bei der Summation von 1 bis 5 wird das Summenzeichen angeschrieben, bei der Summation von 1 bis 4 dagegen unterdrückt.

Mit diesen Bezeichnungen finden wir

$$\left.\begin{aligned} \gamma_{ik} &= g_{ik} + a_i a_k; \quad g_{i5} = 0 \\ \gamma &= \|\gamma_{ik}\| = a_5^2\, g \end{aligned}\right\} (i, k = 1, 2, 3, 4, 5) \qquad \begin{aligned} (14) \\ (15) \end{aligned}$$

$$\left.\begin{aligned} \gamma^{lk} &= g^{lk} \\ \gamma^{5k} &= -\frac{1}{a_5}\, g^{ik} a_i = -\frac{a^i}{a_5} \\ \gamma^{55} &= \frac{1}{(a_5)^2}\cdot(1 + a_i a^i) \end{aligned}\right\} (i, k, l = 1, 2, 3, 4). \tag{16}$$

Die der Gleichung (8) entsprechende Wellengleichung lautet:

$$\sum_{i,k=1}^{5} \frac{\partial}{\partial x_i}\left(\sqrt{-\gamma}\, \gamma^{ik}\, \frac{\partial \psi}{\partial x_k}\right) = 0 \tag{17}$$

oder, ausführlicher geschrieben,

$$\frac{1}{\sqrt{-g}}\frac{\partial}{\partial x_i}\left(\sqrt{-g}\,g^{ik}\frac{\partial\psi}{\partial x_k}\right)-\frac{2}{a_5}a^i\frac{\partial^2\psi}{\partial x_i\,\partial x_5}+\frac{1}{(a_5)^2}(1+a_i a^i)\frac{\partial^2\psi}{\partial x_5^2}=0. \quad (18)$$

Führt man endlich die Funktion ψ_0 und die Potentiale q_i ein, so läßt sich diese Gleichung schreiben:

$$\frac{1}{\sqrt{-g}}\frac{\partial}{\partial x_i}\left(\sqrt{-g}\,g^{ik}\frac{\partial\psi_0}{\partial x_k}\right)-\frac{4\pi}{h}\sqrt{-1}\,c\,e\,q^i\frac{\partial\psi_0}{\partial x_i}-\frac{4\pi^2c^2}{h^2}(m^2+e^2q_iq^i)\psi_0=0. \quad (19)$$

B. Bewegungsgleichungen. Wir wollen nun die Bewegungsgleichungen eines geladenen Massenpunktes als diejenigen einer geodätischen Linie in R_5 aufstellen.

Zu diesem Zwecke müssen wir zunächst die Christoffelschen Klammersymbole berechnen. Wir bezeichnen die fünfdimensionalen Klammersymbole mit $\left\{ {kl \atop r} \right\}_5$ und die vierdimensionalen mit $\left\{ {kl \atop r} \right\}_4$. Wir führen ferner die kovariante Ableitung des Viererpotentials ein:

$$A_{lk}=\frac{\partial a_l}{\partial x_k}-\left\{ {kl \atop r} \right\}_4 a_r, \quad (20)$$

und spalten den Tensor $2A_{ik}$ in seinen symmetrischen und antisymmetrischen Teil:

$$\left.\begin{aligned} B_{lk}&=A_{lk}+A_{kl}\\ M_{lk}&=A_{lk}-A_{kl}=\frac{\partial a_l}{\partial x_k}-\frac{\partial a_k}{\partial x_l}.\end{aligned}\right\} \quad (21)$$

Wir haben dann

$$\left.\begin{aligned}
\left\{ {kl \atop r} \right\}_5 &= \left\{ {kl \atop r} \right\}_4+\frac{1}{2}(a_k g^{ir}M_{il}+a_l g^{ir}M_{ik}),\\
\left\{ {kl \atop 5} \right\}_5 &= \frac{1}{2a_5}B_{lk}-\frac{1}{2a_5}(a_k a^i M_{il}+a_l a^i M_{ik}),\\
\left\{ {k5 \atop 5} \right\}_5 &= -\frac{1}{2}a^iM_{ik},\\
\left\{ {55 \atop k} \right\}_5 &= 0,\\
\left\{ {55 \atop 5} \right\}_5 &= 0.
\end{aligned}\right\} \quad (22)$$

Die Gleichungen der geodätischen Linie in R_5 lauten dann:

$$\frac{d^2 x_r}{d s^2} + \left\{ {k l \atop r} \right\}_4 \frac{d x_k}{d s} \frac{d x_l}{d s} + \frac{d' \Omega}{d s} \cdot g^{ir} M_{il} \frac{d x_l}{d s} = 0, \tag{23}$$

$$\frac{d^2 x_5}{d s^2} + \frac{1}{2 a_5} B_{lk} \frac{d x_k}{d s} \frac{d x_l}{d s} - \frac{1}{a_5} \frac{d' \Omega}{d s} a^i M_{il} \frac{d x_l}{d s} = 0. \tag{24}$$

Hier bezeichnet $d' \Omega$, wie früher, die Linearform

$$d' \Omega = a_i d x_i + a_5 d x_5. \tag{25}$$

Multipliziert man die vier Gleichungen (23) mit a_r, die fünfte (24) mit a_5 und addiert, so erhält man eine Gleichung, die man in der Form

$$\frac{d}{d s} \left(\frac{d' \Omega}{d s} \right) = 0 \tag{26}$$

schreiben kann. Es gilt also

$$\frac{d' \Omega}{d s} = \text{const} \tag{27}$$

Multipliziert man (23) mit $g_{r\alpha} \frac{d x_\alpha}{d s}$ und summiert über r und α, so folgt wegen der Antisymmetrie von M_{ik}

$$\frac{d}{d s} \left(g_{r\alpha} \frac{d x_r}{d s} \frac{d x_\alpha}{d s} \right) = 0 \tag{28}$$

oder, wenn man die Eigenzeit τ durch die Formel

$$g_{ik} d x_i d x_k = -c^2 d \tau^2 \tag{29}$$

einführt,

$$\frac{d}{d s} \left(\frac{d \tau}{d s} \right)^2 = 0. \tag{30}$$

Die Gleichung (28) oder (30) ist übrigens wegen

$$d s^2 = - c^2 d \tau^2 + (d' \Omega)^2 \tag{31}$$

eine Folge von (26).

Aus dem Gesagten folgt, daß die Gleichung (24) eine Folge von (23) ist; wir können sie also beiseite lassen. Führen wir in (23) die Eigenzeit als unabhängige Veränderliche ein, so kommt der fünfte Parameter ganz zum Fortfall; wir unterdrücken noch das Zeichen 4 am Klammersymbol:

$$\frac{d^2 x_r}{d \tau^2} + \left\{ {k l \atop r} \right\} \frac{d x_k}{d \tau} \frac{d x_l}{d \tau} + \frac{d' \Omega}{d \tau} g^{ir} M_{il} \frac{d x_l}{d \tau} = 0. \tag{32}$$

Das letzte Glied auf der linken Seite stellt die Lorentzsche Kraft dar. In der speziellen Relativitätstheorie läßt sich die erste dieser Gleichungen schreiben

$$m \frac{d}{dt} \frac{d x}{d \tau} + \frac{1}{c} \frac{d' \Omega}{d \tau} \left[\frac{e}{c} \left(\dot{z} H_y - \dot{y} H_z + \frac{\partial \mathfrak{A}_x}{\partial t} \right) + e \frac{\partial \varphi}{\partial x} \right] = 0. \tag{33}$$

Um Übereinstimmung mit der Erfahrung zu erhalten, muß der Faktor der eckigen Klammer den Wert 1 haben. Es gilt also

$$\frac{d'\Omega}{d\tau} = c \tag{34}$$

und

$$ds^2 = 0. \tag{35}$$

Die Bahnen des Massenpunktes sind also geodätische Nullinien im fünfdimensionalen Raum.

Um die Hamilton-Jacobische Gleichung zu erhalten, setzen wir das Quadrat des fünfdimensionalen Gradienten einer Funktion ψ gleich Null.

$$g^{ik}\frac{\partial \psi}{\partial x_i}\frac{\partial \psi}{\partial x_k} - \frac{2}{a_5}\frac{\partial \psi}{\partial x_5} a^i \frac{\partial \psi}{\partial x_i} + (1 + a_i a^i)\left(\frac{1}{a_5}\frac{\partial \psi}{\partial x_5}\right)^2 = 0. \tag{36}$$

Setzen wir hier

$$m c a_5 \frac{\dfrac{\partial \psi}{\partial x_i}}{\dfrac{\partial \psi}{\partial x_5}} = \frac{\partial W}{\partial x_i}, \tag{37}$$

und führen wir statt a_i die Potentiale q_i ein, so erhalten wir eine Gleichung

$$g^{ik}\frac{\partial W}{\partial x_i}\frac{\partial W}{\partial x_k} - 2ecq^i\frac{\partial W}{\partial x_i} + c^2(m^2 + e^2 q_i q^i) = 0, \tag{38}$$

die als Verallgemeinerung unserer Gleichung (2), die uns als Ausgangspunkt diente, gelten kann.

Leningrad, Physikalisches Institut der Universität, 24. Juli 1926.

On the Invariant Form of the Wave Equations and the Equations of Motion for a Charged Point Mass*

By **V. Fock** in Leningrad.
(Received on 30 July 1926.)

Abstract

The Schrödinger wave equation is written as an invariant Laplacian equation, and the equations of motion as those of a geodesic, in five-dimensional space. The redundant fifth coordinate parameter is closely related to the linear differential form of the electromagnetic potentials.

In his as yet unpublished work, H. Mandel[1] uses the idea of a five-dimensional space to consider gravitation and the electromagnetic field from a unified point of you. To us, the introduction of a fifth coordinate parameter seems well suited for the invariant formulation of both Schrödinger's wave equation and the mechanical equations.

---

*The idea for this work was formed in a talk with Prof. V. Fréedericksz, from whom I have also received some valuable advice.
Remark during correction. While this note was being printed, the beautiful work by Oskar Klein (ZS. f. Phys. **37**, 895, 1926) has arrived in Leningrad, in which the author arrives at results which coincide in principle with the ones contained in this note. Due to the importance of these results, however, their derivation in a different way (a generalization of an ansatz used in my previous work) may be of interest.

[1]The author was generous enough to give me the chance to read his work in manuscript form.

1 Special Relativity

The Lagrangian for the movement of a charged point mass, in obvious notation, is

$$L = -mc^2\sqrt{1-\frac{v^2}{c^2}} + \frac{e}{c}\mathcal{A}.v - e\phi, \tag{1}$$

and the corresponding Hamilton-Jacobi equation (H.P.) is

$$(\text{grad}W)^2 - \frac{1}{c^2}\left(\frac{\partial W}{\partial t}\right)^2 - \frac{2e}{c}\left(\mathcal{A}.\text{grad}W + \frac{\phi}{c}\frac{\partial W}{\partial t}\right)$$
$$+m^2c^2 + \frac{e^2}{c^2}(\mathcal{A}^2 - \phi^2) = 0. \tag{2}$$

In analogy to the ansatz used in our previous work[2] we shall set

$$\text{grad}W = \frac{\text{grad}\psi}{\frac{\partial\psi}{\partial p}}; \quad \frac{\partial W}{\partial t} = \frac{\frac{\partial\psi}{\partial t}}{\frac{\partial\psi}{\partial p}}, \tag{3}$$

where p designates a new parameter with the dimensions of the quantum of action. After multiplication by $\left(\frac{\partial\psi}{\partial p}\right)^2$ we get a quadratic form

$$Q = (\text{grad}\psi)^2 - \frac{1}{c^2}\left(\frac{\partial\psi}{\partial t}\right)^2 - \frac{2e}{c}\frac{\partial\psi}{\partial p}\left(\mathcal{A}.\text{grad}\psi + \frac{\phi}{c}\frac{\partial\psi}{\partial t}\right)$$
$$+\left[m^2c^2 + \frac{e^2}{c^2}(\mathcal{A}^2 - \phi^2)\right]\left(\frac{\partial\psi}{\partial p}\right)^2. \tag{4}$$

We note that the coefficients of the zeroth, first and second power of $\frac{\partial\psi}{\partial p}$ are four-dimensional invariants. Further, the form Q remains invariant if one puts

$$\left.\begin{aligned} \mathcal{A} &= \mathcal{A}_1 + \text{grad}f, \\ \phi &= \phi_1 - \tfrac{1}{c}\tfrac{\partial f}{\partial t}, \\ p &= p_1 - \tfrac{e}{c}f \end{aligned}\right\} \tag{5}$$

where f denotes an arbitrary function of the coordinates and time. The latter transformation also leaves the linear differential form

$$d'\Omega = \frac{e}{mc^2}(\mathcal{A}_x dx + \mathcal{A}_y dy + \mathcal{A}_z dz) - \frac{e}{mc}\phi dt + \frac{1}{mc}dp \tag{6}$$

[2] V. Fock, Zur Schrödingerschen Wellenmechanik, ZS. f. Phys. **38**, 242, 1926.

invariant[3].

We now want to interpret the form Q as the square of the gradient of the function ψ in five-dimensional space (R_5), and look for the corresponding line element. One easily finds

$$ds^2 = dx^2 + dy^2 + dz^2 - c^2dt^2 + (d'\Omega)^2. \qquad (7)$$

The Laplacian equation in R_5 is

$$\nabla\psi - \frac{1}{c^2}\frac{\partial^2\psi}{\partial t^2} - \frac{2e}{c}\left(\mathcal{A}.\text{grad}\frac{\partial\psi}{\partial p} + \frac{\phi}{c}\frac{\partial^2\psi}{\partial t\partial p}\right)$$

$$-\frac{e}{c}\frac{\partial\psi}{\partial p}\left(\text{div}\mathcal{A} + \frac{1}{c}\frac{\partial\phi}{\partial t}\right) + \left[m^2c^2 + \frac{e^2}{c^2}(\mathcal{A}^2 - \phi^2)\right]\frac{\partial^2\psi}{\partial p^2} = 0. \qquad (8)$$

Like (7) and (4), it remains invariant under Lorentz transformations and transformations (5).

Since the coefficients of equation (8) do not contain the parameter p, we can cast the dependence of the function ψ on p in the form of an exponential factor. To get agreement with experiment we have to set [4]

$$\psi = \psi_0 e^{2\pi i \frac{p}{h}} \qquad (9)$$

The equation for ψ_0 is invariant under Lorentz transformations, but not under transformations (5). The significance of the redundant coordinate parameter p seems to lie exactly in this fact: it causes the invariance of the equations with respect to the addition of an arbitrary gradient to the four-potential.

Note here that, in general, the coefficients in the equation for ψ_0 are complex.

If one further assumes that those coefficients do not depend on t and sets

$$\psi_0 = e^{-\frac{2\pi i}{h}(E+mc^2)t}\psi_1, \qquad (10)$$

one obtains a time-free equation for ψ_1, which is identical to the generalization of Schrödinger's wave equation that we established in our previous work. Consequently, those values of E for which there is a function ψ_1 that satisfies

[3]The symbol d' shall denote the fact that $d'\Omega$ is not a total differential.

[4]The fact that the parameter p, which is associated with the linear form, appears in an exponential function, might be connected to some relations remarked on by E. Schrödinger (ZS. f. Phys. **12**, 13, 1923).

certain requirements of finiteness and continuity, are the Bohr energy levels. It follows from these observations that the addition of a gradient to the four-potential cannot have an influence on the energy levels. The functions ψ_1 and $\bar{\psi}_1$, obtained using the vector potentials $\mathcal{A}$ and $\bar{\mathcal{A}} = \mathcal{A} - \mathrm{grad} f$ respectively, would only differ by a factor of $e^{\frac{2\pi i e}{ch} f}$ from the absolute value 1 and would therefore (under very general assumptions about the function f) have the same continuity properties.

2 General Relativity

[The remainder of the paper consists of a generalization to curved space-time.]

Leningrad, Physical Institute of the University, 24 July 1926.

THE

PHYSICAL REVIEW

A journal of experimental and theoretical physics established by E. L. Nichols in 1893

SECOND SERIES, VOL. 115, No. 3 AUGUST 1, 1959

Significance of Electromagnetic Potentials in the Quantum Theory

Y. AHARONOV AND D. BOHM
H. H. Wills Physics Laboratory, University of Bristol, Bristol, England

(Received May 28, 1959; revised manuscript received June 16, 1959)

In this paper, we discuss some interesting properties of the electromagnetic potentials in the quantum domain. We shall show that, contrary to the conclusions of classical mechanics, there exist effects of potentials on charged particles, even in the region where all the fields (and therefore the forces on the particles) vanish. We shall then discuss possible experiments to test these conclusions; and, finally, we shall suggest further possible developments in the interpretation of the potentials.

1. INTRODUCTION

IN classical electrodynamics, the vector and scalar potentials were first introduced as a convenient mathematical aid for calculating the fields. It is true that in order to obtain a classical canonical formalism, the potentials are needed. Nevertheless, the fundamental equations of motion can always be expressed directly in terms of the fields alone.

In the quantum mechanics, however, the canonical formalism is necessary, and as a result, the potentials cannot be eliminated from the basic equations. Nevertheless, these equations, as well as the physical quantities, are all gauge invariant; so that it may seem that even in quantum mechanics, the potentials themselves have no independent significance.

In this paper, we shall show that the above conclusions are not correct and that a further interpretation of the potentials is needed in the quantum mechanics.

2. POSSIBLE EXPERIMENTS DEMONSTRATING THE ROLE OF POTENTIALS IN THE QUANTUM THEORY

In this section, we shall discuss several possible experiments which demonstrate the significance of potentials in the quantum theory. We shall begin with a simple example.

Suppose we have a charged particle inside a "Faraday cage" connected to an external generator which causes the potential on the cage to alternate in time. This will add to the Hamiltonian of the particle a term $V(x,t)$ which is, for the region inside the cage, a function of time only. In the nonrelativistic limit (and we shall assume this almost everywhere in the following discussions) we have, for the region inside the cage, $H=H_0+V(t)$ where H_0 is the Hamiltonian when the generator is not functioning, and $V(t)=e\phi(t)$. If $\psi_0(x,t)$ is a solution of the Hamiltonian H_0, then the solution for H will be

$$\psi=\psi_0 e^{-iS/\hbar},\quad S=\int V(t)dt,$$

which follows from

$$i\hbar\frac{\partial\psi}{\partial t}=\left(i\hbar\frac{\partial\psi_0}{\partial t}+\psi_0\frac{\partial S}{\partial t}\right)e^{-iS/\hbar}=[H_0+V(t)]\psi=H\psi.$$

The new solution differs from the old one just by a phase factor and this corresponds, of course, to no change in any physical result.

Now consider a more complex experiment in which a single coherent electron beam is split into two parts and each part is then allowed to enter a long cylindrical metal tube, as shown in Fig. 1.

After the beams pass through the tubes, they are combined to interfere coherently at F. By means of time-determining electrical "shutters" the beam is chopped into wave packets that are long compared with the wavelength λ, but short compared with the length of the tubes. The potential in each tube is determined by a time delay mechanism in such a way that the potential is zero in region I (until each packet is well inside its tube). The potential then grows as a function of time, but differently in each tube. Finally, it falls back to zero, before the electron comes near the

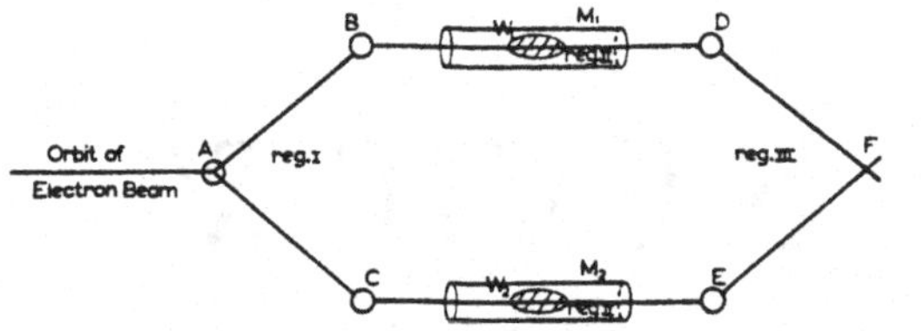

FIG. 1. Schematic experiment to demonstrate interference with time-dependent scalar potential. A, B, C, D, E: suitable devices to separate and divert beams. W_1, W_2: wave packets. M_1, M_2: cylindrical metal tubes. F: interference region.

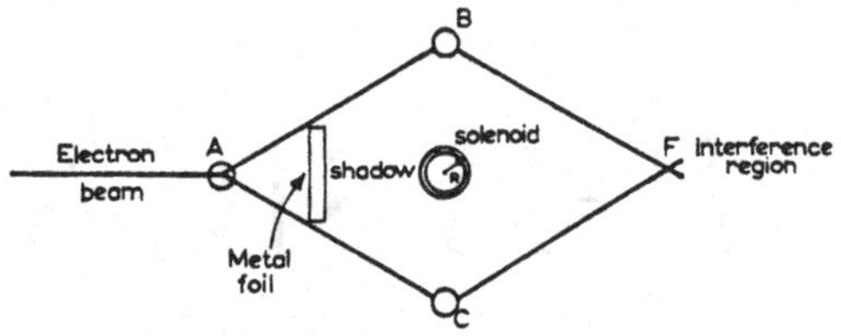

FIG. 2. Schematic experiment to demonstrate interference with time-independent vector potential.

other edge of the tube. Thus the potential is nonzero only while the electrons are well inside the tube (region II). When the electron is in region III, there is again no potential. The purpose of this arrangement is to ensure that the electron is in a time-varying potential without ever being in a field (because the field does not penetrate far from the edges of the tubes, and is nonzero only at times when the electron is far from these edges).

Now let $\psi(x,t)=\psi_1^0(x,t)+\psi_2^0(x,t)$ be the wave function when the potential is absent (ψ_1^0 and ψ_2^0 representing the parts that pass through tubes 1 and 2, respectively). But since V is a function only of t wherever ψ is appreciable, the problem for each tube is essentially the same as that of the Faraday cage. The solution is then

$$\psi=\psi_1^0e^{-iS_1/\hbar}+\psi_2^0e^{-iS_2/\hbar},$$

where

$$S_1=e\int\varphi_1dt,\quad S_2=e\int\varphi_2dt.$$

It is evident that the interference of the two parts at F will depend on the phase difference $(S_1-S_2)/\hbar$. Thus, there is a physical effect of the potentials even though no force is ever actually exerted on the electron. The effect is evidently essentially quantum-mechanical in nature because it comes in the phenomenon of interference. We are therefore not surprised that it does not appear in classical mechanics.

From relativistic considerations, it is easily seen that the covariance of the above conclusion demands that there should be similar results involving the vector potential, **A**.

The phase difference, $(S_1-S_2)/\hbar$, can also be expressed as the integral $(e/\hbar)\oint\varphi dt$ around a closed circuit in space-time, where φ is evaluated at the place of the center of the wave packet. The relativistic generalization of the above integral is

$$\frac{e}{\hbar}\oint\left(\varphi dt-\frac{\mathbf{A}}{c}\cdot d\mathbf{x}\right),$$

where the path of integration now goes over any closed circuit in space-time.

As another special case, let us now consider a path in space only (t=constant). The above argument suggests that the associated phase shift of the electron wave function ought to be

$$\Delta S/\hbar=-\frac{e}{c\hbar}\oint\mathbf{A}\cdot d\mathbf{x},$$

where $\oint\mathbf{A}\cdot d\mathbf{x}=\int\mathbf{H}\cdot d\mathbf{s}=\phi$ (the total magnetic flux inside the circuit).

This corresponds to another experimental situation. By means of a current flowing through a very closely wound cylindrical solenoid of radius R, center at the origin and axis in the z direction, we create a magnetic field, **H**, which is essentially confined within the solenoid. However, the vector potential, **A**, evidently, cannot be zero everywhere outside the solenoid, because the total flux through every circuit containing the origin is equal to a constant

$$\phi_0=\int\mathbf{H}\cdot d\mathbf{s}=\int\mathbf{A}\cdot d\mathbf{x}.$$

To demonstrate the effects of the total flux, we begin, as before, with a coherent beam of electrons. (But now there is no need to make wave packets.) The beam is split into two parts, each going on opposite sides of the solenoid, but avoiding it. (The solenoid can be shielded from the electron beam by a thin plate which casts a shadow.) As in the former example, the beams are brought together at F (Fig. 2).

The Hamiltonian for this case is

$$H=\frac{[\mathbf{P}-(e/c)\mathbf{A}]^2}{2m}.$$

In singly connected regions, where $\mathbf{H}=\nabla\times\mathbf{A}=0$, we can always obtain a solution for the above Hamiltonian by taking $\psi=\psi_0e^{-iS/\hbar}$, where ψ_0 is the solution when $\mathbf{A}=0$ and where $\nabla S/\hbar=(e/c)\mathbf{A}$. But, in the experiment discussed above, in which we have a multiply connected region (the region outside the solenoid), $\psi_0e^{-iS/\hbar}$ is a non-single-valued function[1] and therefore, in general, not a permissible solution of Schrödinger's equation. Nevertheless, in our problem it is still possible to use such solutions because the wave function splits into two parts $\psi=\psi_1+\psi_2$, where ψ_1 represents the beam on

[1] Unless $\phi_0=nhc/e$, where n is an integer.

one side of the solenoid and ψ_2 the beam on the opposite side. Each of these beams stays in a simply connected region. We therefore can write

$$\psi_1=\psi_1^0 e^{-iS_1/\hbar},\quad \psi_2=\psi_2^0 e^{-iS_2/\hbar},$$

where S_1 and S_2 are equal to $(e/c)\int \mathbf{A}\cdot d\mathbf{x}$ along the paths of the first and second beams, respectively. (In Sec. 4, an exact solution for this Hamiltonian will be given, and it will confirm the above results.)

The interference between the two beams will evidently depend on the phase difference,

$$(S_1-S_2)/\hbar=(e/\hbar c)\oint \mathbf{A}\cdot d\mathbf{x}=(e/\hbar c)\phi_0.$$

This effect will exist, even though there are no magnetic forces acting in the places where the electron beam passes.

In order to avoid fully any possible question of contact of the electron with the magnetic field we note that our result would not be changed if we surrounded the solenoid by a potential barrier that reflects the electrons perfectly. (This, too, is confirmed in Sec. 4.)

It is easy to devise hypothetical experiments in which the vector potential may influence not only the interference pattern but also the momentum. To see this, consider a periodic array of solenoids, each of which is shielded from direct contact with the beam by a small plate. This will be essentially a grating. Consider first the diffraction pattern without the magnetic field, which will have a discrete set of directions of strong constructive interference. The effect of the vector potential will be to produce a shift of the relative phase of the wave function in different elements of the gratings. A corresponding shift will take place in the directions, and therefore the momentum of the diffracted beam.

3. A PRACTICABLE EXPERIMENT TO TEST FOR THE EFFECTS OF A POTENTIAL WHERE THERE ARE NO FIELDS

As yet no direct experiments have been carried out which confirm the effect of potentials where there is no field. It would be interesting therefore to test whether such effects actually exist. Such a test is, in fact, within the range of present possibilities.[2] Recent experiments[3,4] have succeeded in obtaining interference from electron beams that have been separated in one case by as much as 0.8 mm.[3] It is quite possible to wind solenoids which are smaller than this, and therefore to place them between the separate beams. Alternatively, we may obtain localized lines of flux of the right magnitude (the magnitude has to be of the order of $\phi_0=2\pi c\hbar/e\sim 4\times 10^{-7}$ gauss cm^2) by means of fine permanently magnetized "whiskers".[5] The solenoid can be used in Marton's device,[3] while the whisker is suitable for another experimental setup[4] where the separation is of the order of microns and the whiskers are even smaller than this.

In principle, we could do the experiment by observing the interference pattern with and without the magnetic flux. But since the main effect of the flux is only to displace the line pattern without changing the interval structure, this would not be a convenient experiment to do. Instead, it would be easier to vary the magnetic flux within the same exposure for the detection of the interference patterns. Such a variation would, according to our previous discussion, alter the sharpness and the general form of the interference bands. This alteration would then constitute a verification of the predicted phenomena.

When the magnetic flux is altered, there will, of course, be an induced electric field outside the solenoid, but the effects of this field can be made negligible. For example, suppose the magnetic flux were suddenly altered in the middle of an exposure. The electric field would then exist only for a very short time, so that only a small part of the beam would be affected by it.

4. EXACT SOLUTION FOR SCATTERING PROBLEMS

We shall now obtain an exact solution for the problem of the scattering of an electron beam by a magnetic field in the limit where the magnetic field region tends to a zero radius, while the total flux remains fixed. This corresponds to the setup described in Sec. 2 and shown in Fig. 2. Only this time we do not split the plane wave into two parts. The wave equation outside the magnetic field region is, in cylindrical coordinates,

$$\left[\frac{\partial^2}{\partial r^2}+\frac{1}{r}\frac{\partial}{\partial r}+\frac{1}{r^2}\left(\frac{\partial}{\partial\theta}+i\alpha\right)^2+k^2\right]\psi=0, \qquad (1)$$

where $\mathbf{k}$ is the wave vector of the incident particle and $\alpha=-e\phi/ch$. We have again chosen the gauge in which $A_r=0$ and $A_\theta=\phi/2\pi r$.

The general solution of the above equation is

$$\psi=\sum_{m+-\infty}^{\infty} e^{im\theta}[a_m J_{m+\alpha}(kr)+b_m J_{-(m+\alpha)}(kr)], \qquad (2)$$

where a_m and b_m are arbitrary constants and $J_{m+\alpha}(kr)$ is a Bessel function, in general of fractional order (dependent on ϕ). The above solution holds only for $r>R$. For $r<R$ (inside the magnetic field) the solution has been worked out.[6] By matching the solutions at $r=R$ it is easily shown that only Bessel functions of positive order will remain, when R approaches zero.

[2] Dr. Chambers is now making a preliminary experimental study of this question at Bristol.

[3] L. Marton, Phys. Rev. **85**, 1057 (1952); **90**, 490 (1953). Marton, Simpson, and Suddeth, Rev. Sci. Instr. **25**, 1099 (1954).

[4] G. Mollenstedt, Naturwissenschaften **42**, 41 (1955); G. Mollenstedt and H. Düker, Z. Physik **145**, 377 (1956).

[5] See, for example, Sidney S. Brenner, Acta Met. **4**, 62 (1956).

[6] L. Page, Phys. Rev. **36**, 444 (1930).

This means that the probability of finding the particle inside the magnetic field region approaches zero with R. It follows that the wave function would not be changed if the electron were kept away from the field by a barrier whose radius also went to zero with R.

The general solution in the limit of R tending to zero is therefore

$$\psi=\sum_{m=-\infty}^{\infty} a_m J_{|m+\alpha|}e^{im\theta}. \tag{3}$$

We must then choose a_m so that ψ represents a beam of electrons that is incident from the right ($\theta=0$). It is important, however, to satisfy the initial condition that the current density,

$$\mathbf{j}=\frac{\hbar(\psi^*\nabla\psi-\psi\nabla\psi^*)}{2im}-\frac{e}{mc}\mathbf{A}\psi^*\psi, \tag{4}$$

shall be constant and in the x direction. In the gauge that we are using, we easily see that the correct incident wave is $\psi_{\mathrm{inc}}=e^{-ikx}e^{-i\alpha\theta}$. Of course, this wave function holds only to the right of the origin, so that no problem of multiple-valuedness arises.

We shall show in the course of this calculation that the above conditions will be satisfied by choosing $a_m=(-i)^{|m+\alpha|}$, in which case, we shall have

$$\psi=\sum_{m=-\infty}^{\infty}(-i)^{|m+\alpha|}J_{|m+\alpha|}e^{im\theta}.$$

It is convenient to split ψ into the following three parts: $\psi=\psi_1+\psi_2+\psi_3$, where

$$\psi_1=\sum_{m=1}^{\infty}(-i)^{m+\alpha}J_{m+\alpha}e^{im\theta},$$

$$\psi_2=\sum_{m=-\infty}^{-1}(-i)^{m+\alpha}J_{m+\alpha}e^{im\theta},$$

$$=\sum_{m=1}^{\infty}(-i)^{m-\alpha}J_{m-\alpha}e^{-im\theta}, \tag{5}$$

$$\psi_3=(-i)^{|\alpha|}J_{|\alpha|}.$$

Now ψ_1 satisfies the simple differential equation

$$\frac{\partial\psi_1}{\partial r'}=\sum_{m=1}^{\infty}(-i)^{m+\alpha}J_{m+\alpha}'e^{im\theta}$$

$$=\sum_{m=1}^{\infty}(-i)^{m+\alpha}\frac{J_{m+\alpha-1}-J_{m+\alpha+1}}{2}e^{im\theta},\quad r'=kr \tag{6}$$

where we have used the well-known formula for Bessel functions:

$$dJ_\gamma(r)/dr=\tfrac{1}{2}(J_{\gamma-1}-J_{\gamma+1}).$$

As a result, we obtain

$$\frac{\partial\psi_1}{\partial r'}=\frac{1}{2}\sum_{m'=0}^{\infty}(-i)^{m'+\alpha+1}J_{m'+\alpha}e^{i(m'+1)\theta}-\frac{1}{2}\sum_{m'=2}^{\infty}(-i)^{m'+\alpha-1}J_{m'+\alpha}e^{i(m'-1)\theta}$$

$$=\frac{1}{2}\sum_{m'=1}^{\infty}(-i)^{m'+\alpha}J_{m'+\alpha}e^{im'\theta}(-ie^{i\theta}+i^{-1}e^{-i\theta})+\tfrac{1}{2}(-i)^{\alpha}[J_{\alpha+1}-ie^{i\theta}J_\alpha]. \tag{7}$$

So

$$\partial\psi_1/\partial r'=-i\cos\theta\psi_1+\tfrac{1}{2}(-i)^\alpha(J_{\alpha+1}-iJ_\alpha e^{i\theta}).$$

This differential equation can be easily integrated to give

$$\psi_1=A\int_0^{r'}e^{ir'\cos\theta}[J_{\alpha+1}-iJ_\alpha e^{i\theta}]dr', \tag{8}$$

where

$$A=\tfrac{1}{2}(-i)^\alpha e^{-ir'\cos\theta}.$$

The lower limit of the integration is determined by the requirement that when r' goes to zero, ψ_1 also goes to to zero because, as we have seen, ψ_1 includes Bessel functions of positive order only.

In order to discuss the asymptotic behavior of ψ_1, let us write it as $\psi_1=A[I_1-I_2]$, where

$$I_1=\int_0^{\infty}e^{ir'\cos\theta}[J_{\alpha+1}-ie^{i\theta}J_\alpha]dr',$$

$$I_2=\int_r^{\infty}e^{ir'\cos\theta}[J_{\alpha+1}-ie^{i\theta}J_\alpha]dr'. \tag{9}$$

The first of these integrals is known[7]:

$$\int_0^{\infty}e^{i\beta r}J_\alpha(kr)=\frac{e^{i[\alpha\arcsin(\beta/k)]}}{(k^2-\beta^2)^{\frac{1}{2}}},\quad 0<\beta<k,\quad -2<\alpha.$$

In our cases, $\beta=\cos\theta$, $k=1$, so that

$$I_1=\left[\frac{e^{i\alpha(\frac{1}{2}\pi-|\theta|)}}{|\sin\theta|}-ie^{i\theta}\frac{e^{i(\alpha+1)(\frac{1}{2}\pi-|\theta|)}}{|\sin\theta|}\right]. \tag{10}$$

Because the integrand is even in θ, we have written the final expression for the above integral as a function of $|\theta|$ and of $|\sin\theta|$. Hence

$$I_1=e^{i\alpha(\frac{1}{2}\pi-|\theta|)}\left[\frac{ie^{-i|\theta|}-ie^{i\theta}}{|\sin\theta|}\right]$$

$$=0\quad\text{for }\theta<0,$$

$$=e^{-i\alpha\theta}2i^\alpha\quad\text{for }\theta>0, \tag{11}$$

where we have taken θ as going from $-\pi$ to π.

[7] See, for example, W. Gröbner and N. Hofreiter, *Integraltafel* (Springer-Verlag, Berlin, 1949).

We shall see presently that I_1 represents the largest term in the asymptotic expansion of ψ_1. The fact that it is zero for $\theta<0$ shows that this part of ψ_1 passes (asymptotically) only on the upper side of the singularity. To explain this, we note that ψ_1 contains only positive values of m, and therefore of the angular momentum. It is quite natural then that this part of ψ_1 goes on the upper side of the singularity. Similarly, since according to (5)

$$\psi_2(r',\theta,\alpha)=\psi_1(r',-\theta,-\alpha),$$

it follows that ψ_2 will behave oppositely to ψ_1 in this regard, so that together they will make up the correct incident wave.

Now, in the limit of $r'\to\infty$ we are allowed to take in the integrand of I_2 the first asymptotic term of J_α,[8] namely $J_\alpha\to(2/\pi r)^{\frac{1}{2}}\cos(r'-\frac{1}{2}\alpha-\frac{1}{4}\pi)$. We obtain

$$I_2=\int_r^\infty e^{ir'\cos\theta}(J_{\alpha+1}-ie^{i\theta}J_\alpha)dr'\to C+D, \quad (12)$$

where

$$C=\int_r^\infty e^{ir'\cos\theta}[\cos(r'-\tfrac{1}{2}(\alpha+1)\pi-\tfrac{1}{4}\pi)]\frac{dr'}{(r')^{\frac{1}{2}}}\left(\frac{2}{\pi}\right)^{\frac{1}{2}},$$
$$D=\int_r^\infty e^{ir'\cos\theta}[\cos(r'-\tfrac{1}{2}\alpha-\tfrac{1}{4}\pi)]\frac{dr'}{(r')^{\frac{1}{2}}}\left(\frac{2}{\pi}\right)^{\frac{1}{2}}(-i)e^{i\theta}. \quad (13)$$

Then

$$C=\int_r^\infty e^{ir'\cos\theta}\Big[e^{i[r'-\frac{1}{2}(\alpha+1)\pi-\frac{1}{4}\pi]}+e^{-i[r'-\frac{1}{2}(\alpha+1)\pi-\frac{1}{4}\pi]}\Big]\frac{dr'}{(2\pi r')^{\frac{1}{2}}}$$
$$=\left(\frac{2}{\pi}\right)^{\frac{1}{2}}\frac{(-i)^{\alpha+\frac{1}{2}}}{(1+\cos\theta)^{\frac{1}{2}}}\int_{[r'(1+\cos\theta)]^{\frac{1}{2}}}^\infty \exp(+iz^2)dz+\left(\frac{2}{\pi}\right)^{\frac{1}{2}}\frac{i^{\alpha+\frac{1}{2}}}{(1-\cos\theta)^{\frac{1}{2}}}\int_{[r'(1-\cos\theta)]^{\frac{1}{2}}}^\infty \exp(-iz^2)dz, \quad (14)$$

where we have put

$$z=[r'(1+\cos\theta)]^{\frac{1}{2}} \quad \text{and} \quad z=[r'(1-\cos\theta)]^{\frac{1}{2}},$$

respectively.

Using now the well-known asymptotic behavior of the error function,[9]

$$\int_a^\infty \exp(iz^2)dz\to\frac{i}{2}\frac{\exp(ia^2)}{a},$$
$$\int_a^\infty \exp(-iz^2)dz\to\frac{-i}{2}\frac{\exp(-ia^2)}{a}, \quad (15)$$

we finally obtain

$$C=\left[\frac{(-i)^{\alpha+\frac{1}{2}}}{(2\pi)^{\frac{1}{2}}}\frac{e^{ir'}}{[r'(1+\cos\theta)^2]^{\frac{1}{2}}}+\frac{i^{\alpha+\frac{1}{2}}}{(2\pi)^{\frac{1}{2}}}\frac{e^{-ir'}}{[r'(1-\cos\theta)^2]^{\frac{1}{2}}}\right]e^{ir'\cos\theta}, \quad (16)$$

$$D=\left[\frac{(-i)^{\alpha-\frac{1}{2}}}{(2\pi)^{\frac{1}{2}}}\frac{e^{ir'}}{[r'(1+\cos\theta)^2]^{\frac{1}{2}}}+\frac{i^{\alpha-\frac{1}{2}}}{(2\pi)^{\frac{1}{2}}}\frac{e^{-ir'}}{[r'(1-\cos\theta)^2]^{\frac{1}{2}}}\right]e^{ir'\cos\theta}(-i)e^{i\theta}. \quad (17)$$

Now adding (16) and (17) together and using (13) and (9), we find that the term of $1/(r')^{\frac{1}{2}}$ in the asymptotic expansion of ψ_1 is

$$\frac{(-i)^{\frac{1}{2}}}{2(2\pi)^{\frac{1}{2}}}\left[(-1)^\alpha\frac{e^{ir'}}{(r')^{\frac{1}{2}}}\frac{1+e^{i\theta}}{1+\cos\theta}+i\frac{e^{-ir'}}{(r')^{\frac{1}{2}}}\frac{1-e^{i\theta}}{1-\cos\theta}\right]. \quad (18)$$

Using again the relation between ψ_1 and ψ_2 we obtain for the corresponding term in ψ_2

$$\frac{(-i)^{\frac{1}{2}}}{2(2\pi)^{\frac{1}{2}}}\left[(-1)^{-\alpha}\frac{e^{ir'}}{(r')^{\frac{1}{2}}}\frac{1+e^{-i\theta}}{1+\cos\theta}+i\frac{e^{-ir'}}{(r')^{\frac{1}{2}}}\frac{1-e^{-i\theta}}{1-\cos\theta}\right]. \quad (19)$$

Adding (18) and (19) and using (11), we finally get

$$\psi_1+\psi_2\to\frac{(-i)^{\frac{1}{2}}}{(2\pi)^{\frac{1}{2}}}\left[\frac{ie^{-ir'}}{(r')^{\frac{1}{2}}}+\frac{e^{ir'}}{(r')^{\frac{1}{2}}}\frac{\cos(\pi\alpha-\frac{1}{2}\theta)}{\cos(\frac{1}{2}\theta)}\right]+e^{-i(r'\cos\theta+\alpha\theta)}. \quad (20)$$

There remains the contribution of ψ_3, whose asymptotic behavior is [see Eq. (12)]

$$(-i)^{|\alpha|}J_{|\alpha|}(r')\to(-i)^{|\alpha|}\left(\frac{2}{\pi r'}\right)^{\frac{1}{2}}\cos(r'-\tfrac{1}{4}\pi-\tfrac{1}{2}|\alpha|\pi).$$

Collecting all terms, we find

$$\psi=\psi_1+\psi_2+\psi_3\to e^{-i(\alpha\theta+r'\cos\theta)}\pm\frac{e^{ir'}}{(2\pi ir')^{\frac{1}{2}}}\sin\pi\alpha\frac{e^{-i\theta/2}}{\cos(\theta/2)}, \quad (21)$$

where the $\pm$ sign is chosen according to the sign of α.

The first term in equation (21) represents the incident wave, and the second the scattered wave.[10] The scattering cross section is therefore

$$\sigma=\frac{\sin^2\pi\alpha}{2\pi}\frac{1}{\cos^2(\theta/2)}. \quad (22)$$

[8] E. Jahnke and F. Emde, *Tables of Functions* (Dover Publications, Inc., New York, 1943), fourth edition, p. 138.

[9] Reference 8, p. 24.

[10] In this way, we verify, of course, that our choice of the a_m for Eq. (3) satisfies the correct boundary conditions.

When $\alpha=n$, where n is an integer, then σ vanishes. This is analogous to the Ramsauer effect.[11] σ has a maximum when $\alpha=n+\frac{1}{2}$.

The asymptotic formula (21) holds only when we are not on the line $\theta=\pi$. The exact solution, which is needed on this line, would show that the second term will combine with the first to make a single-valued wave function, despite the non-single-valued character of the two parts, in the neighborhood of $\theta=\pi$. We shall see this in more detail presently for the special case $\alpha=n+\frac{1}{2}$.

In the interference experiment discussed in Sec. 2, diffraction effects, represented in Eq. (21) by the scattered wave, have been neglected. Therefore, in this problem, it is adequate to use the first term of Eq. (21). Here, we see that the phase of the wave function has a different value depending on whether we approach the line $\theta=\pm\pi$ from positive or negative angles, i.e., from the upper or lower side. This confirms the conclusions obtained in the approximate treatment of Sec. 2.

We shall discuss now the two special cases that can be solved exactly. The first is the case where $\alpha=n$. Here, the wave function is $\psi=e^{-ikx}e^{-i\alpha\theta}$, which is evidently single-valued when α is an integer. (It can be seen by direct differentiation that this is a solution.)

The second case is that of $\alpha=n+\frac{1}{2}$. Because $J_{(n+\frac{1}{2})}(r)$ is a closed trigonometric function, the integrals for ψ can be carried out exactly.

The result is

$$\psi=\frac{i^{\frac{1}{2}}}{\sqrt{2}}e^{-i(\frac{1}{2}\theta+r'\cos\theta)}\int_0^{[r'(1+\cos\theta)]^{\frac{1}{2}}}\exp(iz^2)dz. \qquad (23)$$

This function vanishes on the line $\theta=\pi$. It can be seen that its asymptotic behavior is the same as that of Eq. (2) with α set equal to $n+\frac{1}{2}$. In this case, the single-valuedness of ψ is evident. In general, however, the behavior of ψ is not so simple, since ψ does not become zero on the line $\theta=\pi$.

5. DISCUSSION OF SIGNIFICANCE OF RESULTS

The essential result of the previous discussion is that in quantum theory, an electron (for example) can be influenced by the potentials even if all the field regions are excluded from it. In other words, in a field-free multiply-connected region of space, the physical properties of the system still depend on the potentials.

It is true that all these effects of the potentials depend only on the gauge-invariant quantity $\oint \mathbf{A}\cdot d\mathbf{x}=\int \mathbf{H}\cdot d\mathbf{s}$, so that in reality they can be expressed in terms of the fields inside the circuit. However, according to current relativistic notions, all fields must interact only locally. And since the electrons cannot reach the regions where the fields are, we cannot interpret such effects as due to the fields themselves.

In classical mechanics, we recall that potentials cannot have such significance because the equation of motion involves only the field quantities themselves. For this reason, the potentials have been regarded as purely mathematical auxiliaries, while only the field quantities were thought to have a direct physical meaning.

In quantum mechanics, the essential difference is that the equations of motion of a particle are replaced by the Schrödinger equation for a wave. This Schrödinger equation is obtained from a canonical formalism, which cannot be expressed in terms of the fields alone, but which also requires the potentials. Indeed, the potentials play a role, in Schrödinger's equation, which is analogous to that of the index of refration in optics, The Lorentz force $[e\mathbf{E}+(e/c)\mathbf{v}\times\mathbf{H}]$ does not appear anywhere in the fundamental theory, but appears only as an approximation holding in the classical limit. It would therefore seem natural at this point to propose that, in quantum mechanics, the fundamental physical entities are the potentials, while the fields are derived from them by differentiations.

The main objection that could be raised against the above suggestion is grounded in the gauge invariance of the theory. In other words, if the potentials are subject to the transformation $\mathbf{A}_\mu \rightarrow A_\mu'=A_\mu+\partial\psi/\partial x_\mu$, where ψ is a continuous scalar function, then all the known physical quantities are left unchanged. As a result, the same physical behavior is obtained from any two potentials, $A_\mu(x)$ and $A_\mu'(x)$, related by the above transformation. This means that insofar as the potentials are richer in properties than the fields, there is no way to reveal this additional richness. It was therefore concluded that the potentials cannot have any meaning, except insofar as they are used mathematically, to calculate the fields.

We have seen from the examples described in this paper that the above point of view cannot be maintained for the general case. Of course, our discussion does not bring into question the gauge invariance of the theory. But it does show that in a theory involving only local interactions (e.g., Schrödinger's or Dirac's equation, and current quantum-mechanical field theories), the potentials must, in certain cases, be considered as physically effective, even when there are no fields acting on the charged particles.

The above discussion suggests that some further development of the theory is needed. Two possible directions are clear. First, we may try to formulate a nonlocal theory in which, for example, the electron could interact with a field that was a finite distance away. Then there would be no trouble in interpreting these results, but, as is well known, there are severe difficulties in the way of doing this. Secondly, we may retain the present local theory and, instead, we may try to give a further new interpretation to the poten-

[11] See, for example, D. Bohm, *Quantum Theory* (Prentice-Hall, Inc., Englewood Cliffs, New Jersey, 1951).

tials. In other words, we are led to regard $A_\mu(x)$ as a physical variable. This means that we must be able to define the physical difference between two quantum states which differ only by gauge transformation. It will be shown in a future paper that in a system containing an undefined number of charged particles (i.e., a superposition of states of different total charge), a new Hermitian operator, essentially an angle variable, can be introduced, which is conjugate to the charge density and which may give a meaning to the gauge. Such states have actually been used in connection with recent theories of superconductivity and superfluidity[12] and we shall show their relation to this problem in more detail.

ACKNOWLEDGMENTS

We are indebted to Professor M. H. L. Pryce for many helpful discussions. We wish to thank Dr. Chambers for many discussions connected with the experimental side of the problem.

[12] See, for example, C. G. Kuper, *Advances in Physics*, edited by N. F. Mott (Taylor and Francis, Ltd., London, 1959), Vol. 8, p. 25, Sec. 3, Par. 3.

SHIFT OF AN ELECTRON INTERFERENCE PATTERN BY ENCLOSED MAGNETIC FLUX

R. G. Chambers
H. H. Wills Physics Laboratory, University of Bristol, Bristol, England
(Received May 27, 1960)

Aharonov and Bohm[1] have recently drawn attention to a remarkable prediction from quantum theory. According to this, the fringe pattern in an electron interference experiment should be shifted by altering the amount of magnetic flux passing between the two beams (e.g., in region a of Fig. 1), even though the beams themselves pass only through field-free regions. Theory predicts a shift of n fringes for an enclosed flux Φ of nhc/e; it is convenient to refer to a natural "flux unit," $hc/e = 4.135\times10^{-7}$ gauss cm^2. It has since been pointed out[2] that the same conclusion had previously been reached by Ehrenberg and Siday,[3] using semiclassical arguments, but these authors perhaps did not sufficiently stress the remarkable nature of the result, and their work appears to have attracted little attention.

Clearly the first problem to consider, experimentally, is the effect on the fringe system of stray fields not localized to region a but extending, e.g., over region a' in Fig. 1. In addition

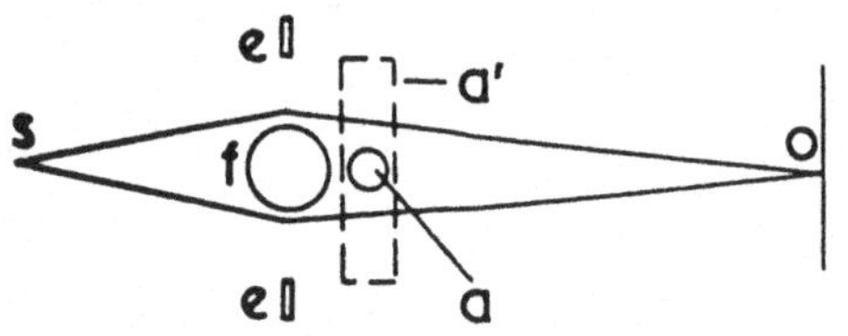

FIG. 1. Schematic diagram of interferometer, with source s, observing plane o, biprism e, f, and confined and extended field regions a and a'.

to the "quantum" fringe shift due to the enclosed flux, there will then be a shift due simply to curvature of the electron trajectories by the field. A straightforward calculation shows that in a "biprism" experiment,[4] such a field should produce a fringe displacement which exactly keeps pace with the deflection of the beams by the field, so that the fringe system appears to remain undisplaced relative to the envelope of the pattern. A field of type a, on the other hand, should leave the envelope undisplaced, and produce a fringe shift within it. In the Marton[5] interferometer, conditions are different, and a field of type a' should leave the fringes undisplaced in space. This explains how Marton et al.[5] were able to observe fringes in the presence of stray 60-cps fields probably large enough to have destroyed them otherwise; this experiment thus constitutes an inadvertent check of the existence of the "quantum" shift.[2]

To obtain a more direct check, a Philips EM100 electron microscope[6] has been modified so that it can be switched at will from normal operation to operation as an interferometer. Fringes are produced by an electrostatic "biprism" consisting of an aluminized quartz fiber f (Fig. 1) flanked by two earthed metal plates e; altering the positive potential applied to f alters the effective angle of the biprism.[4] The distances s-f and f-o (Fig. 1) are about 6.7 cm and 13.4 cm, respectively. With this microscope it was not possible to reduce the virtual source diameter below about 0.2 μ, so that it was necessary to use a fiber f only about 1.5 μ in diameter and a

VOLUME 5, NUMBER 1 PHYSICAL REVIEW LETTERS JULY 1, 1960

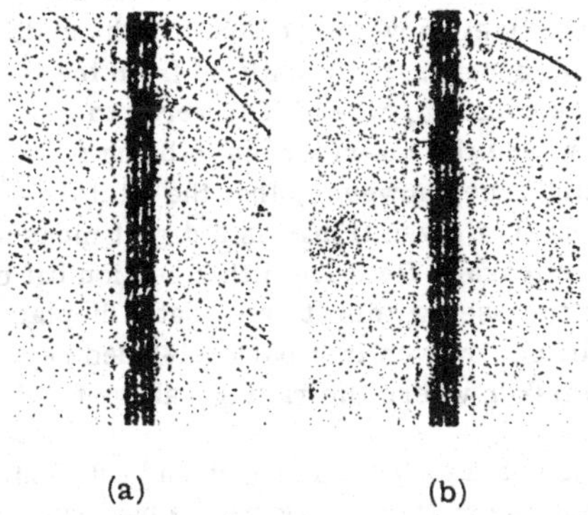

FIG. 2. (a) Fringe pattern due to biprism alone. (b) Pattern displaced by 2.5 fringe widths by field of type a'.

very small biprism angle, to produce a wide pattern of fringes which would not be blurred out by the finite source size. The fringe pattern obtained is shown in Fig. 2(a); the fringe width in the observing plane o is about 0.6 μ.

We first examined the effect of a field of type a', produced by a Helmholtz pair of single turns 3 mm in diameter just behind the biprism. Fields up to 0.3 gauss were applied, sufficient to displace the pattern by up to 30 fringe widths, and as predicted the appearance of the pattern was completely unchanged. Figure 2(b), for instance, shows the pattern in a field producing a displacement of about 2.5 fringe widths. In the absence of the "quantum" shift due to the enclosed flux, this pattern would have had the light and dark fringes interchanged. We also verified that with this interferometer, unlike Marton's, a small ac field suffices to blur out the fringe system completely. These results confirm the presence of the quantum shift in fields of type a'.

Of more interest is the effect predicted for a field of type a, where intuition might expect no effect. Such a field was produced by an iron whisker,[7] about 1 μ in diameter and 0.5 mm long, placed in the shadow of the fiber f. Whiskers as thin as this are expected theoretically[8] and found experimentally[9] to be single magnetic domains; moreover they are found to taper[9] with a slope of the order of 10^{-3}, which is extremely convenient for the present purpose. An iron whisker 1 μ in diameter will contain about 400 flux units; if it tapers uniformly with a slope of 10^{-3}, the flux content will change along the length at a rate $d\Phi/dz$ of about 1 flux unit per micron. Thus if such a whisker is placed in position a (Fig. 1), we expect to see a pattern in which the envelope is undisplaced, but the fringe system within the envelope is inclined at an angle of the order of one fringe width per micron. Since the fringe width in the observing plane is 0.6 μ, and there is a "pin-hole" magnification of $\times 3$ between the biprism-fiber assembly and the observing plane, we thus expect the fringes to show a tilt of order 1 in 5 relative to the envelope of the pattern. Precisely this is observed experimentally, as shown in Fig. 3(a). It will be seen that the whisker taper is not uniform, but in this example becomes very small in the upper part of the picture.

In fact the biprism is an unnecessary refinement for this experiment: Fresnel diffraction into the shadow of the whisker is strong enough to produce a clear fringe pattern from the whisker alone. Thus Fig. 3(b) shows the same section of whisker as Fig. 3(a), moved just out of the shadow of the biprism fiber. The biprism fringes are now unperturbed; the Fresnel fringes in the shadow of the whisker show exactly the same pattern of fringe shifts along their length as in Fig. 3(a). Figure 3(c) shows a further example of these fringes, from a different part of the same whisker, with the biprism moved out of the way. The whisker here is tapering more rapidly.

These fringe shifts cannot be attributed to direct interaction between the electrons and the surface of the whisker, since in Fig. 3(a) the whisker

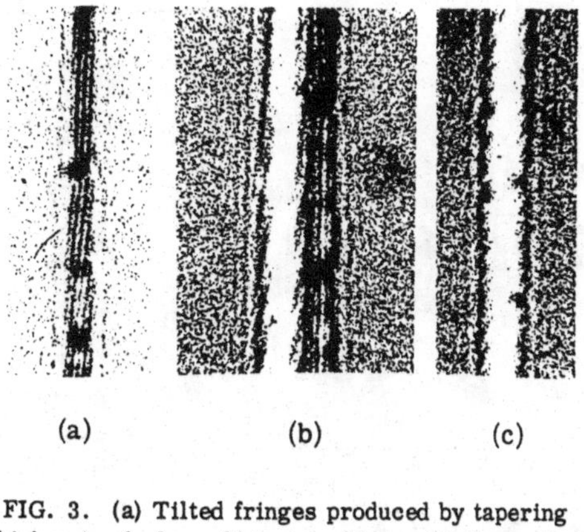

FIG. 3. (a) Tilted fringes produced by tapering whisker in shadow of biprism fiber. (b) Fresnel fringes in the shadow of the whisker itself, just outside shadow of fiber. (c) Same as (b), but from a different part of the whisker, and with fiber out of the field of view.

is completely in the shadow of the fiber. Nor can they be attributed to a return field H_z parallel to the whisker in the region a' outside it, due to the flux emerging from the sides and ends, for two reasons. An estimate of the magnitude of this field shows that it might be strong enough to displace the pattern by perhaps one fringe width, but not more, and observation confirms that the displacement of the envelope is very small; secondly, we have seen experimentally that an extended field H_z in fact produces a completely different effect (Fig. 2). Thus the patterns of Fig. 3 might be taken to demonstrate the existence of the predicted quantum shift. Indeed they do; nevertheless the tilt of the fringes can be attributed to a leakage field, as Pryce has pointed out to me, and it is illuminating to consider this.[10] Immediately outside a tapering whisker, the leakage field is in fact primarily radial and is given by $H_r = (d\Phi/dz)/2\pi r$. This field exerts a force on the electron and gives it a momentum $p_z = \pm \frac{1}{2}(e/c)d\Phi/dz$, the different signs applying to paths on either side of the whisker. The two beams which converge to interfere at o are thus tilted one above and one below the plane of Fig. 1, thus skewing the interference fringes. There is a progressive change in the phase difference between the two beams as one moves in the z direction. This is easily calculated from p_z by de Broglie's relation, and amounts to a phase-difference gradient of $(e/\hbar c)d\Phi/dz$. This is precisely the rate of change of the "quantum" phase difference $e\Phi/\hbar c$ calculated by Aharonov and Bohm. One thus sees fairly intuitively how the "quantum" phase difference is progressively built up from the free end of the whisker, where it is zero, to any section where the interference is being observed. It remains true, however, that the total displacement of a given fringe is a direct measure not of the leakage field from that section but of the flux enclosed within it, and that a displacement will occur even in a parallel-sided region of the whisker where the radial leakage field is zero.

I am indebted to Mr. Aharonov and Dr. Bohm for telling me of their work before publication, and to them and to Professor Pryce for many discussions.

---

[1]Y. Aharonov and D. Bohm, Phys. Rev. 115, 485 (1959).

[2]F. G. Werner and D. R. Brill, Phys. Rev. Letters 4, 344 (1960).

[3]W. Ehrenberg and R. E. Siday, Proc. Phys. Soc. (London) B62, 8 (1949).

[4]G. Möllenstedt and H. Düker, Z. Physik 145, 377 (1956).

[5]L. Marton, J. A. Simpson, and J. A. Suddeth, Rev. Sci. Instr. 25, 1099 (1954).

[6]A. C. van Dorsten, H. Nieuwdorp, and A. Verhoeff, Philips Tech. Rev. 12, 33 (1950).

[7]The use of a whisker was first suggested by Dr. J. W. Mitchell. The whiskers were grown by the method of S. S. Brenner, Acta Met. 4, 62 (1956).

[8]C. Kittel, Phys. Rev. 70, 965 (1946).

[9]R. W. DeBlois, J. Appl. Phys. 29, 459 (1958).

[10]The following analysis is due to Professor Pryce.

PHYSICAL REVIEW VOLUME 96, NUMBER 1 OCTOBER 1, 1954

Conservation of Isotopic Spin and Isotopic Gauge Invariance*

C. N. YANG † AND R. L. MILLS
Brookhaven National Laboratory, Upton, New York
(Received June 28, 1954)

It is pointed out that the usual principle of invariance under isotopic spin rotation is not consistant with the concept of localized fields. The possibility is explored of having invariance under local isotopic spin rotations. This leads to formulating a principle of isotopic gauge invariance and the existence of a **b** field which has the same relation to the isotopic spin that the electromagnetic field has to the electric charge. The **b** field satisfies nonlinear differential equations. The quanta of the **b** field are particles with spin unity, isotopic spin unity, and electric charge $\pm e$ or zero.

INTRODUCTION

THE conservation of isotopic spin is a much discussed concept in recent years. Historically an isotopic spin parameter was first introduced by Heisenberg[1] in 1932 to describe the two charge states (namely neutron and proton) of a nucleon. The idea that the neutron and proton correspond to two states of the same particle was suggested at that time by the fact that their masses are nearly equal, and that the light stable even nuclei contain equal numbers of them. Then in 1937 Breit, Condon, and Present pointed out the approximate equality of $p-p$ and $n-p$ interactions in the ${}^1S$ state.[2] It seemed natural to assume that this equality holds also in the other states available to both the $n-p$ and $p-p$ systems. Under such an assumption one arrives at the concept of a total isotopic spin[3] which is conserved in nucleon-nucleon interactions. Experi-

* Work performed under the auspices of the U. S. Atomic Energy Commission.

† On leave of absence from the Institute for Advanced Study, Princeton, New Jersey.

[1] W. Heisenberg, Z. Physik **77**, 1 (1932).

[2] Breit, Condon, and Present, Phys. Rev. **50**, 825 (1936). J. Schwinger pointed out that the small difference may be attributed to magnetic interactions [Phys. Rev. **78**, 135 (1950)].

[3] The total isotopic spin **T** was first introduced by E. Wigner, Phys. Rev. **51**, 106 (1937); B. Cassen and E. U. Condon, Phys. Rev. **50**, 846 (1936).

ments in recent years[4] on the energy levels of light nuclei strongly suggest that this assumption is indeed correct, An implication of this is that all strong interactions such as the pion-nucleon interaction, must also satisfy the same conservation law. This and the knowledge that there are three charge states of the pion, and that pions can be coupled to the nucleon field *singly*, lead to the conclusion that pions have isotopic spin unity. A direct verification of this conclusion was found in the experiment of Hildebrand[5] which compares the differential cross section of the process $n+p\rightarrow\pi^0+d$ with that of the previously measured process $p+p\rightarrow\pi^++d$.

The conservation of isotopic spin is identical with the requirement of invariance of all interactions under isotopic spin rotation. This means that when electromagnetic interactions can be neglected, as we shall hereafter assume to be the case, the orientation of the isotopic spin is of no physical significance. The differentiation between a neutron and a proton is then a purely arbitrary process. As usually conceived, however, this arbitrariness is subject to the following limitation: once one chooses what to call a proton, what a neutron, at one space-time point, one is then not free to make any choices at other space-time points.

It seems that this is not consistent with the localized field concept that underlies the usual physical theories. In the present paper we wish to explore the possibility of requiring all interactions to be invariant under *independent* rotations of the isotopic spin at all space-time points, so that the relative orientation of the isotopic spin at two space-time points becomes a physically meaningless quantity (the electromagnetic field being neglected).

We wish to point out that an entirely similar situation arises with respect to the ordinary gauge invariance of a charged field which is described by a complex wave function ψ. A change of gauge[6] means a change of phase factor $\psi\rightarrow\psi'$, $\psi'=(\exp i\alpha)\psi$, a change that is devoid of any physical consequences. Since ψ may depend on x, y, z, and t, the relative phase factor of ψ at two different space-time points is therefore completely arbitrary. In other words, the arbitrariness in choosing the phase factor is local in character.

We define *isotopic gauge* as an arbitrary way of choosing the orientation of the isotopic spin axes at all space-time points, in analogy with the electromagnetic gauge which represents an arbitrary way of choosing the complex phase factor of a charged field at all space-time points. We then propose that all physical processes (not involving the electromagnetic field) be invariant under an isotopic gauge transformation, $\psi\rightarrow\psi'$, $\psi'=S^{-1}\psi$, where S represents a space-time dependent isotopic spin rotation.

To preserve invariance one notices that in electrodynamics it is necessary to counteract the variation of α with x, y, z, and t by introducing the electromagnetic field A_μ which changes under a gauge transformation as

$$A_\mu'=A_\mu+\frac{1}{e}\frac{\partial\alpha}{\partial x_\mu}.$$

In an entirely similar manner we introduce a B field in the case of the isotopic gauge transformation to counteract the dependence of S on x, y, z, and t. It will be seen that this natural generalization allows for very little arbitrariness. The field equations satisfied by the twelve independent components of the B field, which we shall call the **b** field, and their interaction with any field having an isotopic spin are essentially fixed, in much the same way that the free electromagnetic field and its interaction with charged fields are essentially determined by the requirement of gauge invariance.

In the following two sections we put down the mathematical formulation of the idea of isotopic gauge invariance discussed above. We then proceed to the quantization of the field equations for the **b** field. In the last section the properties of the quanta of the **b** field are discussed.

ISOTOPIC GAUGE TRANSFORMATION

Let ψ be a two-component wave function describing a field with isotopic spin $\frac{1}{2}$. Under an isotopic gauge transformation it transforms by

$$\psi=S\psi', \tag{1}$$

where S is a 2×2 unitary matrix with determinant unity. In accordance with the discussion in the previous section, we require, in analogy with the electromagnetic case, that all derivatives of ψ appear in the following combination:

$$(\partial_\mu-i\epsilon B_\mu)\psi.$$

B_μ are 2×2 matrices such that[7] for $\mu=1$, 2, and 3, B_μ is Hermitian and B_4 is anti-Hermitian. Invariance requires that

$$S(\partial_\mu-i\epsilon B_\mu')\psi'=(\partial_\mu-i\epsilon B_\mu)\psi. \tag{2}$$

Combining (1) and (2), we obtain the isotopic gauge transformation on B_μ:

$$B_\mu'=S^{-1}B_\mu S+\frac{i}{\epsilon}S^{-1}\frac{\partial S}{\partial x_\mu}. \tag{3}$$

The last term is similar to the gradiant term in the gauge transformation of electromagnetic potentials. In analogy to the procedure of obtaining gauge invariant field strengths in the electromagnetic case, we

---

[4] T. Lauritsen, Ann. Rev. Nuclear Sci. 1, 67 (1952); D. R. Inglis, Revs. Modern Phys. 25, 390 (1953).
[5] R. H. Hildebrand, Phys. Rev. 89, 1090 (1953).
[6] W. Pauli, Revs. Modern Phys. 13, 203 (1941).

[7] We use the conventions $\hbar=c=1$, and $x_4=it$. Bold-face type refers to vectors in isotopic space, not in space-time.

define now

$$F_{\mu\nu}=\frac{\partial B_\mu}{\partial x_\nu}-\frac{\partial B_\nu}{\partial x_\mu}+i\epsilon(B_\mu B_\nu-B_\nu B_\mu). \tag{4}$$

One easily shows from (3) that

$$F_{\mu\nu}'=S^{-1}F_{\mu\nu}S \tag{5}$$

under an isotopic gauge transformation.‡ Other simple functions of B than (4) do not lead to such a simple transformation property.

The above lines of thought can be applied to any field ψ with arbitrary isotopic spin. One need only use other representations S of rotations in three-dimensional space. It is reasonable to assume that different fields with the same total isotopic spin, hence belonging to the same representation S, interact with the same matrix field B_μ. (This is analogous to the fact that the electromagnetic field interacts in the same way with any charged particle, regardless of the nature of the particle. If different fields interact with different and independent B fields, there would be more conservation laws than simply the conservation of total isotopic spin.) To find a more explicit form for the B fields and to relate the B_μ's corresponding to different representations S, we proceed as follows.

Equation (3) is valid for any S and its corresponding B_μ. Now the matrix $S^{-1}\partial S/\partial x_\mu$ appearing in (3) is a linear combination of the isotopic spin "angular momentum" matrices T^i ($i=1, 2, 3$) corresponding to the isotopic spin of the ψ field we are considering. So B_μ itself must also contain a linear combination of the matrices T^i. But any part of B_μ in addition to this, $\bar{B}_\mu$, say, is a scalar or tensor combination of the T's, and must transform by the homogeneous part of (3), $\bar{B}_\mu'=S^{-1}\bar{B}_\mu S$. Such a field is extraneous; it was allowed by the very general form we assumed for the B field, but is irrelevant to the question of isotopic gauge. Thus the relevant part of the B field is of the form

$$B_\mu=2\mathbf{b}_\mu\cdot\mathbf{T}. \tag{6}$$

(Bold-face letters denote three-component vectors in isotopic space.) To relate the $\mathbf{b}_\mu$'s corresponding to different representations S we now consider the product representation $S=S^{(a)}S^{(b)}$. The B field for the combination transforms, according to (3), by

$$B_\mu'=[S^{(b)}]^{-1}[S^{(a)}]^{-1}BS^{(a)}S^{(b)}$$

$$+\frac{i}{\epsilon}[S^{(a)}]^{-1}\frac{\partial S^{(a)}}{\partial x_\mu}+\frac{i}{\epsilon}[S^{(b)}]^{-1}\frac{\partial S^{(b)}}{\partial x_\mu}.$$

‡ *Note added in proof.*—It may appear that B_μ could be introduced as an auxiliary quantity to accomplish invariance, but need not be regarded as a field variable by itself. It is to be emphasized that such a procedure violates the principle of invariance. Every quantity that is not a pure numeral (like 2, or M, or any definite representation of the γ matrices) should be regarded as a dynamical variable, and should be varied in the Lagrangian to yield an equation of motion. Thus the quantities B_μ must be regarded as independent fields.

But the sum of $B_\mu^{(a)}$ and $B_\mu^{(b)}$, the B fields corresponding to $S^{(a)}$ and $S^{(b)}$, transforms in exactly the same way, so that

$$B_\mu=B_\mu^{(a)}+B_\mu^{(b)}$$

(plus possible terms which transform homogeneously, and hence are irrelevant and will not be included). Decomposing $S^{(a)}S^{(b)}$ into irreducible representations, we see that the twelve-component field $\mathbf{b}_\mu$ in Eq. (6) is the same for all representations.

To obtain the interaction between any field ψ of arbitrary isotopic spin with the $\mathbf{b}$ field one therefore simply replaces the gradiant of ψ by

$$(\partial_\mu-2i\epsilon\mathbf{b}_\mu\cdot\mathbf{T})\psi, \tag{7}$$

where T^i ($i=1, 2, 3$), as defined above, are the isotopic spin "angular momentum" matrices for the field ψ.

We remark that the nine components of $\mathbf{b}_\mu$, $\mu=1, 2, 3$ are real and the three of $\mathbf{b}_4$ are pure imaginary. The isotopic-gauge covariant field quantities $F_{\mu\nu}$ are expressible in terms of $\mathbf{b}_\mu$:

$$F_{\mu\nu}=2\mathbf{f}_{\mu\nu}\cdot\mathbf{T}, \tag{8}$$

where

$$\mathbf{f}_{\mu\nu}=\frac{\partial\mathbf{b}_\mu}{\partial x_\nu}-\frac{\partial\mathbf{b}_\nu}{\partial x_\mu}-2\epsilon\mathbf{b}_\mu\times\mathbf{b}_\nu. \tag{9}$$

$\mathbf{f}_{\mu\nu}$ transforms like a vector under an isotopic gauge transformation. Obviously the same $\mathbf{f}_{\mu\nu}$ interact with all fields ψ irrespective of the representation S that ψ belongs to.

The corresponding transformation of $\mathbf{b}_\mu$ is cumbersome. One need, however, study only the infinitesimal isotopic gauge transformations,

$$S=1-2i\mathbf{T}\cdot\delta\boldsymbol{\omega}.$$

Then

$$\mathbf{b}_\mu'=\mathbf{b}_\mu+2\mathbf{b}_\mu\times\delta\boldsymbol{\omega}+\frac{1}{\epsilon}\frac{\partial}{\partial x_\mu}\delta\boldsymbol{\omega}. \tag{10}$$

FIELD EQUATIONS

To write down the field equations for the $\mathbf{b}$ field we clearly only want to use isotopic gauge invariant quantities. In analogy with the electromagnetic case we therefore write down the following Lagrangian density:[8]

$$-\tfrac{1}{4}\mathbf{f}_{\mu\nu}\cdot\mathbf{f}_{\mu\nu}.$$

Since the inclusion of a field with isotopic spin $\frac{1}{2}$ is illustrative, and does not complicate matters very much, we shall use the following total Lagrangian density:

$$\mathcal{L}=-\tfrac{1}{4}\mathbf{f}_{\mu\nu}\cdot\mathbf{f}_{\mu\nu}-\bar{\psi}\gamma_\mu(\partial_\mu-i\epsilon\boldsymbol{\tau}\cdot\mathbf{b}_\mu)\psi-m\bar{\psi}\psi. \tag{11}$$

One obtains from this the following equations of motion:

$$\partial\mathbf{f}_{\mu\nu}/\partial x_\nu+2\epsilon(\mathbf{b}_\nu\times\mathbf{f}_{\mu\nu})+\mathbf{J}_\mu=0,$$
$$\gamma_\mu(\partial_\mu-i\epsilon\boldsymbol{\tau}\cdot\mathbf{b}_\mu)\psi+m\psi=0, \tag{12}$$

[8] Repeated indices are summed over, except where explicitly stated otherwise. Latin indices are summed from 1 to 3, Greek ones from 1 to 4.

where

$$\mathbf{J}_\mu = i\epsilon\bar{\psi}\gamma_\mu\boldsymbol{\tau}\psi. \tag{13}$$

The divergence of $\mathbf{J}_\mu$ does not vanish. Instead it can easily be shown from (13) that

$$\partial\mathbf{J}_\mu/\partial x_\mu = -2\epsilon\mathbf{b}_\mu\times\mathbf{J}_\mu. \tag{14}$$

If we define, however,

$$\mathfrak{J}_\mu = \mathbf{J}_\mu + 2\epsilon\mathbf{b}_\nu\times\mathbf{f}_{\mu\nu}, \tag{15}$$

then (12) leads to the equation of continuity,

$$\partial\mathfrak{J}_\mu/\partial x_\mu = 0. \tag{16}$$

$\mathfrak{J}_{1,2,3}$ and $\mathfrak{J}_4$ are respectively the isotopic spin current density and isotopic spin density of the system. The equation of continuity guarantees that the total isotopic spin

$$\mathbf{T} = \int \mathfrak{J}_4 d^3x$$

is independent of time and independent of a Lorentz transformation. It is important to notice that $\mathfrak{J}_\mu$, like $\mathbf{b}_\mu$, does not transform exactly like vectors under isotopic space rotations. But the total isotopic spin,

$$\mathbf{T} = -\int \frac{\partial \mathbf{f}_{4i}}{\partial x_i} d^3x,$$

is the integral of the divergence of $\mathbf{f}_{4i}$, which transforms like a true vector under isotopic spin space rotations. Hence, under a general isotopic gauge transformation, if $S\to S_0$ on an infinitely large sphere, $\mathbf{T}$ would transform like an isotopic spin vector.

Equation (15) shows that the isotopic spin arises both from the spin-$\frac{1}{2}$ field ($\mathbf{J}_\mu$) and from the $\mathbf{b}_\mu$ field itself. Inasmuch as the isotopic spin is the source of the $\mathbf{b}$ field, this fact makes the field equations for the $\mathbf{b}$ field nonlinear, even in the absence of the spin-$\frac{1}{2}$ field. This is different from the case of the electromagnetic field, which is itself chargeless, and consequently satisfies linear equations in the absence of a charged field.

The Hamiltonian derived from (11) is easily demonstrated to be positive definite in the absence of the field of isotopic spin $\frac{1}{2}$. The demonstration is completely identical with the similar one in electrodynamics.

We must complete the set of equations of motion (12) and (13) by the supplementary condition,

$$\partial\mathbf{b}_\mu/\partial x_\mu = 0, \tag{17}$$

which serves to eliminate the scalar part of the field in $\mathbf{b}_\mu$. This clearly imposes a condition on the possible isotopic gauge transformations. That is, the infinitesimal isotopic gauge transformation $S = 1 - i\boldsymbol{\tau}\cdot\delta\boldsymbol{\omega}$ must satisfy the following condition:

$$2\mathbf{b}_\mu\times\frac{\partial}{\partial x_\mu}\delta\boldsymbol{\omega} + \frac{1}{\epsilon}\frac{\partial^2}{\partial x_\mu^2}\delta\boldsymbol{\omega} = 0. \tag{18}$$

This is the analog of the equation $\partial^2\alpha/\partial x_\mu^2 = 0$ that must be satisfied by the gauge transformation $A_\mu' = A_\mu + e^{-1}(\partial\alpha/\partial x_\mu)$ of the electromagnetic field.

QUANTIZATION

To quantize, it is not convenient to use the isotopic gauge invariant Lagrangian density (11). This is quite similar to the corresponding situation in electrodynamics and we adopt the customary procedure of using a Lagrangian density which is not obviously gauge invariant:

$$\mathcal{L} = -\frac{1}{2}\frac{\partial\mathbf{b}_\mu}{\partial x_\nu}\cdot\frac{\partial\mathbf{b}_\mu}{\partial x_\nu} + 2\epsilon(\mathbf{b}_\mu\times\mathbf{b}_\nu)\frac{\partial\mathbf{b}_\mu}{\partial x_\nu}$$
$$-\epsilon^2(\mathbf{b}_\mu\times\mathbf{b}_\nu)^2 + \mathbf{J}_\mu\cdot\mathbf{b}_\mu - \bar{\psi}(\gamma_\mu\partial_\mu + m)\psi. \tag{19}$$

The equations of motion that result from this Lagrangian density can be easily shown to imply that

$$\frac{\partial^2}{\partial x_\nu^2}\mathbf{a} + 2\epsilon\mathbf{b}_\nu\times\frac{\partial}{\partial x_\nu}\mathbf{a} = 0,$$

where

$$\mathbf{a} = \partial\mathbf{b}_\mu/\partial x_\mu.$$

Thus if, consistent with (17), we put on one space-like surface $\mathbf{a} = 0$ together with $\partial\mathbf{a}/\partial t = 0$, it follows that $\mathbf{a} = 0$ at all times. Using this supplementary condition one can easily prove that the field equations resulting from the Lagrangian densities (19) and (11) are identical.

One can follow the canonical method of quantization with the Lagrangian density (19). Defining

$$\mathbf{\Pi}_\mu = -\partial\mathbf{b}_\mu/\partial x_4 + 2\epsilon(\mathbf{b}_\mu\times\mathbf{b}_4),$$

one obtains the equal-time commutation rule

$$[b_\mu{}^i(x), \Pi_\nu{}^j(x')]_{t=t'} = -\delta_{ij}\delta_{\mu\nu}\delta^3(x-x'), \tag{20}$$

where $b_\mu{}^i$, $i=1, 2, 3$, are the three components of $\mathbf{b}_\mu$. The relativistic invariance of these commutation rules follows from the general proof for canonical methods of quantization given by Heisenberg and Pauli.[9]

The Hamiltonian derived from (19) is identical with the one from (11), in virtue of the supplementary condition. Its density is

$$H = H_0 + H_{\text{int}},$$

$$H_0 = -\tfrac{1}{2}\mathbf{\Pi}_\mu\cdot\mathbf{\Pi}_\mu + \frac{1}{2}\frac{\partial\mathbf{b}_\mu}{\partial x_j}\cdot\frac{\partial\mathbf{b}_\mu}{\partial x_j} + \bar{\psi}(\gamma_j\partial_j + m)\psi,$$
$$H_{\text{int}} = 2\epsilon(\mathbf{b}_i\times\mathbf{b}_4)\cdot\mathbf{\Pi}_i - 2\epsilon(\mathbf{b}_\mu\times\mathbf{b}_j)\cdot(\partial\mathbf{b}_\mu/\partial x_j)$$
$$+\epsilon^2(\mathbf{b}_i\times\mathbf{b}_j)^2 - \mathbf{J}_\mu\cdot\mathbf{b}_\mu. \tag{21}$$

The quantized form of the supplementary condition is the same as in quantum electrodynamics.

[9] W. Heisenberg and W. Pauli, Z. Physik 56, 1 (1929).

PROPERTIES OF THE *b* QUANTA

The quanta of the **b** field clearly have spin unity and isotopic spin unity. We know their electric charge too because all the interactions that we proposed must satisfy the law of conservation of electric charge, which is exact. The two states of the nucleon, namely proton and neutron, differ by charge unity. Since they can transform into each other through the emission or absorption of a **b** quantum, the latter must have three charge states with charges $\pm e$ and 0. Any measurement of electric charges of course involves the electromagnetic field, which necessarily introduces a preferential direction in isotopic space at all space-time points. Choosing the isotopic gauge such that this preferential direction is along the z axis in isotopic space, one sees that for the nucleons

$$Q = \text{electric charge} = e(\tfrac{1}{2} + \epsilon^{-1}T^z),$$

and for the **b** quanta

$$Q = (e/\epsilon)T^z.$$

The interaction (7) then fixes the electric charge up to an additive constant for all fields with any isotopic spin:

$$Q = e(\epsilon^{-1}T^z + R). \qquad (22)$$

The constants R for two charge conjugate fields must be equal but have opposite signs.[10]

FIG. 1. Elementary vertices for **b** fields and nucleon fields. Dotted lines refer to **b** field, solid lines with arrow refer to nucleon field.

We next come to the question of the mass of the **b** quantum, to which we do not have a satisfactory answer. One may argue that without a nucleon field the Lagrangian would contain no quantity of the dimension of a mass, and that therefore the mass of the **b** quantum in such a case is zero. This argument is however subject to the criticism that, like all field theories, the **b** field is beset with divergences, and dimensional arguments are not satisfactory.

One may of course try to apply to the **b** field the methods for handling infinities developed for quantum electrodynamics. Dyson's approach[11] is best suited for the present case. One first transforms into the interaction representation in which the state vector Ψ satisfies

$$i\partial\Psi/\partial t = H_{\text{int}}\Psi,$$

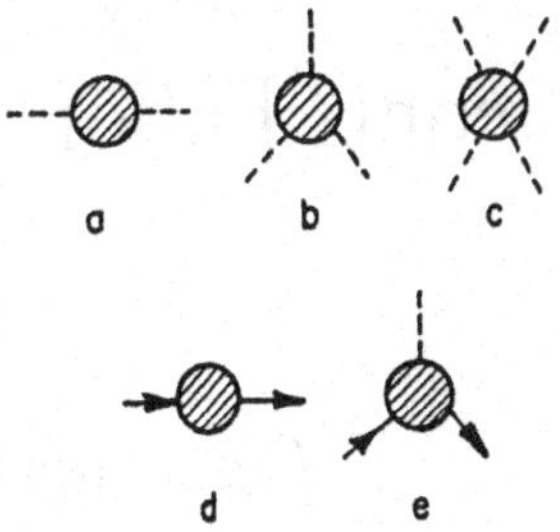

FIG. 2. Primitive divergences.

where H_{int} was defined in Eq. (21). The matrix elements of the scattering matrix are then formulated in terms of contributions from Feynman diagrams. These diagrams have three elementary types of vertices illustrated in Fig. 1, instead of only one type as in quantum electrodynamics. The "primitive divergences" are still finite in number and are listed in Fig. 2. Of these, the one labeled a is the one that effects the propagation function of the **b** quantum, and whose singularity determines the mass of the **b** quantum. In electrodynamics, by the requirement of electric charge conservation,[12] it is argued that the mass of the photon vanishes. Corresponding arguments in the **b** field case do not exist[13] even though the conservation of isotopic spin still holds. We have therefore not been able to conclude anything about the mass of the **b** quantum.

A conclusion about the mass of the **b** quantum is of course very important in deciding whether the proposal of the existence of the **b** field is consistent with experimental information. For example, it is inconsistent with present experiments to have their mass less than that of the pions, because among other reasons they would then be created abundantly at high energies and the charged ones should live long enough to be seen. If they have a mass greater than that of the pions, on the other hand, they would have a short lifetime (say, less than 10^{-20} sec) for decay into pions and photons and would so far have escaped detection.

[10] See M. Gell-Mann, Phys. Rev. **92**, 833 (1953).

[11] F. J. Dyson, Phys. Rev. **75**, 486, 1736 (1949).

[12] J. Schwinger, Phys. Rev. **76**, 790 (1949).

[13] In electrodynamics one can formally prove that $G_{\mu\nu}k_\nu = 0$, where $G_{\mu\nu}$ is defined by Schwinger's Eq. (A12). ($G_{\mu\nu}A_\nu$ is the current generated through virtual processes by the arbitrary external field A_ν.) No corresponding proof has been found for the present case. This is due to the fact that in electrodynamics the conservation of charge is a consequence of the equation of motion of the electron field alone, quite independently of the electromagnetic field itself. In the present case the **b** field carries an isotopic spin and destroys such general conservation laws.

The problem of particle types and other contributions to the theory of elementary particles

Ronald Shaw

Cambridge University Ph.D. thesis, September 1955

1 Representations of the improper Lorentz group and the problem of particle types

2 Some contributions to the theory of elementary particles

2.1 Some results in the theory of interacting fields

2.2 The associated production of θ^0 and Λ^0 particles

2.3 Invariance under general isotopic spin transformations

2.3.1 Introduction

Many general results follow from postulating that the Lagrangian of the system of fields should be invariant under certain classes of transformations. For example, if the Lagrangian is invariant under translations and rotations in space-time, then it is well known that a momentum 4-vector P_α and an angular momentum tensor $M_{\alpha\beta}$ exist and are conserved. These two invariances seem to exhaust the useful possibilities for transformations in space-time. However one can also postulate the invariance of the Lagrangian under rotations in a Euclidean space E_n of dimension n which has no relation to space-time, and useful results still follow.

The invariance of the Lagrangian under special gauge transformations can be looked upon from this point of view. The space involved is then E_2, and invariance under rotations in this space results in the existence of a divergenceless current vector s_α and of a charge Q which is conserved. Neutral fields are scalars in E_2, while fields with charge one are vectors in E_2. For instance, taking the case of two real fermion fields, ψ_1 and ψ_2 with free Lagrangian[1]

$$L_0 = \frac{1}{2}i(\psi_i B\gamma_\alpha \partial^\alpha \psi_i - \partial^\alpha \psi_i B\gamma_\alpha \psi_i) + im\psi_i B\psi_i \tag{1}$$

it is seen that L_0 is invariant under the infinitesimal rotation

$$\left.\begin{aligned} \psi'_1 &= \psi_1 - c\psi_2 \\ \psi'_2 &= \psi_2 + c\psi_1 \end{aligned}\right\} \tag{2}$$

(In equation (1) B is the matrix defined by $B\gamma_\alpha B^{-1} = -\gamma_\alpha^T$; it then follows that $\psi B\psi$ is a scalar, $\psi B\gamma^\alpha\psi$ a vector, etc.) The current density and total charge are in this case

$$s^\alpha = ie(\psi_1 B\gamma_\alpha\psi_2 - \psi_2 B\gamma_\alpha\psi_1) \tag{3}$$

$$Q = \int S^\alpha d\sigma_\alpha \tag{4}$$

In the special gauge transformations (2), c was a constant independent of the space-time argument of ψ. If now the Lagrangian is required to be invariant under general gauge transformations (i.e. those with c a function of position) then it is found necessary to introduce the electromagnetic field. This method was adopted by Schwinger (1954). For under general gauge transformations, L_0 is no longer invariant but becomes

$$L_0' = L_0 - i\partial_\alpha c(\psi_1 B\gamma^\alpha\psi_2 - \psi_2 B\gamma^\alpha\psi_1) \tag{5}$$

However the last term of (5) can be compensated by taking an interaction Lagrangian

$$L_1 = -A_\alpha s^\alpha \tag{6}$$

and requiring that when ψ undergoes the transformation (2), A_α should undergo the transformation

$$eA'_\alpha = eA_\alpha + \partial_\alpha c \tag{7}$$

[1] Lorentz suffices are represented by Greek letter, and suffices in E_n by Latin letters. The summation convention is employed with respect to both kinds of suffices.

Finally. defining as usual the field strengths

$$f_{\alpha\beta} = \partial_\alpha A_\beta - \partial_\beta A_\alpha \tag{8}$$

and a free Lagrangian L_2 for the electromagnetic field

$$L_2 = -\frac{1}{4} f_{\alpha\beta} f^{\alpha\beta} \tag{9}$$

the total Lagrangian

$$L = L_0 + L_1 + L_2 \tag{10}$$

is seen to be invariant under general gauge transformations.

The purpose of the above treatment of the transition from special gauge transformations to general gauge transformations was to present the introduction of the electromagnetic field in a form suitable for generalization to spaces E_n with $n > 2$. For in the usual symmetrical meson theory (Kemmer, 1938), invariance under rotations in a space E_3, the so-called isotopic spin space, is assumed. An infinitesimal rotation in this space involves three real parameters c_i instead of the one parameter c of the transformation (2). In the usual theory the c_i do not depend on position, or in other words, only special isotopic spin transformations are contemplated. However, in analogy with the case of E_2, it is natural to postulate invariance of the Lagrangian under general isotopic spin transformations, the c_i being functions of position. The consequences[2] of the postulate are worked out in Section 2.3.2. In Section 2.3.3, the Salam-Polkinghorne scheme, in which all the strong interactions are assumed to be invariant under rotations in a space E_4, is generalised in a similar manner.

2.3.2 The invariance in the space E_3

The fermion case is taken once more, with free field Lagrangian

$$L_0 = \overline{\psi}(\gamma_\alpha \partial^\alpha + m)\psi \tag{11}$$

where ψ is a 4-component spinor in space-time and a 2-component spinor in E_3. Note that this is in contrast to the case of E_2, when ψ was taken to be

[2]The work described in this chapter was completed, except for its extension in Section 2.3.3, in January 1954, but was not published. In October 1954, Yang and Mills adopted independently the same postulate and derived similar consequences.

a vector in E_2. In (11) the two isotopic spin components of ψ refer of course to the proton and neutron. Under a rotation

$$\phi_i' = \phi_i + c_{ij}\phi_j \tag{12}$$

in isotopic spin space, ψ undergoes the transformation

$$\psi' = \left(1 + \frac{1}{2}ic_i\tau_i\right)\psi \tag{13}$$

where $c_i = \frac{1}{2}\epsilon_{ijk}c_{jk}$ and where τ_i are the usual isotopic spin matrices. In analogy with (3) and (4) the isotopic spin current density $\vec{s}_\alpha$ and the total isotopic spin $\vec{T}$ can be defined

$$\vec{s}_\alpha = +\frac{1}{2}iq\bar{\psi}\gamma_\alpha\vec{\tau}\psi \tag{14}$$

and

$$\vec{T} = \int \vec{s}_\alpha d\sigma^\alpha; \tag{15}$$

$\vec{s}_\alpha$ is seen to be divergenceless and hence $\vec{T}$ is conserved. In (14) q is the unit of 'isotopic charge' and corresponds to e in (3).

Now under general transformations (13) with $\vec{c}$ a function of position, L_0 is no longer invariant but becomes

$$L_0' = L_0 + \frac{1}{2}i\partial^\alpha\vec{c}.\bar{\psi}\gamma_\alpha\vec{\tau}\psi. \tag{16}$$

The additional term can be compensated by introducing an interaction

$$L_1 = -\vec{B}_\alpha.\vec{s}^\alpha \tag{17}$$

and by requiring that when ψ transforms under (13) the field $\vec{B}^\alpha$, which is the generalization to E_3 of the electromagnetic field, transforms

$$q\vec{B}'^\alpha = q(\vec{B}^\alpha - \vec{c}\times\vec{B}^\alpha) + \partial^\alpha\vec{c}. \tag{18}$$

In order to construct a theory invariant under general isotopic spin transformations it only remains to construct an invariant free field Lagrangian for the $\vec{B}$-field. It is natural, in analogy with the electromagnetic case, to define

$$\vec{f}^{\alpha\beta} = \partial^\alpha\vec{B}^\beta - \partial^\beta\vec{B}^\alpha \tag{19}$$

and to take the Lagrangian to be

$$L_2 = -\frac{1}{4}\vec{f}^{\alpha\beta}.\vec{f}_{\alpha\beta} \tag{20}$$

However (20) is not invariant under the transformation (18). For $\vec{f}^{\alpha\beta}$ transforms as follows

$$\vec{f}'^{\alpha\beta} = \vec{f}^{\alpha\beta} - \vec{c} \times \vec{f}^{\alpha\beta} - 2(\partial^\alpha \vec{c} \times \vec{B}^\beta - \partial^\beta \vec{c} \times \vec{B}^\alpha) \tag{21}$$

The second order derivatives of $\vec{c}$ have cancelled as in the electromagnetic case, but the first order derivatives still remain and spoil the invariance of (20). A properly invariant Lagrangian can be obtained by defining

$$\vec{F}^{\alpha\beta} = \vec{f}^{\alpha\beta} - q(\vec{B}^\alpha \times \vec{B}^\beta - \vec{B}^\beta \times \vec{B}^\alpha) \tag{22}$$

and by taking L_2 to be

$$L_2 = -\frac{1}{4}\vec{F}^{\alpha\beta}.\vec{F}_{\alpha\beta} \tag{23}$$

instead of (20). L_2 is now invariant under (18), as $\vec{F}^{\alpha\beta}$ transforms solely as a vector in E_3.

The filed equations which follow from the Lagrangian

$$L = L_0 + L_1 + L_2 \tag{24}$$

are found to be

$$(\gamma_\alpha \partial^\alpha + m)\psi = -\frac{1}{2} iq\vec{B}^\alpha.\vec{\tau}\gamma_\alpha\psi \tag{25}$$

and

$$\partial_\beta \vec{F}^{\alpha\beta} + 2q\vec{B}_\beta \times \vec{F}^{\alpha\beta} = \vec{s}^\alpha \tag{26}$$

The divergence of $\vec{s}^\alpha$ no longer vanishes, but a divergenceless vector $\vec{j}^\alpha$, say, follows immediately from (26)

$$\vec{j}^\alpha = \vec{s}^\alpha - 2q\vec{B}_\beta \times \vec{F}^{\alpha\beta} \tag{27}$$

and instead of (15), the total isotopic spin is now

$$\vec{T} = \int \vec{j}^\alpha d\sigma_\alpha. \tag{28}$$

$\vec{j}^{\alpha}$ replaces $\vec{s}^{\alpha}$ as the isotopic spin current density of the system, and is composed of the isotopic spin $\vec{s}^{\alpha}$ of the nucleons and the isotopic spin $-2q\vec{B}^{\beta}\times\vec{F}^{\alpha\beta}$ of the B-field. The fact that the B-field itself carries isotopic spin accounts for the greater complexity of equation (26) compared with the corresponding equation $-\partial_{\beta} f_{\alpha\beta} = s^{\alpha}$ in the electromagnetic case, and it is not clear from (26) whether the mass of the B-field is zero or not. This means that it is difficult to determine whether or not the existence of the B-field is consistent with experiment. It is clear that these particles, if they exist, will have spin 1 and isotopic spin 1, and so exist in three states with charge $\pm e$ and 0, like the π-meson. Zero mass would appear to be ruled out since otherwise the neutron would have a rapid decay

$$N \to P + B^{-} \tag{29}$$

In fact one would have expected the B-particles to have been observed unless their mass was quite large or unless the coupling constant q was very small.

Finally it should be pointed out that the the invariance postulated in this section does not hold when the electromagnetic field is taken into account. The status of invariances like this, which apply only to part of the Lagrangian, is obscure and is to be contrasted with invariance under gauge transformations, where the ensuing conservation law, that of electric charge, seems never to be violated. The existence of this contrast was the motive for the above work, but the conclusion is that gauge invariance and isotopic spin invariance do not appear to differ essentially as far as generalization to space-dependent transformations is concerned.

2.3.3 Invariance in the space E_4

In this section the 4-dimensional theory of Salam and Polkinghorne (1955) is generalised, in a way similar to the 3-dimensional isotopic spin case of section 2.3.2. The resulting equations are slightly more cumbersome as vector notation cannot be employed.

The fermion free field Lagrangian is now

$$L_0 = \overline{\Psi}_i(\gamma_\alpha \partial^\alpha + m)\Psi_i \tag{30}$$

where Ψ_i is a 4-vector in E_4 and describes both nucleons and cascade particles. The connection between Ψ_i and these fields can be taken as

$$\left.\begin{array}{ll} 2^{\frac{1}{2}}\psi_P = \Psi_0 + i\Psi_3, & 2^{\frac{1}{2}}\psi_N = \Psi_1 - i\Psi_2 \\ 2^{\frac{1}{2}}\chi_{\Xi^-} = \Psi_0 - i\Psi_3, & 2^{\frac{1}{2}}\chi_{\Xi_0} = \Psi_1 + i\Psi_2 \end{array}\right\} \tag{31}$$

In this case the interaction with the electromagnetic field is

$$L' = -e(\overline{\Psi}_0\gamma_\alpha\Psi_3 - \overline{\Psi}_3\gamma_\alpha\Psi_0)A^\alpha \tag{32}$$

In this scheme, the π-meson field transforms under the representation $R(1,0)$ of the rotation group in E_4, and so is a self-dual antisymmetric tensor. Its interaction with nucleons and cascade particles is taken to be

$$L'' = ig\overline{\Psi}_i\gamma_5\Psi_j\phi_{ij} \tag{33}$$

This can be written in the form

$$L'' = ig\overline{\Psi}\gamma_5\vec{T}\Psi.\vec{\phi} \tag{34}$$

where $\vec{T}$ is the self-dual part of the infinitesimal generators M_{ij} of the vector representation in E_4; that is $T_1 = \frac{1}{2}(M_{01} + M_{23})$ etc. Since the T_i satisfy the commutation relations for generators of the spin $\frac{1}{2}$ representation in E_3, we can take

$$-2i\vec{T} = \begin{pmatrix} \vec{\tau} & 0 \\ 0 & \vec{\tau} \end{pmatrix} \tag{35}$$

in a suitable representation, and so (34) can be written in the form

$$L'' = g\vec{\phi}.(\overline{\psi}\gamma_5\vec{\tau}\psi + \overline{\chi}\gamma_5\vec{\tau}\chi) \tag{36}$$

where ψ is the nucleon wave function with components ψ_P and ψ_N, and where χ is the cascade particle wave function with components χ_{Ξ^0} and χ_{Ξ^-}. Thus the above scheme gives the usual symmetrical interaction between the π-mesons and the nucleons.

Under infinitesimal rotations in E_4, Ψ transforms

$$\Psi'_i = \Psi_i + c_{ij}\Psi_j. \tag{37}$$

If invariance is postulated when the c_{ij} are functions of position, then it is necessary to add to L_0 an interaction Lagrangian

$$L_1 = -\frac{1}{2}qB^\alpha_{ij}(\Psi_i\gamma_\alpha\Psi_j - \Psi_j\gamma_\alpha\Psi_i) \tag{38}$$

where B^α_{ij} transforms under (37) as follows

$$qB'^\alpha_{ij} = q(B^\alpha_{ij} + c_{ik}B^\alpha_{kj} - c_{jk}B^\alpha_{ki}) + \partial^\alpha c_{ij} \tag{39}$$

The invariant free Lagrangian for the B-field is

$$L_2 = -\frac{1}{4} F_{ij}^{\alpha\beta} F_{ij,\alpha\beta} \tag{40}$$

where the 'field strengths' $F_{ij}^{\alpha\beta}$ are defined by

$$F_{ij}^{\alpha\beta} = \partial^\alpha B_{ij}^\beta - \partial^\beta B_{ij}^\alpha - q(B_{ik}^\alpha B_{kj}^\beta - B_{jk}^\alpha B_{ki}^\beta) \tag{41}$$

The field equations are easily derived and are

$$(\gamma_\alpha \partial^\alpha + m)\Psi_i = q B_{ij}^\alpha \gamma_\alpha \Psi_j \tag{42}$$

and

$$-\partial_\beta F_{ij}^{\alpha\beta} + q(B_{ik,\beta} F_{kj}^{\alpha\beta} - B_{jk,\beta} F_{ki,\alpha\beta}) = s_{ij}^\alpha \tag{43}$$

where the fermion isotopic spin current density s_{ij}^α is given by

$$s_{ij}^\alpha = q(\overline{\Psi}_i \gamma^\alpha \Psi_j - \overline{\Psi}_j \gamma^\alpha \Psi_i) \tag{44}$$

The total isotopic spin is

$$T_{ij} = \int j_{ij}^\alpha d\sigma_\alpha \tag{45}$$

where

$$j_{ij}^\alpha = s_{ij}^\alpha - q(B_{ik,\beta} F_{kj}^{\alpha\beta} - B_{jk,\beta} F_{ki}^{\alpha\beta}) \tag{46}$$

The divergence of j_{ij}^α vanishes, and so T_{ij} is conserved.

An interesting result of the above theory is obtained on comparing the interactions (32) and (38). It is seen that the electromagnetic interaction L' can be considered as part of the B-field interaction L_1 as long as $q = e$ and $A^\alpha = B_{03}^\alpha$. The total charge is then $Q = T_{03}$. In fact if the B-field were non-quantized, then one could start from the Lagrangian $L = L_0 + L_1 + L_2$, invariant under general transformations (38), and apply such a transformation at each point in space-time, in such a way as to make all components of B_{ij}^α vanish[3] except B_{03}^α. With the above identification it is seen that that the field equations (42) and (43) would then be exactly those for nucleons and cascade particles in interaction with the electromagnetic field[4], so that

[3]Editor's note: This is a mathematical error, as Shaw has confirmed to me.

[4]Except that the mass difference between nucleons and cascade particles is not taken into account. This large mass difference seems to be the only real objection to the Salam-Polkinghorne theory.

while apparently starting with a more general theory, the usual theory is still obtained. Of course the above reduction of B_{ij} to B_{03} could not be carried out in a quantized theory, and so one would expect the existence of six B-particles of spin 1. Since B_{ij} is an antisymmetric tensor, it transforms under the reducible representation $R(1,0) + R(0,1)$, and so it is natural to suppose that the above six particles should exist in two sets of three with wave functions $\vec{C}^\alpha$ and $\vec{D}^\alpha$, say, where $C_1^\alpha = \frac{1}{2}(B_{01}^\alpha + B_{23}^\alpha)$, $D_1^\alpha = \frac{1}{2}(B_{01}^\alpha - B_{23}^\alpha)$ etc. The interaction (38) can be written in terms of the nucleon, cascade, C and D fields. The interaction with the C-field is

$$q(\overline{\psi}\gamma^\alpha\vec{\tau}\psi + \overline{\chi}\gamma_\alpha\vec{\tau}\chi).\vec{C}^\alpha$$

and with the D-filed is

$$q(\overline{\psi}\gamma_\alpha\chi + \overline{\chi}\gamma_\alpha\psi)D_1^\alpha + iq(-\overline{\psi}\gamma_\alpha\chi + \overline{\chi}\gamma_\alpha\psi)D_2^\alpha + q(\overline{\psi}\gamma_\alpha\psi - \overline{\chi}\gamma_\alpha\chi)D_3^\alpha$$

The C-field has the π-meson type interaction in the Salam scheme while the D-field has an interaction of the τ-meson type (except that a vector interaction rather than a pseudoscalar one has to be taken). However, if the B-field does in fact exist, it is not certain that it will be observed in its C and D forms separately, for one would hope that the combination $C_3 + D_3$ would be the electromagnetic field.

REFERENCES

Kemmer, N. 1938. Proc. Camb. Phil. Soc., **34**,354.
Salam, A. & Polkinghorne, J.C., 1955. To be published. [Nuovo Cimento **2** 685.]
Schwinger, J. 1953. Phys. Rev. **91**, 713.
Yang, C.N., & Mills, R.L. 1954. Phys. Rev. **96**, 191.

JOURNAL OF MATHEMATICAL PHYSICS VOLUME 2, NUMBER 2 MARCH–APRIL, 1961

Lorentz Invariance and the Gravitational Field

T. W. B. KIBBLE*
Department of Mathematics, Imperial College, London, England

(Received August 19, 1960)

An argument leading from the Lorentz invariance of the Lagrangian to the introduction of the gravitational field is presented. Utiyama's discussion is extended by considering the 10-parameter group of inhomogeneous Lorentz transformations, involving variation of the coordinates as well as the field variables. It is then unnecessary to introduce *a priori* curvilinear coordinates or a Riemannian metric, and the new field variables introduced as a consequence of the argument include the vierbein components $h_k{}^\mu$ as well as the "local affine connection" $A^i{}_{j\mu}$. The extended transformations for which the 10 parameters become arbitrary functions of position may be interpreted as general coordinate transformations and rotations of the vierbein system. The free Lagrangian for the new fields is shown to be a function of two covariant quantities analogous to $F_{\mu\nu}$ for the electromagnetic field, and the simplest possible form is just the usual curvature scalar density expressed in terms of $h_k{}^\mu$ and $A^i{}_{j\mu}$. This Lagrangian is of first order in the derivatives, and is the analog for the vierbein formalism of Palatini's Lagrangian. In the absence of matter, it yields the familiar equations $R_{\mu\nu}=0$ for empty space, but when matter is present there is a difference from the usual theory (first pointed out by Weyl) which arises from the fact that $A^i{}_{j\mu}$ appears in the matter field Lagrangian, so that the equation of motion relating $A^i{}_{j\mu}$ to $h_k{}^\mu$ is changed. In particular, this means that, although the covariant derivative of the metric vanishes, the affine connection $\Gamma^\lambda{}_{\mu\nu}$ is nonsymmetric. The theory may be reexpressed in terms of the Christoffel connection, and in that case additional terms quadratic in the "spin density" $S^k{}_{ij}$ appear in the Lagrangian. These terms are almost certainly too small to make any experimentally detectable difference to the predictions of the usual metric theory.

1. INTRODUCTION

It has long been realized that the existence of certain fields, notably the electromagnetic field, can be related to invariance properties of the Lagrangian.[1] Thus, if the Lagrangian is invariant under phase transformations $\psi\rightarrow e^{ie\lambda}\psi$, and if we wish to make it invariant under the general gauge transformations for which λ is a function of x, then it is necessary to introduce a new field A_μ which transforms according to $A_\mu\rightarrow A_\mu-\partial_\mu\lambda$, and to replace $\partial_\mu\psi$ in the Lagrangian by a "covariant derivative" $(\partial_\mu+ieA_\mu)\psi$. A similar argument has been applied by Yang and Mills[2] to isotopic spin rotations, and in that case yields a triplet of vector fields. It is thus an attractive idea to relate the existence of the gravitational field to the Lorentz invariance of the Lagrangian. Utiyama[3] has proposed a method which leads to the introduction of 24 new field variables $A^{ij}{}_\mu$ by considering the homogeneous Lorentz transformations specified by six parameters ϵ^{ij}. However, in order to do this it was necessary to introduce *a priori* curvilinear coordinates and a set of 16 parameters $h_k{}^\mu$. Initially, the $h_k{}^\mu$ were treated as given functions of x, but at a later stage they were regarded as field variables and interpreted as the components of a vierbein system in a Riemannian space. This is a rather unsatisfactory procedure since it is the purpose of the discussion to supply an argument for introducing the gravitational field variables, which include the metric as well as the affine connection. The new field variables $A^{ij}{}_\mu$ were subsequently related to the Christoffel connection $\Gamma^\lambda{}_{\mu\nu}$ in the Riemannian space, but this could only be done uniquely by making the *ad hoc* assumption that the quantity $\Gamma'^\lambda{}_{\mu\nu}$ calculated from $A^{ij}{}_\mu$ was symmetric.

It is the purpose of this paper to show that the vierbein components $h_k{}^\mu$, as well as the "local affine connection" $A^{ij}{}_\mu$, can be introduced as new field variables analogous to A_μ if one considers the full 10-parameter group of inhomogeneous Lorentz transformations in place of the restricted six-parameter group. This implies that one must consider transformations of the coordinates as well as the field variables, which will necessitate some changes in the argument, but it also means that only one system of coordinates is required, and that a Riemannian metric need not be introduced *a priori*. The interpretation of the theory in terms of a Riemannian space may be made later if desired. The starting point of the discussion is the ordinary formulation of Lorentz invariance (including translational invariance) in terms of rectangular coordinates in flat space. We shall follow the analogy with gauge transformations as far as possible, and for purposes of comparison we give in Sec. 2 a brief discussion of linear transformations of the field variables. This is essentially a summary of Utiyama's argument, though the emphasis is rather different, particularly with regard to the covariant and noncovariant conservation laws.

In Sec. 3 we discuss the invariance under Lorentz transformations, and in Sec. 4 we extend the discussion to the corresponding group in which the ten parameters become arbitrary functions of position. We show that to maintain invariance of the Lagrangian, it is necessary to introduce 40 new variables so that a suitable covariant derivative may be constructed. To make the action integral invariant, one actually requires the Lagrangian to be an invariant density rather than an invariant, and one must, therefore, multiply the invariant by a suitable (and uniquely determined) function of the new fields. In Sec. 5 we consider the possible forms of the free Lagrangian for the new fields. As in the case of the

* NATO Research Fellow.

[1] See, for example, H. Weyl, *Gruppentheorie und Quantenmechanik* (S. Hirzel, Leipzig, 1931), 2nd ed., Chap. 2, p. 89; and earlier references cited there.

[2] C. N. Yang and R. L. Mills, Phys. Rev. **96**, 191 (1954).

[3] Ryoyu Utiyama, Phys. Rev. **101**, 1597 (1956).

electromagnetic field, we choose the Lagrangian of lowest degree which satisfies the invariance requirements.

The geometrical interpretation in terms of a Riemannian space is discussed in Sec. 6, where we show that the free Lagrangian we have obtained is just the usual curvature scalar density, though expressed in terms of an affine connection $\Gamma^\lambda{}_{\mu\nu}$ which is not necessarily symmetric. In fact, when no matter is present it is symmetric as a consequence of the equations of motion, but otherwise it has an antisymmetric part expressible in terms of the "spin density" $\mathfrak{S}^\mu{}_{ij}$. Thus there is a difference between this theory and the usual metric theory of gravitation. This difference was first pointed out by Weyl,[4] and has more recently been discussed by Sciama.[5] It arises from the fact that our free Lagrangian is of first order in the derivatives, with the $h_k{}^\mu$ and $A^{ij}{}_\mu$ as independent variables. It is possible to re-express the theory in terms of the Christoffel connection ${}^0\Gamma^\lambda{}_{\mu\nu}$ or its local analog ${}^0A^{ij}{}_\mu$, and this is done in Sec. 7. In that case, additional terms quadratic in $\mathfrak{S}^\mu{}_{ij}$, and multiplied by the gravitational constant, appear in the Lagrangian.

2. LINEAR TRANSFORMATIONS

We consider a set of field variables $\chi_A(x)$, which we regard as the elements of a column matrix $\chi(x)$, with the Lagrangian

$$L(x)\equiv L\{\chi(x),\chi_{,\mu}(x)\},$$

where $\chi_{,\mu}=\partial_\mu\chi$. We also consider linear transformations of the form

$$\delta\chi=\epsilon^a T_a\chi, \tag{2.1}$$

where the ϵ^a are n constant infinitesimal parameters, and the T_a are n given matrices satisfying commutation rules appropriate to the generators of a Lie group,

$$[T_a,T_b]=f_a{}^c{}_b T_c.$$

The Lagrangian is invariant under these transformations if the n identities

$$(\partial L/\partial\chi)T_a\chi+(\partial L/\partial\chi_{,\mu})T_a\chi_{,\mu}\equiv 0, \tag{2.2}$$

are satisfied, and we shall assume that this is so. Note that $\partial/\partial\chi$ must be regarded as a row matrix. The equations of motion imply n conservation laws

$$J^\mu{}_{a,\mu}=0,$$

where the "currents" are defined by[6]

$$J^\mu{}_a\equiv-(\partial L/\partial\chi_{,\mu})T_a\chi. \tag{2.3}$$

[4] H. Weyl, Phys. Rev. 77, 699 (1950).

[5] D. W. Sciama, *Festschrift for Infeld* (Pergamon Press, New York), to be published.

[6] We have defined $J^\mu{}_a$ with the opposite sign to that used by Utiyama.[3] This is because with this choice of sign the analogous quantity for translations is $T^\mu{}_\nu$ rather than $-T^\mu{}_\nu$. The change may be considered as a change of sign of ϵ^a and T_a, and there is a corresponding change of sign in (2.6). This convention has the additional advantage that the "local affine connection" $A^i{}_{j\mu}$ defined in Sec. 4 specifies covariant derivatives according to the same rule as $\Gamma^\lambda{}_{\mu\nu}$.

Now, under the more general transformations of the form (2.1), but in which the parameters ϵ^a become arbitrary functions of position, the Lagrangian is no longer invariant, because the derivatives transform according to

$$\delta\chi_{,\mu}=\epsilon^a T_a\chi_{,\mu}+\epsilon^a{}_{,\mu}T_a\chi, \tag{2.4}$$

and the terms in $\epsilon^a{}_{,\mu}$ do not cancel. In fact, one finds

$$\delta L\equiv-\epsilon^a{}_{,\mu}J^\mu{}_a.$$

However, one can obtain a modified Lagrangian which is invariant by replacing $\chi_{,\mu}$ in L by a quantity $\chi_{;\mu}$ which transforms according to

$$\delta\chi_{;\mu}=\epsilon^a T_a\chi_{;\mu}. \tag{2.5}$$

To do this[7] it is necessary to introduce $4n$ new field variables $A^a{}_\mu$ whose transformation properties involve $\epsilon^a{}_{,\mu}$. In fact, if one takes

$$\chi_{;\mu}\equiv\chi_{,\mu}+A^a{}_\mu T_a\chi, \tag{2.6}$$

then the condition (2.5) determines the transformation properties of the new fields uniquely. They are

$$\delta A^a{}_\mu=\epsilon^b f_b{}^a{}_c A^c{}_\mu-\epsilon^a{}_{,\mu}. \tag{2.7}$$

In this way one obtains the invariant Lagrangian

$$L'\{\chi,\chi_{,\mu},A^a{}_\mu\}\equiv L\{\chi,\chi_{;\mu}\}.$$

The expression $\chi_{;\mu}$ may be called the covariant derivative of χ with respect to the transformations (2.1). One may define covariant currents by

$$J'^\mu{}_a\equiv-(\partial L'/\partial A^a{}_\mu)\equiv-(\partial L/\partial\chi_{;\mu})T_a\chi, \tag{2.8}$$

where L is regarded as a function of χ and $\chi_{;\mu}$. They transform linearly according to

$$\delta J'^\mu{}_a=-\epsilon^b f_b{}^c{}_a J'^\mu{}_c,$$

and their covariant divergences vanish in virtue of the equations of motion and the identities (2.2):

$$J'^\mu{}_{a;\mu}\equiv J'^\mu{}_{a,\mu}-A^b{}_\mu f_b{}^c{}_a J'^\mu{}_c$$
$$=0.$$

Two covariant differentiations do not in general commute. From (2.6) one finds

$$\chi_{;\mu\nu}-\chi_{;\nu\mu}=F^a{}_{\mu\nu}T_a\chi,$$

where

$$F^a{}_{\mu\nu}\equiv A^a{}_{\mu,\nu}-A^a{}_{\nu,\mu}-f_b{}^a{}_c A^b{}_\mu A^c{}_\nu. \tag{2.9}$$

Unlike $A^a{}_\mu$, the expression $F^a{}_{\mu\nu}$ is a covariant quantity transforming according to

$$\delta F^a{}_{\mu\nu}=\epsilon^b f_b{}^a{}_c F^c{}_{\mu\nu},$$

and one may, therefore, define its covariant derivative in an obvious manner. It satisfies the cyclic identity

$$F^a{}_{\mu\nu;\rho}+F^a{}_{\nu\rho;\mu}+F^a{}_{\rho\mu;\nu}\equiv 0.$$

[7] For a full discussion, see footnote 3.

It remains to find a free Lagrangian L_0 for the new fields. Clearly L_0 must be separately invariant, and it is easy to see[3] that this implies that it must contain $A^a{}_\mu$ only through the covariant combination $F^a{}_{\mu\nu}$. The simplest such Lagrangian is[8]

$$L_0 = -\tfrac{1}{4}F^a{}_{\mu\nu}F_a{}^{\mu\nu}, \tag{2.10}$$

where the tensor indices are raised with the flat-space metric $\eta^{\mu\nu}$ with diagonal elements $(1, -1, -1, -1)$, and the index a is lowered with the metric[8a]

$$g_{ab} \equiv f_a{}^c{}_d f_c{}^d{}_b$$

associated with the Lie group (except of course for a one-parameter group). It is clear that this Lagrangian is not unique. All that is required is that it should be a scalar both in coordinate space and in the Lie-group space, and one could add to it terms of higher degree in $F^a{}_{\mu\nu}$. However, it seems reasonable to choose the Lagrangian of lowest degree which satisfies the invariance requirements.

With the choice (2.10) of L_0, the equations of motion for the new fields are

$$F_a{}^{\mu\nu}{}_{;\nu} = J'^\mu{}_a.$$

Because of the antisymmetry of $F_a{}^{\mu\nu}$ one can define another current which is conserved in the strict sense:

$$(J'^\mu{}_a + j^\mu{}_a)_{,\mu} = 0, \tag{2.11}$$

where

$$j^\mu{}_a \equiv A^b{}_\nu f_b{}^c{}_a F_c{}^{\mu\nu}.$$

This extra current $j^\mu{}_a$ may be regarded as the current of the new field $A^a{}_\mu$ itself, since it is expressible in the form

$$j^\mu{}_a \equiv -(\partial L_0/\partial A^a{}_\mu) \equiv -(\partial L_0/\partial A^b{}_{\nu,\mu}) f_a{}^b{}_c A^c{}_\nu, \tag{2.12}$$

which should be compared with (2.8). Note, however, that it is not a covariant quantity. To obtain a strict conservation law one must sacrifice the covariance of the current.

3. LORENTZ TRANSFORMATIONS

We now wish to consider infinitesimal variations of both the coordinates and the field variables,

$$\begin{gathered} x^\mu \to x'^\mu = x^\mu + \delta x^\mu, \\ \chi(x) \to \chi'(x') = \chi(x) + \delta\chi(x). \end{gathered} \tag{3.1}$$

It will be convenient to allow for the possibility that the Lagrangian may depend on x explicitly. Then, under a variation (3.1), the change in L is

$$\delta L \equiv (\partial L/\partial\chi)\delta\chi + (\partial L/\partial\chi_{,\mu})\delta\chi_{,\mu} + (\partial L/\partial x^\mu)\delta x^\mu,$$

where $\partial L/\partial x^\mu$ denotes the partial derivative with fixed χ. It is sometimes useful to consider also the variation at a fixed value of x,

$$\delta_0\chi = \chi'(x) - \chi(x) = \delta\chi - \delta x^\mu \chi_{,\mu}. \tag{3.2}$$

In particular, it is obvious that δ_0 commutes with ∂_μ, whence

$$\delta\chi_{,\mu} = (\delta\chi)_{,\mu} - (\delta x^\nu)_{,\mu}\chi_{,\nu}. \tag{3.3}$$

The action integral

$$I(\Omega) \equiv \int_\Omega L(x) d_4x$$

over a space-time region Ω is transformed under (3.1) into

$$I'(\Omega) \equiv \int_\Omega L'(x')\|\partial_\nu x'^\mu\| d_4x.$$

Thus the action integral over an arbitrary region is invariant if[9]

$$\delta L + L(\delta x^\mu)_{,\mu} \equiv \delta_0 L + (L\delta x^\mu)_{,\mu} \equiv 0. \tag{3.4}$$

This is of course the typical transformation law of an invariant density.

We now consider the specific case of Lorentz transformations,

$$\delta x^\mu = \epsilon^\mu{}_\nu x^\nu + \epsilon^\mu, \quad \delta\chi = \tfrac{1}{2}\epsilon^{\mu\nu}S_{\mu\nu}\chi, \tag{3.5}$$

where ϵ^μ and $\epsilon^{\mu\nu} = -\epsilon^{\nu\mu}$ are 10 real infinitesimal parameters, and the $S_{\mu\nu}$ are matrices satisfying

$$S_{\mu\nu} + S_{\nu\mu} = 0,$$

$$[S_{\mu\nu}, S_{\rho\sigma}] = \eta_{\nu\rho}S_{\mu\sigma} + \eta_{\mu\sigma}S_{\nu\rho} - \eta_{\nu\sigma}S_{\mu\rho} - \eta_{\mu\rho}S_{\nu\sigma} \equiv \tfrac{1}{2}f_{\mu\nu}{}^{\kappa\lambda}{}_{\rho\sigma}S_{\kappa\lambda}.$$

From (3.3) one has

$$\delta\chi_{,\mu} = \tfrac{1}{2}\epsilon^{\rho\sigma}S_{\rho\sigma}\chi_{,\mu} - \epsilon^\rho{}_\mu\chi_{,\rho}. \tag{3.6}$$

Moreover, since $(\delta x^\mu)_{,\mu} = \epsilon^\mu{}_\mu = 0$, the condition (3.4) for invariance of the action integral again reduces to $\delta L \equiv 0$, and yields the 10 identities[10]

$$\partial L/\partial x^\rho \equiv L_{,\rho} - (\partial L/\partial\chi)\chi_{,\rho} - (\partial L/\partial\chi_{,\mu})\chi_{,\mu\rho} \equiv 0, \tag{3.7}$$

$$\begin{aligned}(\partial L/\partial\chi)S_{\rho\sigma}\chi + (\partial L/\partial\chi_{,\mu})(S_{\rho\sigma}\chi_{,\mu} \\ + \eta_{\mu\rho}\chi_{,\sigma} - \eta_{\mu\sigma}\chi_{,\rho}) \equiv 0. \end{aligned} \tag{3.8}$$

These are evidently the analogs of the identities (2.2), and we shall assume that they are satisfied. Note that (3.7), which express the conditions for translational invariance, are equivalent to the requirement that L be explicitly independent of x, as might be expected.

As before, the equations of motion may be used to obtain 10 conservation laws which follow from these identities, namely,

$$T^\mu{}_{\rho,\mu} = 0, \quad (S^\mu{}_{\rho\sigma} - x_\rho T^\mu{}_\sigma + x_\sigma T^\mu{}_\rho)_{,\mu} = 0,$$

[8] There could of course be a constant factor multiplying (2.10), but this can be absorbed by a trivial change of definition of $A^a{}_\mu$ and T_a.

[8a] The discussion here applies only to semisimple groups since otherwise g_{ab} is singular. (I am indebted to the referee for this remark.)

[9] See L. Rosenfeld, Ann. Physik 5, 113 (1930).

[10] Compare L. Rosenfeld, Ann. inst. Henri Poincaré 2, 25 (1931).

where

$$T^{\mu}{}_{\rho}\equiv(\partial L/\partial\chi_{,\mu})\chi_{,\rho}-\delta^{\mu}{}_{\rho}L,\quad S^{\mu}{}_{\rho\sigma}\equiv-(\partial L/\partial\chi_{,\mu})S_{\rho\sigma}\chi.$$

These are the conservation laws of energy, momentum, and angular momentum.

It is instructive to examine these transformations in terms of the variation $\delta_0\chi$ also, which in this case is

$$\delta_0\chi=-\epsilon^{\rho}\partial_{\rho}\chi+\tfrac{1}{2}\epsilon^{\rho\sigma}(S_{\rho\sigma}+x_{\rho}\partial_{\sigma}-x_{\sigma}\partial_{\rho})\chi.$$

On comparing this with (2.1), one sees that the role of the matrices T_a is played by the differential operators $-\partial_{\mu}$ and $S_{\rho\sigma}+x_{\rho}\partial_{\sigma}-x_{\sigma}\partial_{\rho}$. Thus, by analogy with the definition (2.3) of the currents $J^{\mu}{}_{a}$, one might expect the currents in this case to be

$$J^{\mu}{}_{\rho}\equiv(\partial L/\partial\chi_{,\mu})\chi_{,\rho},\quad J^{\mu}{}_{\rho\sigma}\equiv S^{\mu}{}_{\rho\sigma}-x_{\rho}J^{\mu}{}_{\sigma}+x_{\sigma}J^{\mu}{}_{\rho},$$

corresponding to the parameters ϵ^{ρ}, $\epsilon^{\rho\sigma}$, respectively. However, in terms of δ_0, the condition for invariance (3.4) is not simply $\delta_0 L\equiv 0$, and the additional term $\delta x^{\rho}L_{,\rho}$ is responsible for the appearance of the term $L_{,\rho}$ in the identities (3.7), and hence for the term $\delta^{\mu}{}_{\rho}L$ in $T^{\mu}{}_{\rho}$.

4. GENERALIZED LORENTZ TRANSFORMATIONS

We now turn to a consideration of the generalized transformations (3.5) in which the parameters ϵ^{μ} and $\epsilon^{\mu\nu}$ become arbitrary functions of position. It is more convenient, and clearly equivalent, to regard as independent functions $\epsilon^{\mu\nu}$ and

$$\xi^{\mu}\equiv\epsilon^{\mu}{}_{\nu}x^{\nu}+\epsilon^{\nu},$$

since this avoids the explicit appearance of x. Moreover, one could consider generalized transformations with $\xi^{\mu}=0$ but nonzero $\epsilon^{\mu\nu}$, so that the coordinate and field transformations can be completely separated. In view of this fact, it is convenient to use Latin indices for ϵ^{ij} (and for the matrices S_{ij}), retaining the Greek ones for ξ^{μ} and x^{μ}. Thus the transformations under consideration are

$$\delta x^{\mu}=\xi^{\mu},\quad \delta\chi=\tfrac{1}{2}\epsilon^{ij}S_{ij}\chi \tag{4.1}$$

or

$$\delta_0\chi=-\xi^{\mu}\chi_{,\mu}+\tfrac{1}{2}\epsilon^{ij}S_{ij}\chi. \tag{4.2}$$

This notation emphasizes the similarity of the ϵ^{ij} transformations to the linear transformations discussed in Sec. 2. These transformations alone were considered by Utiyama.[3] Evidently, the four functions ξ^{μ} specify a general coordinate transformation. The geometrical significance of the ϵ^{ij} will be discussed in Sec. 6.

According to our convention, the differential operator ∂_{μ} must have a Greek index. However, in the Lagrangian function L it would be inconvenient to have two kinds of indices, and we shall, therefore, regard L as a given function of χ and χ_k (no comma),[11] satisfying the identities (3.7) and (3.8). The original Lagrangian is then obtained by setting

$$\chi_k=\delta_k{}^{\mu}\chi_{,\mu}.$$

It is of course not invariant under the generalized transformations (4.1), but we shall later obtain an invariant expression by replacing χ_k by a suitable quantity $\chi_{;k}$.

The transformation of $\chi_{,\mu}$ is given by

$$\delta\chi_{,\mu}=\tfrac{1}{2}\epsilon^{ij}S_{ij}\chi_{,\mu}+\tfrac{1}{2}\epsilon^{ij}{}_{,\mu}S_{ij}\chi-\xi^{\nu}{}_{,\mu}\chi_{,\nu}, \tag{4.3}$$

and so the original Lagrangian transforms according to

$$\delta L\equiv-\xi^{\rho}{}_{,\mu}J^{\mu}{}_{\rho}-\tfrac{1}{2}\epsilon^{ij}{}_{,\mu}S^{\mu}{}_{ij}.$$

Note that it is $J^{\mu}{}_{\rho}$ rather than $T^{\mu}{}_{\rho}$ which appears here. The reason for this is that we have not included the extra term $L(\delta x^{\mu})_{,\mu}$ in (3.4). The left-hand side of (3.4) actually has the value

$$\delta L+L(\delta x^{\mu})_{,\mu}\equiv-\xi^{\rho}{}_{,\mu}T^{\mu}{}_{\rho}-\tfrac{1}{2}\epsilon^{ij}{}_{,\mu}S^{\mu}{}_{ij}.$$

We now look for a modified Lagrangian which makes the action integral invariant. The additional term just mentioned is of a different kind to those previously encountered, in that it involves L and not $\partial L/\partial\chi_k$. In particular, it includes contributions from terms in L which do not contain derivatives. Thus it is clear that we cannot remove it by replacing the derivative by a suitable covariant derivative. For this reason, we shall consider the problem in two stages. We first eliminate the noninvariance arising from the fact that $\chi_{,\mu}$ is not a covariant quantity, and thus obtain an expression L' satisfying

$$\delta L'\equiv 0. \tag{4.4}$$

Then, because the condition (3.4) for invariance of the action integral requires the Lagrangian to be an invariant density rather than an invariant, we make a further modification, replacing L' by $\mathfrak{L}'$, which satisfies

$$\delta\mathfrak{L}'+\xi^{\mu}{}_{,\mu}\mathfrak{L}'\equiv 0. \tag{4.5}$$

The first part of this program can be accomplished by replacing χ_k in L by a "covariant derivative" $\chi_{;k}$ which transforms according to

$$\delta\chi_{;k}=\tfrac{1}{2}\epsilon^{ij}S_{ij}\chi_{;k}-\epsilon^{i}{}_{k}\chi_{;i}. \tag{4.6}$$

The condition (4.4) then follows from the identities (3.8). To do this it is necessary to introduce forty new field variables. We consider first the ϵ^{ij} transformations, and eliminate the $\epsilon^{ij}{}_{,\mu}$ term in (4.3) by setting[12]

$$\chi_{|\mu}\equiv\chi_{,\mu}+\tfrac{1}{2}A^{ij}{}_{\mu}S_{ij}\chi, \tag{4.7}$$

where the $A^{ij}{}_{\mu}=-A^{ji}{}_{\mu}$ are 24 new field variables. We can then impose the condition

$$\delta\chi_{|\mu}=\tfrac{1}{2}\epsilon^{ij}S_{ij}\chi_{|\mu}-\xi^{\nu}{}_{,\mu}\chi_{|\nu} \tag{4.8}$$

which determines the transformation properties of $A^{ij}{}_{\mu}$

[11] Note that since we are using Latin indices for S_{ij} the various tensor components of χ must also have Latin indices, and for spinor components the Dirac matrices must be γ^k.

[12] Our $A^{ij}{}_{\mu}$ differs in sign from that of Utiyama.[3] Compare footnote 6.

uniquely. They are

$$\delta A^{ij}{}_{\mu}=\epsilon^{i}{}_{k}A^{kj}{}_{\mu}+\epsilon^{j}{}_{k}A^{ik}{}_{\mu}-\xi^{\nu}{}_{,\mu}A^{ij}{}_{\nu}-\epsilon^{ij}{}_{,\mu}. \qquad (4.9)$$

The position with regard to the last term in (4.3) is rather different. The term involving $\epsilon^{ij}{}_{,\mu}$ is inhomogeneous in the sense that it contains χ rather than $\chi_{,\mu}$, just like the second term of (2.4), but this is not true of the last term.[13] Correspondingly, the transformation law (4.8) of $\chi_{|\mu}$ is already homogeneous. This means that to obtain an expression $\chi_{;k}$ transforming according to (4.6) we should add to $\chi_{|\mu}$ not a term in χ but rather a term in $\chi_{|\mu}$ itself. In other words, we can merely multiply by a new field:

$$\chi_{;k}\equiv h_{k}{}^{\mu}\chi_{|\mu}. \qquad (4.10)$$

Here the $h_k{}^{\mu}$ are 16 new field variables with transformation properties determined by (4.6) to be

$$\delta h_{k}{}^{\mu}=\xi^{\mu}{}_{,\nu}h_{k}{}^{\nu}-\epsilon^{i}{}_{k}h_{i}{}^{\mu}. \qquad (4.11)$$

It should be noted that the fields $h_k{}^{\mu}$ and $A^{ij}{}_{\mu}$ are quite independent and unrelated at this stage, though of course they will be related by equations of motion.

We have now found an invariant L'. We can easily obtain an invariant density $\mathfrak{L}'$ by multiplying by a suitable function of the fields already introduced:

$$\mathfrak{L}'\equiv\mathfrak{H}L'.$$

Then (4.5) is satisfied provided that $\mathfrak{H}$ is itself an invariant density,

$$\delta\mathfrak{H}+\xi^{\mu}{}_{,\mu}\mathfrak{H}=0.$$

It is easy to see that the only function of the new fields which obeys this transformation law, and does not involve derivatives, is

$$\mathfrak{H}=[\det(h_{k}{}^{\mu})]^{-1},$$

where the arbitrary constant factor has been chosen so that $\mathfrak{H}$ reduces to 1 when $h_k{}^{\mu}$ is set equal to $\delta_k{}^{\mu}$.[14]

The final form of our modified Lagrangian is

$$\mathfrak{L}\{\chi,\chi_{,\mu},h_{k}{}^{\mu},A^{ij}{}_{\mu}\}\equiv\mathfrak{H}L\{\chi,\chi_{;k}\}.$$

(We can drop the prime without risk of confusion.) It may be asked whether this Lagrangian is unique in the same sense as the modified Lagrangian L' of Sec. 2, and in fact it is easy to see that it is not. The reason for this is that if one starts with two Lagrangians L_1 and L_2 which differ by an explicit divergence, and are therefore equivalent, then the modified Lagrangians $\mathfrak{L}_1$ and $\mathfrak{L}_2$ are not necessarily equivalent. Consider for example the Lagrangian for a real scalar field written in its first-order form

$$L_1=\pi^{k}\varphi_{,k}-\tfrac{1}{2}\pi^{k}\pi_{k}-\tfrac{1}{2}m^{2}\varphi^{2}. \qquad (4.12)$$

This is equivalent to

$$L_2=-\pi^{k}{}_{,k}\varphi-\tfrac{1}{2}\pi^{k}\pi_{k}-\tfrac{1}{2}m^{2}\varphi^{2}, \qquad (4.13)$$

but the corresponding modified Lagrangians differ by

$$\begin{aligned}\mathfrak{L}_1-\mathfrak{L}_2&\equiv\mathfrak{H}(\pi^{k}\varphi)_{;k}\\&\equiv\mathfrak{H}h_{k}{}^{\mu}[(\pi^{k}\varphi)_{,\mu}+A^{k}{}_{i\mu}\pi^{i}\varphi]\end{aligned} \qquad (4.14)$$

which is not an explicit divergence. Thus in order to define the modified Lagrangian $\mathfrak{L}$ completely it would be necessary to specify which of the possible equivalent forms of the original Lagrangian is to be chosen. The reasons for this situation and the problem of choosing the correct form are discussed in the Appendix.

As in Sec. 2, one may define modified "currents" in terms of $L\equiv L\{\chi,\chi_{;k}\}$ by

$$\mathfrak{T}^{k}{}_{\mu}\equiv\partial\mathfrak{L}/\partial h_{k}{}^{\mu}\equiv\mathfrak{H}b^{i}{}_{\mu}\{(\partial L/\partial\chi_{;k})\chi_{;i}-\delta^{k}{}_{i}L\}, \qquad (4.15)$$

$$\mathfrak{S}^{\mu}{}_{ij}\equiv-2(\partial\mathfrak{L}/\partial A^{ij}{}_{\mu})\equiv-\mathfrak{H}h_{k}{}^{\mu}(\partial L/\partial\chi_{;k})S_{ij}\chi, \qquad (4.16)$$

where $b^i{}_{\mu}$ is the inverse of $h_i{}^{\mu}$, satisfying

$$b^{i}{}_{\mu}h_{i}{}^{\nu}=\delta_{\mu}{}^{\nu},\quad b^{i}{}_{\mu}h_{j}{}^{\mu}=\delta^{i}{}_{j}.$$

To express the "conservation laws" which these currents satisfy in a simple form, it is convenient to extend the definition of the covariant derivative $\chi_{|\mu}$ (not $\chi_{;k}$). Originally, it is defined for χ and, therefore, by a trivial extension for any other quantity which is invariant under ξ^{μ} transformations, and transforms linearly under ϵ^{ij} transformations. We wish to extend it to any quantity which transforms linearly under ϵ^{ij} transformations, by simply ignoring the ξ^{μ} transformation properties altogether. Thus, for example, we would have

$$h_{i}{}^{\mu}{}_{|\nu}\equiv h_{i}{}^{\mu}{}_{,\nu}-A^{k}{}_{i\nu}h_{k}{}^{\mu}, \qquad (4.17)$$

according to the ϵ^{ij} transformation law of $h_i{}^{\mu}$. We shall call this the ϵ covariant derivative. Later we shall define another covariant derivative which takes account of ξ^{μ} transformations also.

One can easily calculate the commutator of two ϵ covariant differentiations.[15] This gives

$$\chi_{|\mu\nu}-\chi_{|\nu\mu}=\tfrac{1}{2}R^{ij}{}_{\mu\nu}S_{ij}\chi, \qquad (4.18)$$

where

$$R^{i}{}_{j\mu\nu}\equiv A^{i}{}_{j\mu,\nu}-A^{i}{}_{j\nu,\mu}-A^{i}{}_{k\mu}A^{k}{}_{j\nu}+A^{i}{}_{k\nu}A^{k}{}_{j\mu}.$$

This quantity is covariant under ϵ^{ij} transformations, and satisfies the cyclic identity

$$R^{i}{}_{j\mu\nu|\rho}+R^{i}{}_{j\nu\rho|\mu}+R^{i}{}_{j\rho\mu|\nu}\equiv0.$$

---

[13] The reason for this may be seen in terms of the variation $\delta_0\chi$ given by (4.2). The analogs of the matrices T_a are clearly $-\partial_{\mu}$ and S_{ij}, so that the presence of the derivative $\chi_{,\mu}$ in the last term of (4.3) is to be expected. By analogy with (2.6) we should expect the covariant derivative to have the form

$$\chi_{;k}=\delta_{k}{}^{\mu}\chi_{,\mu}+\tfrac{1}{2}A^{ij}{}_{k}S_{ij}\chi-A^{\mu}{}_{k}\partial_{\mu}\chi.$$

Because of the appearance of derivatives, the first and last terms can be combined in the form $h_k{}^{\mu}\chi_{,\mu}$, where $h_k{}^{\mu}=\delta_k{}^{\mu}-A^{\mu}{}_k$. If we then set $A^{ij}{}_k=h_k{}^{\mu}A^{ij}{}_{\mu}$, we arrive at the same form for $\chi_{;k}$ as that obtained in the text.

[14] Multiplication of the entire Lagrangian by a constant factor is of course unimportant.

[15] Note that this could not be done without extending the definition, since one must know how to treat the index on $\chi_{|\rho}$. Here, as in Sec. 2, we simply ignore it.

It is thus closely analogous to $F^a{}_{\mu\nu}$. Note that $R^{ij}{}_{\mu\nu}$ is antisymmetric in both pairs of indices.

In terms of the ϵ covariant derivative, the "conservation laws" can be expressed in the form[5]

$$(\mathfrak{T}^k{}_\nu h_k{}^\mu)_{|\mu}+\mathfrak{T}^k{}_\mu h_k{}^\mu{}_{|\nu}=\tfrac{1}{2}\mathfrak{S}^\mu{}_{ij}R^{ij}{}_{\mu\nu}, \qquad (4.19)$$

$$\mathfrak{S}^\mu{}_{ij|\mu}=\mathfrak{T}_{i\mu}h_j{}^\mu-\mathfrak{T}_{j\mu}h_i{}^\mu. \qquad (4.20)$$

5. FREE GRAVITATIONAL LAGRANGIAN

We now wish to examine the quantity $\chi_{;k}$, rather than $\chi_{|\mu}$. As before, the covariant derivative of any quantity which transforms in a similar way to χ may be defined analogously. Now in particular $\chi_{;k}$ itself (unlike $\chi_{|\mu}$) is such a quantity, and therefore without extending the definition of covariant derivative one can evaluate the commutator $\chi_{;kl}-\chi_{;lk}$. However, this quantity is not simply obtained by multiplying $\chi_{|\mu\nu}-\chi_{|\nu\mu}$ by $h_k{}^\mu h_l{}^\nu$, as one might expect. The reason for this is that in evaluating $\chi_{;kl}$ one differentiates the $h_k{}^\mu$ in $\chi_{;k}$, and moreover adds an extra $A^i{}_{k\mu}$ term on account of the index k. Thus one finds

$$\chi_{;kl}-\chi_{;lk}=\tfrac{1}{2}R^{ij}{}_{kl}S_{ij}\chi-C^i{}_{kl}\chi_{;i}, \qquad (5.1)$$

where

$$R^{ij}{}_{kl}\equiv h_k{}^\mu h_l{}^\nu R^{ij}{}_{\mu\nu}, \qquad (5.2)$$

$$C^i{}_{kl}\equiv (h_k{}^\mu h_l{}^\nu-h_l{}^\mu h_k{}^\nu)b^i{}_{\mu|\nu}. \qquad (5.3)$$

Note that (5.1) is not simply proportional to χ, but involves $\chi_{;i}$ also.[16]

We now look for a free Lagrangian $\mathfrak{L}_0$ for the new fields. Clearly $\mathfrak{L}_0$ must be an invariant density, and if we set

$$\mathfrak{L}_0\equiv\mathfrak{H}L_0,$$

then it is easy to see, as in the case of linear transformations, that the invariant L_0 must be a function only of the covariant quantities $R^{ij}{}_{kl}$ and $C^i{}_{kl}$. As before, there are many possible forms for $\mathfrak{L}_0$, but there is a difference between this case and the previous one in that all the indices on these expressions are of the same type (unlike $F^a{}_{\mu\nu}$), and one can, therefore, contract the upper indices with the lower. In fact, the condition that L_0 be a scalar in two separate spaces is now reduced to the condition that it be a scalar in one space. In particular, this means that there exists a linear invariant which has no analog in the previous case, namely,

$$R\equiv R^{ij}{}_{ij}.$$

There are in addition several quadratic invariants. However, if we again choose for L_0 the form of lowest possible degree, then we are led to the free Lagrangian[17]

$$\mathfrak{L}_0=\tfrac{1}{2}\mathfrak{H}R \qquad (5.4)$$

which differs from (2.10) in being only linear in the derivatives.

[16] This is another example of the fact that for ξ^μ transformations derivatives play the role of the matrices T_a. Compare footnote 13.

[17] We choose units in which $\kappa=1$ (as well as $c=\hbar=1$).

With this choice of Lagrangian, the equations of motion for the new fields are

$$\mathfrak{H}(R^{ik}{}_{jk}-\tfrac{1}{2}\delta^i{}_jR)=-\mathfrak{T}^i{}_\mu h_j{}^\mu, \qquad (5.5)$$

$$-[\mathfrak{H}(h_i{}^\mu h_j{}^\nu-h_j{}^\mu h_i{}^\nu)]_{|\nu}$$
$$\equiv\mathfrak{H}(h_k{}^\mu C^k{}_{ij}-h_j{}^\mu C^k{}_{ik}-h_i{}^\mu C^k{}_{kj})=\mathfrak{S}^\mu{}_{ij}. \qquad (5.6)$$

From Eq. (5.6) one can immediately obtain a strict conservation law

$$(\mathfrak{S}^\mu{}_{ij}+\mathfrak{s}^\mu{}_{ij})_{,\mu}=0, \qquad (5.7)$$

where

$$\mathfrak{s}^\mu{}_{ij}\equiv\mathfrak{H}A^k{}_{i\nu}(h_j{}^\mu h_k{}^\nu-h_k{}^\mu h_j{}^\nu)-\mathfrak{H}A^k{}_{j\nu}(h_i{}^\mu h_k{}^\nu-h_k{}^\mu h_i{}^\nu).$$

This quantity is expressible in the form

$$\mathfrak{s}^\mu{}_{ij}\equiv-2(\partial\mathfrak{L}_0/\partial A^{ij}{}_\mu)\equiv-\tfrac{1}{2}(\partial\mathfrak{L}_0/\partial A^{mn}{}_{\nu,\mu})f_{ij}{}^{mn}{}_{kl}A^{kl}{}_\nu,$$

which is closely analogous to (2.12), and should be compared with (4.16). Equation (5.7) is a rather surprising result, since $\mathfrak{S}^\mu{}_{ij}$ may very reasonably be interpreted as the spin density of the matter field,[18] so that it appears to be a law of conservation of spin with no reference to the orbital angular momentum. In fact, however, the orbital angular momentum appears in the corresponding "covariant conservation law" (4.20), and therefore part of the "spin" of the gravitational field, $\mathfrak{s}^\mu{}_{ij}$, may be regarded as arising from this source. Nevertheless, Eq. (5.7) differs from other statements of angular momentum conservation in that the coordinates do not appear explicitly.

It would also be possible to deduce from Eq. (5.5) a strict conservation law

$$[h_k{}^\nu(\mathfrak{T}^k{}_\mu+\mathfrak{t}^k{}_\mu)]_{,\nu}=0, \qquad (5.8)$$

but there is a considerable amount of freedom in choosing $\mathfrak{t}^k{}_\mu$. The most natural definition, by analogy with (4.15) would be

$$\mathfrak{t}^k{}_\mu\equiv\partial\mathfrak{L}_0/\partial h_k{}^\mu,$$

and this quantity does indeed satisfy (5.8). However, in this case the expression within the parentheses itself vanishes, so that (5.8) is rather trivial. We shall not discuss the question of the correct choice of $\mathfrak{t}^k{}_\mu$ further, as this lies beyond the scope of the present paper.[19]

It should be noted that Eq. (5.6) can be solved, at least in principle, for $A^{ij}{}_\mu$. In the simple case when $\mathfrak{S}^\mu{}_{ij}$ vanishes, one finds[20]

$$A_{ij\mu}={}^0A_{ij\mu}\equiv\tfrac{1}{2}b^k{}_\mu(c_{kij}-c_{ijk}-c_{jki}),$$
$$c^k{}_{ij}\equiv(h_i{}^\mu h_j{}^\nu-h_j{}^\mu h_i{}^\nu)b^k{}_{\mu,\nu}. \qquad (5.9)$$

[18] See H. J. Belinfante, Physica 6, 887 (1939), and footnote 5.

[19] It is well known in the case of the ordinary metric theory of gravitation that many definitions of the energy pseudotensor are possible. See, for example, P. G. Bergmann, Phys. Rev. 112, 287 (1958).

[20] The ${}^0A^{ij}{}_\mu$ are Ricci's coefficients of rotation. See for instance V. Fock, Z. Physik 57, 261 (1929).

In general, if we write

$$\mathfrak{S}^{\mu}{}_{ij} \equiv \mathfrak{H} h_k{}^{\mu} S^k{}_{ij},$$

then

$$A_{ij\mu} = {}^0A_{ij\mu} - \tfrac{1}{2} b^k{}_\mu (S_{kij} - S_{ijk} - S_{jki} - \eta_{ki} S^l{}_{lj} - \eta_{kj} S^l{}_{il}). \quad (5.10)$$

If the original Lagrangian L is of first order in the derivatives, then $S^k{}_{ij}$ is independent of $A^{ij}{}_\mu$ so that (5.10) is an explicit solution. Otherwise, however, $A^{ij}{}_\mu$ also appears on the right-hand side of this equation.

We conclude this section with a discussion of the Lagrangian for the fields $A^a{}_\mu$ introduced in Sec. 2 when the "gravitational" fields $h_k{}^\mu$ and $A^{ij}{}_\mu$ are also introduced. The fields $A^a{}_\mu$ should not be regarded merely as components of χ when dealing with Lorentz transformations, since one must preserve the invariance under the linear transformations. To find the correct form of the Lagrangian, one should consider simultaneously Lorentz transformations and these linear transformations. This can be done provided that the matrices T^a commute with the S_{ij}, a condition which is always fulfilled in practice. Then one finds that χ_k in L should be replaced by a derivative which is covariant under both (2.1) and (4.1), namely,

$$\chi_{;k} = h_k{}^\mu(\chi_{,\mu} + \tfrac{1}{2} A^{ij}{}_\mu S_{ij}\chi + A^a{}_\mu T_a \chi).$$

The commutator $\chi_{;kl} - \chi_{;lk}$ then contains the extra term

$$F^a{}_{kl} T^a \chi,$$

where

$$F^a{}_{kl} \equiv h_k{}^\mu h_l{}^\nu F^a{}_{\mu\nu},$$

with $F^a{}_{\mu\nu}$ given by (2.9). It is important to notice that the derivatives of $A^a{}_\mu$ in $F^a{}_{\mu\nu}$ are ordinary derivatives, not covariant ones. (We shall see in the next section that the ordinary and covariant curls are not equal, because the affine connection is in general nonsymmetric.) As before, one can see that any invariant function of $A^a{}_\mu$ must be a function of $F^a{}_{kl}$ only, and the simplest free Lagrangian for $A^a{}_\mu$ is, therefore,

$$-\tfrac{1}{4} \mathfrak{H} F^a{}_{kl} F_a{}^{kl}. \quad (5.11)$$

6. GEOMETRICAL INTERPRETATION

Up to this point, we have not given any geometrical significance to the transformations (4.1), or to the new fields $h_k{}^\mu$ and $A^{ij}{}_\mu$, but it is useful to do so in order to be able to compare the theory with the more familiar metric theory of gravitation.

Now the ξ^μ transformations are general coordinate transformations, and according to (4.11) $h_k{}^\mu$ transforms like a contravariant vector under these transformations, while $b^k{}_\mu$ and $A^{ij}{}_\mu$ transform like covariant vectors. Thus the quantity

$$g_{\mu\nu} \equiv b^k{}_\mu b_{k\nu} \quad (6.1)$$

is a symmetric covariant tensor, and may therefore be interpreted as the metric tensor of a Riemannian space. It is moreover invariant under the ϵ^{ij} transformations. Evidently, the Greek indices may be regarded as world tensor indices, and we must of course abandon for them the convention that all indices are to be raised or lowered with the flat-space metric $\eta_{\mu\nu}$, and use $g_{\mu\nu}$ instead. It is easy to see that the scalar density $\mathfrak{H}$ is equal to $(-g)^{\frac{1}{2}}$, where $g = \det(g_{\mu\nu})$.

Now, in view of the relation (6.1), $h_k{}^\mu$ and $b^k{}_\mu$ are the contravariant and covariant components, respectively, of a vierbein system in the Riemannian space.[21] Thus the ϵ^{ij} transformations should be interpreted as vierbein rotations, and the Latin indices as local tensor indices with respect to this vierbein system. The original field χ may be decomposed into local tensors and spinors,[22] and from the tensors one can form corresponding world tensors by multiplying by $h_k{}^\mu$ or $b^k{}_\mu$. For example, from a local vector v^i one can form

$$v^\mu = h_i{}^\mu v^i, \quad v_\mu = b^i{}_\mu v_i. \quad (6.2)$$

No confusion can be caused by using the same symbol v for the local and world vectors, since they are distinguished by the type of index, and indeed we have already used this convention in (5.2). Note that $v_\mu = g_{\mu\nu} v^\nu$, so that (6.2) is consistent with the choice of metric (6.1). We shall frequently use this convention of associating world tensors with given local tensors without explicit mention on each occasion.

The field $A^i{}_{j\mu}$ may reasonably be called a "local affine connection" with respect to the vierbein system, since it specifies the covariant derivatives of local tensors or spinors.[23] For a local vector, this takes the form

$$v^i{}_{|\nu} = v^i{}_{,\nu} + A^i{}_{j\nu} v^j, \quad v_{j|\nu} = v_{j,\nu} - A^i{}_{j\nu} v_i. \quad (6.3)$$

It may be noticed that the relation (4.10) between $\chi_{|\mu}$ and $\chi_{;k}$ is of the same type as (6.2) and could be written simply as

$$\chi_{;\mu} = \chi_{|\mu} \quad (6.4)$$

according to our convention. However, we shall retain the use of two separate symbols because we wish to extend the definition of covariant derivative in a different way to that of Sec. 4. It seems natural to define the covariant derivative of a world tensor in terms of the covariant derivative of the associated local tensor. Thus, for instance, to define the covariant derivatives of the world vectors (6.2) one would form the world tensors corresponding to (6.3). This gives

$$v^\lambda{}_{;\nu} \equiv h_i{}^\lambda v^i{}_{|\nu} = v^\lambda{}_{,\nu} + \Gamma^\lambda{}_{\mu\nu} v^\mu,$$

$$v_{\mu;\nu} \equiv b^i{}_\mu v_{i|\nu} = v_{\mu,\nu} - \Gamma^\lambda{}_{\mu\nu} v_\lambda,$$

where

$$\Gamma^\lambda{}_{\mu\nu} \equiv h_i{}^\lambda b^i{}_{\mu|\nu} \equiv -b^i{}_\mu h_i{}^\lambda{}_{|\nu}. \quad (6.5)$$

Note that this definition of $\Gamma^\lambda{}_{\mu\nu}$ is equivalent to the

[21] See for instance H. Weyl, Z. Physik **56**, 330 (1929).
[22] H. J. Belinfante, Physica **7**, 305 (1940).
[23] Compare J. A. Schouten, J. Math. and Phys. **10**, 239 (1931).

requirement that the covariant derivatives of the vierbein components should vanish,

$$h_i{}^\lambda{}_{;\nu}\equiv 0, \quad b^i{}_{\mu;\nu}\equiv 0. \tag{6.6}$$

For a generic quantity α transforming according to

$$\delta\alpha=\tfrac{1}{2}\epsilon^{ij}S_{ij}\alpha+\xi^\lambda{}_{,\mu}\Sigma_\lambda{}^\mu\alpha, \tag{6.7}$$

the covariant derivative is defined by[21]

$$\alpha_{;\nu}\equiv\alpha_{,\nu}+\tfrac{1}{2}A^{ij}{}_\nu S_{ij}\alpha+\Gamma^\lambda{}_{\mu\nu}\Sigma_\lambda{}^\mu\alpha, \tag{6.8}$$

whereas the ϵ covariant derivative defined in Sec. 4 is obtained by simply omitting the last term of (6.8). Note that the two derivatives are equal for purely local tensors or spinors, but not otherwise. One easily finds that the commutator of two covariant differentiations is given by

$$\alpha_{;\mu\nu}-\alpha_{;\nu\mu}=\tfrac{1}{2}R^{ij}{}_{\mu\nu}S_{ij}\alpha+R^\rho{}_{\sigma\mu\nu}\Sigma_\rho{}^\sigma\alpha-C^\lambda{}_{\mu\nu}\alpha_{;\lambda},$$

where $R^\rho{}_{\sigma\mu\nu}$ and $C^\lambda{}_{\mu\nu}$ are defined in the usual way in terms of $R^i{}_{j\mu\nu}$ and $C^i{}_{kl}$. They are both world tensors, and can easily be expressed in terms of $\Gamma^\lambda{}_{\mu\nu}$, in the form[24]

$$R^\rho{}_{\sigma\mu\nu}=\Gamma^\rho{}_{\sigma\mu,\nu}-\Gamma^\rho{}_{\sigma\nu,\mu}-\Gamma^\rho{}_{\lambda\mu}\Gamma^\lambda{}_{\sigma\nu}+\Gamma^\rho{}_{\lambda\nu}\Gamma^\lambda{}_{\sigma\nu}, \tag{6.9}$$

$$C^\lambda{}_{\mu\nu}=\Gamma^\lambda{}_{\mu\nu}-\Gamma^\lambda{}_{\nu\mu}. \tag{6.10}$$

Thus one sees that $R^\rho{}_{\sigma\mu\nu}$ is just the Riemann tensor formed from the affine connection $\Gamma^\lambda{}_{\mu\nu}$.

From (6.6) it follows that

$$g_{\mu\nu;\rho}\equiv 0, \tag{6.11}$$

so that it is consistent to interpret $\Gamma^\lambda{}_{\mu\nu}$ as an affine connection in the Riemannian space. However, the definition (6.5) evidently does not guarantee that it is symmetric, so that in general it is not the Christoffel connection. The curvature scalar R has the usual form

$$R\equiv R^\mu{}_\mu, \quad R_{\mu\nu}\equiv R^\lambda{}_{\mu\lambda\nu},$$

so that the free gravitational Lagrangian is just the usual one except for the nonsymmetry of $\Gamma^\lambda{}_{\mu\nu}$. It should be remarked that it would be incorrect to treat the 64 components of $\Gamma^\lambda{}_{\mu\nu}$ as independent variables, since there are only 24 components of $A^{ij}{}_\mu$. In fact the $\Gamma^\lambda{}_{\mu\nu}$ are restricted by the 40 identities (6.11). Thus there is no contradiction with the well-known fact that the first-order Palatini Lagrangian with nonsymmetric $\Gamma^\lambda{}_{\mu\nu}$ does not yield (6.11) as equations of motion.[25]

The equations of motion (5.5) and (5.6) can be rewritten in the form

$$\mathfrak{H}(R_{\mu\nu}-\tfrac{1}{2}g_{\mu\nu}R)=-\mathfrak{T}_{\mu\nu}, \tag{6.12}$$

$$\mathfrak{H}C^\lambda{}_{\mu\nu}=\mathfrak{S}^\lambda{}_{\mu\nu}-\tfrac{1}{2}\delta^\lambda{}_\mu\mathfrak{S}^\rho{}_{\rho\nu}-\tfrac{1}{2}\delta^\lambda{}_\nu\mathfrak{S}^\rho{}_{\mu\rho}. \tag{6.13}$$

From Eqs. (6.10) and (6.13) one sees that in the absence of matter the affine connection $\Gamma^\lambda{}_{\mu\nu}$ is symmetric, and therefore equal to the Christoffel connection ${}^0\Gamma^\lambda{}_{\mu\nu}$. (This is the analog for world tensors of ${}^0A^i{}_{j\mu}$.) Then $R_{\mu\nu}$ is symmetric, and Eq. (6.12) yields Einstein's familiar equations for empty space,

$$R_{\mu\nu}=0.$$

However, when matter is present, $\Gamma^\lambda{}_{\mu\nu}$ is no longer symmetric, and its antisymmetric part is given by (6.13). Then the tensor $R_{\mu\nu}$ is also nonsymmetric, and correspondingly the energy tensor density $\mathfrak{T}_{\mu\nu}$ is in general nonsymmetric, because $h_k{}^\mu$ does not appear in $\mathfrak{L}$ only through the symmetric combination $g^{\mu\nu}$. Thus the theory differs slightly from the usual one, in a way first noted by Weyl.[4] In the following section, we shall investigate this difference in more detail.[5]

Finally, we can rewrite the covariant conservation laws in terms of world tensors. It is convenient to define the contraction

$$C_\mu\equiv C^\lambda{}_{\mu\lambda},$$

since the covariant divergence of a vector density $\mathfrak{f}^\mu$ is then

$$\mathfrak{f}^\mu{}_{;\mu}=\mathfrak{f}^\mu{}_{,\mu}+C_\mu\mathfrak{f}^\mu. \tag{6.14}$$

The conservation laws become

$$\mathfrak{T}^\nu{}_{\mu;\nu}-C_\nu\mathfrak{T}^\nu{}_\mu+C^\lambda{}_{\mu\nu}\mathfrak{T}^\nu{}_\lambda=\tfrac{1}{2}R^{\rho\sigma}{}_{\mu\nu}\mathfrak{S}^\nu{}_{\rho\sigma},$$

$$\mathfrak{S}^\mu{}_{\rho\sigma;\mu}-C_\mu\mathfrak{S}^\mu{}_{\rho\sigma}=\mathfrak{T}_{\rho\sigma}-\mathfrak{T}_{\sigma\rho}.$$

It may be noticed that these are slightly more complicated than the expressions in terms of the ϵ covariant derivative.

7. COMPARISON WITH METRIC THEORY

For simplicity, we shall assume in this section that L is only of first order in the derivatives, so that (5.10) is an explicit solution for $A^{ij}{}_\mu$. The difference between the theory presented here and the usual one arises because we are using a Lagrangian $\mathfrak{L}_0$ of first order, in which $h_k{}^\mu$ and $A^{ij}{}_\mu$ are independent variables. The situation is entirely analogous to that which obtains for any theory with "derivative" interaction. In first-order form, the "momenta" $A^{ij}{}_\mu$ are not just equal to derivatives of the "coordinates" $h_k{}^\mu$, or in other words to ${}^0A^{ij}{}_\mu$. Thus an interaction which appears simple in first-order form will be more complicated if a second-order Lagrangian is used, and vice versa.

The second-order form of the Lagrangian may be obtained by substituting for $A^{ij}{}_\mu$ the expression (5.10). This gives

$$\mathfrak{L}'={}^0\mathfrak{L}+{}^0\mathfrak{L}_0+{}^1\mathfrak{L},$$

where ${}^0\mathfrak{L}$ and ${}^0\mathfrak{L}_0$ are obtained from $\mathfrak{L}$ and $\mathfrak{L}_0$ by replacing $A^{ij}{}_\mu$ by ${}^0A^{ij}{}_\mu$ (or equivalently $\Gamma^\lambda{}_{\mu\nu}$ by ${}^0\Gamma^\lambda{}_{\mu\nu}$), and ${}^1\mathfrak{L}$ is an additional term quadratic in $S^k{}_{ij}$, namely,

$${}^1\mathfrak{L}=\tfrac{1}{8}\mathfrak{H}(2S_{ijk}S^{jki}-S_{ijk}S^{ijk}+2S^i{}_{ik}S_j{}^{jk}). \tag{7.1}$$

In this Lagrangian, only $h_k{}^\mu$ and χ are treated as inde-

[24] This is a generalization to nonsymmetric affinities of the result proved in the appendix to footnote 3. See also footnotes 4 and 5.

[25] See for instance E. Schrödinger, *Space-time Structure* (Cambridge University Press, New York, 1950).

pendent variables. The equations of motion are equivalent to those previously obtained if the variables $A^{ij}{}_{\mu}$ are eliminated from the latter by using (5.10).

The usual metric theory, on the other hand, is given by the Lagrangian

$$\mathfrak{L}''={}^0\mathfrak{L}+{}^0\mathfrak{L}_0,$$

without the extra terms (7.1). If this Lagrangian were written in a first-order form by introducing additional independent variables $A^{ij}{}_{\mu}$, then one would arrive at a form identical to the one given here except for the appearance of extra terms equal to (7.1) with a negative sign.

Thus we see that the only difference between the two theories is the presence or absence of these "direct-interaction" terms. Now if we had not set $\kappa=1$, then $\mathfrak{L}_0$ would have a factor κ^{-1}, whereas the terms (7.1) would appear with the factor κ. They are, therefore, extremely small in comparison to other interaction terms. In particular, for a Dirac field, they would be proportional to (see Appendix)

$$\kappa\bar{\psi}\gamma_k\gamma_5\psi\bar{\psi}\gamma^k\gamma_5\psi.$$

Thus they are similar in form to the Fermi interaction terms, but much smaller in magnitude, so that it seems impossible that they would lead to any observable difference between the predictions of the two theories. Hence we must conclude that for all practical purposes the theory presented here is equivalent to the usual one.

ACKNOWLEDGMENTS

The author is indebted to Drs. J. L. Anderson, P. W. Higgs, and D. W. Sciama for helpful discussions and comments.

APPENDIX

In this appendix we shall discuss the remaining ambiguity in the modified Lagrangian. It was pointed out in Sec. 4 that the generally covariant Lagrangians obtained from two equivalent Lagrangians L_1 and L_2 are in general inequivalent. One can now see that in fact they differ by a covariant divergence. Thus (4.14) can be written in the form

$$\mathfrak{L}_1-\mathfrak{L}_2=(\mathfrak{H}h_k{}^{\mu}\pi^k\varphi)_{;\mu},$$

but in view of (6.14) this is not equal to the ordinary divergence. It is clear that quite generally changing L by a divergence must change $\mathfrak{L}$ by the covariant divergence of a quantity which is a vector density under coordinate transformations, and invariant under all other transformations. This is the reason for the difference between this case and that of the linear transformations of Sec. 2.

We now wish to investigate the possibility of choosing a criterion which will select a particular form of L, and thus specify $\mathfrak{L}$ completely. There does not seem to be any really compelling reason for one choice rather than another, but there are plausible arguments for a particular choice.

The most obvious criterion would be to require that the Lagrangian should be written in the symmetrized first-order form suggested by Schwinger,[26] which in the case of the scalar field discussed in Sec. 4 is

$$L=\tfrac{1}{2}(L_1+L_2).$$

This corresponds to treating φ and π^k on a symmetrical footing. However, this may not in fact be the correct choice, because for some purposes φ and π^k should not be treated in this way. In fact, the two Lagrangians differ in one important respect: $\mathfrak{L}_1$ is independent of $A^{ij}{}_{\mu}$, whereas $\mathfrak{L}_2$ is not. Correspondingly, for L_1 the quantity $S^k{}_{ij}$ vanishes, whereas for L_2 one finds

$$S^k{}_{ij}=(\delta^k{}_i\pi_j-\delta^k{}_j\pi_i)\varphi.$$

The conservation laws in the two cases are of course the same, because the quantities $T^k{}_i$ also differ. Now the tensor $S^k{}_{ij}$ has often been interpreted as the spin density,[18] so that the two cases differ with regard to the separation of the total angular momentum into orbital and spin terms. The scalar field is normally regarded as a field of spinless particles, so that one would naturally expect $S^k{}_{ij}$ to vanish. This, therefore, furnishes a possible criterion, which would select L_1 rather than L_2. With this choice, a preferred position is assigned to the "wave function" φ rather than the "momenta" π^k, and the derivatives are written on φ only. In this way one achieves a vanishing spin tensor, because the matrices S_{ij} are zero for the scalar field φ, but not for the vector π^k. It may be noticed that L_1 is automatically selected if one writes the Lagrangian in its second-order form in terms of φ only:

$$L_1'=\tfrac{1}{2}\varphi_{,k}\varphi^{,k}-\tfrac{1}{2}m^2\varphi^2,$$

which yields the modified Lagrangian

$$\mathfrak{L}_1'=\tfrac{1}{2}\mathfrak{H}(g^{\mu\nu}\varphi_{,\mu}\varphi_{,\nu}-m^2\varphi^2),$$

equivalent to $\mathfrak{L}_1$.[27] This should be contrasted with the second-order form of $\mathfrak{L}_2$, which is

$$\mathfrak{L}_2'=\tfrac{1}{2}\mathfrak{H}^{-1}(\mathfrak{H}h_i{}^{\mu}\varphi)_{;\mu}(\mathfrak{H}h^{i\nu}\varphi)_{;\nu}-\tfrac{1}{2}\mathfrak{H}m^2\varphi^2,$$

and clearly differs from $\mathfrak{L}_1'$ by a covariant divergence.

This seems to be a resonable criterion, but the arguments for it cannot be regarded as conclusive. For, although it is true that the spin tensor obtained from L_2 is nonzero, it is still true that the three space-space components of the total spin

$$\mathrm{S}_{ij}=\int d_3x\, S^0{}_{ij}$$

are zero. Thus L_1 and L_2 differ only in the values of the

[26] J. Schwinger, Phys. Rev. 91, 713 (1953).

[27] Here $\mathfrak{L}_1$ is a "linearization" of $\mathfrak{L}_1'$ in the sense of T. W. B. Kibble and J. C. Polkinghorne, Nuovo cimento 8, 74 (1958).

spin part of the $(0i)$ components of angular momentum. Indeed, one easily sees that it is true in general that adding a divergence to L will change only the $(0i)$ components of S_{ij}. Since it is not at all clear what significance should be attached to the separation of these components into "orbital" and "spin" terms, it might be questioned whether one should expect the spin terms to vanish even for a spinless particle. Even so, the choice of L_1 seems in this case to be the most reasonable.

For a field of spin 1, the corresponding choice would be

$$L_1 = -\tfrac{1}{2} f^{ij}(a_{i,j} - a_{j,i}) + \tfrac{1}{4} f^{ij} f_{ij} + \tfrac{1}{2} m^2 a_i a^i,$$

which is again equivalent to the choice of the second-order Lagrangian in terms of a_i only. It yields

$$S^k{}_{ij} = a_i f_j{}^k - a_j f_i{}^k,$$

which is a reasonable definition of the spin density.[28] The modified Lagrangian may be expressed in terms of the world vector a_μ as

$$\mathfrak{L} = -\tfrac{1}{4}\mathfrak{H} g^{\mu\rho} g^{\nu\sigma}(a_{\mu;\nu} - a_{\nu;\mu})(a_{\rho;\sigma} - a_{\sigma;\rho}) + \tfrac{1}{2}\mathfrak{H} m^2 g^{\mu\nu} a_\mu a_\nu. \quad \text{(A.1)}$$

It should be noticed that the electromagnetic Lagrangian is not obtained simply by putting $m=0$ in (A.1). The difference is that the derivatives in (A.1) are covariant derivatives, and since $\Gamma^\lambda{}_{\mu\nu}$ is nonsymmetric the covariant curl is not equal to the ordinary curl (though both are of course tensors). In fact, (A.1) with $m=0$ would not be gauge invariant. The reason for the difference is that a_i is here treated simply as a component of χ, whereas A_μ is introduced along with the gravitational variables to ensure gauge invariance.[29]

For a spinor field ψ, symmetry between ψ and $\bar\psi$ appears to demand that one should choose the symmetrized Lagrangian

$$L = \tfrac{1}{2}(\bar\psi i\gamma^k \psi_{,k} - \bar\psi_{,k} i\gamma^k \psi) - m\bar\psi\psi,$$

which yields the spin density

$$S_{kij} = \tfrac{1}{2}\epsilon_{kijl}\bar\psi i\gamma^l\gamma_5\psi.$$

Since the Lagrangian $\mathfrak{L}$ must be Hermitian, one could not write the derivative on ψ alone. There remains, however, another possible choice: We could introduce a distinction between the left- and right-handed components, $\psi_\pm = \tfrac{1}{2}(1 \pm i\gamma_5)\psi$, treating one of them line φ and the other like π^k. This gives the Lagrangian

$$L = \tfrac{1}{2}\bar\psi i\gamma^k(1+i\gamma_5)\psi_{,k} - \tfrac{1}{2}\bar\psi_{,k} i\gamma^k(1-i\gamma_5)\psi - m\bar\psi\psi.$$

This form of Lagrangian may seem rather unnatural, but it should be mentioned because there are other grounds for treating ψ_+ and ψ_- on a nonsymmetrical footing.[30]

[28] Compare footnote 18.

[29] This has the rather strange consequence that for the electromagnetic field the "spin" tensor $S^k{}_{ij}$ vanishes, since the Lagrangian is independent of $A^i{}_\mu$.

[30] See R. P. Feynman and M. Gell-Mann, Phys. Rev. **109**, 193 (1958).

VOLUME 33, NUMBER 7 PHYSICAL REVIEW LETTERS 12 AUGUST 1974

Integral Formalism for Gauge Fields

C. N. Yang
Institute of Theoretical Physics, University of Wrocław, Wrocław, Poland, and
Institute of Theoretical Physics, State University of New York, Stony Brook, New York 11990
(Received 10 June 1974)

A new integral formalism for gauge fields is described. Further developments are presented, including gravitation equations related to, but not identical with, Einstein's equations.

It was pointed out by Weyl many years ago that the electromagnetic field can be formulated in terms of an Abelian gauge transformation. This idea was extended[1] in 1954 to the concept of gauge fields for non-Abelian groups. That formulation, like the Weyl formulation for electromagnetism, was based on the replacement of ∂_μ by $\partial_\mu - ieB_\mu$. One might call such formulations differential formulations. It is the purpose of the present paper to reformulate the concept of gauge fields in an *integral formalism*. The new formalism is conceptually superior to the differential formalism and allows for natural developments of additional concepts. It further allows a mathematical and physical discussion of the gravitational field *as a gauge field*, resulting in equations related, but not identical, to Einstein's.

The basic point is the fact that *electromagnetism is a nonintegrable phase factor*, a fact discussed many years ago by Dirac, Peierls, and others, and more recently by many authors.[2] This fact is now generalized as follows:

Definition of a gauge field.—Consider a manifold with points on it labeled by x^μ ($\mu = 1, 2, \ldots, n$) and consider a gauge G which is a Lie group with generators X_k ($k = 1, 2, \ldots, m$). [For $G = U(1)$ we have electromagnetism; for G non-Abelian we have non-Abelian gauge fields.] Define a path-dependent (i.e., nonintegrable) phase factor φ_{AB} as an element of the group G associated with path AB between two points A and B on the manifold. The association is to have the group property: $\varphi_{ABC} = \varphi_{AB}\varphi_{BC}$, where the paths AB and BC are segments of ABC. Furthermore for an infinitiesimal path A to $A + dx^\mu$ the phase factor is close to the identity I of G, so that[3]

$$\varphi_{A(A+dx)} = I + b_\mu{}^k(x) X_k dx^\mu . \tag{1}$$

The function $b_\mu{}^k(x)$ defined on the manifold will be called a *gauge potential;* φ_{AB} will be called a *gauge phase factor.*

With this definition additional concepts and theorems are naturally developed. We summarize some of these below. Details will be published elsewhere.

Gauge field strength.—Consider a path $ABCDA$ forming the border of an infinitesimal parallelogram with sides dx and dx'. φ_{ABCDA} can be computed by multiplying four phase factors like (1) together, resulting in

$$\varphi_{ABCDA} = I + f_{\mu\nu}{}^k X_k dx^\mu dx^{\nu\prime} , \tag{2}$$

where

$$f_{\mu\nu}{}^k = \frac{\partial b_\mu{}^k}{\partial x^\nu} - \frac{\partial b_\nu{}^k}{\partial x^\mu} - b_\mu{}^i b_\nu{}^j C_{ij}{}^k = -f_{\nu\mu}{}^k \tag{3}$$

in which $C_{ij}{}^k$ is the structure constant of G:

$$X_k X_j - X_j X_k = C_{kj}{}^i X_i . \tag{4}$$

$f_{\mu\nu}{}^k$ will be called a *gauge field,* or gauge field strength. They are the Faraday-Maxwell fields when $G = U(1)$.

Gauge transformation.—A gauge transformation in the integral formalism is defined by a transformation

$$\varphi_{AB} \to \varphi_{AB}' = \xi_A \varphi_{AB} \xi_B^{-1} , \tag{5}$$

where ξ_A is an element of G which depends on the point A. It is clear that under (5)

$$\varphi_{ABCDA} \to \varphi_{ABCDA}' = \xi_A \varphi_{ABCDA} \xi_A^{-1} . \tag{6}$$

Thus

$$f_{\mu\nu}{}^{k\prime} = \langle k | R_{\rm adj} | j \rangle f_{\mu\nu}{}^j , \tag{7}$$

where $R_{\rm adj}$ is the adjoint representation for the element ξ_A. The simple transformation property (7) is the definition for the concept that $f_{\mu\nu}{}^k$ is *gauge covariant.* Generalization to other representations R of G for a gauge-covariant quantity $\psi_{\alpha\beta\gamma}{}^K$ is immediate[3]:

$$\psi_{\alpha\beta\gamma}{}^{K\prime} = \langle K | R(\xi_A) | J \rangle \psi_{\alpha\beta\gamma}{}^J . \tag{8}$$

$b_\mu{}^k$ is not gauge covariant; $f_{\mu\nu}{}^k$ is.

Gauge-covariant differentiation.—To *retain*

gauge covariance in differentiation we define

$$\psi^K|_\mu = \frac{\partial\psi^K}{\partial x^\mu} + b_\mu{}^k\langle K|Z_k|J\rangle\psi^J, \tag{9}$$

where Z_k is the matrix representation of X_k. Generalization to other cases is obvious. An interesting theorem is that

$$f_{\mu\nu}|_\lambda{}^k + f_{\nu\lambda}|_\mu{}^k + f_{\lambda\mu}|_\nu{}^k = 0, \tag{10}$$

which is the gauge-Bianchi identity.

Introduction of a Riemannian metric.—So far we need no metric for the manifold. Now we introduce a metric for it and discuss arbitrary coordinate transformations. We come then naturally to *Riemannian covariant* quantities and *doubly covariant derivatives.* $b_\mu{}^k$ is Riemannian covariant, since φ_{AB} is coordinate-system independent. $f_{\mu\nu}{}^k$ is doubly covariant. We have

$$\begin{aligned}
\psi^K{}_{\|\mu} &= \psi^K|_\mu,\\
\psi^{K\nu}{}_{\|\mu} &= \psi^{K\nu}|_\mu + \left\{ {\nu \atop \mu\alpha} \right\}\psi^{K\alpha},\\
f_{\mu\nu\|\lambda}{}^k &= f_{\mu\nu}|_\lambda{}^k - \left\{ {\alpha \atop \mu\lambda} \right\} f_{\alpha\nu}{}^k - \left\{ {\alpha \atop \nu\lambda} \right\} f_{\mu\alpha}{}^k,
\end{aligned} \tag{11}$$

etc. It is easily shown that

$$f_{\mu\nu\|\lambda}{}^k + f_{\nu\lambda\|\mu}{}^k + f_{\lambda\mu\|\nu}{}^k = 0 \tag{12}$$

which is satisfied by *all* gauge fields on *all* Riemannian manifolds.

Source of gauge fields.— We *define,* in analogy with electromagnetism, a source four-vector $J_\mu{}^k$ for a gauge field:

$$J_\mu{}^k = g^{\nu\lambda} f_{\mu\nu\|\lambda}{}^k = f_{\mu\nu}{}^{k\|\nu}. \tag{13}$$

After some computation one derives a theorem:

$$g^{\mu\lambda}J_{\mu\|\lambda}{}^k = 0 \quad \text{(conserved current)}, \tag{14}$$

which in electromagnetism states charge conservation. In Ref. 1 this was Eq. (14). One can also generalize Eqs. (15) and (16) of Ref. 1, leading to the concept of "total charge."

Parallel-displacement gauge field.—For any Riemannian manifold, the important concept of parallel displacement defines, along any path AB, a *linear* relationship between any vector V_A at A and its parallel vector V_B at B. Thus parallel displacement is defined by an $n\times n$ matrix M_{AB} which gives this linear relationship. M_{AB} is a representation of an element of GL(n). Thus we have the following:

Theorem.—Parallel displacement defines a gauge field with G being GL(n). The index k has n^2 values and we write $k = (\alpha\beta)$. The gauge potential and gauge fields are respectively

$$b_\mu{}^{(\alpha\beta)} = \left\{ {\alpha \atop \beta\mu} \right\}, \quad f^{(\alpha\beta)}{}_{\mu\nu} = -R^\alpha{}_{\beta\mu\nu}. \tag{15}$$

It is important to recognize that in this definition we have chosen a fixed coordinate system. A coordinate transformation would generate a linear transformation in the vector spaces V_A and V_B. In other words $M_{AB} \to N_A M_{AB} N_B{}^{-1}$. Comparison with (5) shows thus that a coordinate transformation generates a simultaneous gauge transformation of the parallel-displacement gauge potential. In fact, the usual nonlinear term in the transformation of $\left\{ {\alpha \atop \alpha\gamma} \right\}$ is precisely the nonlinear term needed in the gauge transformation of the gauge noncovariant quantity $b_\mu{}^{(\alpha\beta)}$. In this connection we observe that for GL(n),

$$C^{(\alpha\beta)}_{(\lambda\mu)(\eta\zeta)} = \delta_{\mu\eta}\delta_{\alpha\lambda}\delta_{\beta\zeta} - \delta_{\lambda\zeta}\delta_{\alpha\eta}\delta_{\beta\mu}. \tag{16}$$

Thus by definitions (9) and (11)

$$\begin{aligned}
\psi^{(\alpha\beta)}{}_{\|\mu} &= \frac{\partial\psi^{(\alpha\beta)}}{\partial x^\mu} + b_\mu{}^{(\lambda\nu)} C^{(\alpha\beta)}_{(\lambda\nu)(\eta\zeta)}\psi^{(\eta\zeta)}\\
&= \psi^\alpha{}_{\beta;\mu},
\end{aligned} \tag{17}$$

where the semicolon represents the usual Riemannian covariant differentiation with α and β treated as usual contravariant and covariant indices. The rule works also in general. E.g.,

$$f^{(\alpha\beta)}{}_{\mu\nu\|\lambda} = -R^\alpha{}_{\beta\mu\nu;\lambda}. \tag{18}$$

Nontrivial sourceless gauge fields.—Gauge fields for which $f_{\mu\nu}{}^k \neq 0$ and $J_\mu{}^k = 0$ are of physical interest. So far only nonanalytic examples are known.[4]

We now can construct two general types of general types of examples.

(a) Consider the natural Riemannian geometry of a semisimple Lie group. Its parallel-displacement gauge field is sourceless and analytic.

(b) Consider the same Riemannian manifold of a group G as above in (a). Define φ_{AB} as that for an infinitesimal path AB, $\varphi_{AB} = (A^{-1}B)^{1/2}$. This gauge phase factor which is itself an element of G gives a gauge field which is analytic and sourceless.

Pure spaces.—A Riemannian manifold for which the parallel-displacement gauge field is sourceless will be called a pure space. A necessary and sufficient condition for a pure space is

$$R_{\mu\alpha;\beta} = R_{\mu\beta;\alpha} \tag{19}$$

A four-dimensional Einstein space, i.e., one for

VOLUME 33, NUMBER 7 PHYSICAL REVIEW LETTERS 12 AUGUST 1974

which $R_{\alpha\beta}=0$, is a pure space.

Gravitational field as a gauge field.—The electromagnetic field and the usual gauge fields are special cases of gauge fields, satisfying (12) and (13). A natural question is whether one should identify these *same equations* for the parallel-displacement gauge field as the equations for the gravitational field. There are advantages in this identification and we shall come back to this topic in a later communication. If one adopts this identification then gravitational equations are third-order differential equations[5] for $g_{\mu\nu}$. A pure gravitational field is then described by a pure space as defined above.

Variational principles.—Equation (13) with $J_\mu{}^k=0$ follows from a variational principle $\delta\int\sqrt{-g}\,d^nx=0$, where

$$L=f_{\mu\nu}{}^k f_{\alpha\beta}{}^j g^{\mu\alpha}g^{\nu\beta}C_{ka}{}^b C_{jb}{}^a. \qquad (20)$$

In the variation $g_{\mu\nu}$ is kept fixed and $b_\mu{}^k$ is varied, and $f_{\mu\nu}{}^k$ is given by (3); $C_{ka}{}^b$ are not varied. One could also find a variational principle which is satisfied by a pure space (19). Choose $C_{ka}{}^b$ to be the structure constants for GL(n), given by (16). Write the L of (20) as a functional of $b_\mu{}^{(\alpha\beta)}$ and $g^{\lambda\nu}$:

$$L=L(b_\mu{}^{(\alpha\beta)},g^{\lambda\nu}), \qquad (21)$$

which of course also contains derivatives of $b_\mu{}^k$ and $g^{\lambda\nu}$. Now form the variation

$$\delta\int\left[L(b_\mu{}^{(\alpha\beta)},g^{\lambda\nu})-L\left(\left\{\begin{matrix}\alpha\\ \beta\mu\end{matrix}\right\},g^{\lambda\nu}\right)\right]\times\sqrt{-g}\,d^nx=0, \qquad (22)$$

in which $b_\mu{}^{(\alpha\beta)}$ and $g^{\lambda\nu}$ are independently varied. The resultant equations are satisfied by (15) and (19).

It is a great pleasure to acknowledge the warm hospitality extended to me during my visit to the Institute of Theoretical Physics at Wrocław where this paper was written.

[1]C. N. Yang and R. L. Mills, Phys. Rev. 96, 191 (1954).

[2]S. Mandelstam, Ann. Phys. (New York) 19 1, 25 (1962); I. Białynicki-Birula, Bull. Acad. Pol. Sci., Ser. Sci. Math. Astron. Phys. 11, 135 (1963).

[3]We use the summation convention for repeated indices. Greek indices run from 1 to n. Lower case Latin indices run from 1 to m. m of course is also the dimension of the adjoint representation of G. Upper case Latin indices run from 1 to M, where M is the dimension of a representation of G.

[4]T. T. Wu and C. N. Yang, in *Properties of Matter under Unusual Conditions*, edited by H. Mark and S. Fernbach (Wiley, New York, 1969), p. 349.

[5]R. Utiyama, Phys. Rev. 101, 1957 (1956), had concluded that Einstein's equations are gauge-field equations. We believe that was an unnatural interpretation of gauge fields.

PHYSICAL REVIEW VOLUME 117, NUMBER 3 FEBRUARY 1, 1960

Quasi-Particles and Gauge Invariance in the Theory of Superconductivity*

YOICHIRO NAMBU
The Enrico Fermi Institute for Nuclear Studies and the Department of Physics, The University of Chicago, Chicago, Illinois
(Received July 23, 1959)

Ideas and techniques known in quantum electrodynamics have been applied to the Bardeen-Cooper-Schrieffer theory of superconductivity. In an approximation which corresponds to a generalization of the Hartree-Fock fields, one can write down an integral equation defining the self-energy of an electron in an electron gas with phonon and Coulomb interaction. The form of the equation implies the existence of a particular solution which does not follow from perturbation theory, and which leads to the energy gap equation and the quasi-particle picture analogous to Bogoliubov's.

The gauge invariance, to the first order in the external electromagnetic field, can be maintained in the quasi-particle picture by taking into account a certain class of corrections to the charge-current operator due to the phonon and Coulomb interaction. In fact, generalized forms of the Ward identity are obtained between certain vertex parts and the self-energy. The Meissner effect calculation is thus rendered strictly gauge invariant, but essentially keeping the BCS result unaltered for transverse fields.

It is shown also that the integral equation for vertex parts allows homogeneous solutions which describe collective excitations of quasi-particle pairs, and the nature and effects of such collective states are discussed.

1. INTRODUCTION

A NUMBER of papers have appeared on various aspects of the Bardeen-Cooper-Schrieffer[1] theory of superconductivity. On the whole, the BCS theory, which leads to the existence of an energy gap, presents us with a remarkably good understanding of the general features of superconducivity. A mathematical formulation based on the BCS theory has been developed in a very elegant way by Bogoliubov,[2] who introduced coherent mixtures of particles and holes to describe a superconductor. Such "quasi-particles" are not eigenstates of charge and particle number, and reveal a very bold departure, inherent in the BCS theory, from the conventional approach to many-fermion problems. This, however, creates at the same time certain theoretical difficulties which are matters of principle. Thus the derivation of the Meissner effect in the original BCS theory is not gauge-invariant, as is obvious from the viewpoint of the quasi-particle picture, and poses a serious problem as to the correctness of the results obtained in such a theory.

This question of gauge invariance has been taken up by many people.[3] In the Meissner effect one deals with a linear relation between the Fourier components of the external vector potential A and the induced current J, which is given by the expression

$$J_i(q)=\sum_{j=1}^{3} K_{ij}(q)A_j(q),$$

with

$$K_{ij}(q)=-\frac{e^2}{m}\langle 0|\rho|0\rangle\delta_{ij}+\sum_n\left(\frac{\langle 0|j_i(q)|n\rangle\langle n|j_j(-q)|0\rangle}{E_n}+\frac{\langle 0|j_j(-q)|n\rangle\langle n|j_i(q)|0\rangle}{E_n}\right). \quad (1.1)$$

ρ and j are the charge-current density, and $|0\rangle$ refers to the superconductive ground state. In the BCS model, the second term vanishes in the limit $q\rightarrow 0$, leaving the first term alone to give a nongauge invariant result. It has been pointed out, however, that there is a significant difference between the transversal and longitudinal current operators in their matrix elements. Namely, there exist collective excited states of quasi-particle pairs, as was first derived by Bogoliubov,[2] which can be excited only by the longitudinal current.

As a result, the second term does not vanish for a longitudinal current, but cancels the first term (the longitudinal sum rule) to produce no physical effect; whereas for a transversal field, the original result will remain essentially correct.

If such collective states are essential to the gauge-invariant character of the theory, then one might argue that the former is a necessary consequence of the latter. But this point has not been clear so far.

Another way to understand the BCS theory and its problems is to recognize it as a generalized Hartree-Fock approximation.[4] We will develop this point a little further here since it is the starting point of what follows later as the main part of the paper.

* This work was supported by the U. S. Atomic Energy Commission.

[1] Bardeen, Cooper, and Schrieffer, Phys. Rev. **106**, 162 (1957); **108**, 1175 (1957).

[2] N. N. Bogoliubov, J. Exptl. Theoret. Phys. U.S.S.R. **34**, 58, 73 (1958) [translation: Soviet Phys. **34**, 41, 51 (1958)]; Bogoliubov, Tolmachev, and Shirkov, *A New Method in the Theory of Superconductivity* (Academy of Sciences of U.S.S.R., Moscow, 1958). See also J. G. Valatin, Nuovo cimento **7**, 843 (1958).

[3] M. J. Buckingam, Nuovo cimento **5**, 1763 (1957). J. Bardeen, Nuovo cimento **5**, 1765 (1957). M. R. Schafroth, Phys. Rev. **111**, 72 (1958). P. W. Anderson, Phys. Rev. **110**, 827 (1958); **112**, 1900 (1958). G. Rickayzen, Phys. Rev. **111**, 817 (1958); Phys. Rev. Letters **2**, 91 (1959). D. Pines and R. Schrieffer, Nuovo cimento **10**, 496 (1958); Phys. Rev. Letters **2**, 407 (1958). G. Wentzel, Phys. Rev. **111**, 1488 (1958); Phys. Rev. Letters **2**, 33 (1959). J. M. Blatt and T. Matsubara, Progr. Theoret. Phys. (Kyoto) **20**, 781 (1958). Blatt, Matsubara, and May, Progr. Theoret. Phys. (Kyoto) **21**, 745 (1959). K. Yosida, *ibid.* **731**.

[4] Recently N. N. Bogoliubov, Uspekhi Fiz. Nauk **67**, 549 (1959) [translation: Soviet Phys.—Uspekhi **67**, 236 (1959)], has also reformulated his theory as a Hartree-Fock approximation, and discussed the gauge invariance collective excitations from this viewpoint. The author is indebted to Prof. Bogoliubov for sending him a preprint.

Take the Hamiltonian in the second quantization form for electrons interacting through a potential V:

$$H=\int\sum_{i=1}^{2}\psi_i^+(x)K_i\psi_i(x)d^3x+\tfrac{1}{2}\int\int\sum_{i,k}\psi_i^+(x)\times\psi_k^+(y)V(x,y)\psi_k(y)\psi_i(x)d^3xd^3y$$

$$\equiv H_0+H_{\text{int}}. \tag{1.2}$$

K is the kinetic energy plus any external field. $i=1, 2$ refers to the two spin states (e.g., spin up and down along the z axis).

The Hartree-Fock method is equivalent to linearizing the interaction H_{int} by replacing bilinear products like $\psi_i^+(x)\psi_k(y)$ with their expectation values with respect to an approximate wave function which, in turn, is determined by the linearized Hamiltonian. We may consider also expectation values $\langle\psi_i(x)\psi_k(y)\rangle$ and $\langle\psi_i^+(x)\psi_k^+(y)\rangle$ although they would certainly be zero if the trial wave function were to represent an eigenstate of the number of particles, as is the case for the true wave function.

We write thus a linearized Hamiltonian

$$H_0'=\int\sum_i\psi_i^+K_i\psi_id^3x+\int\int\sum_{i,k}[\psi_i^+(x)\chi_{ik}(xy)\psi_k(y)+\psi_i^+(x)\phi_{ik}(xy)\psi_k^+(y)+\psi_k(x)\phi_{ki}^+(xy)\psi_i(y)]d^3xd^3y$$

$$\equiv H_0+H_s, \tag{1.3}$$

where

$$\chi_{ik}(xy)=\delta_{ik}\delta^3(x-y)\int V(xz)\sum_j\langle\psi_j^+(z)\psi_j(z)\rangle d^3z-V(xy)\langle\psi_k^+(y)\psi_i(x)\rangle, \tag{1.4}$$

$$\phi_{ik}(xy)=\tfrac{1}{2}V(xy)\langle\psi_k(y)\psi_i(x)\rangle,$$

$$\phi_{ik}^+(xy)=\tfrac{1}{2}V(xy)\langle\psi_k^+(y)\psi_i^+(x)\rangle.$$

We diagonalize H_0' and take, for example, the ground-state eigenfunction which will be a Slater-Fock product of individual particle eigenfunctions. The defining equations (1.4) then represent just generalized forms of Hartree-Fock equations to be solved for the self-consistent fields χ and ϕ.

The justification of such a procedure may be given by writing the original Hamiltonian as

$$H=(H_0+H_s)+(H_{\text{int}}-H_s)\equiv H_0'+H_{\text{int}}',$$

and demanding that H_{int}' shall have no matrix elements which would cause single-particle transitions; i.e., no matrix elements which would effectively modify the starting H_0': to put it more precisely, we demand our approximate eigenstates to be such that

$$\langle n|H_{\text{int}}'|0\rangle=\langle n|H|0\rangle=0, \tag{1.5}$$

if in $|n\rangle$ more than one particle change their states from those in $|0\rangle$. This condition is contained in Eq. (1.4).[5] Since in many-body problems, as in relativistic field theory, we often take a picture in which particles and holes can be created and annihilated, the condition (1.5) should also be interpreted to include the case where $|n\rangle$ and $|0\rangle$ differ only by such pairs. The significance of the BCS theory lies in the recognition that with an essentially attractive interaction V, a non-vanishing ϕ is indeed a possible solution, and the corresponding ground state has a lower energy than the normal state. It is also separated from the excited states by an energy gap $\sim 2\phi$.

The condition (1.5) was first invoked by Bogoliubov[2] in order to determine the transformation from the ordinary electron to the quasi-particle representation. He derived this requirement from the observation that H_{int}' contains matrix elements which spontaneously create virtual pairs of particles with opposite momenta, and cause the breakdown of the perturbation theory as the energy denominators can become arbitrarily small. Equation (1.5), as applied to such pair creation processes, determines only the nondiagonal part (in quasi-particle energy) of H_s in the representation in which H_0+H_s is diagonal. The diagonal part of H_s is still arbitrary. We can fix it by requiring that

$$\langle 1'|H_{\text{int}}'|1\rangle=0, \tag{1.6}$$

namely, the vanishing of the diagonal part of H_{int} for the states where one more particle (or hole) having a Hamiltonian H_0' is added to the ground state. In this way we can interpret H_0' as describing single particles (or excitations) moving in the "vacuum," and the diagonal part of H_s represents the self-energy (or the Hartree potential) for such particles arising from its interaction with the vacuum.

The distinction between Eqs. (1.5) and (1.6) is not so clear when applied to normal states. On the one hand, particles and holes (negative energy particles) are not separated by an energy gap; on the other hand, there is little difference when one particle is added just above the ground state.

In the above formulation of the generalized Hartree fields, χ and ϕ will in general depend on the external field as well as the interaction between particles. There is a complication due to the fact that they are gauge dependent. This is because a phase transformation $\psi_i(x)\rightarrow e^{i\lambda(x)}\psi_i(x)$ applied on Eq. (1.3) will change χ and ϕ according to

$$\chi(xy)\rightarrow e^{-i\lambda(x)+i\lambda(y)}\chi(xy),$$
$$\phi(xy)\rightarrow e^{i\lambda(x)+i\lambda(y)}\phi(xy), \tag{1.6}$$
$$\phi^+(xy)\rightarrow e^{-i\lambda(x)-i\lambda(y)}\phi^+(xy).$$

[5] Equation (1.5) refers only to the transitions from occupied states to unoccupied states. Transitions between occupied states or unoccupied states are given by Eq. (1.6). These two together then are equivalent to Eq. (1.4). For the analysis of the Hartree approximation in terms of diagrams, see J. Goldstone, Proc.

It is especially serious for ϕ (and ϕ^+) since, even if $\phi(xy)=\delta^3(x-y)$ times a constant in some gauge, it is not so in other gauges. Therefore, unless we can show explicitly that physical quantities do not depend on the gauge, any calculation based on a particular ϕ is open to question. It would not be enough to say that a longitudinal electromagnetic potential produces no effect because it can be transformed away before making the Hartree approximation. A natural way to reconcile the existence of ϕ, which we want to keep, with gauge invariance would be to find the dependence of ϕ on the external field explicitly. If the gauge invariance can be maintained, the dependence must be such that for a longitudinal potential $A=-\text{grad}\lambda$, it reduces to Eq. (1.6). This should not be done in an arbitrary manner, but by studying the actual influence of H_{int} on the primary electromagnetic interaction when ϕ is first determined without the external field.

After these preliminaries, we are going to study the points raised here by means of the techniques developed in quantum electrodynamics. We will first develop the Feynman-Dyson formulation adapted to our problem, and write down an integral eauation for the self-energy part which corresponds to the Hartree approximation. It is observed that it can possess a nonperturbational solution, and the existence of an energy gap is immediately recognized.

Next we will introduce external fields. Guided by the well-known theorems about gauge invariance, we are led to consider the so-called vertex parts, which include the "radiative corrections" to the primary charge-current operator. When an integral equation for the general vertex part is written down, certain exact solutions are obtained in terms of the assumed self-energy part, leading to analogs of the Ward identity.[6] They are intimately related to inherent invariance properties of the theory. Among other things, the gauge invariance is thus strictly established insofar as effects linear in the external field are concerned, including the Meissner effect.

Later we look into the collective excitations. A very interesting result emerges when we observe that one of the exact solutions to the vertex part equations becomes a homogeneous solution if the external energy-momentum is zero, and expresses a bound state of a pair with zero energy-momentum. Then by perturbation, other bound states with nonzero energy-momentum are obtained, and their dispersion law determined. Thus the existence of the bound state is a logical consequence of the existence of the special self-energy ϕ and the gauge invariance, which are seemingly contradictory to each other.

When the Coulomb interaction is taken into account, the bound pair states are drastically modified, turning into the plasma modes due to the same mechanism as in the normal case. This situation will also be studied.

Roy. Soc. (London) **A239**, 267 (1957). Compare also T. Kinoshita and Y. Nambu, Phys. Rev. **94**, 598 (1953).

[6] J. C. Ward, Phys. Rev. **78**, 182 (1950).

2. FEYNMAN-DYSON FORMULATION

We start from the Lagrangian for the electron-phonon system, which is supposed to be uniform and isotropic.[7]

$$\mathcal{L}=\sum_p\sum_i[i\psi_i^+(p)\dot{\psi}_i(p)-\psi_i^+(p)\epsilon_p\psi_i(p)]$$
$$+\sum_k\tfrac{1}{2}[\dot{\varphi}(k)\dot{\varphi}(-k)-c^2\varphi(k)\varphi(-k)]$$
$$-g\frac{1}{\sqrt{\mathcal{V}}}\sum_{p,k}\psi_i^+(p+k)\psi_i(p)h(k)\varphi(k). \quad (2.1)$$

p is the phonon field, with the momentum k (energy $\omega_k=ck$) running up to a cutoff value $k_m(\omega_m)$; c is the phonon velocity. ϵ_p is the electron kinetic energy relative to the Fermi energy; $gh(k)$ represents the strength of coupling.[8] ($\mathcal{V}$ is the volume of the system.)

The Coulomb interaction between the electrons is not included for the moment in order to avoid complication. Later we will make remarks whenever necessary about the modifications when the Coulomb interaction is taken into account.

It will turn out to be convenient to introduce a two-component notation[9] for the electrons

$$\Psi(x)=\begin{pmatrix}\psi_1(x)\\ \psi_2^+(x)\end{pmatrix} \quad\text{or}\quad \Psi(p)=\begin{pmatrix}\psi_1(p)\\ \psi_2^+(-p)\end{pmatrix}, \quad (2.2)$$

and the corresponding 2×2 Pauli matrices

$$\tau_1=\begin{pmatrix}0&1\\1&0\end{pmatrix},\quad \tau_2=\begin{pmatrix}0&-i\\i&0\end{pmatrix},\quad \tau_3=\begin{pmatrix}1&0\\0&-1\end{pmatrix}. \quad (2.3)$$

The Lagrangian then becomes:

$$\mathcal{L}=\sum_p\Psi^+(p)\left(i\frac{\partial}{\partial t}-\epsilon_p\tau_3\right)\Psi(p)$$
$$+\sum_k\tfrac{1}{2}[\dot{\varphi}(k)\dot{\varphi}(-k)-c^2\varphi(k)\varphi(-k)]$$
$$-g\frac{1}{\sqrt{\nabla}}\sum_{p,k}\Psi^+(p+k)\tau_3\Psi(p)h(k)\varphi(k)+\sum_p\epsilon_p$$
$$=\mathcal{L}_0+\mathcal{L}_{\text{int}}+\text{const}.$$

The last infinite c-number term comes from the rearrangement of the kinetic energy term. This is certainly uncomfortable, but will not be important except for the calculation of the total energy.

The fields obey the standard commutation relations.

[7] We use the units $\hbar=1$.

[8] For convenience, we have included in $h(k)$ the frequency factor: $h(k)=h_1(k)k_0$.

[9] P. W. Anderson [Phys. Rev. **112**, 1900 (1958)], has also introduced this two-component wave function.

Especially for Ψ, we have

$$\{\Psi_i(x),\Psi_j^+(y)\}\equiv\Psi_i(x)\Psi_j^+(y)+\Psi_j^+(y)\Psi_i(x) =\delta_{ij}\delta^3(x-y), \tag{2.5}$$
$$\{\Psi_i(p),\Psi_j^+(p')\}=\delta_{ij}\delta_{pp'}.$$

We may now formally treat $H_{\rm int}$ as perturbation, using the formulation of Feynman and Dyson.[10] The unperturbed ground state (vacuum) is then the state where all individual electron states $\epsilon_p<0(>0)$ are occupied (unoccupied) in the representation where $\psi_i^+(p)\psi_i(p)$ is the occupation number.

Having defined the vacuum, the time-ordered Green's functions for free electrons and phonons

$$\langle T(\Psi_i(xt),\Psi_j^+(x't'))\rangle=[G_0(x-x', t-t')]_{ij},$$
$$\langle T(\varphi(xt),\varphi(x't'))\rangle=\Delta_0(x-x', t-t') \tag{2.6}$$

are easily determined. We get for their Fourier representation (in the limit $\mathcal{V}\to\infty$)[10a]

$$G_0(xt)=(1/(2\pi)^4)\int G_0(pp_0)e^{i\mathbf{p}\cdot\mathbf{x}-ip_0t}d^3pdp_0,$$

$$\Delta_0(xt)=\frac{1}{(2\pi)^4}\int_{|k|<k_m}\Delta_0(kk_0)e^{i\mathbf{k}\cdot\mathbf{x}-ik_0t}d^3kdk_0,$$

$$G_0(pp_0)=i\left[P\frac{1}{p_0-\epsilon_p\tau_3}-i\pi\,\mathrm{sgn}(\tau_3\epsilon_p)\delta(p_0-\tau_3\epsilon_p)\right] \tag{2.7}$$
$$=i(p_0+\epsilon_p\tau_3)/(p_0^2-\epsilon_p^2+i\epsilon),$$
$$\Delta_0(kk_0)=i\left[P\frac{1}{k_0^2-c^2k^2}-i\pi\delta(k_0^2-c^2k^2)\right]$$
$$=i/(k_0^2-c^2k^2+i\epsilon).$$

With the aid of these Green's functions, we are able to calculate the S matrix and other quantities according to a well-defined set of rules in perturbation theory.

We will analyze in particular the self-energies of the electron and the phonon. In the many-particle system, these energies express (apart from the self-interaction of the electron) the average interaction of a single particle or phonon placed in the medium. Because the phonon spectrum is limited, there will be no ultraviolet divergences, unlike the case of quantum electrodynamics.

These self-energies may be obtained in a perturbation expansion with respect to $H_{\rm int}$. We are, however, interested in the Hartree method which proposes to take account of them in an approximate but nonperturbational way. It is true that the self-energies are in general complex due to the instability of single particles. But to the extent that the single-particle picture makes physical sense, we will ignore the small imaginary part of the self-energies in the following considerations.

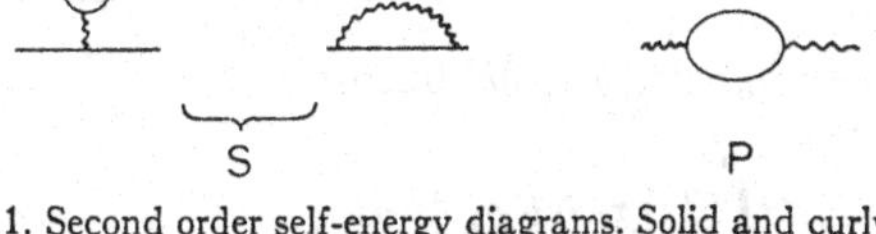

FIG. 1. Second order self-energy diagrams. Solid and curly lines represent electron and phonons, respectively, themselves being under the influence of the self-energies Σ and Π. All diagrams are to be interpreted in the sense of Feynman, lumping together all topologically equivalent processes.

Let us thus introduce the approximate self-energy Lagrangian $\mathcal{L}_s$, and write

$$\mathcal{L}=(\mathcal{L}_0+\mathcal{L}_s)+(\mathcal{L}_{\rm int}-\mathcal{L}_s)$$
$$\equiv\mathcal{L}_0'+\mathcal{L}_{\rm int}',$$
$$\mathcal{L}_0=\sum_p\Psi_p^+L_0\Psi_p+\sum_k\tfrac{1}{2}\varphi_kM_0\varphi_{-k}, \tag{2.8}$$
$$\mathcal{L}_s=-\sum_p\Psi_p^+\Sigma\Psi_p-\sum_k\tfrac{1}{2}\varphi_k\Pi\varphi_{-k},$$
$$L_0-\Sigma\equiv L,\quad M_0-\Pi\equiv M.$$

The free electrons with "spin" functions u and phonons obey the dispersion law

$$L_0(\mathbf{p}, p_0=\epsilon_p)u_p=0,\quad M_0(\mathbf{k}, k_0=\omega_k)=0, \tag{2.9}$$

whereas they obey in the medium

$$L(\mathbf{p}, p_0=E_p)u_p=0,\quad M(\mathbf{k}, k_0=\Omega_k)=0. \tag{2.9'}$$

Σ will be a function of momentum p and "spin." Π will consist of two parts: $\Pi(k_0k)=\Pi_1(k)k_0^2+\Pi_2(k)$ in conformity with the second order character (in time) of the phonon wave equation.[11]

The propagators corresponding to these modified electrons and phonons are

$$G(pp_0)=i/(L(pp_0)+i\,\mathrm{sgn}(p_0)\epsilon),$$
$$\Delta(kk_0)=i/(M(kk_0)+i\epsilon). \tag{2.10}$$

We now determine Σ and Π self-consistently to the second order in the coupling g. Namely the second order self-energies coming from the phonon-electron interaction have to be cancelled by the first order effect of $\mathcal{L}_{\rm int}$.

These second order self-energies are represented by the nonlocal operators[12] (Fig. 1)

$$\mathcal{S}((t+t')/2)=\int\int\int\Psi^+(xt)S(x-x', t-t')\times\Psi(x't')d^3xd^3x'd(t-t'),$$
$$\mathcal{P}((t+t')/2)=\tfrac{1}{2}\int\int\int\varphi(xt)P(x-x', t-t')\times\varphi(x't')d^3xd^3x'd(t-t'), \tag{2.11}$$

[10] F. J. Dyson, Phys. Rev. **75**, 486, 1736 (1949); R. P. Feynman, Phys. Rev. **76**, 769 (1949); J. Schwinger, Phys. Rev. **74**, 1439 (1948). Although we followed here the perturbation theory of Dyson, there is no doubt that the relations obtained in this paper can be derived by a nonperturbational formulation such as J. Schwinger's: Proc. Natl. Acad. Sci. U. S. **37**, 452, 455 (1951).

[10a] P stands for the principal value; $i\epsilon$ in the denominator is a small positive imaginary quantity.

[11] In the same spirit Σ should actually be in the form $\Sigma_1(p)p_0+\Sigma_2(p)$. Here we neglect the renormalization term Σ_1 since the two conditions (2.13) can be met without it.

[12] We use the word nonlocal here for nonlocality in time.

where S and P have the Fourier representation

$$S(pp_0)=-ig^2\tau_3\delta^3(p)\delta(p_0)h^2(0)\Delta(0)$$
$$\times\int \mathrm{Tr}[\tau_3 G(p'p_0)]d^3p dp_0$$
$$-ig^2\int \tau_3 G(p-k,\, p_0-k_0)\tau_3 h^2(kk_0)$$
$$\times\Delta(kk_0)d^3k dk_0, \quad (2.12)$$
$$P(kk_0)=ig^2 h(kk_0)^2\int \mathrm{Tr}[\tau_3 G(pp_0)$$
$$\times G(p+k,\, p_0+k_0)]d^3p dp_0.$$

In Eq. (2.11) we have chosen more or less arbitrarily $(t+t')/2$ as the fixed time to which we refer the nonlocal operators $\mathcal{S}$ and $\mathcal{P}$. The self-consistency requirements (1.5) and (1.6) mean in the present case that Σ, Π must be identical with S, P (a): for the diagonal elements [on the energy shell, Eq. (2.9)], and (b): for the non-diagonal matrix elements for creating a pair out of the vacuum.

The pair creation of electrons is possible because Ψ, being a two-component wave function, can have in general two eigenfunctions u_{ps} ($s=1, 2$) with different energies E_{ps} for a fixed momentum, p, only one of which is occupied in the ground state.

Thus taking particular plane waves $u_{ps}{}^*e^{-ip\cdot x+ip_0t}$, $u_{p's'}e^{ip'\cdot x'-ip_0't'}$ for Ψ^+ and Ψ in (2.11), we easily find that the diagonal matrix element of Σ corresponds to $u_{ps}{}^*S(p,E_{ps})u_{ps}$, while the nondiagonal part corresponds to $u_{ps}{}^*S(p,0)u_{ps}'$, $s\neq s'$ $(p_0'=-p_0)$.

A similar situation holds also for the photon self-energy Π. Since Π consists of two parts, the diagonal and off-diagonal conditions will fix these.

With this understanding, the self-consistency relations may be written

$$\begin{aligned}\Sigma(\mathbf{p}E_p)_{\mathrm{D}}=S(\mathbf{p},E_p)_{\mathrm{D}},\quad \Sigma(\mathbf{p}0)_{\mathrm{ND}}=S(\mathbf{p}0)_{\mathrm{ND}},\\ \Pi(\mathbf{k}\Omega_k)=P(\mathbf{k}\Omega_k),\quad \Pi(\mathbf{k}0)=P(\mathbf{k}0),\end{aligned}\quad (2.13)$$

where D, ND signify the diagonal and nondiagonal parts in the "spin" space. As stated before, we have agreed to omit possible imaginary parts in S and P. (The nondiagonal components, however, will turn out to be real.)

Before discussing the general solutions, let us consider the meaning of Eq. (2.13) in terms of perturbation theory. Suppose we expand G occurring in Eq. (2.12), with respect to Σ:

$$G=G_0-iG_0\Sigma G_0-G_0\Sigma G_0\Sigma G_0+\cdots,$$

and expand Σ itself with respect to g^2, then we easily realize that Eq. (2.13) defines an infinite sum of a particular class of diagrams, which are illustrated in Fig. 2. The first term in S of Eq. (2.12) corresponds to the ordinary Hartree potential which is just a constant,

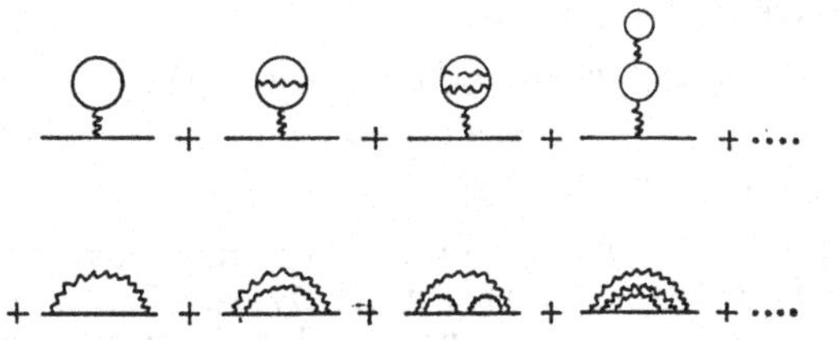

FIG. 2. Expansion of the self-consistent self-energy $\Sigma\sim S$ in terms of bare electron diagrams.

whereas the second term gives an exchange effect. In the latter, the approximation is characterized by the fact that no phonon lines cross each other.

It must be said that the Hartree approximation does not really sum the series of Fig. 2 completely since we equate in Eq. (2.13) only special matrix elements of both sides. For in the perturbation series the Σ obtained to any order is a function of p_0, whereas in Eq. (2.13) it is replaced by a p_0-independent quantity. Hence there will be a correction left out in each order (analogous to the radiative correction after mass renormalization in quantum electrodynamics).

In this perturbation expansion, S in Eq. (2.13) is always proportional to τ_3 on the energy shell since $H_0\propto\tau_3$. Accordingly Σ will be $\propto\tau_3$ and commute with H_0, so that no off-diagonal part exists.[11]

It is important, however, to note the possibility of a nonperturbational solution by assuming that Σ contains also a term proportional to τ_1 or τ_2. Thus, take

$$\begin{aligned}\Sigma(p)&=\chi(p)\tau_3+\phi(p)\tau_1,\\ H_0'&=(\epsilon+\chi)\tau_3+\phi\tau_1\\ &\equiv\bar{\epsilon}\tau_3+\phi\tau_1.\end{aligned}\quad (2.14)$$

This form bears a resemblance to the Dirac equation. Its eigenvalues are

$$E=\pm E_p\equiv\pm(\bar{\epsilon}_p{}^2+\phi_p{}^2)^{\frac{1}{2}}. \quad (2.15)$$

Since H_0' describes by definition excited states, we have to adopt the hole picture and conclude that the ground state (vacuum) is the state where all negative energy "quasi-particles" ($E<0$) are occupied and no positive energy particles exist. If ϕ remains finite on the Fermi surface, the positive and negative states are separated by a gap $\sim2|\phi|$. The corresponding Green's function G now has the representation

$$G(pp_0)=i\frac{p_0+\bar{\epsilon}_p\tau_3+\phi_p\tau_1}{p_0{}^2-E_p{}^2+i\epsilon}. \quad (2.16)$$

In order to extract the diagonal and nondiagonal parts in spin space, we will use the trick

$$\begin{aligned}O_{\mathrm{D}}&=\tfrac{1}{2}\,\mathrm{Tr}\,(\Lambda O),\\ O_{\mathrm{ND}}&=-(i/2)\,\mathrm{Tr}\,(\Lambda O\tau_2),\\ \Lambda&=[E_p+H_0'(p)]/2E_p.\end{aligned}\quad (2.17)$$

Applying this to Eq. (2.13a) with Eqs. (2.12), (2.14), and (2.15), we finally obtain the following equations for χ and ϕ

$$\frac{\epsilon_p\chi_p+\phi_p^2}{E_p}=\frac{g^2\pi}{(2\pi)^4}P\int\left[\frac{E_p}{\Omega_k}+\frac{E_{p-k}+\Omega_k}{E_pE_{p-k}\Omega_k}\right.$$
$$\left.\times(\bar{\epsilon}_p\bar{\epsilon}_{p-k}-\phi_p\phi_{p-k})\right]\frac{h(k)^2}{E_p^2-(E_{p-k}-\Omega_k)^2}d^3k,$$
$$\epsilon_p\phi_p=\frac{g^2\pi}{(2\pi)^4}\int(\bar{\epsilon}_{p-k}\phi_p+\bar{\epsilon}_p\phi_{p-k})$$
$$\times\frac{h(k)^2d^3k}{E_{p-k}\Omega_k(E_{p-k}+\Omega_k)}. \quad (2.18)$$

The second equation, coming from the nondiagonal condition, has a trivial solution $\phi=0$. If a finite solution ϕ exists, it cannot follow from perturbation treatment since there is no inhomogeneous term to start with.

Equation (2.18) is equivalent to, but slightly different from, the corresponding conditions of Bogoliubov because of a slightly different definition of the nondiagonal part of the self-energy operator, which is actually due to an inherent ambiguity in approximating nonlocal operators by local ones. (This is the same kind of ambiguity as one encounters in the derivation of a potential from field theory. The difference between the local operator Σ and the nonlocal one S shows up in a situation like that in Fig. 3, and the compensation between Σ and S is not complete.) We may avoid this unpleasant situation, by extending the Hartree self-consistency conditions to all virtual matrix elements, but this would mean that ϕ (and χ) must be treated as nonlocal. We will discuss this situation in a separate section since such a generalization brings simplification in dealing with the problem of gauge invariance and collective excitations.

For the moment we consider the second equation of (2.18) and rewrite it

$$\phi_p=A_p\frac{g^2\pi}{(2\pi)^4}\int\frac{\phi_{p-k}}{E_{p-k}}\frac{h(k)d^3k}{\Omega_k(E_{p-k}+\Omega_k)},$$
$$A_p=\bar{\epsilon}_p\Big/\left(\epsilon_p-\frac{g^2\pi}{(2\pi)^4}\int\frac{\bar{\epsilon}_ph(k)^2d^3k}{E_{p-k}\Omega_k(E_{p-k}+\Omega_k)}\right). \quad (2.19)$$

This is essentially the energy gap equation of BCS if $g^2A_ph(k)^2/\Omega_k(E_{p-k}+\Omega_k)$ is identified with the effective interaction potential V, and if $\bar{\epsilon}_p\sim\epsilon_p(\chi_p\sim0)$. It has a solution

$$\phi\sim\Omega_m\exp(-1/VN),$$

if $VN\ll1$, N being the density of states: $N=dn/d\epsilon_p$ on the Fermi surface.

The phonon self-energy Π may be studied similarly from Eq. (2.13), which should determine the renormalization of the phonon field. It does not play an essential role in superconductivity, though it gives rise to an important correction when the Coulomb effect is taken into account. (See the following section.)

FIG. 3. An example of the situation where the cancellation of Σ_{ND} *versus* S_{ND} is not complete. The two self-energy parts overlap in time, and their centers of time t_1 and t_2 are such that $t_1>t_2$. If calculated according to the usual perturbation theory, this process will not be eliminated by the condition $\Sigma_{ND}=(S_1)_{ND}$.

From the nature of Eq. (2.12), it is clear that $\tau_1\phi$ can actually be pointed in any direction in the 1–2 plane of the τ space: $\tau_1\phi_1+\tau_2\phi_2$. It was thus sufficient to take $\phi_1\neq0$, $\phi_2=0$. Any other solution is obtained by a transformation

$$\Psi\rightarrow\exp(i\alpha\tau_3/2)\Psi,$$
$$(\phi,0)\rightarrow(\phi\cos\alpha,\phi\sin\alpha). \quad (2.20)$$

In view of the definition of Ψ, Eq. (2.20) is a gauge transformation with a constant phase. Thus the arbitrariness in the direction of ϕ is the 1–2 plane is a reflection of the gauge invariance.

For later use, we also mention here the particle-antiparticle conjugation C of the quasi-particle field Ψ. This is defined by

$$C:\ \Psi\rightarrow\Psi^C=C\Psi^+=\tau_2\Psi^+,$$

or

$$\begin{pmatrix}\psi_1{}^C\\ \psi_2{}^C\end{pmatrix}=\begin{pmatrix}-i\psi_2{}^+\\ i\psi_1{}^+\end{pmatrix}, \quad (2.21)$$

and changes quasi-particles of energy-momentum (p_0,p) into holes of energy-momentum $(-p_0,-p)$, or interchanges up-spin and down-spin electrons. Under C, the τ operators transform as

$$C:\ \tau_i\rightarrow C^{-1}\tau_iC=-\tau_i{}^T,\quad i=1,2,3 \quad (2.22)$$

where T means transposition.

As a consequence, we have also

$$C:\ L(p)\rightarrow L^C(-p)=-L(-p)^T. \quad (2.23)$$

Finally we make a remark about the Coulomb interaction. When this is taken into account, the phonon interaction factor $g^2h(k)^2\Delta(k,k_0)$ in Eq. (2.12a) has to be replaced by

$$[g^2h(k)^2\Delta(kk_0)+ie^2/k^2]/$$
$$\{1-i\Pi(kk_0)[\Delta(kk_0)+ie^2/g^2h(k)^2k^2]\}.$$

As is well known, the denominator represents the screening of the Coulomb interaction. Discussion about this point will be made later in connection with the plasma oscillations.

3. NONLOCAL (ENERGY-DEPENDENT) SELF-CONSISTENCY CONDITIONS

In the last section we remarked that the self-consistency conditions Eq. (2.13) may be extended to all virtual matrix elements, namely, not only on the energy shell (diagonal) and for the virtual pair creation out of the vacuum, but also for the self-energy effects which appear in intermediate states of any process.

This simply means that ϕ and π are now nonlocal; i.e., depend both on energy and momentum arbitrarily, and are to be completely equated with S and P, respectively,

$$\Sigma(pp_0)=S(pp_0), \quad \Pi(kk_0)=P(kk_0). \tag{3.1}$$

Actually, these self-energies can no more be incorporated in H_0' as the zeroth order Lagrangian since they contain infinite orders of time derivatives.[13] Nevertheless, Eq. (3.1) has a precise meaning in the bare particle perturbation theory. It defines the (proper) self-energy parts (in the sense of Dyson) as an infinite sum of the special class of diagrams illustrated in Fig. 2.

The earlier condition of Eq. (2.13) represented, as was noted there, only an approximation to this sum. In other words, Eqs. (2.13) and (3.1) are not exactly identical even on the energy shell.

The Hartree-Fock approximation based on Eq. (3.1) could be interpreted as a nonperturbation approximation to determine the "dressed" single particles (together with the "dressed vacuum") or the Green's function $\langle 0|T(\Psi(xt),\Psi^+(x't'))|0\rangle$ for the true interacting system. Such single particles will satisfy

$$L(p,p_0)u\cong 0, \quad M(k,k_0)\cong 0. \tag{3.2}$$

We use the approximate equality since a really stable single particle may not exist.

Let us assume that these determine the approximate renormalized dispersion law

$$p_0^2=E_r(p)^2, \quad k_0^2=\Omega_r(k)^2. \tag{3.3}$$

If we write for Σ

$$\Sigma(pp_0)=p_0\zeta(pp_0)+\chi(pp_0)\tau_3+\phi(pp_0)\tau_1, \tag{3.4}$$

where ζ, χ, ϕ are even functions of p_0, then

$$\begin{aligned} E_r^2(p)&=[\bar{\epsilon}(pp_0)^2+\phi(pp_0)^2]/[1-\zeta(pp_0)]^2|_{p_0^2=E_r(p)^2} \\ &\equiv E(pp_0)^2/Z(pp_0)^2|_{p_0^2=E_r(p)^2}. \end{aligned} \tag{3.5}$$

The Green's functions G and Δ will be given by

$$\begin{aligned} G(pp_0)&=i/L(pp_0) \\ &=i\int_0^\infty \frac{dx}{p_0^2-x+i\epsilon} \\ &\qquad\times \mathrm{Im}\frac{p_0Z(px)+\bar{\epsilon}(px)\tau_3+\phi(px)\tau_1}{x^2Z(px)^2-E(px)^2}, \end{aligned} \tag{3.6}$$

$$\begin{aligned} \Delta(kk_0)&=i/M(kk_0) \\ &=i\int_0^\infty \frac{dx}{k_0^2-x+i\epsilon}\,\mathrm{Im}\frac{1}{M(kx)}. \end{aligned}$$

This representation assumes that $G(p_0)[\Delta(k_0)]$ is analytic except for a branch cut on the real axis. The imaginary part in the integrand is expected to have a delta function or a sharp peak at $x=E_r^2(p)$ $[\Omega_r(k^2)]$. These properties are necessary in order that the vacuum is stable and the quasi-particles and phonons have a valid physical meaning as excitations.[14] In the following, we will generally consider this quasi-particle peak only, and write

$$G(pp_0)=i\frac{p_0Z(pp_0)+\bar{\epsilon}(pp_0)\tau_3+\phi(pp_0)\tau_1}{p_0Z(pp_0)^2-E(pp_0)^2+i\epsilon}, \quad \text{etc.}$$

The Hartree equations now take the form

$$\Sigma(pp_0)=-i\frac{g^2}{(2\pi)^4}\int \tau_3G(p-k,\,p_0-k_0)\tau_3h(kk_0)^2d^3kdk_0,$$

$$\begin{aligned} \Pi(kk_0)=i\frac{g^2}{(2\pi)^4}\int \mathrm{Tr}\,[\tau_3G(k-p,\,k_0-p_0) \\ \times\tau_3G(pp_0)]d^3pdp_0. \end{aligned} \tag{3.7}$$

This equation for Σ is much simpler than the previous one (2.18) since we may just equate the coefficients of 1, τ_3, τ_1 on both sides. In particular, we get the energy gap equation

$$\begin{aligned} \phi(pp_0)=-\frac{ig^2}{(2\pi)^4}\int \frac{\phi(p'p_0')}{p_0^2Z(p'p_0')^2-E(p'p_0')^2+i\epsilon} \\ \times h(\mathbf{p}-\mathbf{p}',\,p_0-p_0')^2\Delta(\mathbf{p}-\mathbf{p}',\,p_0-p_0')d^3p'dp_0', \end{aligned} \tag{3.8}$$

which is to be compared with Eq. (2.19).

Although the existence of a solution to Eq. (3.6) may be difficult to establish, the solution, if it exists, should not be much different from the older solution to Eq. (2.19). At any rate, our assumption about the analyiticity of G and Π is consistent with Eq. (3.6) or (3.7) which implies that Σ and Π are also analytic except for a cut on the real axis.

In later calculations we shall encounter various integrals which we may classify into three types regarding their sensitivity to the energy gap. First, a normal self-energy part, for example, represents the effect of the bulk of the surrounding electrons on a particular electron, and is insensitive to the change of the small fraction $\sim\phi/E_F$ of the electrons near the Fermi surface in a superconductor. Such a quantity is

[13] It would seem then that we lose the advantage of the generalization since we cannot find the Bogoliubov transformation. However, we could still start from the older solution (2.13) as the zeroth approximation to Eq. (3.1), and then calculate the correction; namely, the "radiative" correction to the Bogoliubov vacuum and the Bogoliubov quasi-particle. These corrections would take account of the single-particle transitions which remain after the Bogoliubov condition (2.13) is imposed.

[14] This is a representation of the Lehmann type [H. Lehmann, Nuovo cimento 11, 342 (1954)] which can be derived by defining the Green's functions in terms of Heisenberg operators. See also V. M. Galizkii and A. B. Migdal, J. Exptl. Theoret. Phys. U.S.S.R. 34, 139 (1958) [translation: Soviet Phys. JETP 7, 96 (1958)].

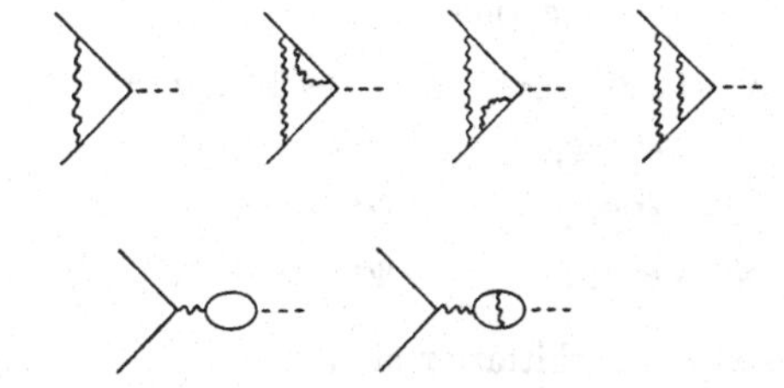

FIG. 4. Construction of the vertex part Γ in bare particle picture. The second line represents the polarization diagrams.

given by an integral like

$$g^2\int\frac{\epsilon_k}{E_k}f(\mathbf{p}-\mathbf{k})d^3k, \tag{3.9}$$

where the region $\epsilon_k \lesssim E_k=(\epsilon_k{}^2+\phi^2)^{\frac{1}{2}}$ makes little contribution if $f(p-k)$ is a smooth function.

Second, the energy gap itself is determined from an equation of the form

$$g^2\int\frac{d^3k}{E_k}f(\mathbf{p}-\mathbf{k})\sim g^2\int_{E_k\lesssim\omega_m}\frac{d^3k}{E_k}f(\mathbf{p}-\mathbf{k})\sim 1, \tag{3.10}$$

which means that even if g^2 is small, such an expression is always of the order 1.

Finally we meet with integrals like

$$g^2\int\frac{\epsilon_k\phi}{E_k{}^3}f(\mathbf{p}-\mathbf{k})d^3k,\quad g^2\int\frac{\phi^2}{E_k{}^3}f(\mathbf{p}-\mathbf{k})d^3k,\quad \text{etc.} \tag{3.11}$$

They have an extra cutoff factor $\sim 1/E$, $1/E^2$, etc., in the integrand which restricts the contribution to an energy interval $\sim 2\phi$ near the Fermi surface. The integrals are thus of the order

$$g^2N\phi/\omega_m,\quad g^2N,\quad \text{etc.}$$

In the following, we will not be primarily concerned with the ordinary self-energy effects. We will assume that proper renormalization has been carried out, or else simply disregard it unless essential. When we carry out perturbation type calculations, we will arrange things so that quantities of the second type are taken into account rigorously, and treat quantities of the third type as small, and hence negligible ($g^2N \ll 1$).

4. INTEGRAL EQUATIONS FOR VERTEX PARTS[15]

In the presence of an electromagnetic potential, the original Lagrangian $\mathfrak{L}$ has to be modified according to the rule

$$i\frac{\partial}{\partial t}\rightarrow i\frac{\partial}{\partial t}+eA_0,\quad \mathbf{p}\rightarrow\mathbf{p}-\frac{e}{c}\mathbf{A}$$

for the electron. Going to the two-component representation, this corresponds to the prescription

$$i\frac{\partial}{\partial t}\rightarrow i\frac{\partial}{\partial t}+e\tau_3A_0,\quad \mathbf{p}\rightarrow\mathbf{p}-\frac{e}{c}\tau_3\mathbf{A} \tag{4.1}$$

acting on Ψ. It can also be inferred from the gauge transformation $\Psi\rightarrow\exp(i\alpha\tau_3)\Psi$ as was observed previously. So the ordinary charge-current operator turns out to be in our form given by

$$\rho(x)=\frac{e}{2}([\Psi^+(x),\tau_3\Psi(x)]+\{\Psi^+(x),\Psi(x)\}),$$

$$\begin{aligned}\mathbf{j}(x)=\frac{-ie}{4m}([\Psi^+(x),\ &(\nabla-ie\tau_3\mathbf{A})\Psi(x)]\\ &+[(-\nabla-ie\tau_3A)\Psi^+(x),\ \Psi(x)]\\ &+\{\Psi^+(x),\ (\nabla\tau_3-ieA)\Psi(x)\}\\ &-\{(\tau_3\nabla-ie\mathbf{A})\Psi^+(x),\ \Psi(x)\}).\end{aligned} \tag{4.2}$$

The second terms on the right-hand side, being infinite C numbers, arise from the rearrangement of ψ and ψ^+, and will actually be compensated for by the first terms.

This expression, however, has to be modified when we go to the quasi-particle picture.

For we have seen that the self-energy ϕ of a quasi-particle is a gauge-dependent quantity. If we want to have the quasi-particle picture and gauge invariance at the same time, then it is clear that the electromagnetic current of a quasi-particle must contain, in addition to the normal terms given by Eq. (4.2), terms which would cause a physically unobservable transformation of ϕ if the electromagnetic potential is replaced by the gradient of a scalar. In other words, the complete charge current of a quasi-particle has to satisfy the continuity equation, which Eq. (4.2) does not, since

$$\partial\rho/\partial t+\nabla\cdot\mathbf{j}=2\Psi^+\phi\tau_2\Psi.$$

In order to find such a conserving expression for charge current, it is instructive to go back to the bare electron picture, in which the self-energy is represented by a particular class of diagrams discussed in the previous sections.

It is well known[16] in quantum electrodynamics that, in any process involving electromagnetic interaction, perturbation diagrams can be grouped into gauge-invariant subsets, such that the invariance is maintained by each subset taken as a whole. Such a subset can be constructed by letting each photon line in a diagram interact with a charge of all possible places along a chain of charge-carrying particle lines. The gauge-invariant interaction of a quasi-particle with an electromagnetic potential should then be obtained by attaching a photon line at all possible places in the diagrams of Fig. 2. The result is illustrated in Fig. 4,

---

[15] Hereafter we will often use the four-dimensional notation $x=(\mathbf{x},t)$, $p=(\mathbf{p},p_0)$, $d^4p=d^3pdp_0$.

[16] Z. Koba, Progr. Theoret. Phys. (Kyoto) 6, 322 (1951).

which consists of the "vertex" part Γ and the self-energy part Σ.

In this way we are led to consider the modification of the vertex due to the phonon interaction in the same approximation as the self-energy effect is included in the quasi-particle. It is not difficult to see that it corresponds to a "ladder approximation" for the vertex part, and we get an integral equation[17]

$$\Gamma_i(p',p)=\gamma_i(p',p)-g^2\int\tau_3G(p'-k)\Gamma_i(p'-k,\,p-k)\times G(p-k)\tau_3h(k)^2\Delta(k)d^4k,\quad(4.3)$$

where γ_i, $i=0, 1, 2, 3$ stand for the free particle charge current $[\tau_3,\,(1/2m)(\mathbf{p}+\mathbf{p}')]$ which follows from Eq. (4.2). Similar equations may be set up for any type of vertex interactions.

Equation (4.3) is the basis of the rest of this paper. It expresses a clear-cut approximation procedure in which the "free" charge-current operator γ_i of a quasi-particle is modified by a special class of "radiative corrections" due to $H_{\rm int}'$.

As the next important step, we observe that there exist exact solutions to Eq. (4.3) for the following four types of vertex interactions

$$\begin{aligned}
(a)\quad &\gamma^{(a)}(p',p)=L_0(p')-L_0(p)\\
&\qquad=(p_0'-p_0)-\tau_3(\epsilon_{p'}-\epsilon_p),\\
&\Gamma^{(a)}(p',p)=L(p')-L(p)\\
&\qquad=\gamma_a(p')-[\Sigma(p')-\Sigma(p)],\\
(b)\quad &\gamma^{(b)}(p',p)=L_0(p')\tau_3-\tau_3L_0(p)\\
&\qquad=(p_0'-p_0)\tau_3-(\epsilon_{p'}-\epsilon_p),\\
&\Gamma^{(b)}(p',p)=L(p')\tau_3-\tau_3L(p),\\
(c)\quad &\gamma^{(c)}(p',p)=L_0(p')\tau_1+\tau_1L_0(p'),\\
&\Gamma^{(c)}(p',p)=L(p')\tau_1+\tau_1L(p),\\
(d)\quad &\gamma^{(d)}(p',p)=L_0(p')\tau_2+\tau_2L(p),\\
&\Gamma^{(d)}(p',p)=L(p')\tau_2+\tau_2L(p).
\end{aligned}\qquad(4.4)$$

The verification is straightforward by noting that $G(p)=i/L(p)$, and making use of Eq. (3.7).

The fact that there are simple solutions is not accidental. These solutions express continuity equations and other relations following from the four types of operations, which do not depend on the presence or absence of the interaction:

$$\begin{aligned}
(a)\quad &\Psi(x)\to e^{i\alpha(x)}\Psi(x), && \Psi^+(x)\to\Psi^+(x)e^{-i\alpha(x)},\\
(b)\quad &\Psi\to e^{i\tau_3\alpha}\Psi, && \Psi^+\to\Psi^+e^{-i\tau_3\alpha},\\
(c)\quad &\Psi\to e^{\tau_1\alpha}\Psi, && \Psi^+\to\Psi^+e^{\tau_1\alpha},\\
(d)\quad &\Psi\to e^{\tau_2\alpha}\Psi, && \Psi^+\to\Psi^+e^{\tau_2\alpha},
\end{aligned}\qquad(4.5)$$

where $\alpha(x)$ is an arbitrary real function.

(a) and (b) correspond, respectively, to the spin rotation around the z axis, and the gauge transformation. The entire Lagrangian is invariant under them, so that we obtain continuity equations for the z component of spin and charge, respectively:

$$\begin{aligned}
&\frac{\partial}{\partial t}\Psi^+\Psi+\nabla\cdot\Psi^+\frac{\mathbf{p}}{m}\tau_3\Psi=0,\\
&\frac{\partial}{\partial t}\Psi^+\tau_3\Psi+\nabla\cdot\Psi^+\frac{\mathbf{p}}{m}\Psi=0,
\end{aligned}\qquad(4.6)$$

where Ψ is the true Heisenberg operator.

These equations are identical with

$$\Psi^+\gamma^{(a)}\Psi=0,\quad\Psi^+\gamma^{(b)}\Psi=0.\qquad(4.7)$$

Taken between two "dressed" quasi-particle states, the left-hand side of Eq. (4.7) will become

$$\begin{aligned}
e^{-i(p'-p)\cdot x}\langle p'|\Psi^+(x)\gamma^{(n)}\Psi(x)|p\rangle
&=u_{p'}^*\Gamma^{(n)}(p',p)u_p\\
&=0,\quad(n=a,b)
\end{aligned}\qquad(4.7')$$

where u_p, $u_{p'}$ are single-particle wave functions satisfying $L(p)u_p=u_{p'}^*L(p')=0$.

In this way we have shown the existence of spin and charge currents $\Gamma_i^{(a)}(p',p)$ and $\Gamma_i^{(b)}(p',p)$ for a quasi-particle, for which the continuity equations

$$(p_0'-p_0)\cdot\Gamma_0^{(n)}-\sum_{i=1}^{3}(p'-p)_i\cdot\Gamma_i^{(n)}=\Gamma^{(n)}(p',p)=0$$

will hold.

The last two transformations of Eq. (4.4) are not unitary, but mix ψ_1 and ψ_2^+ in such a way as to keep $\Psi^+\tau_3\Psi$ invariant. From infinitesimal transformations of these kinds we get

$$\begin{aligned}
&i\Psi^+\tau_1\left(\frac{\vec\partial}{\partial t}-\frac{\overleftarrow\partial}{\partial t}\right)\Psi+\nabla\cdot\Psi^+\tau_2\left(\frac{\vec{\mathbf{p}}}{m}+\frac{\overleftarrow{\mathbf{p}}}{m}\right)\Psi=0,\\
&-i\Psi^+\tau_2\left(\frac{\vec\partial}{\partial t}-\frac{\overleftarrow\partial}{\partial t}\right)\Psi+\nabla\cdot\Psi^+\tau_1\left(\frac{\vec{\mathbf{p}}}{m}+\frac{\overleftarrow{\mathbf{p}}}{m}\right)\Psi=0,
\end{aligned}\qquad(4.8)$$

which bear the same relations to $\gamma^{(c),(d)}$ and $\Gamma^{(c),(d)}$ as Eq. (4.6) did to $\gamma^{(a),(b)}$ and $\Gamma^{(a),(b)}$. Note that the above equations are unaffected by the presence of the phonon interaction.

The fact that we can find a conserved charge-current

---

[17] This equation may also be derived simply by considering the self-energy equation (3.7) in the presence of an external field, and expanding Σ in A. Σ should be now a function of initial and final momenta, and we define

$$\Sigma^{(A)}(p',p)\equiv\Sigma(p)\delta^4(p'-p)+\sum_{i=0}^{3}(\Gamma_i(p',p)-\gamma_i(p',p))\times A^i(p'-p)+O(A^2).$$

In the limit $p'-p=0$, $\Gamma_i-\gamma_i=\partial\Sigma/\partial A^i$, which is the content of the Ward identity.[6] Investigation of the higher order terms in A is beyond the scope of this paper.

FIG. 5. The diagram for the kernel $K^{(2)}$.

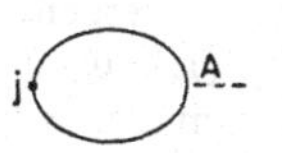

for a quasi-particle is rather surprising. A quasi-particle cannot be an eigenstate of charge since it is a linear combination of an electron and a hole, tending to an electron well above the Fermi surface, and to a hole well below. We must conclude than that an accelerated wave packet of quasi-particles, whose energy is confined to a finite region of space, continuously picks up charge from, or deposits it with, the surrounding medium which extends to infinity. This situation will be studied in Sec. 7, where we will derive the charge current operators Γ_i explicitly.

5. GAUGE INVARIANCE IN THE MEISSNER EFFECT

We will next discuss how the gauge invariance is maintained in the problem of the Meissner effect when the external magnetic field is static. We calculate the Fourier component of the current $J(q)$ induced in the superconducting ground state by an external vector potential $A(q)$:

$$J_i(q)=\sum_{j=1}^{3} K_{ij}(q)A_j(q), \tag{5.1}$$

where q is kept finite.

For free electrons, K is represented by

$$K_{ij}=K_{ij}^{(1)}+K_{ij}^{(2)},\quad K_{ij}^{(1)}=-\delta_{ij}ne^2/m, \tag{5.2}$$

where n is the number of electrons inside the Fermi sphere. $K^{(1)}$ comes from the expectation value of the current operator Eq. (4.2), whereas $K^{(2)}$ corresponds to the diagram in Fig. 5. [Compare also Eq. (1.1).] It is well known that in this case K_{ij} is of the form

$$K_{ij}(q)=(\delta_{ij}q^2-q_iq_j)K(q^2), \tag{5.3}$$

so that for a longitudinal vector potential $A_i(q)\sim q_i\lambda(q)$, we have

$$J_i(q)=K_{ij}q_j\lambda(q)=0, \tag{5.4}$$

establishing the unphysical nature of such a potential.

In the case of a superconducting state, the free electron lines in Fig. 5 will be replaced by quasi-particle lines. But then we have $K^{(2)}(q)\to 0$ as $q\to 0$ since the intermediate pair formation is suppressed due to the finite energy gap, whereas $K^{(1)}$ is essentially unaltered. Thus Eq. (5.2) takes the form of the London equation, except that even a longitudinal field creates a current.

According to our previous argument, this lack of gauge invariance should be remedied by taking account of the vertex corrections. Starting again from the free electron picture, and inserting the phonon interaction

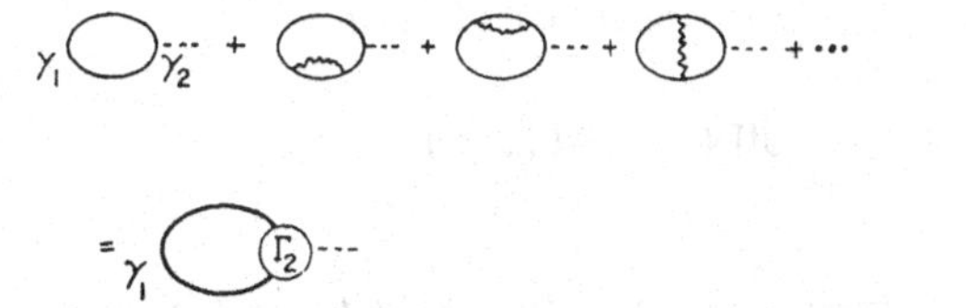

FIG. 6. Graphical derivation of Eq. (5.5). The thick lines represent quasi-particles.

effects, as indicated in Fig. 6, we arrive at the conclusion that either one of the vertices γ in Fig. 5 has to be replaced by the full Γ^{10}. In addition, there is the polarization correction represented by a string of bubbles. Let us, however, first neglect this correction. $K_{ij}^{(2)}$ is then

$$K_{ij}^{(2)}(q)=\frac{-ie^2}{(2\pi)^4}\int \mathrm{Tr}[\gamma_i(p-q/2,\,p+q/2)G(p+q/2)\times\Gamma_j(p+q/2,\,p-q/2)G(p-q/2)]d^4p. \tag{5.5}$$

Although we do not know $\Gamma_j(p,p)$ explicitly, we can establish Eq. (5.4) easily. For

$$-\sum_{j=1}^{3}\Gamma_j(p+q,\,q)q_j$$

is exactly the solution $\Gamma^{(b)}(p+q,\,p)$ of Eq. (4.4) where q_0 is equal to zero. Substituting this solution in Eq. (4.5) we find

$$\begin{aligned}K_{ij}^{(2)}(q)q_j
&=\frac{-1}{(2\pi)^4}\int \mathrm{Tr}\{\gamma_i(p-q/2,\,p+q/2)\times[\tau_3G(p+q/2)-G(p-q/2)\tau_3]\}d^4p\\
&=\frac{-1}{(2\pi)^4}\int \mathrm{Tr}\{[\gamma_i(p-q,\,p)-\gamma_i(p,\,p+q)]\times\tau_3G(p)\}d^4p\\
&=\frac{1}{(2\pi)^4}\frac{q_i}{m}\int \mathrm{Tr}[\tau_3G(p)]d^4p,\end{aligned} \tag{5.6}$$

where the properties of γ_i and G under particle conjugation and a translation in p space were utilized in going from the first to the second line.

On the other hand, the part $K^{(1)}$ is, according to Eq. (4.2) given by

$$\begin{aligned}K_{ij}^{(1)}&=-\delta_{ij}\frac{e^2}{2m}(\langle 0|[\Psi^+(x),\tau_3\Psi(x)]|0\rangle+\{\Psi^+(x),\Psi(x)\})\\
&=K_{ij}^{(1a)}+K_{ij}^{(1b)}.\end{aligned} \tag{5.7}$$

The first term becomes further

$$-\delta_{ij}\frac{e^2}{2m}\langle 0|[\Psi^+(x),\tau_3\Psi(x)]|0\rangle$$

$$=-\delta_{ij}\frac{e^2}{m}\tfrac{1}{2}\lim_{\epsilon\to 0}\sum_{\pm}\langle 0|T(\Psi^+(x,t\pm\epsilon)\tau_3\Psi(x,t))|0\rangle$$

$$=-\delta_{ij}\frac{e}{m}\mathrm{Tr}[\tau_3 G(xt=0)]$$

$$=-\delta_{ij}\frac{e}{m}\frac{1}{(2\pi)^4}\int\mathrm{Tr}[\tau_3 G(p)]d^4p. \quad (5.8)$$

Thus

$$[K_{ij}^{(1a)}(q)+K_{ij}^{(2)}(q)]q_j=0. \quad (5.9)$$

The second term $K^{(1b)}$ comes from the c-number term of the current operator (4.2), and is just the anticommutator of the electron field, which does not depend on the quasi-particle picture, nor on the presence of interaction. Therefore we may write for this contribution

$$K_{ij}^{(1b)}(q)A_j(q)=\frac{-ie}{2m}\frac{1}{(2\pi)^3}\int e^{-iqx}d^3x$$

$$\times\{\Psi^+(x),(\tau_3\nabla-ieA(x))_i\Psi(x)\} \quad (5.10)$$

to show its formal gauge invariance since $\tau_3\nabla-ieA(x)$ is certainly a gauge-invariant combination for free electron field.

As for the polarization correction, we can easily show in a similar way that it vanishes for the static case ($q_0=0$) because

$$\int\mathrm{Tr}\,\Gamma_i(p-q/2,p+q/2)G(p+q/2)$$

$$\times\gamma_0(p+q/2,p-q/2)G(p-q/2)d^4p=0.$$

Thus the above proof is complete and independent of the Coulomb interaction which profoundly influences the polarization effect. Although the proof is thus rigorous, it is still somewhat disturbing since $K^{(1a)}$, $K^{(1b)}$ and $K^{(2)}$ are all infinite. Actually there is a certain ambiguity in the evaluation of $K^{(2)}$, Eq. (5.6), which is again similar to the one encountered in quantum electrodynamics.[18] An alternative way would be to expand quantities in q without making translations in p space. In this case we may write

$$-\Gamma^{(b)}(p+q/2,p-q/2)=\bar{\epsilon}(p+q/2)-\bar{\epsilon}(p-q/2)$$

$$-i\tau_2[\phi(p+q/2)+\phi(p-q/2)]$$

$$\approx p\cdot q/m-2i\tau_2\phi. \quad (5.11)$$

The first term then gives

$$\frac{e^2}{(2\pi)^3}\int\frac{\phi^2}{4E_p^5}\left(\frac{\mathbf{p}\cdot\mathbf{q}}{m}\right)^2 p_i\mathbf{p}\cdot\mathbf{q}d^3p\propto q^2q_i, \quad (5.12)$$

which is convergent and the same as the one obtained from Eq. (1.1) using the bare quasi-particle states. The second term also is finite and equal to

$$\frac{e^2}{(2\pi)^3}\int\frac{\phi^2}{E_p^3}\frac{p^2}{3m^2}q_id^3p+O(q^2)q_i\approx N\alpha^2q_i=n(e^2/m)q_i. \quad (5.13)$$

The last line follows from Eqs. (6.11) and (6.11′) below.

The calculation of $K^{(1)}$ from Eqs. (5.7) and (5.10), gives, on the other hand, the same value as Eq. (5.2), so that we get $(K_{ij}^{(1)}+K_{ij}^{(2)})q_j=0$ in the limit of small q. (The polarization correction is again zero.)

Since Eq. (5.13) is a contribution from the collective intermediate state (see Secs. 6 and 7), we may say that the collective state saves gauge invariance, as has been claimed by several people.[3,19]

It goes without saying that the effect of the vertex correction on K_{ij} will be felt also for real magnetic field. But as we shall see later, it is a small correction of order g^2N (except for the renormalization effects), and not as drastic as for the longitudinal case.

6. THE COLLECTIVE EXCITATIONS

In order to understand the mechanism by which gauge invariance was restored in the calculation of the Meissner effect, and also to solve the integral equations for general vertex interactions, it is necessary to examine the collective excitations of the quasi-particles. In fact, people[3] have shown already that the essential difference between the transversal and longitudinal vector potentials in inducing a current is due to the fact that the latter can excite collective motions of quasi-particle pairs.

We see that the existence of such collective excitations follows naturally from our vertex solutions Eq. (4.4). For taking $p=p'$, the second solution $\Gamma^{(b)}(p',p)$ becomes

$$\Gamma^{(b)}(p,p)=L(p)\tau_3-\tau_3L(p)$$
$$=2i\tau_2\phi, \quad (6.1)$$
$$\gamma^{(b)}=0.$$

In other words $\tau_2\phi(p)\equiv\Phi_0(p)$ satisfies a homogeneous integral equation:

$$\Phi_0(p)=-\frac{g^2}{(2\pi)^4}\int\tau_3G(p')\Phi(p')G(p')$$

$$\times\tau_3h(p-p')^2\Delta(p-p')d^4p. \quad (6.2)$$

We interpret this as describing a pair of a particle and an antiparticle interacting with each other to form a bound state with zero energy and momentum $q=p'-p=0$.

[18] H. Fukuda and T. Kinoshita, Progr. Theoret. Phys. (Kyoto) 5, 1024 (1950).

[19] On the other hand, the way in which the collective mode accomplishes this end seems to differ from one paper to another. We will not attempt to analyze this situation here.

In fact, by defining

$$F(p,-p)\equiv -G(p)\Phi_0(p)G(p), \tag{6.3}$$

Eq. (6.2) becomes

$$L(p)F(p,-p)L(p)=-\frac{g^2}{(2\pi)^4}\int \tau_3F(p',-p')\times\tau_3h(p-p')^2\Delta(p-p')d^4p',$$

or

$$\sum_{j,l=1}^{2} L(p)_{ij}L^C(-p)_{kl}F(p,-p)_{jl}=\frac{-g^2}{(2\pi)^4}\int\sum_{j,l}(\tau_3)_{ij}(\tau_3)_{kl}F(p',-p')_{jl}\times h(p-p')^2\Delta(p-p')d^4p. \tag{6.4}$$

The particle-conjugate quantity L^C was defined in Eq. (2.23).

Equations (6.2) and (6.4) are the analog of the so-called Bethe-Salpeter equation[20] for the bound pair of quasi-particles with zero total energy-momentum. $F_{ij}(p,-p)$ is the four-dimensional wave function with the spin variables i, j and the relative energy-momentum $(p_0,\mathbf{p})$.

Since there, thus, exists a bound pair of zero momentum, there will also be pairs moving with finite momentum and kinetic energy. In other words, there will be a continuum of pair states with energies going up from zero. We have to determine their dispersion law.

For a finite total energy-momentum q, the homogeneous integral equation takes the form

$$\Phi_q(p)\equiv L(\tfrac{1}{2}q+p)F(\tfrac{1}{2}q+p,\tfrac{1}{2}q-p)L(p-\tfrac{1}{2}q)=-g^2\frac{1}{(2\pi)^4}\int\tau_3F(\tfrac{1}{2}q+p',\tfrac{1}{2}q-p')\times\tau_3h(p-p')^2\Delta(p-p')d^4p. \tag{6.5}$$

From here on we carry out perturbation calculation. Let us expand F and L in terms of the small change $L(p\pm q/2)-L(p)$, thus

$$\begin{aligned}F(\tfrac{1}{2}q+p,\tfrac{1}{2}q-p)&=F^{(0)}(p)+F^{(1)}(p,q/2)+\cdots,\\ L(p\pm q/2)&=L(p)+\Delta L(p,\pm q/2).\end{aligned} \tag{6.6}$$

Collecting terms of the first order, we get

$$\begin{aligned}L(p)F^{(1)}(p,q/2)L(p)+U^{(1)}(p,q/2)&=-g^2\frac{1}{(2\pi)^4}\int\tau_3F^{(1)}(p',q/2)\times\tau_3h(p-p')^2\Delta(p-p')d^4p',\\ U^{(1)}(p,q/2)&=\Delta L(p,q/2)F^{(0)}(p)L(p)+L(p)F^{(0)}(p)\Delta L(p,-q/2).\end{aligned} \tag{6.7}$$

[20] E. E. Salpeter and H. A. Bethe, Phys. Rev. 84, 1232 (1951).

This is an inhomogeneous integral equation for $F^{(1)}$. In order that it has a solution, the inhomogeneous term $U(p)$ must be orthogonal to the solution $\Phi_0(p)$ of the homogeneous equation. This condition can be derived as follows:

We multiply Eq. (6.7) by $F^{(0)}(p)=-G(p)\Phi_0(p)G(p)$, and integrate thus:

$$\begin{aligned}&\int \mathrm{Tr}\,F^{(0)}(p)L(p)F^{(1)}(p,q/2)L(p)d^4p+\int\mathrm{Tr}\,F^{(0)}(p)U^{(1)}(p,q/2)d^4p\\ &=-g^2\frac{1}{(2\pi)^4}\int\int\mathrm{Tr}\,F^{(0)}(p)\tau_3F^{(1)}(p',q/2)\times\tau_3h(p-p')^2\Delta(p-p')d^4pd^4p'.\end{aligned}$$

In view of Eq. (6.5) the last line is

$$=\int\mathrm{Tr}\,L(p')F^{(0)}(p')L(p')F^{(1)}(p',q/2)d^4p',$$

so that

$$(F^{(0)},U^{(1)})\equiv\int\mathrm{Tr}\,F^{(0)}(p)U^{(1)}(p,q/2)d^4p=0. \tag{6.8}$$

This is the desired condition.

For the evaluation of Eq. (6.8), we will neglect the p dependence of the self-energy terms. Thus

$$\begin{aligned}F^{(0)}(p)&=\tau_2\phi/(p_0^2-E_p^2+i\epsilon),\quad E_p^2=\epsilon_p^2+\phi^2,\\ \Delta L(p,q/2)&=q_0/2-\tau_3(\mathbf{p}\cdot\mathbf{q}/2m+(q/2)^2/2m).\end{aligned} \tag{6.9}$$

We then obtain

$$(F^{(0)},U^{(1)})=2\pi i\int\frac{\phi^2}{E_p^3}\left[\left(\frac{q_0}{2}\right)^2-\left(\frac{\mathbf{p}\cdot\mathbf{q}}{2m}\right)^2-\frac{\epsilon_p}{m^2}\left(\frac{q}{2}\right)^2\right]d^3p=0,$$

or

$$\left(\frac{q_0}{2}\right)^2-\left(\frac{q}{2}\right)^2\left[\frac{1}{3}\frac{\bar{p}^2}{m^2}-\frac{\bar{\epsilon}_p}{m}\right]=0, \tag{6.10}$$

where the average $\bar{f}$ is defined

$$\bar{f}=\int f(p)\frac{\phi^2}{E^3}d^3p\Big/\int\frac{\phi^2}{E^3}d^3p. \tag{6.10'}$$

The weight function $\phi^2/E_p^3=\phi^2/(\epsilon_p^2+\phi^2)^{\frac{3}{2}}$ peaks around the Fermi momentum, so that $p^2\sim p_F^2$, $\epsilon_p\sim 0$. Thus

$$q_0^2\approx q^2\frac{1}{3}\frac{\bar{p}^2}{m^2}\equiv\alpha^2q^2,\quad \alpha^2\approx p_F^2/3m^2, \tag{6.11}$$

which is the dispersion law for the collective excitations.[2,3] We also note, incidentally, that

$$\frac{1}{(2\pi)^3}\int\frac{\phi^2}{E^3}d^3p\approx N=mp_F/\pi^2, \qquad (6.11')$$
$$\alpha^2N\approx p_F^3/3\pi^2m=n.$$

We would like to emphasize here that these collective excitations are based on Eq. (6.2), which takes account of the phonon-Coulomb scattering of the quasi-particle pairs, but does not take into account the annihilation-creation process of the pair due to the same interaction.

It is well known that this annihilation-creation process is very important in the case of the Coulomb interaction, and plays the role of creating the plasma mode of collective oscillations. We will consider it in a later section.

As for the wave function $F^{(1)}$ itself, we have still to solve the integral equation (6.7). But this can be done by perturbation because on substituting $U^{(1)}$ in the integrand, we find that all the terms are of the type (3.11). In other words, to the zeroth order we may neglect the integral entirely and so

$$F^{(1)}(p,q/2)=-G(p)U^{(1)}(p,q/2)G(p). \qquad (6.12)$$

The original function

$$\Phi_q(p)=-L(p+q/2)F(p,q/2)L(p-q/2)$$

is even simpler. We get

$$\Phi_q(p)\approx\Phi_0(p) \qquad (6.13)$$

to this order.

7. CALCULATION OF THE CHARGE-CURRENT VERTEX FUNCTIONS

In this section we determine explicitly the charge-current vertex functions Γ_i, $(i=0, 1, 2, 3)$ from their integral equations. Only the particular combination $\Gamma^{(b)}$ of these was given before.

Let us first go back to the integral equation for Γ_0 generated by τ_3:

$$\Gamma_0(p+q/2,\, p-q/2)$$
$$=\tau_3-g^2\int\tau_3G(p'+q/2)\Gamma_0(p'+q/2,\, p'-q/2)\times G(p'-q/2)\tau_3h(p-p')^2\Delta(p-p')d^4p',$$

or

$$L(p+q/2)F_0(p+q/2,\, p-q/2)L(p-q/2)$$
$$=\tau_3+g^2\int\tau_3F_0(p'+q/2,\, p'-q/2)\times\tau_3h(p-p')^2\Delta(p-p')d^4p'. \qquad (7.1)$$

For small g^2, the standard approach to solve the equation would be the perturbation expansion in powers of g^2. We know, however, that there are low-lying collective excitations, discussed before, to which τ_3 can be coupled, and these excitations do not follow from perturbation.[21]

Fortunately, if we assume $q=0$, $q_0\neq 0$, then we have an exact solution to Eq. (7.1) in terms of $\Gamma^{(b)}$ of Eq. (4.4). Namely,

$$\begin{aligned}\Gamma_0(p+q/2,\, p-q/2)&=\Gamma^{(b)}(p+q/2,\, p-q/2)/q_0\\&=\tau_3\{[Z(p+q/2)+Z(p-q/2)\}/2\\&\quad+(p_0/q_0)[Z(p+q/2)-Z(p-q/2)]\}\\&\quad-[\chi(p+q/2)-\chi(p-q/2)]/q_0\\&\quad+i\tau_2[\phi(p+q/2)+\phi(p-q/2)]/q_0,\end{aligned} \qquad (7.2)$$

which can readily be verified.

The second term is the result of the coupling of τ_3 to the collective mode. This can be understood in the following way. Γ_0 contains matrix elements for creation or annihilation of a pair out of the vacuum. These processes can go through the collective intermediate state with the dispersion law (6.11), so that Γ will contain terms of the form

$$R_\pm/(q_0\pm\alpha q).$$

The residues $R_\pm$ can be obtained by taking the limit

$$R_\pm=\lim_{q_0\pm\alpha q\to 0}\Gamma_0(p+q/2,\, p-q/2)(q_0\pm\alpha q). \qquad (7.3)$$

Applying this procedure to the integral equation (7.1) for Γ_0, we find that $R_\pm$ must be a solution of the homogeneous equation; namely,

$$R_\pm=C_\pm\Phi_q(p), \qquad (7.4)$$

under the condition $q_0\pm\alpha q=0$.

For the particular case $q=0$, $\Phi_q(p)$ reduces to $\tau_2\phi(p)$, which in fact agrees with Eq. (7.2) if

$$C_\pm=-2i. \qquad (7.5)$$

This observation enables us to write down Γ_0 for $q\neq 0$. According to the results of Sec. 6, $\Phi_q(p)=\Phi_0(p)$ in the zeroth order in g^2N. Since corrections to the non-collective part of Γ_0 also turn out to be calculable by perturbation, we may now put

$$\Gamma_0(p+q/2,\, p-q/2)\approx\tau_3\bar{Z}+2i\tau_2\bar{\phi}q_0/(q_0^2-\alpha^2q^2),$$
$$\bar{\phi}\equiv[\phi(p+q/2)+\phi(p-q/2)]/2,$$
$$\bar{Z}\equiv[Z(p+q/2)+Z(p-q/2)\}/2$$

to the extent that terms of order g^2N and/or the p-dependence of the renormalization constants are neglected.

In quite a similar way the current vertex $\mathbf{\Gamma}$ may be constructed. This time we start from the longitudinal

[21] If we proceeded by perturbation theory, we would find in each order terms of order 1.

component for $q_0=0$, $q\neq 0$, which has the exact solution

$$\Gamma(p+q/2,p-q/2)\cdot\mathbf{q}/q=-\Gamma^{(b)}(p+q/2,p+q/2)/q$$
$$=\frac{\mathbf{p}\cdot\mathbf{q}}{mq}\left\{1+\frac{\chi(p+q/2)-\chi(p-q/2)}{\mathbf{p}\cdot\mathbf{q}/m}\right\}$$
$$-\tau_3p_0[\zeta(p+q/2)-\zeta(p-q/2)]/q$$
$$-2i\tau_2\frac{\phi(p+q/2)+\phi(p-q/2)}{2q}. \quad (7.7)$$

For $q_0\neq 0$, then, we get

$$\Gamma(p+q/2,p-q/2)\cdot\mathbf{q}/q$$
$$\approx(\mathbf{p}\cdot\mathbf{q}/q)\bar{Y}+2i\tau_2\phi\alpha^2q/(q_0^2-\alpha^2q^2), \quad (7.8)$$
$$\bar{Y}\equiv 1+[\chi(p+q/2)-\chi(p-q/2)]/(\mathbf{p}\cdot\mathbf{q}/m).$$

Combining (7.6) and (7.8), the continuity equation takes the form

$$q_0\Gamma_0-\mathbf{q}\cdot\Gamma=q_0\tau_3\bar{Z}+(\mathbf{p}\cdot\mathbf{q}/m)\bar{Y}+2i\tau_2\bar{\phi}$$
$$\approx\Gamma^{(b)},$$

which is indeed zero on the energy shell.

The transversal part of Γ, on the other hand, is not coupled with the collective mode because the latter is a scalar wave.[22] We may, therefore, write instead of Eq. (7.8)

$$\Gamma(p+q/2,p-q/2)\approx(\mathbf{p}/m)\bar{Y}$$
$$+2i\tau_2\bar{\phi}\alpha^2q/(q_0^2-\alpha^2q^2). \quad (7.10)$$

Equations (7.6) and (7.10) for Γ_i have a very interesting structure. The noncollective part is essentially the same as the charge current for a free quasi-particle except for the renormalization $\bar{Z}$ and $\bar{Y}$, whereas the collective part is spread out both in space and time. Neglecting the momentum dependence of $\bar{Z}$, $\bar{Y}$, and ϕ, we may thus write the charge-current density (ρ,j) as

$$\rho(x,t)\cong e\Psi^+\tau_3Z\Psi(x,t)+\frac{1}{\alpha^2}\frac{\partial f(x,t)}{\partial t}\equiv\rho_0+\frac{1}{\alpha^2}\frac{\partial f}{\partial t}, \quad (7.11)$$
$$\mathbf{j}(x,t)\cong e\Psi^+(\mathbf{p}/m)Y\Psi(x,t)-\nabla f(x,t)\equiv\mathbf{j}_0-\nabla f,$$

where f satisfies the wave equation

$$\left(\Delta-\frac{1}{\alpha^2}\frac{\partial^2}{\partial t^2}\right)f\approx-2e\Psi^+\tau_2\phi\Psi. \quad (7.12)$$

$(\rho_0,\mathbf{j}_0)$ is the charge-current residing in the "core" of a quasi-particle. The latter is surrounded by a cloud of the excitation field f. In a static situation, for example, f will fall off like $1/r$ from the core. When the particle is accelerated, a fraction of the charge is exchanged between the core and the cloud.

The total charge residing in a finite volume around a core is not constant because the current $-\nabla f$ reaches out to infinity.

[22] There may be transverse collective excitations (Bogoliubov, reference 2), but they do not automatically follow from the self-energy equation nor affect the energy gap structure.

8. THE PLASMA OSCILLATIONS

The inclusion of the annihilation-creation processes in the equations of the previous sections means that the vertex parts get multiplied by a string of closed loops, which represent the polarization (or shielding effect) of the surrounding medium. We will call the new quantities Λ, which now satisfy the following type of integral equations

$$\Lambda(p',p)=\gamma-i\int\tau_3G(p'-k)\Lambda(p'-k,p-k)$$
$$\times G(p-k)\tau_3D(k)d^4k$$
$$+iD(p'-p)\tau_3\int\mathrm{Tr}[\tau_3G(p'-k)$$
$$\times\Lambda(p'-k,p-k)G(p-k)]d^4k, \quad (8.1)$$
$$D(q)\equiv-ig^2h(q)^2\Delta(q)+e^2/q^2.$$

$D(q)$ includes the effect of the Coulomb interaction [see Eq. (2.24)]. Putting

$$\bar{X}(p'-p)\equiv i\int\mathrm{Tr}[\tau_3G(p'-k)\Lambda(p'-k,p'-k)$$
$$\times G(p-k)]d^4k, \quad (8.2)$$

Eq. (8.1) takes the same form as Eq. (4.3) for Γ with the inhomogeneous term replaced by $\gamma+\tau_3D\bar{X}$, so that Λ is a linear combination of the Γ corresponding to γ and Γ_0:

$$\Lambda=\Gamma+\Gamma_0D\bar{X}. \quad (8.3)$$

Substitution in Eq. (8.2) then yields

$$\bar{X}(p'-p)=i\int\mathrm{Tr}[\tau_3G(p'-k)$$
$$\times\Gamma(p'-k,p-k)G(p-k)]d^4k$$
$$+iD(p'-p)\bar{X}(p'-p)\int\mathrm{Tr}[\tau_3G(p'-k)$$
$$\times\Gamma_0(p'-k,p-k)G(p-k)]d^4k,$$

or

$$\bar{X}(p'-p)=i\int\mathrm{Tr}[\tau_3G(p'-k)$$
$$\times\Gamma(p'-k,p-k)G(p-k)]d^4k$$
$$\times\left\{1-iD(p'-p)\int\mathrm{Tr}[\tau_3G(p'-k)\right.$$
$$\left.\times\Gamma_0(p'-k,p-k)G(p-k)]d^4k\right\}^{-1}$$
$$\equiv X(p'-p)/[1-D(p'-p)X_0(p'-p)]. \quad (8.4)$$

Especially for $\gamma=\tau_3$, we get

$$\bar{X}_0(p'-p)=X_0(p'-p)/[1-D(p'-p)X_0(p'-p)],$$
$$\Lambda_0(p',p)=\Gamma_0(p',p)/[1-D(p'-p)X_0(p',p)]. \quad (8.5)$$

To obtain the collective excitations, let us next write down the homogeneous integral equation:

$$\Theta_q(p)=-i\int \tau_3 G(p'+q/2)\Theta_q(p')\times G(p'-q/2)\tau_3 D(p-p')d^4p' +i\tau_3 D(q)\int \mathrm{Tr}\,[\tau_3 G(p'+q/2)\Theta_q(p')\times G(p'-q/2)]d^4p', \quad (8.6)$$

which means

$$\Theta_q(p)=\Gamma_0(p+q/2,\,p-q/2)D(q)\chi(q),$$

$$\chi(q)\equiv i\int \mathrm{Tr}[\tau_3 G(p'+q/2)\times\Theta_q(p')G(p'-q/2)\tau_3]d^4p'. \quad (8.7)$$

Substituting Θ_q in the second equation from the first, we get

$$1=D(q)X_0(q), \quad (8.8)$$

where $X_0(q)$ is defined in Eq. (8.4).

The solutions to Eq. (8.8) determine the new dispersion law $q_0=f(q)$ for the collective excitations.

With the solution (7.6), the quantity X_0 in Eq. (8.8) can be calculated. After some simplifications using Eq. (6.11), we obtain

$$X_0=\frac{1}{(2\pi)^3}\left[\frac{\alpha^2q^2}{q_0^2-\alpha^2q^2}\int\frac{\phi^2d^3p}{E_p(E_p^2-\alpha^2q^2/4)} +\frac{q^2}{4}\int\frac{\phi^2d^3p}{E_p(q_0^2/4-E_p^2)}\left(\frac{p^2}{3m^2E_p^2}-\frac{\alpha^2}{E_p^2-\alpha^2q^2/4}\right)\right]+O(q^4). \quad (8.9)$$

For $\alpha q\ll\phi$, and $q_0\gg\phi$ or $\ll\phi$, the second integral may be dropped and

$$X_0\cong\alpha^2q^2N/(q_0^2-\alpha^2q^2). \quad (8.10)$$

For small q^2, the dominant part of $D(q)$ in Eq. (8.8) is the Coulomb interaction e^2/q^2. Equation (8.8) then becomes

$$q_0^2=e^2\alpha^2N=e^2n \quad (q^2\rightarrow 0), \quad (8.11)$$

where n is the number of electrons per unit volume. This agrees with the ordinary plasma frequency for free electron gas.

We see thus that the previous collective state with $q_0^2=\alpha^2q^2$ has shifted its energy to the plasma energy as a result of the Coulomb interaction.

On the other hand, if Coulomb interaction is neglected, Eq. (8.8) leads to[23]

$$q_0^2=\alpha^2q^2[1-ig^2\Delta(q,q_0)h(q,q_0)^2N]. \quad (8.12)$$

The correction term, however, is of the order g^2N, hence should be neglected to be consistent with our approximation.

We can also study the behavior of X_0 in the limit $q_0\rightarrow 0$ for small but finite q^2:

$$X_0\approx\frac{1}{(2\pi)^3}\int\frac{\phi^2}{E^3}d^3p\approx N, \quad (8.13)$$

which comes entirely from the noncollective part of Γ_0, but again agrees with the free electron value.

Another observation we can make regarding $\bar{X}_0(q,q_0)$ is the following. $\bar{X}_0$ represents the charge density correlation in the ground state:

$$\bar{X}_0(q,q_0)=\int\langle 0|T(\rho(xt),\rho(0))|0\rangle e^{-i\mathbf{q}\cdot\mathbf{x}+iq_0t}d^3xdt.$$

If $|0\rangle$ is an eigenstate of charge, $\bar{X}_0$ should vanish for $q\rightarrow 0$, $q_0\neq 0$ since the right-hand side then consists of the nondiagonal matrix elements of the total charge operator Q:

$$\bar{X}_0(0,q_0)\propto\sum_n\left(\frac{1}{q_0-E_n}-\frac{1}{q_0+E_n}\right)|\langle n|Q|0\rangle|^2.$$

The converse is also true if $E_n>|q_0|$, $n\neq 0$ for some $q_0\neq 0$. Our result for $\bar{X}_0$, as is clear from Eqs. (8.5) and (8.9), has indeed the correct property in spite of the fact that the "bare" vacuum, from which we started, is not an eigenstate of charge.

9. CONCLUDING REMARKS

We have discussed here formal mathematical structure of the BCS-Bogoliubov theory. The nature of the approximation is characterized essentially as the Hartree-Fock method, and can be given a simple interpretation in terms of perturbation expansion. In the presence of external fields, the corresponding approximation insures, if treated properly, that the gauge invariance is maintained. It is interesting that the quasi-particle picture and charge conservation (or gauge invariance) can be reconciled at all. This is possible because we are taking account of the "radiative corrections" to the bare quasi-particles which are not eigenstates of charge. These corrections manifest themselves primarily through the existence of collective excitations.

There are some questions which have been left out. We would like to know, for one thing, what will happen if we seek corrections to our Hartree-Fock approximation by including processes (or diagrams) which have not been considered here. Even within our ap-

[23] Compare Anderson, reference 7.

proximation, there is an additional assumption of the weak coupling ($g^2N \ll 1$), and the importance of the neglected terms (of order g^2N and higher) is not known.

Experimentally, there has been some evidence[24] regarding the presence of spin paramagnetism in superconductors. This effect has to do with the spin density induced by a magnetic field and can be derived by means of an appropriate vertex solution. However, this does not seem to give a finite spin paramagnetism at 0°K.[25] The collective excitations do not play an important role here as they are not excited by spin density. [$\Gamma^{(a)}$, Eq. (4.4), does not have the characteristic pole.]

It is desirable that both experiment and theory about spin paramagnetism be developed further since this may be a crucial test of the fundamental ideas underlying the BCS theory.

ACKNOWLEDGMENT

We wish to thank Dr. R. Schrieffer for extremely helpful discussions throughout the entire course of the work.

[24] Knight, Androes, and Hammond, Phys. Rev. **104**, 852 (1956); F. Reif, Phys. Rev. **106**, 208 (1957); G. M. Androes and W. D. Knight, Phys. Rev. Letters **2**, 386 (1959).

[25] K. Yosida, Phys. Rev. **110**, 769 (1958).

Volume 1, number 7 PHYSICS LETTERS 1 July 1962

POSSIBLE NEW EFFECTS IN SUPERCONDUCTIVE TUNNELLING *

B. D. JOSEPHSON
Cavendish Laboratory, Cambridge, England

Received 8 June 1962

We here present an approach to the calculation of tunnelling currents between two metals that is sufficiently general to deal with the case when both metals are superconducting. In that case new effects are predicted, due to the possibility that electron pairs may tunnel through the barrier leaving the quasi-particle distribution unchanged.

Our procedure, following that of Cohen et al. [1], is to treat the term in the Hamiltonian which transfers electrons across the barrier as a perturbation. We assume that in the absence of the transfer term there exist quasi-particle operators of definite energies, whose corresponding number operators are constant.

A difficulty, due to the fact that we have a system containing two disjoint superconducting regions, arises if we try to describe quasi-particles by the usual Bogoliubov operators [2]. This is because states defined as eigenfunctions of the Bogoliubov quasi-particle number operators contain phase-coherent superpositions of states with the same total number of electrons but different numbers in the two regions. However, if the regions are independent these states must be capable of superposition with arbitrary phases. On switching-on the transfer term the particular phases chosen will affect the predicted tunnelling current. This behaviour is of fundamental importance to the argument that follows. The neglect, in the quasi-particle approximation, of the collective excitations of zero energy [3] results in an unphysical restriction in the free choice of phases, but may be avoided by working with the projected states with definite numbers of electrons on both sides of the barrier. Corresponding to these projections we use operators which alter the numbers of electrons on the two sides by definite numbers **. In particular, corresponding to the Bogoliubov operators α_k^+ we use quasi-particle creation operators α_{ek}^+, α_{hk}^+ which respectively add or remove an electron from the same side as the quasi-particle and leave the number on the other side unchanged, and pair creation operators S_k^+ † which add a pair of electrons on one side leaving the quasi-particle distribution unchanged. The Hermitean conjugate destruction operators have similar definitions. The S operators, referring to macroscopically occupied states, may be treated as time dependent c-numbers ††, and we normalise them to have unit amplitude. Relations expressing electron operators in terms of quasi-particle operators, equal-time anticommutation relations and number operator relations may be derived from those of the Bogoliubov theory by requiring both sides of the equations to have the same effect on N_l and N_r, the numbers of electrons on the two sides of the barrier. For example,

$$a_{k\uparrow}^+ = u_k\, \alpha_{ek0}^+ + v_k\, \alpha_{hk1}\,, \qquad \alpha_{ek}^+\, \alpha_{ek} = n_k\,, \qquad a_{ek}^+\, \alpha_{hk} = S_k^+\, n_k\,. \tag{1}$$

Noting that the Bogoliubov Hamiltonian is $H - \lambda N$ (λ = chemical potential), we take our unperturbed Hamiltonian to be

$$H_0 = \sum_k n_k E_k + \lambda_l N_l + \lambda_r N_r\,,$$

where E_k is the quasi-particle energy in the Bogoliubov theory, and derive the commutation relations

$$[H_0, \alpha_{ek}^+] = (E_k + \lambda_k)\, \alpha_{ek}^+\,, \qquad [H_0, \alpha_{hk}^+] = (E_k - \lambda_k)\, \alpha_{hk}^+\,, \qquad [H_0, S_k^+] = 2\, \lambda_k\, S_k^+\,. \tag{2}$$

In the presence of tunnelling the Hamiltonian is $H_0 + H_T$, where H_T expressed in electron operators is

* Work supported by Trinity College, Cambridge, and the Department of Scientific and Industrial Research.

** We shall use subscripts l and r to distinguish operators on the two sides, and k to denote an operator referring to either side.

† These are equivalent to the operators which change $|N\rangle$ to $|N+2\rangle$ in the theory of Gor'kov [4].

†† Cf. N. N. Bogoliubov et al. [5]. The phase of an S operator is related to the orientation of the plane containing the pseudospin operators [3]. Physical observables cannot depend on the phase of a single S operator, but they can depend on the relative phases of the S operators associated with two superconducting regions, as in the phenomena dealt with here.

Volume 1, number 7 PHYSICS LETTERS 1 July 1962

$$\sum_{l,r} (T_{lr}\, a_l^+ a_r + T_{rl}\, a_r^+ a_l) .$$

If we describe the time dependence of operators by the interaction picture [6], equations (2) imply that the α and S operators have exponential time dependence. The current operator in the Heisenberg picture is related to that in the interaction picture according to

$$J_H(t) = U^{-1}(t)\, J_{\text{int}}(t)\, U(t) ,$$

where

$$U(t) = \lim_{\epsilon \to 0+} \{T \exp(-\, i/\hbar \int_{-\infty}^{t} e^{\epsilon t'} H_T(t')\, dt')\} .$$

Here H_T is expressed in the interaction picture and $U(t)$ can be evaluated by writing H_T in terms of quasi-particle operators and using the method of Goldstone [6]. We also express

$$J_{\text{int}}(t) = ie/\hbar\, [H_T, N_r] \quad 1)$$

in terms of quasi-particle operators, and by retaining only those terms in $J_H(t)$ which can be expressed in accordance with (1) as products of S and number operators obtain an expression equivalent to the usual one, of the form

$$J_H = J_0 + \tfrac{1}{2} J_1\, S_l^+ S_r + \tfrac{1}{2} J_1^*\, S_r^+ S_l . \qquad (3)$$

To second order in H_T, J_0 is similar to the expression of Cohen et al. [1], and reduces for the same reasons to the usual one obtained by neglecting coherence factors. The remaining terms oscillate with frequency $\nu = 2eV/h$ ($V = \lambda_l - \lambda_r$ being the applied voltage), owing to the time dependence of the S operators. J_1 is proportional to the effective matrix element for the transfer of electron pairs across the barrier without affecting the quasi-particle distribution, and typical terms are of the form

$$2\, i\, e\, u_l\, v_l\, u_r\, v_r\, T_{lr}\, T_{-l,-r} [(1 - n_{l0} - n_{r1})$$
$$\times \{P \frac{1}{eV - E_l - E_r} + \pi i \delta(eV - E_l - E_r)\}$$
$$- (n_{l0} - n_{r0}) \{P \frac{1}{eV + E_l - E_r} + \pi i \delta(eV + E_l - E_r)\}] \qquad (4)$$

where $-k$ denotes the state paired with k. The second and fourth terms result from processes involving real intermediate states, and can be regarded as fluctuations in the normal current due to coherence effects. We note that the first term remains finite at zero temperature and zero applied voltage. From (3) our theory predicts that

(i) At finite voltages the usual DC current occurs, but there is also an AC supercurrent of amplitude $|J_1|$ and frequency $2\, eV/h$ (1 μV corresponds to 483.6 Mc/s).

(ii) at zero voltages, J_0 is zero, but a DC supercurrent of up to a maximum of $|J_1|$ can occur.

Applied r.f. fields can be treated by noting that the oscillations in V frequency-modulate the supercurrent. Thus if a DC voltage V on which is superimposed an AC voltage of frequency ν is applied across the barrier, the current has Fourier components at frequencies $2eV/h \pm n\nu$, where n is an integer. If for some n, $2eV/h = n\nu$, the supercurrent has a DC component dependent on the magnitude and phase of the AC voltage. Hence the DC characteristic has a zero slope resistance part over a range of current dependent on the magnitude of the AC voltage.

Equivalent quantum-mechanical explanations of these effects can be given. For example (i) is due to the transfer of an electron pair across the barrier with photon emission, leaving the quasi-particle distribution unchanged. Consequently the photon frequency is not broadened by the finite quasi-particle lifetimes occurring in real superconductors. (ii) is due to pair transfer without photon emission. The linear dependence of the current on the matrix element is due to the fact that the process involves macroscopically occupied states between which phase relationships can occur.

The possibility of observing these effects depends on the value of $|J_1|$. At low temperatures and voltages the first term of (4) dominates, and in the presence of time-reversal symmetry all contributions to it are in phase. $|J_1|$ is then equal to the current flowing in the normal state at an applied voltage equal to π times the energy gap, assumed to be the same on both sides. At higher temperatures the third term reduces $|J_1|$, and at high frequencies the effects are reduced by the capacitance across the barrier. Magnetic fields, and currents in the films destroy the time-reversal symmetry and reduce $|J_1|$. The effects may be taken into account approximately by replacing (3) by

$$j = j_0 + \tfrac{1}{2} j_1\, \psi_l^*\, \psi_r + \tfrac{1}{2} j_1^*\, \psi_r^*\, \psi_l ,$$

where j is the tunnelling current density, and ψ_l, ψ_r the effective superconducting wave functions [7] in the films on the two sides. This formula predicts that in very weak fields diamagnetic currents will screen the field from the space between the films, but with a large penetration depth owing to the smallness of j_1. In larger fields, owing to the existence of a critical current density, screening will not occur; the phases of the supercurrents will vary rapidly over the barrier, causing the maximum total supercurrent to drop off rapidly with increasing field. Anderson [8] has suggested that the absence of tunnelling supercurrents in most experiments hitherto performed may be due to the earth's field acting in this way. Cancellation of supercurrents would start to occur when the amount of flux between the films, including that in the penetration regions, became of the order of a quantum of flux $hc/2e$. This would occur for typical films in a field of about 0.1 gauss. Such a field would not be appreciably excluded by the critical currents obtainable in specimens of all but the highest conductivity.

When two superconducting regions are separated by a thin normal region, effects similar to those considered here should occur and may be relevant to the theory of the intermediate state.

I am indebted to Dr. P. W. Anderson and Prof. A. B. Pippard for stimulating discussions.

References

1) M.H. Cohen, L.M. Falicov and J.C. Phillips, Phys. Rev. Letters 8 (1962) 316.
2) N.N. Bogoliubov, J. Exptl. Theoret. Phys. (USSR) 34 (1958) 58; translation: Soviet Phys. JETP 7 (1958) 41.
3) P.W. Anderson, Phys. Rev. 112 (1958) 1900.
4) L.P. Gor'kov, J. Exptl. Theoret. Phys. (USSR) 34 (1958) 735; translation: Soviet Phys. JETP 7 (1958) 505.
5) N.N. Bogoliubov, V.V. Tolmachev and D.V. Shirkov, A new method in the theory of superconductivity (Consultants Bureau, Inc., New York, 1959), p. 6.
6) J. Goldstone, Proc. Roy. Soc. (London) A 239 (1957) 267.
7) L.P. Gor'kov, J. Exptl. Theoret. Phys. (USSR) 36 (1959) 1918; translation: Soviet Phys. JETP 9 (1959) 1364.
8) P.W. Anderson, private discussion.

* * * * *

PHYSICAL REVIEW VOLUME 125, NUMBER 1 JANUARY 1, 1962

Gauge Invariance and Mass

JULIAN SCHWINGER
Harvard University, Cambridge, Massachusetts, and University of California, Los Angeles, California

(Received July 20, 1961)

It is argued that the gauge invariance of a vector field does not necessarily imply zero mass for an associated particle if the current vector coupling is sufficiently strong. This situation may permit a deeper understanding of nucleonic charge conservation as a manifestation of a gauge invariance, without the obvious conflict with experience that a massless particle entails.

DOES the requirement of gauge invariance for a vector field coupled to a dynamical current imply the existence of a corresponding particle with zero mass? Although the answer to this question is invariably given in the affirmative,[1] the author has become convinced that there is no such necessary implication, once the assumption of weak coupling is removed. Thus the path to an understanding of nucleonic (baryonic) charge conservation as an aspect of a gauge invariance, in strict analogy with electric charge,[2] may be open for the first time.

One potential source of error should be recognized at the outset. A gauge-invariant system is not the continuous limit of one that fails to admit such an arbitrary function transformation group. The discontinuous change of invariance properties produces a corresponding discontinuity of the dynamical degrees of freedom and of the operator commutation relations. No reliable conclusions about the mass spectrum of a gauge-invariant system can be drawn from the properties of an apparently neighboring system, with a smaller invariance group. Indeed, if one considers a vector field coupled to a divergenceless current, where gauge invariance is destroyed by a so-called mass term with parameter m_0, it is easily shown[3] that the mass spectrum must extend below m_0. The lowest mass value will therefore become arbitrarily small as m_0 approaches zero. Nevertheless, if m_0 is exactly zero the commutation relations, or equivalent properties, upon which this conclusion is based become entirely different and the argument fails.

If invariance under arbitrary gauge transformations is asserted, one should distinguish sharply between numerical gauge functions and operator gauge functions, for the various operator gauges are not on the same quantum footing. In each coordinate frame there is a unique operator gauge, characterized by three-dimensional transversality (radiation gauge), for which one has the standard operator construction in a vector space of positive norm, with a physical probability interpretation. When the theory is formulated with the aid of vacuum expectation values of time-ordered operator products, the Green's functions, the freedom of formal gauge transformation can be restored.[4] The Green's functions of other gauges have more complicated operator realizations, however, and will generally lack the positiveness properties of the radiation gauge.

Let us consider the simplest Green's function associated with the field $A_\mu(x)$, which can be derived from the unordered product

$$\langle A_\mu(x)A_\nu(x')\rangle = \int \frac{(dp)}{(2\pi)^3} e^{ip(x-x')} dm^2\, \eta_+(p)\delta(p^2+m^2)A_{\mu\nu}(p),$$

where the factor $\eta_+(p)\delta(p^2+m^2)$ enforces the spectral restriction to states with mass $m\geq 0$ and positive energy. The requirement of non-negativeness for the matrix $A_{\mu\nu}(p)$ is satisfied by the structure associated with the radiation gauge, in virtue of the gauge-dependent asymmetry between space and time (the time axis is specified by the unit vector n_μ):

$$A_{\mu\nu}{}^R(p)=B(m^2)\left[g_{\mu\nu}-\frac{(p_\mu n_\nu+p_\nu n_\mu)(np)+p_\mu p_\nu}{p^2+(np)^2}\right].$$

Here $B(m^2)$ is a real non-negative number. It obeys the sum rule

$$1=\int_0^\infty dm^2\, B(m^2),$$

which is a full expression of all the fundamental equal-time commutation relations.

The field equations supply the analogous construction for the vacuum expectation value of current products $\langle j_\mu(x)j_\nu(x')\rangle$, in terms of the non-negative matrix

$$j_{\mu\nu}(p)=m^2B(m^2)(p_\mu p_\nu-g_{\mu\nu}p^2).$$

The factor m^2 has the decisive consequence that $m=0$ is not contained in the current vector's spectrum of vacuum fluctuations. The latter determines $B(m^2)$ for $m>0$, but leaves unspecified a possible delta function contribution at $m=0$,

$$B(m^2)=B_0\delta(m^2)+B_1(m^2).$$

The non-negative constant B_0 is then fixed by the sum rule,

$$1=B_0+\int_0^\infty dm^2\, B_1(m^2).$$

[1] For example, J. Schwinger, Phys. Rev. 75, 651 (1949).
[2] T. D. Lee and C. N. Yang, Phys. Rev. 98, 1501 (1955).
[3] K. Johnson, Nuclear Phys. 25, 435 (1961).
[4] J. Schwinger, Phys. Rev. 115, 721 (1959).

We have now recognized that the vacuum fluctuations of the vector A_μ are composed of two parts. One, with $m>0$, is directly related to corresponding current fluctuations, while the other part, with $m=0$, can be associated with a pure radiation field, which is transverse in both three- and four-dimensional senses and has no accompanying current. Imagine that the current vector contains a variable numerical factor. If this is set equal to zero, we have $B_1(m^2)=0$ and $B_0=1$ or, just the radiation field. For a sufficiently small nonzero value of the parameter, B_0 will be slightly less than unity, which may be the situation for the electromagnetic field. Or it may be that the electrodynamic coupling is quite considerable and gives rise to a small value of B_0, which has the appearance of a fairly weak coupling. Can we increase further the magnitude of the variable parameter until $\int dm^2 B_1(m^2)$ attains its limiting value of unity, at which point $B_0=0$, and $m=0$ disappears from the spectrum of A_μ? The general requirement of gauge invariance no longer seems to dispose of this essentially dynamical question.

Would the absence of a massless particle imply the existence of a stable, unit spin particle of nonzero mass? Not necessarily, since the vacuum fluctuation spectrum of A_μ becomes identical with that of j_μ, which is governed by all of the dynamical properties of the fields that contribute to this current. For the particularly interesting situation of a vector field that is coupled to the current of nucleonic charge, the relevant spectrum, in the approximate strong-interaction framework, is that of the states with $N=Y=T=0$, $R_T=-1$, $J=1$, and odd parity. This is a continuum, beginning at three pion masses.[5] It is entirely possible, of course, that $B(m^2)$ shows a more or less pronounced maximum which could be characterized approximately as an unstable particle.[6] But the essential point is embodied in the view that the observed physical world is the outcome of the dynamical play among underlying primary fields, and the relationship between these fundamental fields and the phenomenological particles can be comparatively remote, in contrast to the immediate correlation that is commonly assumed.

---

[5] The very short range of the resulting nuclear interaction together with the qualitative inference that like nucleonic charges are thereby repelled suggests that the vector field which defines nucleonic charge is also the ultimate instrument of nuclear stability.

[6] *Note added in proof.* Experimental evidence for an unstable particle of this type has recently been announced by B. C. Maglić, L. W. Alvarez, A. H. Rosenfeld, and M. L. Stevenson, in Phys. Rev. Letters 7, 178 (1961).

PHYSICAL REVIEW VOLUME 130, NUMBER 1 1 APRIL 1963

Plasmons, Gauge Invariance, and Mass

P. W. ANDERSON
Bell Telephone Laboratories, Murray Hill, New Jersey
(Received 8 November 1962)

Schwinger has pointed out that the Yang-Mills vector boson implied by associating a generalized gauge transformation with a conservation law (of baryonic charge, for instance) does not necessarily have zero mass, if a certain criterion on the vacuum fluctuations of the generalized current is satisfied. We show that the theory of plasma oscillations is a simple nonrelativistic example exhibiting all of the features of Schwinger's idea. It is also shown that Schwinger's criterion that the vector field $m\neq 0$ implies that the matter spectrum before including the Yang-Mills interaction contains $m=0$, but that the example of superconductivity illustrates that the physical spectrum need not. Some comments on the relationship between these ideas and the zero-mass difficulty in theories with broken symmetries are given.

RECENTLY, Schwinger[1] has given an argument strongly suggesting that associating a gauge transformation with a local conservation law does not necessarily require the existence of a zero-mass vector boson. For instance, it had previously seemed impossible to describe the conservation of baryons in such a manner because of the absence of a zero-mass boson and of the accompanying long-range forces.[2] The problem of the mass of the bosons represents the major stumbling block in Sakurai's attempt to treat the dynamics of strongly interacting particles in terms of the Yang-Mills gauge fields which seem to be required to accompany the known conserved currents of baryon number and hypercharge.[3] (We use the term "Yang-Mills" in Sakurai's sense, to denote any generalized gauge field accompanying a local conservation law.)

The purpose of this article is to point out that the familiar plasmon theory of the free-electron gas exemplifies Schwinger's theory in a very straightforward manner. In the plasma, transverse electromagnetic waves do not propagate below the "plasma frequency," which is usually thought of as the frequency of long-wavelength longitudinal oscillation of the electron gas. At and above this frequency, three modes exist, in close analogy (except for problems of Galilean invariance implied by the inequivalent dispersion of longitudinal and transverse modes) with the massive vector boson mentioned by Schwinger. The plasma frequency is equivalent to the mass, while the finite density of electrons leading to divergent "vacuum" current fluctuations resembles the strong renormalized coupling of Schwinger's theory. In spite of the absence of low-frequency photons, gauge invariance and particle conservation are clearly satisfied in the plasma.

In fact, one can draw a direct parallel between the dielectric constant treatment of plasmon theory[4] and Schwinger's argument. Schwinger comments that the commutation relations for the gauge field A give us one sum rule for the vacuum fluctuations of A, while those for the matter field give a completely independent value for the fluctuations of matter current j. Since j is the source for A and the two are connected by field equations, the two sum rules are normally incompatible unless there is a contribution to the A rule from a free, homogeneous, weakly interacting, massless solution of the field equations. If, however, the source term is large enough, there can be no such contribution and the massless solutions cannot exist.

The usual theory of the plasmon does not treat the electromagnetic field quantum-mechanically or discuss vacuum fluctuations; yet there is a close relationship between the two arguments, and we, therefore, show that the quantum nature of the gauge field is irrelevant. Our argument is as follows:

The equation for the electromagnetic field is

$$p^2 A_\mu = (k^2-\omega^2)A_\mu(\mathbf{k},\omega) = 4\pi j_\mu(\mathbf{k},\omega).$$

[1] J. Schwinger, Phys. Rev. **125**, 397 (1962).
[2] T. D. Lee and C. N. Yang, Phys. Rev. **98**, 1501 (1955).
[3] J. J. Sakurai, Ann. Phys. (N. Y.) **11**, 1 (1961).
[4] P. Nozières and D. Pines, Phys. Rev. **109**, 741 (1958).

A given distribution of current j_μ will, therefore, lead to a response A_μ given by

$$A_\mu=\frac{4\pi}{k^2-\omega^2}j_\mu=\frac{4\pi}{p^2}j_\mu. \tag{1}$$

(1) is merely the statement that only the electromagnetic current can be a source of the field; it is required for general gauge invariance and charge conservation according to the usual arguments.

The dynamics of the matter system—of the plasma in that case, of the vacuum in the elementary particle problem—determine a second response function, the response of the current to a given electromagnetic or Yang-Mills field. Let us call this response function

$$j_\mu(\mathbf{k},\omega)=-K_{\mu\nu}(\mathbf{k},\omega)A_\nu(\mathbf{k},\omega). \tag{2}$$

By well-known arguments of gauge invariance, $K_{\mu\nu}$ must have a certain form: Schwinger points out that in the relativistic case it must be proportional to $p_\mu p_\nu - g_{\mu\nu}p^2$, and equivalent arguments give one the same form in superconductivity.[5] It will be convenient to consider, for simplicity, only the gauge

$$p_\mu A_\mu=0. \tag{3}$$

Then the response is diagonal: $K_{\mu\nu}=-g_{\mu\nu}K$. For a plasma with n carriers of charge e and mass M it is simply (in the limit $p\to 0$)

$$K=ne^2/M. \tag{4}$$

In an insulator the response is not relativistically invariant. If the insulator has magnetic polarizability α_m and electric α_e, the response equations may be written, in the gauge (3),

$$j_\mu=-\alpha_e p^2 A_\mu \quad \text{(longitudinal and time components)},$$
$$\mathbf{j}=-\alpha_m p^2 \mathbf{A} \quad \text{(transverse components)}.$$

In a truly relativistic situation such as our normal picture of a vacuum, we expect

$$j_\mu=-\alpha p^2 A_\mu \tag{5}$$

to describe normal polarizable behavior.

Since we cannot turn off the interactions, we do not actually observe the responses (1), (2), or (5). If we insert a test particle, its field A_μ^e induces a current j_μ which in turn acts as the source for an internal field A_μ^i:

$$j_\mu=-K(A_\mu^i+A_\mu^e),\quad A_\mu^i=+4\pi j_\mu/p^2,$$

or, the total field is modified to

$$A_\mu=[p^2/(p^2+4\pi K)]A_\mu^e. \tag{6}$$

The pole at which A propagates freely occurs at a mass (frequency)

$$m^2=-p^2=4\pi K, \tag{7}$$

which in a conductor is

$$m^2=\omega^2-k^2=\omega_p^2. \tag{8}$$

ω_p is the usual plasma frequency $(4\pi ne^2/M)^{1/2}$.

It is not necessary here to go in detail into the relationship between longitudinal and transverse behavior of the plasmon. In the limit $p\to 0$ both waves propagate according to (8). The longitudinal plasmon is generally thought of as entirely an attribute of the plasma, while the transverse ones are considered to result from modification of the propagation of real photons by the medium. This is reasonable in the classical case because the longitudinal plasmon disappears at a certain cutoff energy and has a different dispersion law; but in a Lorentz-covariant theory of the vacuum it would be indistinguishable from the third component of a massive vector boson of which the transverse photons are the two transverse components.

How, then, if we were confined to the plasma as we are to the vacuum and could only measure renormalized quantities, might we try to determine whether, before turning on the effects of electromagnetic interaction, A had been a massless gauge field and K had been finite? As far as we can see, this is not possible; it is, nonetheless, interesting to see what the criterion is in terms of the actual current response function to a perturbation in the Lagrangian

$$\delta L=j_\mu\delta A_\mu. \tag{9}$$

This will turn out to be identical to Schwinger's criterion. The original "bare" response function was K:

$$j_\mu=-K_{\mu\nu}\delta A_\mu.$$

Taking into account the interaction, however, we must correct for the induced fields and currents, and we get

$$j_\mu=-K'\delta A_\mu^e=-K[p^2/(p^2+4\pi K)]\delta A_\mu^e\to -(p^2/4\pi)\delta A_\mu^e,\quad p^2\to 0. \tag{10}$$

Thus, the new response to an applied perturbing field (9) is very like that of an ordinary polarizable medium. The only difference from an ordinary polarizable "vacuum" with bare response (5) is that in that case as $p\to 0$

$$K'\to -[\alpha/(1+4\pi\alpha)]p^2, \tag{11}$$

so that the coefficient of $p^2/4\pi$ is less than unity.

This criterion is precisely the same as Schwinger's criterion

$$\int B_1(m^2)dm^2=1,$$

where $B_1 m^2$ is the weight function for the current vacuum fluctuations. This can be shown by a simple dispersion argument. Schwinger expresses the unordered

[5] M. R. Schafroth, Helv. Phys. Acta 24, 645 (1951).

product expectation value of the current as

$$\langle j_\mu(x)j_\nu(x')\rangle=\int dm^2\, m^2B_1(m^2)\int\frac{dp}{(2\pi)^3}e^{ip(x-x')}$$
$$\times\eta_+(p)\delta(p^2+m^2)(p_\mu p_\nu-g_{\mu\nu}p^2).$$

The Fourier transform of the corresponding retarded Green's function is our response function:

$$K'(p)=\int\frac{dm^2\, m^2B_1(m^2)}{p^2-m^2}[p_\mu p_\nu-g_{\mu\nu}p^2],$$

and

$$\lim_{p\to 0}K'(p)=(p_\mu p_\nu-g_{\mu\nu}p^2)\int dm^2\, B_1(m^2).$$

Thus, (aside from a factor 4π which Schwinger has not used in his field equation) his criterion is also that the polarizability α', here expressed in terms of a dispersion integral, have its maximum possible value, 1.

The polarizability of the vacuum is not generally considered to be observable[6] except in its p dependence (terms of order p^4 or higher in K). In fact, we can remove (11) entirely by the conventional renormalization of the field and charge

$$A_r=AZ^{-1/2},\quad e_r=eZ^{1/2},\quad j_r=jZ^{1/2}.$$

Z, here, can be shown to be precisely

$$Z=1-4\pi\alpha'=1-\int_0^\infty dm^2\, B_1(m^2).$$

Thus, the renormalization procedure is possible for any merely polarizable "vacuum," but not for the special case of the conducting "plasma" type of vacuum. In this case, no net true charge remains localized in the region of the dressed particle; all of the charge is carried "at infinity" corresponding to the fact, well known in the theory of metals, that all the charge carried by a quasi-particle in a plasma is actually on the surface. Nonetheless, conservation of particles, if not of bare charge, is strictly maintained. Note that the situation does not resemble the case of "infinite" charge renormalization because the infinity in the vacuum polarizability need only occur at $p^2=0$.

Either in the case of the polarizable vacuum or of the "conducting" one, no low-energy experiment, and even possibly no high-energy one, seems capable of directly testing the value of the vacuum polarizability prior to renormalization. Thus, we conclude that the plasmon is a physical example demonstrating Schwinger's contention that under some circumstances the Yang-Mills type of vector boson need not have zero mass. In addition, aside from the short range of forces and the finite mass, which we might interpret without resorting to Yang-Mills, it is not obvious how to characterize such a case mathematically in terms of observable, renormalized quantities.

We can, on the other hand, try to turn the problem around and see what other conclusions we can draw about possible Yang-Mills models of strong interactions from the solid-state analogs. What properties of the vacuum are needed for it to have the analog of a conducting response to the Yang-Mills field?

Certainly the fact that the polarizability of the "matter" system, without taking into account the interaction with the gauge field, is infinite need not bother us, since that is unobservable. In physical conductors we can see it, but only because we can get outside them and apply to them true electromagnetic fields, not only internal test charges.

More serious is the implication—obviously physically from the fact that α has a pole at $p^2=0$—that the "matter" spectrum, at least for the "undressed" matter system, must extend all the way to $m^2=0$. In the normal plasma even the final spectrum extends to zero frequency, the coupling rather than the spectrum being affected by the screening. Is this necessarily always the case? The answer is no, obviously, since the superconducting electron gas has no zero-mass excitations whatever. In that case, the fermion mass is finite because of the energy gap, while the boson which appears as a result of the theorem of Goldstone[7,8] and has zero unrenormalized mass is converted into a finite-mass plasmon by interaction with the appropriate gauge field, which is the electromagnetic field. The same is true of the charged Bose gas.

It is likely, then, considering the superconducting analog, that the way is now open for a degenerate-vacuum theory of the Nambu type[9] without any difficulties involving either zero-mass Yang-Mills gauge bosons or zero-mass Goldstone bosons. These two types of bosons seem capable of "canceling each other out" and leaving finite mass bosons only. It is not at all clear that the way for a Sakurai[3] theory is equally uncluttered. The only mechanism suggested by the present work (of course, we have not discussed non-Abelian gauge groups) for giving the gauge field mass is the degenerate vacuum type of theory, in which the original symmetry is not manifest in the observable domain. Therefore, it needs to be demonstrated that the necessary conservation laws can be maintained.

I should like to close with one final remark on the Goldstone theorem. This theorem was initially conjectured, one presumes, because of the solid-state analogs, via the work of Nambu[10] and of Anderson.[11] The theorem states, essentially, that if the Lagrangian

[6] We follow here, as elsewhere, the viewpoint of W. Thirring, *Principles of Quantum Electrodynamics* (Academic Press Inc., New York, 1958), Chap. 14.

[7] J. Goldstone, Nuovo Cimento **19**, 154 (1961).
[8] J. Goldstone, A. Salam, and S. Weinberg, Phys. Rev. **127**, 965 (1962).
[9] Y. Nambu and G. Jona-Lasinio, Phys. Rev. **122**, 345 (1961).
[10] Y. Nambu, Phys. Rev. **117**, 648 (1960).
[11] P. W. Anderson, Phys. Rev. **110**, 827 (1958).

possesses a continuous symmetry group under which the ground or vacuum state is not invariant, that state is, therefore, degenerate with other ground states. This implies a zero-mass boson. Thus, the solid crystal violates translational and rotational invariance, and possesses phonons; liquid helium violates (in a certain sense only, of course) gauge invariance, and possesses a longitudinal phonon; ferro-magnetism violates spin rotation symmetry, and possesses spin waves; superconductivity violates gauge invariance, and would have a zero-mass collective mode in the absence of long-range Coulomb forces.

It is noteworthy that in most of these cases, upon closer examination, the Goldstone bosons do indeed become tangled up with Yang-Mills gauge bosons and, thus, do not in any true sense really have zero mass. Superconductivity is a familiar example, but a similar phenomenon happens with phonons; when the phonon frequency is as low as the gravitational plasma frequency, $(4\pi G\rho)^{1/2}$ (wavelength$\sim 10^4$ km in normal matter) there is a phonon-graviton interaction: in that case, because of the peculiar sign of the gravitational interaction, leading to instability rather than finite mass.[12] Utiyama[13] and Feynman have pointed out that gravity is also a Yang-Mills field. It is an amusing observation that the three phonons plus two gravitons are just enough components to make up the appropriate tensor particle which would be required for a finite-mass graviton.

Spin waves also are known to interact strongly with magnetostatic forces at very long wavelengths,[14] for rather more obscure and less satisfactory reasons. We conclude, then, that the Goldstone zero-mass difficulty is not a serious one, because we can probably cancel it off against an equal Yang-Mills zero-mass problem. What is not clear yet, on the other hand, is whether it is possible to describe a truly strong conservation law such as that of baryons with a gauge group and a Yang-Mills field having finite mass.

I should like to thank Dr. John R. Klauder for valuable conversations and, particularly, for correcting some serious misapprehensions on my part, and Dr. John G. Taylor for calling my attention to Schwinger's work.

[12] J. H. Jeans, Phil. Trans. Roy. Soc. London **101**, 157 (1903).
[13] R. Utiyama, Phys. Rev. **101**, 1597 (1956); R. P. Feynman (unpublished).
[14] L. R. Walker, Phys. Rev. **105**, 390 (1957).

BROKEN SYMMETRY AND THE MASS OF GAUGE VECTOR MESONS*

F. Englert and R. Brout
Faculté des Sciences, Université Libre de Bruxelles, Bruxelles, Belgium
(Received 26 June 1964)

It is of interest to inquire whether gauge vector mesons acquire mass through interaction[1]; by a gauge vector meson we mean a Yang-Mills field[2] associated with the extension of a Lie group from global to local symmetry. The importance of this problem resides in the possibility that strong-interaction physics originates from massive gauge fields related to a system of conserved currents.[3] In this note, we shall show that in certain cases vector mesons do indeed acquire mass when the vacuum is degenerate with respect to a compact Lie group.

Theories with degenerate vacuum (broken symmetry) have been the subject of intensive study since their inception by Nambu.[4-6] A characteristic feature of such theories is the possible existence of zero-mass bosons which tend to restore the symmetry.[7,8] We shall show that it is precisely these singularities which maintain the gauge invariance of the theory, despite the fact that the vector meson acquires mass.

We shall first treat the case where the original fields are a set of bosons φ_A which transform as a basis for a representation of a compact Lie group. This example should be considered as a rather general phenomenological model. As such, we shall not study the particular mechanism by which the symmetry is broken but simply assume that such a mechanism exists. A calculation performed in lowest order perturbation theory indicates that those vector mesons which are coupled to currents that "rotate" the original vacuum are the ones which acquire mass [see Eq. (6)].

We shall then examine a particular model based on chirality invariance which may have a more fundamental significance. Here we begin with a chirality-invariant Lagrangian and introduce both vector and pseudovector gauge fields, thereby guaranteeing invariance under both local phase and local γ_5-phase transformations. In this model the gauge fields themselves may break the γ_5 invariance leading to a mass for the original Fermi field. We shall show in this case that the pseudovector field acquires mass.

In the last paragraph we sketch a simple argument which renders these results reasonable.

(1) Lest the simplicity of the argument be shrouded in a cloud of indices, we first consider a one-parameter Abelian group, representing, for example, the phase transformation of a charged boson; we then present the generalization to an arbitrary compact Lie group.

The interaction between the φ and the A_μ fields is

$$H_{\rm int} = ieA_\mu \varphi^* \overleftrightarrow{\partial}_\mu \varphi - e^2 \varphi^* \varphi A_\mu A_\mu, \qquad (1)$$

where $\varphi = (\varphi_1 + i\varphi_2)/\sqrt{2}$. We shall break the symmetry by fixing $\langle\varphi\rangle \neq 0$ in the vacuum, with the phase chosen for convenience such that $\langle\varphi\rangle = \langle\varphi^*\rangle = \langle\varphi_1\rangle/\sqrt{2}$.

We shall assume that the application of the

theorem of Goldstone, Salam, and Weinberg[7] is straightforward and thus that the propagator of the field φ_2, which is "orthogonal" to φ_1, has a pole at $q=0$ which is not isolated.

We calculate the vacuum polarization loop $\Pi_{\mu\nu}$ for the field A_μ in lowest order perturbation theory about the self-consistent vacuum. We take into consideration only the broken-symmetry diagrams (Fig. 1). The conventional terms do not lead to a mass in this approximation if gauge invariance is carefully maintained. One evaluates directly

$$\Pi_{\mu\nu}(q)=(2\pi)^4 ie^2[g_{\mu\nu}\langle\varphi_1\rangle^2-(q_\mu q_\nu/q^2)\langle\varphi_1\rangle^2]. \quad (2)$$

Here we have used for the propagator of φ_2 the value $[i/(2\pi)^4]/q^2$; the fact that the renormalization constant is 1 is consistent with our approximation.[9] We then note that Eq. (2) both maintains gauge invariance ($\Pi_{\mu\nu}q_\nu=0$) and causes the A_μ field to acquire a mass

$$\mu^2=e^2\langle\varphi_1\rangle^2. \quad (3)$$

We have not yet constructed a proof in arbitrary order; however, the similar appearance of higher order graphs leads one to surmise the general truth of the theorem.

Consider now, in general, a set of boson-field operators φ_A (which we may always choose to be Hermitian) and the associated Yang-Mills field $A_{a,\mu}$. The Lagrangian is invariant under the transformation[10]

$$\delta\varphi_A=\textstyle\sum_{a,A}\epsilon_a(x)T_{a,AB}\varphi_B,$$

$$\delta A_{a,\mu}=\textstyle\sum_{c,b}\epsilon_c(x)c_{acb}A_{b,\mu}+\partial_\mu\epsilon_a(x), \quad (4)$$

where c_{abc} are the structure constants of a compact Lie group and $T_{a,AB}$ the antisymmetric generators of the group in the representation defined by the φ_B.

Suppose that in the vacuum $\langle\varphi_{B'}\rangle\neq 0$ for some B'. Then the propagator of $\sum_{A,B'}T_{a,AB'}\varphi_A \times\langle\varphi_{B'}\rangle$ is, in the lowest order,

$$\left[\frac{i}{(2\pi)^4}\right]\sum_{A,B',C'}\frac{T_{a,AB'}\langle\varphi_{B'}\rangle T_{a,AC'}\langle\varphi_{C'}\rangle}{q^2}$$

$$\equiv\left[\frac{-i}{(2\pi)^4}\right]\frac{(\langle\varphi\rangle T_a T_a\langle\varphi\rangle)}{q^2}.$$

With λ the coupling constant of the Yang-Mills field, the same calculation as before yields

$$\Pi_{\mu\nu}^{a}(q)=-i(2\pi)^4\lambda^2(\langle\varphi\rangle T_a T_a\langle\varphi\rangle)$$

$$\times[g_{\mu\nu}-q_\mu q_\nu/q^2],$$

giving a value for the mass

$$\mu_a^2=-(\langle\varphi\rangle T_a T_a\langle\varphi\rangle). \quad (6)$$

(2) Consider the interaction Hamiltonian

$$H_{\text{int}}=-\eta\bar\psi\gamma_\mu\gamma_5\psi B_\mu-\epsilon\bar\psi\gamma_\mu\psi A_\mu, \quad (7)$$

where A_μ and B_μ are vector and pseudovector gauge fields. The vector field causes attraction whereas the pseudovector leads to repulsion between particle and antiparticle. For a suitable choice of ϵ and η there exists, as in Johnson's model,[11] a broken-symmetry solution corresponding to an arbitrary mass m for the ψ field fixing the scale of the problem. Thus the fermion propagator $S(p)$ is

$$S^{-1}(p)=\gamma p-\Sigma(p)=\gamma p[1-\Sigma_2(p^2)]-\Sigma_1(p^2), \quad (8)$$

with

$$\Sigma_1(p^2)\neq 0$$

and

$$m[1-\Sigma_2(m^2)]-\Sigma_1(m^2)=0.$$

We define the gauage-invariant current J_μ^5 by using Johnson's method[12]:

$$J_\mu^5=-\eta\lim_{\xi\to 0}\bar\psi'(x+\xi)\gamma_\mu\gamma_5\psi'(x),$$

$$\psi'(x)=\exp[-i\textstyle\int_{-\infty}^{x}\eta B_\mu(y)dy^\mu\gamma_5]\psi(x). \quad (9)$$

This gives for the polarization tensor of the

(a) (b)

FIG. 1. Broken-symmetry diagram leading to a mass for the gauge field. Short-dashed line, $\langle\varphi_1\rangle$; long-dashed line, φ_2 propagator; wavy line, A_μ propagator. (a)→ $(2\pi)^4ie^2g_{\mu\nu}\langle\varphi_1\rangle^2$, (b)→ $-(2\pi)^4ie^2(q_\mu q_\nu/q^2)\times\langle\varphi_1\rangle^2$.

pseudovector field

$$\Pi_{\mu\nu}{}^5(q)=\eta^2\frac{i}{(2\pi)^4}\int \mathrm{Tr}\{S(p-\tfrac{1}{2}q)\Gamma_{\nu5}(p-\tfrac{1}{2}q;p+\tfrac{1}{2}q)\times S(p+\tfrac{1}{2}q)\gamma_\mu\gamma_5 - S(p)[\partial S^{-1}(p)/\partial p_\nu]S(p)\gamma_\mu\}d^4p, \quad (10)$$

where the vertex function $\Gamma_{\nu5}=\gamma_\nu\gamma_5+\Lambda_{\nu5}$ satisfies the Ward identity[5]

$$q_\nu\Lambda_{\nu5}(p-\tfrac{1}{2}q;p+\tfrac{1}{2}q)=\Sigma(p-\tfrac{1}{2}q)\gamma_5+\gamma_5\Sigma(p+\tfrac{1}{2}q), \quad (11)$$

which for low q reads

$$q_\nu\Gamma_{\nu5}=q_\nu\gamma_\nu\gamma_5[1-\Sigma_2]+2\Sigma_1\gamma_5 -2(q_\nu p_\nu)(\gamma_\lambda p_\lambda)(\partial\Sigma_2/\partial p^2)\gamma_5. \quad (12)$$

The singularity in the longitudinal $\Gamma_{\nu5}$ vertex due to the broken-symmetry term $2\Sigma_1\gamma_5$ in the Ward identity leads to a nonvanishing gauge-invariant $\Pi_{\mu\nu}{}^5(q)$ in the limit $q\to 0$, while the usual spurious "photon mass" drops because of the second term in (10). The mass of the pseudovector field is roughly η^2m^2 as can be checked by inserting into (10) the lowest approximation for $\Gamma_{\nu5}$ consistant with the Ward identity.

Thus, in this case the general feature of the phenomenological boson system survives. We would like to emphasize that here the symmetry is broken through the gauge fields themselves. One might hope that such a feature is quite general and is possibly instrumental in the realization of Sakurai's program.[3]

(3) We present below a simple argument which indicates why the gauge vector field need not have zero mass in the presence of broken symmetry. Let us recall that these fields were introduced in the first place in order to extend the symmetry group to transformations which were different at various space-time points. Thus one expects that when the group transformations become homogeneous in space-time, that is $q\to 0$, no dynamical manifestation of these fields should appear. This means that it should cost no energy to create a Yang-Mills quantum at $q=0$ and thus the mass is zero. However, if we break gauge invariance of the first kind and still maintain gauge invariance of the second kind this reasoning is obviously incorrect. Indeed, in Fig. 1, one sees that the A_μ propagator connects to intermediate states, which are "rotated" vacua. This is seen most clearly by writing $\langle\varphi_1\rangle=\langle[Q\varphi_2]\rangle$ where Q is the group generator. This effect cannot vanish in the limit $q\to 0$.

*This work has been supported in part by the U. S. Air Force under grant No. AFEOAR 63-51 and monitored by the European Office of Aerospace Research.

[1]J. Schwinger, Phys. Rev. 125, 397 (1962).

[2]C. N. Yang and R. L. Mills, Phys. Rev. 96, 191 (1954).

[3]J. J. Sakurai, Ann. Phys. (N.Y.) 11, 1 (1960).

[4]Y. Nambu, Phys. Rev. Letters 4, 380 (1960).

[5]Y. Nambu and G. Jona-Lasinio, Phys. Rev. 122, 345 (1961).

[6]"Broken symmetry" has been extensively discussed by various authors in the Proceedings of the Seminar on Unified Theories of Elementary Particles, University of Rochester, Rochester, New York, 1963 (unpublished).

[7]J. Goldstone, A. Salam, and S. Weinberg, Phys. Rev. 127, 965 (1962).

[8]S. A. Bludman and A. Klein, Phys. Rev. 131, 2364 (1963).

[9]A. Klein, reference 6.

[10]R. Utiyama, Phys. Rev. 101, 1597 (1956).

[11]K. A. Johnson, reference 6.

[12]K. A. Johnson, reference 6.

VOLUME 13, NUMBER 16 PHYSICAL REVIEW LETTERS 19 OCTOBER 1964

BROKEN SYMMETRIES AND THE MASSES OF GAUGE BOSONS

Peter W. Higgs
Tait Institute of Mathematical Physics, University of Edinburgh, Edinburgh, Scotland
(Received 31 August 1964)

In a recent note[1] it was shown that the Goldstone theorem,[2] that Lorentz-covariant field theories in which spontaneous breakdown of symmetry under an internal Lie group occurs contain zero-mass particles, fails if and only if the conserved currents associated with the internal group are coupled to gauge fields. The purpose of the present note is to report that, as a consequence of this coupling, the spin-one quanta of some of the gauge fields acquire mass; the longitudinal degrees of freedom of these particles (which would be absent if their mass were zero) go over into the Goldstone bosons when the coupling tends to zero. This phenomenon is just the relativistic analog of the plasmon phenomenon to which Anderson[3] has drawn attention: that the scalar zero-mass excitations of a superconducting neutral Fermi gas become longitudinal plasmon modes of finite mass when the gas is charged.

The simplest theory which exhibits this behavior is a gauge-invariant version of a model used by Goldstone[2] himself: Two real[4] scalar fields φ_1, φ_2 and a real vector field A_μ interact through the Lagrangian density

$$L = -\tfrac{1}{2}(\nabla\varphi_1)^2 - \tfrac{1}{2}(\nabla\varphi_2)^2 - V(\varphi_1^2 + \varphi_2^2) - \tfrac{1}{4}F_{\mu\nu}F^{\mu\nu}, \qquad (1)$$

where

$$\nabla_\mu\varphi_1 = \partial_\mu\varphi_1 - eA_\mu\varphi_2,$$

$$\nabla_\mu\varphi_2 = \partial_\mu\varphi_2 + eA_\mu\varphi_1,$$

$$F_{\mu\nu} = \partial_\mu A_\nu - \partial_\nu A_\mu,$$

e is a dimensionless coupling constant, and the metric is taken as $-+++$. L is invariant under simultaneous gauge transformations of the first kind on $\varphi_1 \pm i\varphi_2$ and of the second kind on A_μ. Let us suppose that $V'(\varphi_0^2) = 0$, $V''(\varphi_0^2) > 0$; then spontaneous breakdown of U(1) symmetry occurs. Consider the equations [derived from (1) by treating $\Delta\varphi_1$, $\Delta\varphi_2$, and A_μ as small quantities] governing the propagation of small oscillations about the "vacuum" solution $\varphi_1(x) = 0$, $\varphi_2(x) = \varphi_0$:

$$\partial^\mu\{\partial_\mu(\Delta\varphi_1) - e\varphi_0 A_\mu\} = 0, \qquad (2a)$$

$$\{\partial^2 - 4\varphi_0^2 V''(\varphi_0^2)\}(\Delta\varphi_2) = 0, \qquad (2b)$$

$$\partial_\nu F^{\mu\nu} = e\varphi_0\{\partial^\mu(\Delta\varphi_1) - e\varphi_0 A^\mu\}. \qquad (2c)$$

Equation (2b) describes waves whose quanta have (bare) mass $2\varphi_0\{V''(\varphi_0^2)\}^{1/2}$; Eqs. (2a) and (2c) may be transformed, by the introduction of new variables

$$B_\mu = A_\mu - (e\varphi_0)^{-1}\partial_\mu(\Delta\varphi_1),$$
$$G_{\mu\nu} = \partial_\mu B_\nu - \partial_\nu B_\mu = F_{\mu\nu}, \qquad (3)$$

into the form

$$\partial_\mu B^\mu = 0, \quad \partial_\nu G^{\mu\nu} + e^2\varphi_0^2 B^\mu = 0. \qquad (4)$$

Equation (4) describes vector waves whose quanta have (bare) mass $e\varphi_0$. In the absence of the gauge field coupling ($e = 0$) the situation is quite different: Equations (2a) and (2c) describe zero-mass scalar and vector bosons, respectively. In passing, we note that the right-hand side of (2c) is just the linear approximation to the conserved current: It is linear in the vector potential, gauge invariance being maintained by the presence of the gradient term.[5]

When one considers theoretical models in which spontaneous breakdown of symmetry under a semisimple group occurs, one encounters a variety of possible situations corresponding to the various distinct irreducible representations to which the scalar fields may belong; the gauge field always belongs to the adjoint representation.[6] The model of the most immediate interest is that in which the scalar fields form an octet under SU(3): Here one finds the possibility of two nonvanishing vacuum expectation values, which may be chosen to be the two $Y = 0$, $I_3 = 0$ members of the octet.[7] There are two massive scalar bosons with just these quantum numbers; the remaining six components of the scalar octet combine with the corresponding components of the gauge-field octet to describe

VOLUME 13, NUMBER 16 PHYSICAL REVIEW LETTERS 19 OCTOBER 1964

massive vector bosons. There are two $I=\frac{1}{2}$ vector doublets, degenerate in mass between $Y=\pm1$ but with an electromagnetic mass splitting between $I_3=\pm\frac{1}{2}$, and the $I_3=\pm1$ components of a $Y=0$, $I=1$ triplet whose mass is entirely electromagnetic. The two $Y=0$, $I=0$ gauge fields remain massless: This is associated with the residual unbroken symmetry under the Abelian group generated by Y and I_3. It may be expected that when a further mechanism (presumably related to the weak interactions) is introduced in order to break Y conservation, one of these gauge fields will acquire mass, leaving the photon as the only massless vector particle. A detailed discussion of these questions will be presented elsewhere.

It is worth noting that an essential feature of the type of theory which has been described in this note is the prediction of incomplete multiplets of scalar and vector bosons.[8] It is to be expected that this feature will appear also in theories in which the symmetry-breaking scalar fields are not elementary dynamic variables but bilinear combinations of Fermi fields.[9]

---

[1]P. W. Higgs, to be published.

[2]J. Goldstone, Nuovo Cimento 19, 154 (1961); J. Goldstone, A. Salam, and S. Weinberg, Phys. Rev. 127, 965 (1962).

[3]P. W. Anderson, Phys. Rev. 130, 439 (1963).

[4]In the present note the model is discussed mainly in classical terms; nothing is proved about the quantized theory. It should be understood, therefore, that the conclusions which are presented concerning the masses of particles are conjectures based on the quantization of linearized classical field equations. However, essentially the same conclusions have been reached independently by F. Englert and R. Brout, Phys. Rev. Letters 13, 321 (1964): These authors discuss the same model quantum mechanically in lowest order perturbation theory about the self-consistent vacuum.

[5]In the theory of superconductivity such a term arises from collective excitations of the Fermi gas.

[6]See, for example, S. L. Glashow and M. Gell-Mann, Ann. Phys. (N.Y.) 15, 437 (1961).

[7]These are just the parameters which, if the scalar octet interacts with baryons and mesons, lead to the Gell-Mann–Okubo and electromagnetic mass splittings: See S. Coleman and S. L. Glashow, Phys. Rev. 134, B671 (1964).

[8]Tentative proposals that incomplete SU(3) octets of scalar particles exist have been made by a number of people. Such a rôle, as an isolated $Y=\pm1$, $I=\frac{1}{2}$ state, was proposed for the κ meson (725 MeV) by Y. Nambu and J. J. Sakurai, Phys. Rev. Letters 11, 42 (1963). More recently the possibility that the σ meson (385 MeV) may be the $Y=I=0$ member of an incomplete octet has been considered by L. M. Brown, Phys. Rev. Letters 13, 42 (1964).

[9]In the theory of superconductivity the scalar fields are associated with fermion pairs; the doubly charged excitation responsible for the quantization of magnetic flux is then the surviving member of a U(1) doublet.

GLOBAL CONSERVATION LAWS AND MASSLESS PARTICLES*

G. S. Guralnik,† C. R. Hagen,‡ and T. W. B. Kibble
Department of Physics, Imperial College, London, England
(Received 12 October 1964)

In all of the fairly numerous attempts to date to formulate a consistent field theory possessing a broken symmetry, Goldstone's remarkable theorem[1] has played an important role. This theorem, briefly stated, asserts that if there exists a conserved operator Q_i such that

$$[Q_i, A_j(x)] = \sum_k t_{ijk} A_k(x),$$

and if it is possible consistently to take $\sum_k t_{ijk} \times \langle 0 | A_k | 0 \rangle \neq 0$, then $A_j(x)$ has a zero-mass particle in its spectrum. It has more recently been observed that the assumed Lorentz invariance essential to the proof[2] may allow one the hope of avoiding such massless particles through the introduction of vector gauge fields and the consequent breakdown of manifest covariance.[3] This, of course, represents a departure from the assumptions of the theorem, and a limitation on its applicability which in no way reflects on the general validity of the proof.

In this note we shall show, within the framework of a simple soluble field theory, that it is possible consistently to break a symmetry (in the sense that $\sum_k t_{ijk} \langle 0 | A_k | 0 \rangle \neq 0$) without requiring that $A(x)$ excite a zero-mass particle. While this result might suggest a general procedure for the elimination of unwanted massless bosons, it will be seen that this has been accomplished by giving up the global conservation law usually implied by invariance under a local gauge group. The consequent time dependence of the generators Q_i destroys the usual global operator rules of quantum field theory (while leaving the local algebra unchanged), in such a way as to preclude the possibility of applying the Goldstone theorem. It is clear that such a modification of the basic operator relations is a far more drastic step than that taken in the usual broken-symmetry theories in which a degenerate vacuum is the sole symmetry-breaking agent, and the operator algebra possesses the full symmetry. However, since superconductivity appears to display a similar behavior, the possibility of breaking such global conservation laws must not be lightly discarded.

Normally, the time independence of

$$Q_i = \int d^3x\, j_i^{\,0}(\vec{x}, t)$$

is asserted to be a consequence of the local conservation law $\partial_\mu j^\mu = 0$. However, the relation

$$\partial_\mu \langle 0 | [j_i^{\,\mu}(x), A_j(x')] | 0 \rangle = 0$$

implies that

$$\int d^3x \langle 0 | [j_i^{\,0}(x), A_j(x')] | 0 \rangle = \text{const}$$

only if the contributions from spatial infinity vanish. This, of course, is always the case in a fully causal theory whose commutators vanish outside the light cone. If, however, the theory is not manifestly covariant (e.g., radiation-gauge electrodynamics), causality is a requirement which must be imposed with caution. Since Q_i consequently may not be time independent, it will not necessarily generate local gauge transformations upon $A_j(x')$ for $x^0 \neq x'^0$ despite the existence of the differential conservation laws $\partial_\mu j^\mu = 0$.

The phenomenon described here has previously been observed by Zumino[4] in the radiation-gauge formulation of two-dimensional electrodynamics where the usual electric charge cannot be conserved. The same effect is not present in the Lorentz gauge where zero-mass excitations which preserve charge conservation are found to occur. (These correspond to gauge parts rather than physical particles.) We shall, however, allow the possibility of the breakdown of such global conservation laws, and seek solutions of our model consistent only with the differential conservation laws.

We consider, as our example, a theory which was partially solved by Englert and Brout,[5] and bears some resemblance to the classical theory of Higgs.[6] Our starting point is the ordinary electrodynamics of massless spin-zero particles, characterized by the Lagrangian

$$\mathcal{L} = -\tfrac{1}{2}F^{\mu\nu}(\partial_\mu A_\nu - \partial_\nu A_\mu) + \tfrac{1}{4}F^{\mu\nu}F_{\mu\nu} + \varphi^\mu \partial_\mu \varphi + \tfrac{1}{2}\varphi^\mu \varphi_\mu + ie_0 \varphi^\mu q \varphi A_\mu,$$

where φ is a two-component Hermitian field, and q is the Pauli matrix σ_2. The broken-symmetry condition

$$ie_0 q \langle 0 | \varphi | 0 \rangle = \eta \equiv \begin{pmatrix} \eta_1 \\ \eta_2 \end{pmatrix}$$

will be imposed by approximating $ie_0 \varphi^\mu q \varphi A_\mu$ in the Lagrangian by $\varphi^\mu \eta A_\mu$. The resulting equations of motion,

$$F^{\mu\nu} = \partial^\mu A^\nu - \partial^\nu A^\mu,$$

$$\partial_\nu F^{\mu\nu} = \varphi^\mu \eta,$$

$$\varphi^\mu = -\partial^\mu \varphi - \eta A^\mu,$$

$$\partial_\mu \varphi^\mu = 0,$$

are essentially those of the Brout-Englert model, and can be solved in either the radiation[7] or Lorentz gauge. The Lorentz-gauge formulation, however, suffers from the fact that the usual canonical quantization is inconsistent with the field equations. (The quantization of A_μ leads to an indefinite metric for one component of φ.) Since we choose to view the theory as being imbedded as a linear approximation in the full theory of electrodynamics, these equations will have significance only in the radiation gauge.

With no loss of generality, we can take $\eta_2 = 0$, and find

$$(-\partial^2 + \eta_1^{\,2})\varphi_1 = 0,$$

$$-\partial^2 \varphi_2 = 0,$$

$$(-\partial^2 + \eta_1^{\,2}) A_k^{\,T} = 0,$$

where the superscript T denotes the transverse part. The two degrees of freedom of $A_k^{\,T}$ combine with φ_1 to form the three components of a massive vector field. While one sees by inspection that there is a massless particle in the theory, it is easily seen that it is completely decoupled from the other (massive) excitations, and has nothing to do with the Goldstone theorem.

It is now straightforward to demonstrate the failure of the conservation law of electric charge. If there exists a conserved charge Q, then the relation expressing Q as the generator of rotations in charge space is

$$[Q, \varphi(x)] = e_0 q \varphi(x).$$

Our broken symmetry requirement is then

$$\langle 0 | [Q, \varphi_1(x)] | 0 \rangle = -i\eta$$

or, in terms of the soluble model considered here,

$$\int d^3x' \, \eta_1 \langle 0 | [\varphi_1^0(x'), \varphi_1(x)] | 0 \rangle = -i\eta_1.$$

From the result

$$\langle 0 | \varphi_1^0(x') \varphi_1(x) | 0 \rangle = \partial_0 \Delta^{(+)}(x'-x; \eta_1^2),$$

one is led to the consistency condition

$$\eta_1 \exp[-i\eta_1(x_0'-x_0)] = \eta_1,$$

which is clearly incompatible with a nontrivial η_1. Thus we have a direct demonstration of the failure of Q to perform its usual function as a conserved generator of rotations in charge space. It is well to mention here that this result not only does not contradict, but is actually required by, the field equations, which imply

$$(\partial_0^2 + \eta_1^2) Q = 0.$$

It is also remarkable that if A_μ is given any bare mass, the entire theory becomes manifestly covariant, and Q is consequently conserved. Goldstone's theorem can therefore assert the existence of a massless particle. One indeed finds that in that case φ_1 has only zero-mass excitations.

In summary then, we have established that it may be possible consistently to break a symmetry by requiring that the vacuum expectation value of a field operator be nonvanishing without generating zero-mass particles. If the theory lacks manifest covariance it may happen that what should be the generators of the theory fail to be time-independent, despite the existence of a local conservation law. Thus the absence of massless bosons is a consequence of the inapplicability of Goldstone's theorem rather than a contradiction of it. Preliminary investigations indicate that superconductivity displays an analogous behavior.

The first named author wishes to thank Dr. W. Gilbert for an enlightening conversation, and two of us (G.S.G. and C.R.H.) thank Professor A. Salam for his hospitality.

---

*The research reported in this document has been sponsored in whole, or in part, by the Air Force Office of Scientific Research under Grant No. AF EOAR 64-46 through the European Office of Aerospace Research (OAR), U. S. Air Force.

†National Science Foundation Postdoctoral Fellow.

‡On leave of absence from the University of Rochester, Rochester, New York.

[1]J. Goldstone, Nuovo Cimento 19, 154 (1961); J. Goldstone, A. Salam, and S. Weinberg, Phys. Rev. 127, 965 (1962); S. A. Bludman and A. Klein, Phys. Rev. 131, 2364 (1963).

[2]W. Gilbert, Phys. Rev. Letters 12, 713 (1964).

[3]P. W. Higgs, Phys. Letters 12, 132 (1964).

[4]B. Zumino, Phys. Letters 10, 224 (1964).

[5]F. Englert and R. Brout, Phys. Rev. Letters 13, 321 (1964).

[6]P. W. Higgs, to be published.

[7]This is an extension of a model considered in more detail in another context by D. G. Boulware and W. Gilbert, Phys. Rev. 126, 1563 (1962).

PHYSICAL REVIEW VOLUME 155, NUMBER 5 25 MARCH 1967

Symmetry Breaking in Non-Abelian Gauge Theories*

T. W. B. KIBBLE
Department of Physics, Imperial College, London, England
(Received 24 October 1966)

According to the Goldstone theorem, any manifestly covariant broken-symmetry theory must exhibit massless particles. However, it is known from previous work that such particles need not appear in a relativistic theory such as radiation-gauge electrodynamics, which lacks manifest covariance. Higgs has shown how the massless Goldstone particles may be eliminated from a theory with broken $U(1)$ symmetry by coupling in the electromagnetic field. The primary purpose of this paper is to discuss the analogous problem for the case of broken non-Abelian gauge symmetries. In particular, a model is exhibited which shows how the number of massless particles in a theory of this type is determined, and the possibility of having a broken non-Abelian gauge symmetry with no massless particles whatever is established. A secondary purpose is to investigate the relationship between the radiation-gauge and Lorentz-gauge formalisms. The Abelian-gauge case is reexamined, in order to show that, contrary to some previous assertions, the Lorentz-gauge formalism, properly handled, is perfectly consistent, and leads to physical conclusions identical with those reached using the radiation gauge.

I. INTRODUCTION

THEORIES with spontaneous symmetry breaking (in which the Hamiltonian but not the ground state is symmetric) have played an important role in our understanding of nonrelativistic phenomena like superconductivity and ferromagnetism. Many authors, beginning with Nambu,[1] have discussed the possibility that some at least of the observed approximate symmetries of relativistic particle physics might be interpreted in a similar way. The most serious obstacle has been the appearance in such theories of unwanted massless particles, as predicted by the Goldstone theorem.[2]

In nonrelativistic theories such as the BCS model, the corresponding zero-energy-gap excitation modes may be eliminated by the introduction of long-range forces. The first indication of a similar effect in relativistic theories was provided by the work of Anderson,[3] who showed that the introduction of a long-range field, like the electromagnetic field, might serve to eliminate massless particles from the theory. More recently, Higgs[4] has exhibited a model which shows explicitly how the massless Goldstone bosons are eliminated by coupling the current associated with the broken symmetry to a gauge field. The reasons for the breakdown of the Goldstone theorem in this case have been analyzed by Guralnik, Hagen, and Kibble.[5] The situation is identical with that in the nonrelativistic domain. In either case the theorem is inapplicable in the presence of long-range forces, essentially because the continuity equation $\partial_\mu j^\mu=0$ no longer implies the time independence of expressions like $\int d^3x\,[j^0(x),\phi(0)]$, since the relevant surface integrals do not vanish in the limit of infinite volume. (In the relativistic case, the theorem does apply if we use the Lorentz gauge; but then it tells us nothing about whether the massless particles are physical.) It should be noted that the extension or corollary of the Goldstone theorem discussed by Streater[6] also fails in precisely this case. If long-range fields are introduced, spontaneous symmetry breaking can lead to mass splitting.

As has been emphasized recently by Higgs,[7] it thus appears that the only way of reconciling spontaneous symmetry breaking in relativistic theories with the absence of massless particles is to couple in gauge fields. Another possibility is that Goldstone bosons may turn out to be completely uncoupled and therefore physically irrelevant. In this case, however, the Hilbert space decomposes into the direct product of a physical Hilbert space and a free-particle Fock space for the Goldstone bosons. The broken symmetry appears only in the latter, and no trace of it remains in any physical quantities. In most simple cases, the symmetry transformations leave the physical Hilbert space completely invariant; and in any case they act unitarily on it. Such theories cannot therefore explain observed approximate symmetries. This decoupling of Goldstone modes does occur in the Lorentz-gauge treatment of models like that discussed by Higgs, in which in fact no trace of the original $U(1)$ symmetry remains in the physical states. However it does not occur in corresponding non-Abelian gauge theories, to which the conventional (i.e., Gupta-Bleuler) Lorentz-gauge formation is inapplicable.

It has been suggested by Fuchs[8] that in the case of non-Abelian gauges the massless particles may persist

* The research reported in this document has been sponsored in part by the U. S. Air Force Office of Scientific Research OAR through the European Office Aerospace Research, U. S. Air Force.

[1] Y. Nambu, Phys. Rev. Letters 4, 380 (1960). Y. Nambu and G. Jona-Lasinio, Phys. Rev. 122, 345 (1961); M. Baker and S. L. Glashow, *ibid.* 128, 2462 (1962); S. L. Glashow, *ibid.* 130, 2132 (1962).

[2] J. Goldstone, Nuovo Cimento 19, 154 (1961); J. Goldstone, A. Salam, and S. Weinberg, Phys. Rev. 127, 965 (1962).

[3] P. W. Anderson, Phys. Rev. 130, 439 (1963).

[4] P. W. Higgs, Phys. Letters 12, 132 (1964).

[5] G. S. Guralnik, C. R. Hagen, and T. W. B. Kibble, Phys. Rev. Letters 13, 585 (1964). See also T. W. B. Kibble, in *Proceedings of the Oxford International Conference on Elementary Particles, 1965* (Rutherford High Energy Laboratory, Harwell, England, 1966), p. 19.

[6] R. F. Streater, Phys. Rev. Letters 15, 475 (1965).

[7] P. W. Higgs, Phys. Rev. 145, 1156 (1966).

[8] N. Fuchs, Phys. Rev. 140, B911, (1965).

even after the introduction of gauge fields. His argument is based on the use of the Lorentz gauge, and Schwinger's extended-operator formalism.[9] His conclusions disagree with those reached by Higgs[4], using the radiation gauge. However, Fuchs has already remarked that his method leads to considerable difficulties of interpretation inasmuch as the energy spectrum is not bounded below. This is not the only difficulty which the method encounters.

In order to bring out clearly the relationships between the Lorentz-gauge and radiation-gauge treatments, we shall begin by re-examining, in Secs. II and III, the simple Abelian-gauge case. The Coulomb-gauge treatment given in Sec. II contains a summary of some of Higgs's results, re-expressed in a form appropriate to the comparison with the Lorentz-gauge treatment given in Sec. III. In particular, we aim to show that the correct treatment of $U(1)$ symmetry breaking in Schwinger's extended-operator formalism does not involve any alteration in the "subsidiary condition" which selects the gauge-invariant physical states. (This condition is changed in the method proposed by Fuchs.)

In Sec. IV we go on to discuss the generalization of the model treated by Higgs to an arbitrary non-Abelian group. Our aim here is not to give a complete discussion of this model but mainly to show how the number of massless fields in the theory is determined, in terms of the "canonical number" introduced by Bludman and Klein.[10] Finally, in Sec. V, we exhibit a model involving spontaneously broken $U(2)$ symmetry which is entirely free of massless particles, and moreover, in which the physical states retain clear indications of the underlying symmetry. As in all such theories, the most obvious indication of symmetry breaking is the appearance of an incomplete multiplet of massive scalar particles.

Our results lead to the following conclusion: If all the currents associated with a broken non-Abelian symmetry group are coupled to gauge-vector fields, the number of massless vector bosons remaining in the theory is just the dimensionality of the subgroup of *unbroken* symmetry transformations. In particular, if there are no unbroken components of the symmetry group, then no massless particles remain.

II. COULOMB GAUGE

It will be useful to begin by summarizing in rather different language some of the results obtained by Higgs.[7]

We start with the Goldstone model: a complex scalar field ϕ described by the Lagrangian

$$L=\phi^{\mu*}\partial_\mu\phi+\phi^\mu\partial_\mu\phi^*-\phi^{\mu*}\phi_\mu-V(\phi^*\phi). \quad (1)$$

This is clearly invariant under the constant gauge transformations

$$\phi(x)\to e^{ie\lambda}\phi(x). \quad (2)$$

[9] J. Schwinger, Phys. Rev. **125**, 1043 (1962); 130, 402 (1963).
[10] S. A. Bludman and A. Klein, Phys. Rev. **131**, 2364 (1963).

Consequently, the current

$$j^\mu=-ie(\phi^{\mu*}\phi-\phi^\mu\phi^*) \quad (3)$$

satisfies

$$\partial_\mu j^\mu=0. \quad (4)$$

If $V(\phi^*\phi)$ has a maximum at $\phi^*\phi=0$ and a minimum elsewhere, then we may expect that the expectation value of ϕ in the vacuum (or ground state) is nonzero. From the equations of motion we obtain the consistency condition

$$\langle\partial V/\partial\phi^*\rangle=\langle\phi V'(\phi^*\phi)\rangle=0, \quad (5)$$

which serves to determine the magnitude of $\langle\phi\rangle$. If $\langle\phi\rangle=\eta$ is a possible solution, then so is $\langle\phi\rangle=\eta e^{i\alpha}$. There is, therefore, an infinitely degenerate set of vacuum states, parametrized by the phase α. Formally, transformations from one to another are generated by the "unitary operator" $e^{i\lambda Q}$ with

$$Q=\int d^3x\, j^0(x). \quad (6)$$

However, when $\langle\phi\rangle\neq0$ the integral here is divergent, and the various degenerate vacuums belong to unitarily inequivalent representations.

The Goldstone theorem requires the existence of massless particles in this theory. They may be exhibited by making the polar decomposition

$$\phi=2^{-1/2}\rho e^{i\vartheta}, \quad (7)$$

introduced by Higgs. (We shall ignore problems of operator ordering.) The canonically conjugate variables are the time components of the vectors

$$\rho^\mu=2^{-1/2}(\phi^{\mu*}e^{i\vartheta}+\phi^\mu e^{-i\vartheta}),$$

and

$$\vartheta^\mu=2^{-1/2}i\rho(\phi^{\mu*}e^{i\vartheta}-\phi^\mu e^{-i\vartheta})=-e^{-1}j^\mu. \quad (8)$$

It should be noted that ρ, ρ^μ, and ϑ^μ are all invariant under the transformations (2) while ϑ transforms according to

$$\vartheta(x)\to\vartheta(x)+e\lambda(x). \quad (9)$$

This shows, incidentally, that there is no fundamental distinction between transformations expressible as rotations or translations of the field variables, since one may be converted into the other by a change of variables.

In terms of the new variables, the Lagrangian becomes

$$L=\rho^\mu\partial_\mu\rho+\vartheta^\mu\partial_\mu\vartheta-\tfrac{1}{2}\rho^\mu\rho_\mu-\vartheta^\mu\vartheta_\mu/2\rho^2-V(\tfrac{1}{2}\rho^2). \quad (10)$$

The broken-symmetry condition is expressed by setting

$$\rho=|\eta|+\rho', \quad \langle\rho'\rangle=0. \quad (11)$$

Clearly, the ρ' field describes particles whose mass is determined (to lowest order) by the second derivative of V at $\rho=|\eta|$, while the ϑ field describes massless particles.

Now let us consider the coupling of the current (3) to the electromagnetic field. We have[11]

$$L=-F^{\mu\nu}\partial_\nu A_\mu+\tfrac{1}{4}F^{\mu\nu}F_{\mu\nu}+\phi^{\mu *}(\partial_\mu+ieA_\mu)\phi+\phi^\mu(\partial_\mu-ieA_\mu)\phi^*-\phi^{\mu *}\phi_\mu-V(\phi^*\phi),\quad(12)$$

which is invariant not only under (2) but also under the gauge transformations

$$\phi(x)\to e^{ie\lambda(x)}\phi(x),\quad A_\mu(x)\to A_\mu(x)-\partial_\mu\lambda(x).\quad(13)$$

In terms of the polar variables introduced in (7) and (8), this Lagrangian becomes

$$L=-F^{\mu\nu}\partial_\nu A_\mu+\tfrac{1}{4}F^{\mu\nu}F_{\mu\nu}+\rho^\mu\partial_\mu\rho+\vartheta^\mu\partial_\mu\vartheta-\tfrac{1}{2}\rho^\mu\rho_\mu-\vartheta^\mu\vartheta_\mu/2\rho^2-V(\tfrac{1}{2}\rho^2)+eA_\mu\vartheta^\mu.\quad(14)$$

We now wish to investigate the relationship between the Coulomb-gauge and Lorentz-gauge treatments of this Lagrangian. Let us consider first the Coulomb gauge. To preserve the analogy with our later treatment of the Lorentz gauge, it will be convenient to impose the Coulomb-gauge condition by adding to L a Lagrange multiplier term

$$-C\partial_k A^k.\quad(15)$$

This destroys the invariance under (13), but not that under the constant gauge transformations (2).

It is now convenient to introduce new variables

$$B_\mu=A_\mu+e^{-1}\partial_\mu\vartheta.\quad(16)$$

Then, using the equation

$$\vartheta_\mu=e\rho^2B_\mu$$

to eliminate ϑ_μ from the Lagrangian,[12] we obtain

$$L=-F^{\mu\nu}\partial_\nu B_\mu+\tfrac{1}{4}F^{\mu\nu}F_{\mu\nu}+\tfrac{1}{2}e^2\rho^2B_\mu B^\mu+\rho^\mu\partial_\mu\rho-\tfrac{1}{2}\rho^\mu\rho_\mu-V(\tfrac{1}{2}\rho^2)-C\partial_k(B^k-e^{-1}\partial^k\vartheta).\quad(17)$$

This Lagrangian is still invariant under (2), or its equivalent

$$\vartheta(x)\to\vartheta(x)+e\lambda,\quad(18)$$

but in a completely trivial way. In fact, ϑ is determined by

$$\nabla^2\vartheta=-e\partial_k B^k,\quad(19)$$

and (18) represents merely the arbitrariness in the solution of this equation. (Explicit dependence on the coordinates is ruled out by the requirement of translational invariance.) It may be noted that the equation obtained by variation of ϑ, namely

$$\nabla^2C=0,\quad(20)$$

shows similarly that C is at most a constant.

From the structure of (17) we see that no massless particles remain in the theory. The mass of the scalar particles described by ρ' is as before determined to lowest order by the second derivative of V at $\rho=|\eta|$. Now, however, the massless particles described by ϑ have become the longitudinal modes of the vector field described by B_μ, whose mass is to a first approximation $e|\eta|$.

It may be worth recalling the reasons for the failure of the Goldstone theorem to apply in this case. The essential point is that the continuity equation (4) no longer implies the time independence of the commutator

$$\int d^3x\langle[j^0(x),\phi(0)]\rangle\quad(21)$$

because the relevant surface integral fails to vanish in the limit of infinite volume. Although the operator (6) does not exist, its commutators do exist in a formal sense provided we perform the space integration after the evaluation of the commutator. In the absence of long-range fields Q is time-independent (in the sense that these commutators are so), but, as was pointed out by Guralnik, Hagen, and Kibble,[5] this is no longer true when gauge fields are present. This is easy to verify for our particular model. We have

$$j^\mu=-e^2\rho^2B^\mu,\quad(22)$$

whence

$$\langle[j^\mu(x),\vartheta(0)]\rangle=-e^3(1/\nabla^2)\partial_k\langle[\rho^2B^\mu(x),B^k(0)]\rangle.\quad(23)$$

This form clearly exhibits the nonlocal structure of $\langle[j^i,\vartheta]\rangle$. Inserting a Lehmann spectral form on the right-hand side of this equation, it is easy to see that $\langle[j^0,\vartheta]\rangle$ is causal but that its space integral is not time-independent. Indeed in lowest order it is $-e\cos(e|\eta|x^0)$.

III. LORENTZ GAUGE

Now let us turn to the description of this model in terms of the Lorentz gauge. Following Schwinger,[9] we may impose the Lorentz-gauge condition by adding to L a Lagrange multiplier term

$$-G\partial_\mu A^\mu+\tfrac{1}{2}\alpha GG,\quad(24)$$

where α is an arbitrary constant introduced to allow direct comparison both with Schwinger's formalism ($\alpha=0$) and with the more conventional formalism ($\alpha=1$) adopted by Fuchs. Note that in second-order form (24) is equivalent to

$$-\tfrac{1}{2}\alpha^{-1}(\partial_\mu A^\mu)^2,\quad(25)$$

so that $\alpha=1$ corresponds to the usual Fermi Lagrangian.

[11] This model, or a closely related one, has been discussed in Refs. 4, 5, and 7 and also by F. Englert and R. Brout, Phys. Rev. Letters, 13, 321 (1964).

[12] This is permissible since it is an algebraic equation for ϑ_μ. One is not allowed to solve an equation of motion and substitute the results in the Lagrangian, but one is allowed to substitute explicit solutions obtained without integration. See, for example, T. W. B. Kibble and J. C. Polkinghorne, Nuovo Cimento 8, 74 (1958).

The advantage of the first-order form (24) lies precisely in the possibility of taking $\alpha=0$.

The generator of the gauge transformations (13) is

$$G(\lambda)=\int d^3x\,[\lambda(x)\partial^0G(x)-G(x)\partial^0\lambda(x)]$$

$$=\int d^3x\,[\lambda(j^0-\partial_kF^{0k})-G\partial^0\lambda].$$

In Schwinger's extended-operator formalism the physical states are distinguished by the gauge-invariance requirement

$$G(\lambda)\Psi=0\,, \tag{26}$$

or equivalently,

$$G\Psi=0\,,\quad (j^0-\partial_kF^{0k})\Psi=0\,. \tag{27}$$

On the other hand, in the more familiar Gupta-Bleuler formalism,[13] only the positive frequency components of G are required to annihilate the physical states. Both these formalisms will be considered in the sequel.

The important difference between the Coulomb and Lorentz gauges lies of course in the number of degrees of freedom. Since (15) involves no time derivatives, C is not a dynamical variable. However, in the Lorentz gauge, the canonically conjugate pairs of variables are (A_i,F^{0i}), (G,A^0), (ρ,ρ^0), and (ϑ,ϑ^0). The ρ-field excitations are essentially irrelevant to our discussion. So for simplicity we shall make the approximation of replacing ρ by $|\eta|$. This should be a good first approximation if the mass determined by the second derivative of V is large. Thus we have to consider the Lagrangian

$$L=-F^{\mu\nu}\partial_\nu A_\mu+\tfrac{1}{4}F^{\mu\nu}F_{\mu\nu}+\vartheta^\mu\partial_\mu\vartheta$$
$$-\vartheta^\mu\vartheta_\mu/2|\eta|^2+eA_\mu\vartheta^\mu-G\partial_\mu A^\mu+\tfrac{1}{2}\alpha GG\,. \tag{28}$$

Since this Lagrangian is only quadratic in the field variables, the resulting theory is exactly soluble.

As before, it is convenient to introduce new variables. We write

$$m^2=e^2|\eta|^2 \tag{29}$$

and make a canonical transformation to the pairs (B_i,F^{0i}), (G,G^0), and (ψ,ψ^0), where

$$B_i=A_i+\frac{1}{e}\partial_i\vartheta+\frac{1}{m^2}\partial_iG\,,$$

$$G^0=A^0+\frac{1}{m^2}\partial_kF^{0k}\,, \tag{30}$$

$$\psi=|\eta|\vartheta\,,$$

and

$$\psi^0=\frac{1}{|\eta|}\vartheta^0+\frac{1}{m}\partial_kF^{0k}.$$

[13] See, for example, J. M. Jauch and F. Rohrlich, *The Theory of Photons and Electrons* (Addison-Wesley Publishing Company, Inc., Reading, Massachusetts, 1955), Chap. 2.

To exhibit the covariance of the Lagrangian it is convenient to introduce also new dependent field variables B_0, G^i, and ψ^i, so that we may write

$$L=-F^{\mu\nu}\partial_\nu B_\mu+\tfrac{1}{4}F^{\mu\nu}F_{\mu\nu}+\tfrac{1}{2}m^2B^\mu B_\mu$$
$$+G^\mu\partial_\mu G+\psi^\mu\partial_\mu\psi-\tfrac{1}{2}\psi^\mu\psi_\mu+m\psi^\mu G_\mu+\tfrac{1}{2}\alpha GG\,. \tag{31}$$

Clearly, we have a vector field describing particles of mass m, and two scalar fields (in addition to field ρ' corresponding to the suppressed modes). The vector particles are, of course, precisely those found earlier in the Coulomb gauge.

To discuss the physical significance of the scalar fields in (31), we have to be more precise about the conditions on physical states. Let us first consider Schwinger's extended-operator formalism. Here the fields are not to be regarded as operators in a Hilbert space; they are extended operators acting on a more general space of functionals in which no scalar product is defined. The states are labeled by the eigenvalues of, for example, ψ, G^0, and B_i, all at one time t_0:

$$\Psi=\langle\psi',G^{0\prime},B_i'|\rangle\,. \tag{32}$$

The canonically conjugate variables ψ^0, G, and F^{0i} are represented as functional derivatives with respect to the appropriate variables. Then the functionals representing physical states are distinguished by the gauge-invariance requirement (26) or (27), which may now be written

$$\begin{aligned}G\Psi&=i(\delta/\delta G^{0\prime})\Psi=0\,,\\ -\psi^0\Psi&=i(\delta/\delta\psi')\Psi=0\,.\end{aligned} \tag{33}$$

It follows that Ψ is actually independent of the two scalar fields, and may be represented by a functional of B_i' alone. This is, of course, in conformity with the conclusion reached in the Coulomb-gauge treatment, that only vector particles appear in the physical states.

It should be remarked that the symmetry breaking corresponds to having a nonzero value of η and therefore of the vector particle mass m, and has nothing whatsoever to do with the conditions (33) which should be imposed whether or not the symmetry is broken. The physical states are still gauge invariant in the sense described by (33) even when the original symmetry (2) is broken. Indeed, in this formalism, the local-gauge transformations (13) do not act on the physical states at all, although of course the global transformations (2) do. This is associated with the well-known fact that while the one-parameter group of gauge transformations (2) yields a conservation law for the electric charge, the infinite-parameter group of local transformations (13) does not yield an infinite set of physical conservation laws.

The symmetry-breaking discussed by Fuchs[8] is the breaking of the conditions (33). This means that more states than usual are admitted as physical. This procedure has a number of grave disadvantages, notably the fact that when (33) is relaxed, the energy spectrum is no longer bounded below. It should also be noted that (33) provides the only guarantee that the equations of motion agree, for physical states, with those obtained from the original gauge-invariant Lagrangian. To relax this condition is not merely to break a symmetry but to change the physical equations of motion. It may be that symmetry breaking of this type has some physical relevance, despite these difficulties. What we are concerned to show here, however, is that it is not the method which is the true Lorentz-gauge analog of the Coulomb-gauge formalism described in the preceding section. To achieve agreement between the two approaches it is necessary (and sufficient) to break the symmetry under the transformations (2) while maintaining the conditions (33) intact.

It is interesting to examine in more detail the unphysical fields G and ψ, which satisfy the field equations

$$\partial_\mu G=-m\psi_\mu,\quad \partial_\mu\psi=\psi_\mu-mG_\mu,\qquad \partial_\mu G^\mu=\alpha G,\quad \partial_\mu\psi^\mu=0,\tag{34}$$

whence

$$\Box G=0,\quad \Box\psi=-\alpha mG.\tag{35}$$

From these equations and the canonical commutation rules, we can easily derive covariant commutation relations. The consistency of the conditions (33) for different times is assured by the relation

$$[G(x),G(0)]=0.\tag{36}$$

We also find

$$[G(x),\psi(0)]=-imD(x)=\frac{im}{2\pi}\epsilon(x^0)\delta(x^2),\tag{37}$$

and finally,

$$[\psi(x),\psi(0)]=iD(x)-i\alpha m^2\left[\frac{\partial}{\partial\kappa^2}\Delta(x,\kappa^2)\right]_{\kappa^2=0}=-\frac{i}{2\pi}\epsilon(x^0)[\delta(x^2)+\tfrac{1}{4}\alpha m^2\vartheta(x^2)].\tag{38}$$

In the extended-operator formalism, these relations are not required to possess a representation in a Hilbert space, and pose no particular problem.

However, let us now examine the formalism of Gupta and Bleuler, in which the fields are operators in a Hilbert space of indefinite metric, and the physical states are selected by the condition

$$G^{(+)}|\rangle=0.\tag{39}$$

Then, taking the vacuum expectation value of the commutator (38), and denoting the diagonal elements (±1) of the metric operator by ρ_n, we find

$$\sum_n \rho_n|\langle n|\psi(0)|0\rangle|^2(2\pi)^4[\delta_4(p_n-k)-\delta_4(p_n+k)]=\epsilon(k^0)2\pi\delta(k^2)-\alpha m^2\left[\frac{\partial}{\partial\kappa^2}\{\epsilon(k^0)2\pi\delta(k^2-\kappa^2)\}\right]_{\kappa^2=0}.$$

For $k\neq0$, we can drop the second term in the square brackets on the left-hand side of this equation, and replace $\epsilon(k^0)$ on the right by $\theta(k^0)$. However, in the neighborhood of $k=0$ this is *not* correct. For, although $(\partial/\partial\kappa^2)[\epsilon(k^0)2\pi\delta(k^2-\kappa^2)]$ yields a well-defined distribution in the limit $\kappa^2\to0$, the corresponding symmetric expression $(\partial/\partial\kappa^2)[2\pi\delta(k^2-\kappa^2)]$ does not.[14] In fact, integrating it over a small volume around $k=0$, one finds a result which diverges like $\ln\kappa^2$ as $\kappa^2\to0$. To obtain a well-defined distribution, one must subtract off this infinite term, and consider for example the expression

$$\frac{\partial}{\partial\kappa^2}[2\pi\delta(k^2-\kappa^2)]-2\pi^2\ln\frac{\kappa^2}{M^2}\delta_4(k),$$

for any constant mass M, which does possess a well-defined limit as $\kappa^2\to0$. Thus, we obtain

$$\sum_n \rho_n|\langle n|\psi(0)|0\rangle|^2(2\pi)^4\delta_4(p_n-k)=\theta(k^0)2\pi\delta(k^2)-\alpha m^2\left[\frac{\partial}{\partial\kappa^2}\{\theta(k^0)2\pi\delta(k^2-\kappa^2)\}-\pi^2\ln\frac{\kappa^2}{M^2}\delta_4(k)\right]_{\kappa^2=0}.\tag{40}$$

The arbitrariness in the constant M arises from the fact that division of a distribution equation by $\epsilon(k^0)$ yields a result arbitrary to the extent of a multiple of $\delta_4(k)$. This arbitrary constant may also be regarded as a manifestation of the arbitrary additive constant in the field ψ, since a field translation of ψ would clearly change the left-hand side of (40) by a multiple of $\delta_4(k)$.

It is evident that unless $\alpha=0$, the matrix elements of ψ must be extremely singular. The basic reason for this may be seen from the equations of motion (35). Since the field G has a singularity at $k^2=0$, like $1/k^2$, we see that ψ must have an even worse singularity, like $(1/k^2)^2$. The structure in the right side of (40) is itself well defined[14] (for given M) but can only be obtained by a cancellation between infinite positive and negative terms on the left. It is in this sense that the conventional Lorentz-gauge treatment (which implies $\alpha=1$) is inconsistent.

It is very interesting that the particular case $\alpha=0$ does not encounter these problems. This case is interesting from another point of view also. Since this theory is manifestly covariant, the Goldstone theorem is certainly applicable. In general, the massless bosons it predicts are described by the field G [since $\partial_\mu G$ is the conserved current; note that the proof of the theorem rests on the commutator function (37)]. However, in the special case $\alpha=0$, there are two independent massless fields corresponding to the fact that in that case the Lagrangian is invariant (up to a divergence) not only under the transformations (2) but also under $G(x)\to G(x)+\lambda$.

[14] J. Gårding and S. Lions, Nuovo Cimento Suppl. 14, 9 (1959).

We note that Schwinger's extended-operator formalism is more flexible, even in the simple Abelian-gauge case, than that of Gupta and Bleuler, which works properly only for $\alpha=0$. In the case of non-Abelian gauges, the Gupta-Bleuler formalism is wholly inapplicable, because the analog to G no longer satisfies the free wave equation, and cannot therefore be resolved covariantly into positive and negative frequency components, so that (39) becomes meaningless.

IV. NON-ABELIAN GAUGE MODELS

Let us now consider a model of an n-component real scalar field ϕ, which transforms according to a given n-dimensional representation of a compact Lie group G of dimension g,

$$\phi(x) \to e^{\lambda\cdot T}\phi(x),$$

and

$$\lambda\cdot T=\lambda^A T_A. \tag{41}$$

Here the λ^A are g real parameters and the T_A are g real antisymmetric $n\times n$ matrices obeying the commutation relations of the associated Lie algebra

$$[T_A,T_B]=T_C t^C{}_{AB}. \tag{42}$$

These relations are satisfied in particular by the matrices $t_A=(t^C{}_{AB})$ of the adjoint representation.

The Lagrangian density is taken to be

$$L=\phi^\mu\partial_\mu\phi-\tfrac{1}{2}\phi^\mu\phi_\mu-V(\phi), \tag{43}$$

where ϕ_μ transforms like ϕ, and $V(\phi)$ is invariant, under (41). (The notation implies the use of an invariant scalar product in the n-dimensional space of the variables ϕ.) From the invariance of the Lagrangian we may infer the existence of conserved currents,

$$j_A{}^\mu=-\phi^\mu T_A\phi. \tag{44}$$

In any finite space-time volume, the transformation (41) is generated by the operators

$$\lambda^A\int d^3x\, j_A{}^0(x). \tag{45}$$

However if ϕ has a nonvanishing expectation value so that the symmetry is broken, then as usual the integrals over all space do not exist, and the transformation (41) is not unitarily implementable.

The expectation value $\langle\phi\rangle$ in a translationally invariant (vacuum) state is restricted by the consistency condition

$$\langle\partial V/\partial\phi\rangle=0. \tag{46}$$

If $\langle\phi\rangle=\eta$ is a consistent broken-symmetry solution then so also is $\langle\phi\rangle=e^{\lambda\cdot T}\eta$, for any λ. If we choose a particular η, then all other physically equivalent solutions may be expressed in this form. [There may, of course, be other physically inequivalent disjoint solutions of (46)]. However, not all these will be independent in general, since there may be a subgroup G_η of G which leave η invariant (the *isotropy group* of G at η[15]). This is the subgroup corresponding to symmetries which are *not* broken. Let ν—the "canonical number" of Bludman and Klein[10]—be the number of algebraically independent invariants constructible from η, or equivalently the number of algebraically independent invariant "Hartree conditions" (46). We assume that η can be brought to a canonical form in which only ν of its components are nonzero. Further, we assume that none of these components is accidentally zero. Then the set of equivalent solutions $\langle\phi\rangle$ forms a manifold of dimension $r=n-\nu$. It is clear that this manifold may be identified with the factor space G/G_η (not in general a group). In fact, we may write the representative of each element of G in the form

$$e^{\mu\cdot T}e^{\nu\cdot T}, \tag{47}$$

where $e^{\nu\cdot T}$ is an element of the $(g-r)$-dimensional subgroup G_η, and the remaining r parameters μ serve to parametrize the solutions $\langle\phi\rangle$ by the identification

$$\langle\phi\rangle=e^{\mu\cdot T}\eta. \tag{48}$$

Since these solutions are physically equivalent (though, of course, unitarily inequivalent) there is no essential loss of generality in choosing

$$\langle\phi\rangle=\eta. \tag{49}$$

It will be convenient to adopt a set of coordinates in which the first r elements of η are zero, while the last ν elements are not. It is then useful to make a corresponding "polar decomposition" of ϕ, analogous to (7). We write

$$\phi=e^{\vartheta\cdot T}\rho=e^{\vartheta\cdot T}(\eta+\rho'), \tag{50}$$

where ϑ, like μ, has r components, while the first r components of ρ are zero. We shall distinguish these components by using indices $a, b, \cdots=1, \cdots, r$ and $\alpha, \beta, \cdots=r+1, \cdots, n$.

Consider the action of the generators T_A on ρ. We note that $T_A\rho=0$ for those indices A belonging to elements of the Lie algebra of G_η. A nonzero result occurs only for T_a, $a=1, \cdots, r$. Moreover, consider the matrix

$$X_a{}^b=(T_a\rho)^b, \tag{51}$$

which consists of the components of these vectors in the subspace in which ρ itself vanishes. We assert that this matrix is nonsingular. For, if not, we can find some linear combination of the generators T_a which gives zero acting on all vectors ρ. But then this linear combination should be an element of the Lie algebra of G_η, which it is not.

[15] See, for example, R. Hermann, *Lie Groups for Physicists* (W. A. Benjamin, Inc., New York, 1966), p. 3.

The canonically conjugate variables to ϑ and ρ are, as in (8), the time components of the 4-vectors

$$\vartheta_a{}^\mu=\phi^\mu e^{\vartheta\cdot T}T_b\rho\Lambda^b{}_a, \qquad (52)$$

$$\rho_a{}^\mu=(\phi^\mu e^{\vartheta\cdot T})_a, \qquad (53)$$

where

$$\Lambda^b{}_a=[(1-e^{-\vartheta\cdot t})/\vartheta\cdot t]^b{}_a. \qquad (54)$$

This follows from the relation

$$e^{-\lambda\cdot T}\frac{\partial}{\partial\lambda_A}e^{\lambda\cdot T}=T_B\left(\frac{1-e^{-\lambda\cdot t}}{\lambda\cdot t}\right)^B{}_A. \qquad (55)$$

Note, however, that we have defined Λ as an $r\times r$ submatrix of the $g\times g$ matrix appearing in (55). This is permissible because, as we have seen, $T_B\rho=0$ unless B is one of the first r indices. Thus, in place of $T_B\rho\Lambda^B{}_a$, we can write $T_b\rho\Lambda^b{}_a$.

The currents (44) may all be expressed in terms of the canonical variables ϑ and ϑ^μ. In fact

$$j^\mu{}_A=-\vartheta_a{}^\mu(\Lambda^{-1})^a{}_b(e^{-\vartheta\cdot t})^b{}_A. \qquad (56)$$

In terms of the new variables the Lagrangian (43) may be written

$$L=\rho^\mu\partial_\mu\rho-\tfrac{1}{2}\rho^\mu\rho_\mu-V(\rho)$$
$$+\vartheta_a{}^\mu\partial_\mu\vartheta^a-\tfrac{1}{2}[\{\vartheta_a{}^\mu(\Lambda^{-1})^a{}_b-\rho^\mu T_b\rho\}(X^{-1})^b{}_c]^2. \qquad (57)$$

Here we have used explicitly the fact that X, as defined by (51), is nonsingular. This Lagrangian, of course, retains the invariance under (41). Clearly, ρ and ρ^μ are invariant. The effect on ϑ of an infinitesimal transformation may be written as

$$\delta\vartheta^a=(\Lambda^{-1})^a{}_b(e^{-\vartheta\cdot t})^b{}_A\delta\lambda^A, \qquad (58)$$

while the effect on ϑ^μ may most easily be expressed in the form

$$\delta\vartheta_a{}^\mu=-\vartheta_b{}^\mu\frac{\partial}{\partial\vartheta^a}[(\Lambda^{-1})^b{}_c(e^{-\vartheta\cdot t})^c{}_A]\delta\lambda^A. \qquad (59)$$

The masses of the particles described by the fields ρ' would be principally determined, if the interaction is weak, by the second derivatives of V near the point $\rho=\eta$. Normally, these will all be positive. The absence from (57) of any terms involving ϑ, but not ϑ^μ, is indicative of the fact that the fields ϑ contain the massless excitations required by the Goldstone theorem.

Now let us consider the coupling of the currents (44) or (56) to a set of g gauge fields $A^A{}_\mu$. Thus, we now take

$$L=-\tfrac{1}{2}F_A{}^{\mu\nu}(\partial_\nu A^A{}_\mu-\partial_\mu A^A{}_\nu-t^A{}_{BC}A^B{}_\mu A^C{}_\nu)$$
$$+\tfrac{1}{4}F_A{}^{\mu\nu}F^A{}_{\mu\nu}+\rho^\mu\partial_\mu\rho-\tfrac{1}{2}\rho^\mu\rho_\mu-V(\rho)$$
$$+\vartheta_a{}^\mu\partial_\mu\vartheta^a-\tfrac{1}{2}[\{\vartheta_a{}^\mu(\Lambda^{-1})^a{}_b-\rho^\mu T_b\rho\}(X^{-1})^b{}_c]^2$$
$$+\vartheta_a{}^\mu(\Lambda^{-1})^a{}_b(e^{-\vartheta\cdot t})^b{}_AA^A{}_\mu. \qquad (60)$$

We shall work implicitly in the Coulomb gauge, but will not explicitly indicate the Lagrange multiplier term analogous to (15). It is now convenient to introduce new variables

$$B^A{}_\mu=(e^{-\vartheta\cdot t})^A{}_BA^B{}_\mu+[(1-e^{-\vartheta\cdot t})/\vartheta\cdot t]^A{}_b\partial_\mu\vartheta^b$$

and (61)

$$G_A{}^{\mu\nu}=F_B{}^{\mu\nu}(e^{\vartheta\cdot t})^B{}_A.$$

Then, elimination of the variables ϑ^μ yields the Lagrangian

$$L=-\tfrac{1}{2}G_A{}^{\mu\nu}(\partial_\nu B^A{}_\mu-\partial_\mu B^A{}_\nu-t^A{}_{BC}B^B{}_\mu B^C{}_\nu)$$
$$+\tfrac{1}{4}G_A{}^{\mu\nu}G^A{}_{\mu\nu}+\rho^\mu\partial_\mu\rho-\tfrac{1}{2}\rho^\mu\rho_\mu-V(\rho)$$
$$+\rho^\mu T_a\rho B^a{}_\mu+\tfrac{1}{2}(X^a{}_bB^b{}_\mu)^2. \qquad (62)$$

It is clear from this form that $g-r$ of the g vector fields have zero mass and do not interact directly with the fields ρ, while the remaining r have masses given in lowest order by the mass matrix

$$M_{ab}=-(\eta T_a)_c(T_b\eta)^c, \qquad (63)$$

which is positive definite because of the antisymmetry of the matrices T_a.

We may summarize the situation as follows. Before introducing the gauge-vector fields, we have r massless scalar fields which may be placed in one-to-one correspondence with the *broken* components of the symmetry group, and $n-r=\nu$ massive scalar fields. When the vector fields are coupled in, the ν massive scalar fields remain, but the massless scalar fields combine with r of the vector fields to yield r massive vector fields. We are left finally with $g-r$ massless vector fields corresponding to the unbroken components of the symmetry group. Thus, in order to avoid the appearance of *any* massless particles, it is necessary to choose a representation for which $r=g$; or, in other words, for which the subgroup G_η of elements of G leaving η invariant is trivial or at most a discrete group.

V. A SIMPLE MODEL

As an illustration of the discussion in the preceding section, we shall consider here a simple model of broken $U(2)$ symmetry in which no massless particles remain.

The model contains a complex three-component field $\phi=(\phi_i)$ and four vector fields A_μ and $\mathbf{A}_\mu=(A_{i\mu})$. It is described by the Lagrangian

$$L=-\tfrac{1}{2}F^{\mu\nu}(\partial_\nu A_\mu-\partial_\mu A_\nu)+\tfrac{1}{4}F^{\mu\nu}F_{\mu\nu}$$
$$-\tfrac{1}{2}\mathbf{F}^{\mu\nu}\cdot(\partial_\nu\mathbf{A}_\mu-\partial_\mu\mathbf{A}_\nu-e\mathbf{A}_\mu\times\mathbf{A}_\nu)+\tfrac{1}{4}\mathbf{F}^{\mu\nu}\cdot\mathbf{F}_{\mu\nu}$$
$$+\phi^{\mu*}\cdot(\partial_\mu\phi+e\mathbf{A}_\mu\times\phi+ieA_\mu\phi)$$
$$+\phi^\mu\cdot(\partial_\mu\phi^*+e\mathbf{A}_\mu\times\phi^*-ieA_\mu\phi^*)$$
$$-\phi^{\mu*}\cdot\phi_\mu-V(\phi^*\cdot\phi;|\phi^2|). \qquad (64)$$

This is invariant under infinitesimal transformations of the form

$$\delta\phi=ie\delta\lambda\phi+e\delta\omega\times\phi,\quad \delta\phi_\mu=ie\delta\lambda\phi_\mu+e\delta\omega\times\phi_\mu,$$
$$\delta A_\mu=-\partial_\mu\delta\lambda,\quad \delta F_{\mu\nu}=0,$$
$$\delta\mathbf{A}_\mu=e\delta\omega\times\mathbf{A}_\mu-\partial_\mu\delta\omega,\quad \delta\mathbf{F}_{\mu\nu}=e\delta\omega\times\mathbf{F}_{\mu\nu},$$

which belong to the group $U(1)\times 0(3)\simeq U(1)\times SU(2) = U(2)$.

We may now write

$$\phi = e^{ie\lambda} e^{\vartheta\cdot t} \varrho\,, \tag{65}$$

where

$$t_j = (t^i{}_{jk}) = (e\epsilon_{ijk})$$

and

$$\varrho = 2^{-1/2}(\varrho_1 + i\varrho_2)\,,$$

and we may choose, for example, to set

$$\varrho_1 = \begin{pmatrix} \rho_1 \\ 0 \\ 0 \end{pmatrix}, \quad \text{and} \quad \varrho_2 = \begin{pmatrix} 0 \\ \rho_2 \\ 0 \end{pmatrix}.$$

Then the transformation (61) leads to the equivalent Lagrangian

$$\begin{aligned} L = &-\tfrac{1}{2}G^{\mu\nu}(\partial_\nu B_\mu - \partial_\mu B_\nu) + \tfrac{1}{4}G^{\mu\nu}G_{\mu\nu} \\ &-\tfrac{1}{2}\mathbf{G}^{\mu\nu}\cdot(\partial_\nu \mathbf{B}_\mu - \partial_\mu \mathbf{B}_\nu - e\mathbf{B}_\mu \times \mathbf{B}_\nu) + \tfrac{1}{4}\mathbf{G}^{\mu\nu}\cdot\mathbf{G}_{\mu\nu} \\ &+\rho_1{}^\mu \partial_\mu \rho_1 + \rho_2{}^\mu \partial_\mu \rho_2 - \tfrac{1}{2}\rho_1{}^\mu \rho_{1\mu} - \tfrac{1}{2}\rho_2{}^\mu \rho_{2\mu} \\ &- V(\tfrac{1}{2}(\rho_1{}^2 + \rho_2{}^2), \tfrac{1}{2}|\rho_1{}^2 - \rho_2{}^2|) \\ &+ \tfrac{1}{2}e^2[\rho_1{}^2 B_2{}^\mu B_{2\mu} + \rho_2{}^2 B_1{}^\mu B_{1\mu} \\ &+ \tfrac{1}{2}(\rho_1 + \rho_2)^2 (B^\mu - B_3{}^\mu)(B_\mu - B_{3\mu}) \\ &\qquad + \tfrac{1}{2}(\rho_1 - \rho_2)^2 (B^\mu + B_3{}^\mu)(B_\mu + B_{3\mu})]. \end{aligned} \tag{66}$$

If

$$\langle \rho_{1,2} \rangle = \eta_{1,2}\,,$$

then we have in lowest-order four vector particles of masses

$$\eta_1{}^2, \eta_2{}^2, \quad \text{and} \quad (\eta_1 \pm \eta_2)^2,$$

in addition to the two scalar particles whose masses are determined by the second derivatives of V. Provided that we choose the form of V so that none of these quantities vanish, no zero-mass particles remain in the theory.

It should be remarked that this model is in no sense unusual. Suppose we construct such a model for any group G, with g vector fields transforming according to the adjoint representation, and n scalar fields ϕ transforming according to some other specified representation. Then, except for a few representations corresponding to small values of n, the number of invariants constructible from ϕ will be exactly $\nu = n - g$. In that case the model will be completely free of massless particles. For example, for $SU(2)$, the *only* two irreducible representations for which G_η is not trivial are the one-dimensional identity representation and the three-dimensional adjoint representation. [The two-dimensional (fundamental) representation must be complex. It therefore has four real components, $(n=4)$, and a single invariant $(r=1)$.]

We note certain characteristic features of our model. It is perfectly possible to describe it without ever introducing the notion of symmetry breaking, merely by writing down the Lagrangian (66). Indeed if the physical world were really described by this model, it is (66) rather than (64) to which we should be led by experiment. The only advantage of (64) is that it is easier to understand the appearance of an exact symmetry than of an approximate one. Experimentally, we would discover the existence of a set of four vector bosons with different masses but whose interactions exhibited a remarkable degree of symmetry. We would also discover a pair of scalar particles forming an apparently incomplete multiplet under the group describing this symmetry. In such circumstances it would surely be regarded as a considerable advance if we could recast the theory into a form described by the symmetric Lagrangian (64).

VI. CONCLUSIONS

In this paper we have tried to establish two main points: Firstly, that it is possible to handle the problem of symmetry-breaking consistently in the Lorentz gauge as well as in the Coulomb gauge, and to reach identical conclusions; and, secondly, that in the case of non-Abelian gauge groups (as well as in the Abelian case) the introduction of gauge-vector fields coupled to currents associated with a broken symmetry can serve to eliminate massless particles completely from the theory. The condition that there be no massless particles is also the condition that no components of the symmetry remain unbroken. For each unbroken component there remains a massless vector field. This is of course precisely the physical situation in regard to groups like $SU(3)$. There is just one unbroken component, generated by the electric charge, and one known massless vector boson, the photon.

Considerable difficulties still face any theory of this type. In particular it is not so easy to give the vector bosons a reasonable mass as to give them some nonzero mass. However, it does at least seem worthy of further study.

ACKNOWLEDGMENTS

I am indebted to Dr. G. S. Guralnik for numerous discussions, and to Dr. C. R. Hagen for commenting on an earlier version of the manuscript.

Vol. XXIV (1963) *ACTA PHYSICA POLONICA* Fasc. 6 (12)

QUANTUM THEORY OF GRAVITATION*

By R. P. Feynman

(Received July 3, 1963)

My subject is the quantum theory of gravitation. My interest in it is primarily in the relation of one part of nature to another. There's a certain irrationality to any work in gravitation, so it's hard to explain why you do any of it; for example, as far as quantum effects are concerned let us consider the effect of the gravitational attraction between an electron and a proton in a hydrogen atom; it changes the energy a little bit. Changing the energy of a quantum system means that the phase of the wave function is slowly shifted relative to what it would have been were no perturbation present. The effect of gravitation on the hydrogen atom is to shift the phase by 43 seconds of phase in every hundred times the lifetime of the universe! An atom made purely by gravitation, let us say two neutrons held together by gravitation, has a Bohr orbit of 10^8 light years. The energy of this system is 10^{-70} rydbergs. I wish to discuss here the possibility of calculating the Lamb correction to this thing, an energy, of the order 10^{-120}. This irrationality is shown also in the strange gadgets of Prof. Weber, in the absurd creations of Prof. Wheeler and other such things, because the dimensions are so peculiar. It is therefore clear that the problem we are working on is not the correct problem; the correct problem is what determines the size of gravitation? But since I am among equally irrational men I won't be criticized I hope for the fact that there is no possible, practical reason for making these calculations.

I am limiting myself to not discussing the questions of quantum geometry nor what happens when the fields are of very short wave length. I am not trying to discuss any problems which we don't already have in present quantum field theory of other fields, not that I believe that gravitation is incapable of solving the problems that we have in the present theory, but because I wish to limit my subject. I suppose that no wave lengths are shorter than one-millionth of the Compton wave length of a proton, and therefore it is legitimate to analyze everything in perturbation approximation; and I will carry out the perturbation approximation as far as I can in every direction, so that we can have as many terms as we want, which means that we can go to ten to the minus two-hundred and something rydbergs.

I am investigating this subject despite the real difficulty that there are no experiments. Therefore there is so real challenge to compute true, physical situations. And so I made

---

* Based on a tape-recording of Professor Feynman's lecture at the *Conference on Relativistic Theories of Gravitation*, Jabłonna, July, 1962. — Ed.

believe that there were experiments; I imagined that there were a lot of experiments and that the gravitational constant was more like the electrical constant and that they were coming up with data on the various gravitating atoms, and so forth; and that it was a challenge to calculate whether the theory agreed with the data. So that in each case I gave myself a specific physical problem; not a question, what happens in a quantized geometry, how do you define an energy tensor *etc.*, unless that question was necessary to the solution of the physical problem, so please appreciate that the plan of the attack is a succession of increasingly complex physical problems; if I could do one, then I was finished, and I went to a harder one imagining the experimenters were getting into more and more complicated situations. Also I decided not to investigate what I would call familiar difficulties. The quantum electrodynamics diverges; if this theory diverges, it's not something to be investigated unless it produces any specific difficulties associated with gravitation. In short, I was looking entirely for unfamiliar (that is, unfamiliar to meson physics) difficulties. For example, it's immediately remarked that the theory is non-linear. This is not at all an unfamiliar difficulty; the theory, for example, of the spin 1/2 particles interacting with the electromagnetic field has a coupling term $\bar{\psi}A\psi$ which involves three fields and is therefore non-linear; that's not a new thing at all. Now, I thought that this would be very easy and I'd just go ahead and do it, and here's what I planned. I started with the Lagrangian of Einstein for the interacting field of gravity and I had to make some definition for the matter since I'm dealing with real bodies and make up my mind what the matter was made of; and then later I would check whether the results that I have depend on the specific choice or they are more powerful. I can only do one example at a time; I took spin zero matter; then, since I'm going to make a perturbation theory, just as we do in quantum electrodynamics, where it is allowed (it is especially more allowed in gravity where the coupling constant is smaller), $g_{\mu\nu}$ is written as flat space as if there were no gravity plus $\varkappa$ times $h_{\mu\nu}$, where $\varkappa$ is the square root of the gravitational constant. Then, if this is substituted in the Lagrangian, one gets a big mess, which is outlined here.

$$\mathcal{L} = \frac{1}{\varkappa^2}\int R\sqrt{g}\,d\tau + \frac{1}{2}\int (\sqrt{g}\,g^{\mu\nu}\varphi_{,\mu}\varphi_{,\nu} - m^2\sqrt{g}\,\varphi^2)\,d\tau \tag{1}$$

$$g_{\mu\nu} = \delta_{\mu\nu} + \varkappa h_{\mu\nu}.$$

Substituting and expanding, and simplifying the results by a notation (a bar over a tensor means

$$\bar{x}_{\mu\nu} \equiv \frac{1}{2}(x_{\mu\nu} + x_{\nu\mu} - \delta_{\mu\nu}x_{\sigma\sigma});$$

notice that if $x_{\mu\nu}$ is symmetric, $\bar{\bar{x}}_{\mu\nu} = x_{\mu\nu}$) we get

$$\mathcal{L} = \int (h_{\mu\nu,\sigma}\bar{h}_{\mu\nu,\sigma} - 2\bar{h}_{\mu\sigma,\sigma}\bar{h}_{\mu\sigma,\sigma}) + \frac{1}{2}\int (\varphi_{,\mu}^2 - m^2\varphi^2)\,d\tau +$$

$$+ \varkappa\int\left(\bar{h}_{\mu\nu}\varphi_{,\mu}\varphi_{,\nu} - m^2\frac{1}{2}h_{\sigma\sigma}\varphi^2\right) + \varkappa\int \text{“}hhh\text{”} + \varkappa^2\int \text{“}hh\varphi\varphi\text{”} + \text{etc.} \tag{2}$$

First, there are terms which are quadratic in h; then there are terms which are quadratic in φ, the spin zero meson field variable; then there are terms which are more complicated than quadratic; for example, here is a term with two φ's and one h, which I will write $h\varphi\varphi$ (I have written that one out, in particular); there are terms with three h's; then there are terms which involve two h's and two φ's; and so on and so on with more and more complicated terms. The first two terms are considered as the free Lagrangian of the gravitational field and of the matter.

Now we look first at what we would want to solve problem classically, we take the variation of this with respect to h, from the first term we produce a certain combination of second derivatives, and on the other side a mess involving higher orders than first. And the same with the φ, of course.

$$h_{\mu\nu,\sigma\sigma}-\bar{h}_{\sigma\nu,\sigma\mu}-\bar{h}_{\sigma\mu,\sigma\nu} = \bar{S}_{\mu\nu}(h, \varphi) \tag{3}$$

$$\varphi_{,\sigma\sigma}-m^2\varphi = \chi(\varphi, h). \tag{4}$$

We will speak in the following way: (3) is a wave equation, of which $S_{\mu\nu}$ is the source, just like (4) is the wave equation of which χ is the source. The problem is to solve those equations in succession, and to use the usual methods of calculation of the quantum theory. Inasmuch as I wanted to get into the minimum of difficulties, I just took a guess that I use the same plan as I do in electricity; and the plan in electricity leads to the following suggestion here: that if you have a source, you divide by the operator on the left side of (3) in momentum space to get the propagator field. So I have to solve this equation (3). But as you all know it is singular; the entire Lagrangian in the beginning was invariant under a complicated transformation of g, which in the form of h is the following; if you add to h a gradient plus more, the entire system is invariant:

$$h'_{\mu\nu} = h_{\mu\nu}+2\xi_{\mu,\nu}+2h_{\mu\sigma}\,\xi_{\sigma,\nu}+\xi_\sigma\,h_{\mu\nu,\sigma}, \tag{5}$$

where ξ_μ is arbitrary, and μ and ν should be made symmetric in all these equations. As a consequence of this same invariance in the complet Lagrangian one can show that the source $S_{\mu\nu}$ must have zero divergence $S_{\mu\nu,\nu} = 0$. In fact equations (3) would not be consistent without this condition as can be seen by barring both sides and taking the divergence — the left side vanishes identically. Now, because of the invariance of the equations, in the same way that the Maxwell equations cannot be solved to get a unique vector potential — so these can't be solved and we can't get a unique propagator. But because of the invariance under the transformation some arbitrary choice of a condition on $h_{\mu\nu}$ can be made, analogous to the Lorentz condition $A_{\mu,\mu}=0$ in quantum electrodynamics. Making the simplest choice which I know, I make choice $\bar{h}_{\mu\sigma,\sigma} = 0$. This is four conditions and I have free the four variables ξ_μ that I can adjust to make the condition satisfied by $h'_{\mu\nu}$. Then this equation (3) is very simple, because two terms in (3) fall away and all we have is that the d'Alemberian of h is equal to S. Therefore the generating field from a source $S_{\mu\nu}$ will equal the $\bar{S}_{\mu\nu}$ times $1/k^2$ in Fourier series, where k^2 is the square of the frequency, wave vector; the time part might be called the frequency ω, the space part $\boldsymbol{k}$. This is the analogue of the equation in electricity that says that the field is $1/k^2$ times the current. In the method of quantum field theory, you have a source which generates something, and that may interact later with something else; the iteraction, of course, is $S_{\mu\nu}\, h_{\mu\nu}$; so that, I say, one source may create a potential which acts on another source. So, to take the very simplest example of two interacting systems, let's say S and S', the result would be the following: h would be generated by $S_{\mu\nu}$, and then it would interact with $S'_{\mu\nu}$, so we would get for the interaction of two systems, of two particles, the fundamental interaction that we investigate

$$\varkappa^2 \bar{S}_{\mu\nu} \frac{1}{k^2} S_{\mu\nu}. \tag{6}$$

This represents the law of gravitational interaction expressed by means of an interchange of a virtual graviton. To understand the theory better and to see how far we already arrived we expand it out in components. Let index 4 represent the time, and 3 the direction of **k**, so that 1 and 2 are transverse. The condition $\mathrm{k}_\mu S_{\mu\nu} = 0$ becomes $\omega S_{4\nu} = \mathrm{k} S_{3\nu}$ where k is the magnitude of **k**. Using this, many of the terms involving number 3 component of S can be replaced by terms in number 4 components. After some rearranging there results

$$-2\bar{S}_{\mu\nu} \frac{1}{k^2} S_{\mu\nu} = \frac{1}{\mathrm{k}^2}[S_{44} S'_{44}] + \frac{1}{\mathrm{k}^2}[S_{44}(S'_{11}+S'_{22}) + S'_{44}(S_{11}+S_{22}) +$$

$$+ S_{43} S'_{43} - 4S_{41} S'_{41} - 4S_{42} S'_{42}] + \frac{1}{\mathrm{k}^2 - \omega^2 + i\varepsilon}[(S_{11}-S_{22})\,(S'_{11}-S'_{22}) + 4S_{12}S'_{12}]. \tag{7}$$

There is a singular point in the last term when $\omega = \mathrm{k}$, and to be precise we put in the $+i\varepsilon$ as is well-known from electrodynamics. You note that in the first two terms instead of one over a four-dimensional $\omega^2 - \mathbf{k}^2$ we have here just $1/\mathbf{k}^2$, the momentum itself. S_{44} is the energy density, so this first term represents the two energy densities interacting with no ω dependence which means, in the Fourier transform an interaction instantaneous in time; and $1/\mathbf{k}^2$ means $1/r$ in space, so there's an instantaneous $1/r$ interaction between masses, Newton's law. In the next term there's another instantaneous term which says that Newton's mass law should be corrected by some other components analogous to a kind of magnetic interaction (not quite analogous because the magnetic interaction in electricity already involves a $\mathrm{k}^2 - \omega^2 + i\varepsilon$ propagator rather than just k^2. But the $\mathrm{k}^2 - \omega^2 + it$ in gravitation comes even later and is a much smaller term which involves velocities to the fourth). So if we really wanted to do problems with atoms that were held together gravitationally it would be very easy; we would take the first term, and possibly even the second as the interaction. Being instantaneous, it can be put directly into a Schrödinger equation, analogous to the e^2/r term for electrical interaction. And that take care of gravitation to a very high accuracy, without a quantized field theory at all. However, for still higher accuracy we have to do the radiative corrections, which come from the last term.

Radiation of free gravitons corresponds to the situation that there is a pole in the propagator. There is a pole in the last term when $\omega = \mathrm{k}$, of course, which means that the wave number and the frequency are related as for a mass zero particle. The residue of the pole, we see, is the product of two terms; which means that there are two kinds of waves, one generated by $S_{11} - S_{22}$ and the other generated by S_{12}, and so we have two kinds of transverse polarized waves, that is there are two polarization states for the graviton. The linear combination $S_{11} - S_{22} \pm 2iS_{12}$ vary with angle Θ of rotation in the $1-2$ plane as $e^{\pm 2i\theta}$ so the gravitaton has spin 2, component ± 2 along direction of polarization. Everything is clear directly from the expression (7); I just wanted to illustrate that the propagator (6) of quantum mechanics and all that we know about the classical situation are in evident coincidence.

In order to proceed to make specific calculations by means of diagrams, beside the propagator we need to know just what the junctions are, in other words just what the S's are for a particular problem; and I shall just illustrate how that's done in one example. It is done by looking at the non-quadratic terms in the Lagrangian I've written one out completely. This one has an h and two φ's in the Lagrangian (2). The rules of the quantum mechanics for writting this thing are to look at the h and two φ's: one φ each refers to the in and out particle, and the one h corresponds to the graviton; so we immediately see in that term a two particle interaction through a graviton (see Fig. 1). And we can immediately

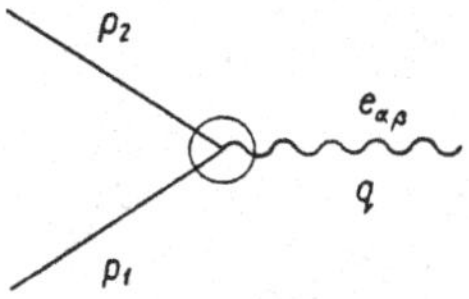

Fig. 1

read off the answer for the interaction this way: if the p_1 and p_2 are the momenta of the particles and q the momentum of the graviton; and $e_{\alpha\beta}$ is the polarization tensor of the plane wave representing the graviton, that is $h_{\alpha\beta} = e_{\alpha\beta}\, e^{iq\cdot x}$, the Fourier expansion of this term gives the amplitude for the coupling of two particles to a graviton

$$p^1_\mu p^2_\nu \bar{e}_{\mu\nu} - \frac{1}{2} m^2 e_{\sigma\sigma}. \tag{8}$$

So this is a coupling of matter to gravity; it is first order, and then there are higher terms; but the point I'm trying to make is that there is no mystery about what to write down — everything is perfectly clear, from the Lagrangian. We have the propagator, we have the couplings, we can write everything. A term like hhh implies a definite formula for the interaction of three gravitons; it is very complicated, and I won't write it down, but you can read it right off directly by substituting momenta for the gradients. That such a term exists is, of course, natural, because gravity interacts with any kind of energy, including its own, so if it interacts with an object-particles it will interact with gravitons; so this is the scattering of a graviton in a gravitational field, which must exist. So that everything is directly readable and all we have to do is proceed to find out if we get a sensible physics. I've already indicated that the physics of direct interactions is sensible; and I go ahead now to compute a number of other things.

To take just one example, we compute the Compton effect, or the analogue rather, of the Compton effect, in which a graviton comes in and out on a particle. The amplitude for this is a sum of terms corresponding to the diagrams of Fig. 2. The amplidute for the first diagram of Fig. 2 is the coupling (8) times the propagator for the intermediate meson which reads $(p^2-m^2)^{-1}$, which is the Fourier transform of the equation (4) which is the propagation of the spin zero particle. Then there is another coupling of the same form as (8). We multiply these together, to get the amplitude for that diagram

$$\left(p^2_\mu p_\nu \bar{e}^b_{\mu\nu} - \frac{1}{2} e_{\mu\mu} m^2\right) \frac{1}{p^2-m^2} \left(p_\sigma p^1_\tau e^a_{\sigma\tau} - \frac{1}{2} \bar{e}_{\sigma\sigma} m^2\right),$$

where we should substitute $p = p^2+q^b = p^1+q^a$. Then you must add similar contributions from the other diagrams.

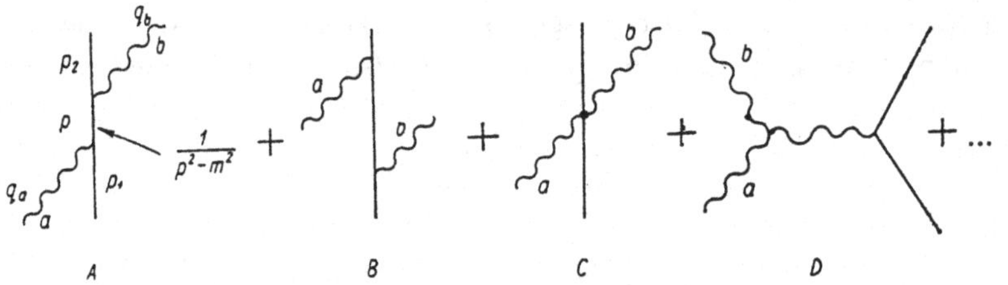

Fig. 2

The third one comes in because there are terms with two h's and two φ's in the Lagrangian. One adds the four diagrams together and gets an answer for the Compton effect. It is rather simple, and quite interesting; that it is simple is what is interesting, because the labour is fantastic in all these things.

But the thing I would like to emphasize is this; in this problem we used a certain wave $e^a_{\alpha\beta}$ for the incoming graviton number "a" say; the question is could we use a different one? According to the theory, it should really be invariant under coordinate transformations and so on, but what it corresponds to here is the analogue of gauge invariance, that you can add to the potential a gradient (see (5)). And therefore it should be that if I changed $e_{\alpha\beta}$ of a particular graviton to $e_{\alpha\beta}+q_\alpha\xi_\beta$ where ξ is arbitrary, and q_α is the momentum of the graviton, there should be no change in the physics. In short, the amplitude should be unchanged; and it is. The amplitude for this particular process is what I call gauge-invariant, or coordinate-transforming invariant. At first sight this is somewhat puzzling, because you would have expected that the invariance law of the whole thing is more complicated, including the last two terms in (5), which I seem to have omitted. But those terms have been included; you see asymptotically all you have to do is worry about the second term, the last two in h's times ξ's are in fact generated by the last diagram, Fig. 2D; when I put a gradient in here for this one, what this means is if I put for the incoming wave a pure gradient, I should get zero. If I put the gradient $q_\alpha\xi_\beta$ in for $e^a_{\alpha\beta}$ on this term D, I get a coupling between ξ and the other field $e^a_{\alpha\beta}$ because of the three graviton coupling. The result, as far as the matter line is concerned is that it is acted on in first order by a resultant field $e^b_{\mu\sigma}\xi_\sigma q^a_\nu +$ $+ \frac{1}{2} q^b_\sigma e_{\mu\nu}\xi_\sigma$ which is just the last two terms in (5). The rule is that the field which acts on the matter itself must be invariant the way described by (5); but here in Fig. 2 I've already calculated all the corrections, the generator and all the necessary non-linear modifications if I take all the diagrams into account. In short, asymptotically far away if I include all kinds of diagrams such as D, the invariance need be checked only for a pure gradient added to an incoming wave. It takes care of the non-linearities by calculating them through the interaction.

I would like, now, to emphasize one more point that is very important for our later discussion. If I add a gradient, I said, the result was zero. Let's call a the one graviton coming in and b the other one in every diagram. The result is zero if I use a gradient for a, only if b is a free graviton with no source; that is if it is either really an honest graviton with $(q^b)^2 = 0$, or a pure potential, which is a solution of the free wave equation. That is unlike electrodynamics, where the field b could have been any potential at all and adding a gradient to a would have made no difference. But in gravity, it must be that b is a pure wave; the reason is very simple. There is no way to avoid this by changing any propagators; this is not a disease — there is a physical reason. The reason can be seen as follows: If this b had a source let me modify my diagrams to show the source of b, suppose some other matter particle made the b, so we add onto each b line a matter line at the end, like Fig. 3a. (*E.g.* Fig. 2a becomes Fig. 3b *etc.*)

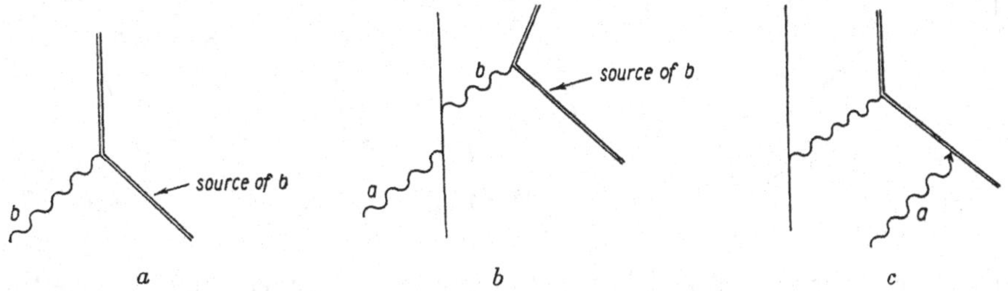

Fig. 3

Now, if b isn't a free wave, but it had a source, the situation is this. If this "a" field is taken as a gradient field which operates everywhere on everything in the diagram it should give zero. But we forgot something; there's another type of diagram, if the "a" is supposed to act on everything, one of which looks like Fig. 3c, in which the "a" itself acts on the source of b and then b comes over to interact with the original matter. In other words, among all the diagrams where there is a source, there's also these of type 3c. The sum of all diagrams is zero; but the sum of those like Fig. 2 without those of type 3c is not zero, and therefore if I were to just calculate the diagrams of Fig. 2 and forget about the source of b and then put a gradient in for "a" the result cannot be zero, but must be getting ready to cancel the terms from the likes of 3c when I do it right. That will turn out to be an important point to emphasize. I have done a lot of problems like this, without closed loops but I won't bore you with all the problems and answers; there's nothing new, I mean nothing interesting, in the sense that no apparent difficulties arise.

However, the next step is to take situations in which we have what we call closed loops, or rings, or circuits, in which not all momenta of the problem are defined. Let me just mention something. I've analyzed this method both by doing a number of problems, and by a mathematical high-class elegant technique — I can do high class mathematics too, but I don't believe in it, that's the difference. I have to check it in a problem. I can prove that no matter how complicated the problem is, if you take it in the order in which there are no rings, in which every momentum is determined, the invariance is satisfied, the system is independent of what choice I made of gauge and of the propagator I made in the beginning; and everything is all right, there are no difficulties. I emphasize that this contains all the classical cases, and so I 'm really saying there are no difficulties in the classical gravitation theory. This is not meant as a grand discovery, because after all, you've been worrying about all these difficulties that I say don't exist, but only for you to get an idea of the calibration — what I mean by difficulties! If we take the next case, let's say the interaction of two particles in a higher order, then you get diagrams of which I'll only begin to write a few of them. One that looks like this in which two gravitons are exchanged,

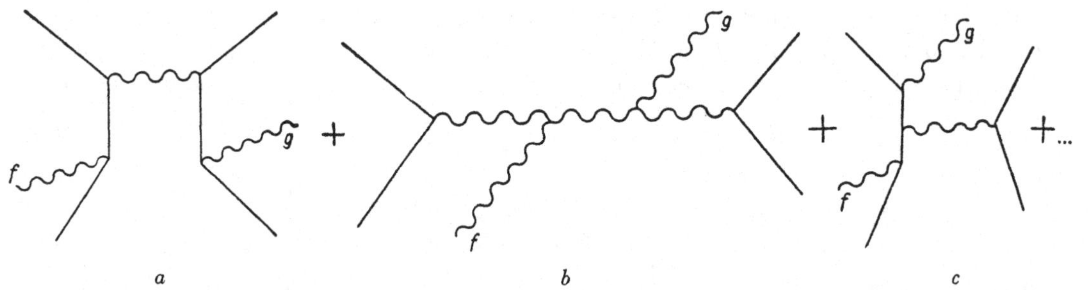

Fig. 4

or, for instance, a graviton gets split into two gravitons and then come back — these are only the beginning of a whole series of frightening-looking pictures, which correspond to the problem of calculating the Lamb shift, or the radiative corrections to the hydrogen atom. When I tried to do this, I did it in a straightforward way, following all the rules, putting in the propagator $1/k^2$, and so on. I had some difficulties, the thing didn't look gauge invariant but that had to do with the way I was making the cutoffs, because the stuff is infinite. Shortage of time does'nt permit me to explain the way I got around all those things, because in spite of getting around all those things the result is nevertheless definitely incorrect. It's gauge-invariant, it's perfectly O.K. looking, but it is definitely incorrect. The reason I knew it was incorrect is the following. In order to get it gauge-invariant, I had to do a lot of pushing and pulling, and I got the feeling that the thing might not be unique. I figured that maybe somebody else could do it another way or something, and I was rather suspicious, so I tried to get more tests for it; and a student of mine, by the name of Yura, tested to see if it was unitary; and what that means is the following: Let me take instead of this scattering problem, a problem of Fig. 4 in which time runs vertically, a problem which gives the same diagrams but in which time is running horizontally, which is the annihilation of a pair, to produce another pair, and we are calculating second order corrections to that problem. Let's suppose for simplicity that in the final state the pair is in the same state as before.

Then, adding all these diagrams gives the amplitude that if you have a pair, particle and antiparticle, they annihilate and recreate themselves; in other words it's the amplitude that the pair is still in the same state as a function of time. The amplitude to remain in the same state for a time T in general is of the form

$$e^{-i\left(E_0-i\frac{\gamma}{2}\right)T}$$

you see that the imaginary part of the phase goes as $e^{-\frac{\gamma}{2}T}$; which means that the probability of being in a state must decrease with time. Why does the probability decrease in time? Because there's another possibility, namely, these two objects could come together, annihilate, and produce a real pair of gravitons. Therefore, it is necessary that this decay rate of the closed loop diagrams in Fig. 4 that I obtain by directly finding the imaginary part of the sum agrees with another thing I can calculate independently, without looking at the closed loop diagrams. Namely, what is the rate at which a particle and antiparticle annihilate into two gravitons? And this is very easy to calculate (same set of diagrams as Fig. 2, only turned on its side). I calculated this rate from Fig. 2, checked whether this rate agrees with the rate at which the probability of the two particles staying the same decreases (imaginary part of Fig. 4), and it does not check. Somethin'gs the matter.

This made me investigate the entire subject in great detail to find out what the trouble is. I discovered in the process two things. First, I discovered a number of theorems, which as far as I know are new, which relate closed loop diagrams and diagrams without closed loop diagrams (I shall call the latter diagrams "trees"). The unitarity relation which I have just been describing, is one connection between a closed loop diagram and a tree; but I found a whole lot of other ones, and this gives me more tests on my machinery. So let me just tell you a little bit about this theorem, which gives other rules. It is rather interesting. As a matter of fact, I proved that if you have a diagram with rings in it there are enough theorems altogether, so that you can express any diagram with circuits completely in terms of diagrams with trees and with all momenta for tree diagrams in physically attainable regions and on the mass shell. The demonstration is remarkably easy. There are several ways of demonstrating it; I'll only chose one. Things propagate from one place to another, as I said, with amplitude $1/k^2$. When translated into space, that's a certain propagation function which you might call $K_+(1, 2)$, a function of two positions, 1, 2, in space-time. It represents, in the past, incoming waves and in the future, it represents outgoing waves; so you have waves come in and out; and that's the conventional propagator, with the $i\varepsilon$ and so on, as usually represented. However, this is only a solution of the propagators's equation, the wave equation I mean; it is a special solution, as you all know. There are other solutions; for instance there is a solution which is purely retarded, which I'll call K_{ret} and which exists only inside the future light-cone. Now, if you have two Green's functions for the same equation they must differ by some solution of the homogeneous equation, say K_x. That means K_x is a solution of the free wave equation and $K_+ = K_{\text{ret}} + K_x$. In a ring like Fig. 4a we have a whole product of these K_+'s. For example, for four points 1, 2, 3, 4 in a ring we have a product like this: $K_+(1, 2)K_+(2, 3)K_+(3, 4)K_+(4, 1)$ (all K's are not the same, some of them belong to the gravitons and some are propagators for the particles and so on).

But now let us see what happens if we were to replace one (or more) of these K_+ by K_x, say $K_+(1, 2)$ is $K_x(1, 2)$? Then between 1, 2 we have just free particles, you've broken the ring; you've got an open diagram, because K_x is free wave solution, and this means it's an integral over all real momenta of free particles, on the mass shell and perfectly honest. Therefore if we replace one of K_+ by K_x then that particular line is opened; and the process is changed to one in which there is a forward scattering of an extra particle; there's a fake particle that belongs to this propagator that has to be integrated over, but it's a free diagram — it is now a tree, and therefore perfectly definite and unique to calculate. But I said that I could open every diagram; the reason is this. First I note that if I put K_{ret} for every K in a ring, I get zero

$$K_{\text{ret}}(1, 2)K_{\text{ret}}(2, 3)K_{\text{ret}}(3, 4)K_{\text{ret}}(4, 1) = 0 \tag{9}$$

for to be non zero t_1 must be greater than t_2, $t_2 > t_3$, $t_3 > t_4$ and $t_4 > t_1$ which is impossible. Now make the substitution $K_{\text{ret}} = K_+ - K_x$ in (9). You get either all K_+ in each factor, which is the closed loop we want; or at least one K_x, which are represented by tree diagrams. Since the sum is zero, closed loops can be represented as integrals over tree diagrams. I was surprised I had never noticed this thing before.

Well, then I checked whether these diagrams of Fig. 4 when opened into trees agreed with the theorem. I mean I hoped that the theorem proved for other meson theories would agree in principle for the gravity case, such that on opening a virtual graviton line the tree would correspond to forward scattering of free graviton waves. And it does not work in the gravity case. But, you say, how could it fail, after you just demonstrated that it ought to work? The reason it fails is the following: This argument has to do with the position of the poles in the propagators; a typical propagator is a factor $1/(k^2-m^2+i\varepsilon)$, the $+i\varepsilon$ due to the poles, and all I'm doing here is changing the rule about the poles and picking up an extra delta function $\delta(k^2-m^2)$ as a consequence, which is the free wave coming in and out. What I want these free waves to represent in the gravity case are physical gravitons and not something wrong. They do represent waves of $q^2 = 0$ of course, but, as it turns out, not with the correct polarization to be free gravitons. I'd like to show it. It has to do with the numerator, not the denominator. You see the propagator that I wrote before, which was $S_{\mu\nu}$ times $1/(k^2+i\varepsilon)$ times $\bar{S}'_{\mu\nu}$, is being replaced by $S_{\mu\nu}\,\delta(q^2)\bar{S}'_{\mu\nu}$. Now when I make $q^2 = 0$ I have a free wave instead of arbitıary momentum. This should be a real graviton or else there's going to be physical trouble. It ins't; although it is of zero momentum, it is not transverse. It does not make any difference in understanding the point so forget one index in $S_{\mu\nu}$ — it's a lot of extra work to carry the other index so just imagine there's one index: $S_\mu S_\mu \delta(q^2)$. This combination $S_\mu S'_\mu$, is $S_4S'_4-S_3S'_3-S_1S'_1-S_2S'_2$, where 4 is the time and 3 is the direction, say, of momentum of the four-vector q. Then 1 and 2 are transverse, and those are the only two we want. (Please appreciate I ıemoved one index — I can make it more elaborate, but it is the same idea.) That is we want only $-S_1S'_1-S_2S'_2$ instead of the sum over four. Now what about this extra term $S_4S'_4-S_3S'_3$? Well, it is S_4-S_3 times $S'_4+S'_3$ plus S_4+S_3 times $S'_4-S'_3$. But S_4-S_3 is proprtional to $q_\mu S_\mu$ (suppressing one index) because q_4 in this notation is the frequency and equals q_3, if we assume the 3-direction is the direction of the momentum. So S_4-S_3 is the response of the system to a gradient potential, which we proved was zero in our invariance discussion. Therefore, we have shown $(S_4-S_3)/(S_4'+S_3') = 0$ and this should be accounted for by purely transverse wave contributions. But it ins't, and it isn't because *the proof that the response to a gradient potential is zero required that the other particle that was interacting was an honest free graviton.* And four plus three in $S_4'+S_3'$ is not honest — it's not transverse, it is not a correct kind of graviton. You see, the only way you can get a polarization 4+5 going in the 4—3 direction is to have what I call longitudinal response; it's not a transverse wave. Such a wave could only be generated by an artificial source here of some silly kind; it is not a free wave. When there's an artificial source for one graviton, even the another is a pure gradient, the sum of all the diagrams does not give zero. If the beam is not exactly that of a free wave, perfectly transverse and everything, the argument that the gradient has to be zero must fail, for the reason outlined previously.

Although this gradient for S_4-S_3 is what I want and I hoped it was going to be zero I forgot that the other end of it — $S_4'+S_3'$ is a funny wave which is not a gradient, and which is not a free wave — and therefore you do not get zero and should not get zero, and something is fundamentally wrong.

Incidentally I investigated further and discovered another very interesting point. There is another theory, more well-known to meson physicists, called the Yang-Mills theory, and I take the one with zero mass; it is a special theory that has never been investigated in great detail. It is very analogous to gravitation; instead of the coordinate transformation group being the source of everything, it's the isotopic spin rotation group that's the source of everything. It is a non-linear theory, that's like the gravitation theory, and so forth. At the suggestion of Gell-Mann I looked at the theory of Yang-Mills with zero mass, which has a kind of gauge group and everything the same; and found exactly the same difficulty. And therefore in meson theory it was not strictly unknown difficulty, because it should have been noticed by meson physicists who had been fooling around the Yang-Mills theory. They had not noticed it because they're practical, and the Yang-Mills theory with zero mass obviously does not exist, because a zero mass field would be obvious; it would come out of nuclei right away. So they didn't take the case of zero mass and investigate it carefully. But this disease which I discovered here is a disease which exist in other theories. So at least there is one good thing: gravity isn't alone in this difficulty. This observation that Yang-Mills was also in trouble was of very great advantage to me; it made everything much easier in trying to straighten out the troubles of the preceding paragraph, for several reasons. The main reason is if you have two examples of the same disease, then there are many things you don't worry about. You see, if there is something different in the two theories it is not caused by that. For example, for gravity, in front of the second derivatives of $g_{\mu\nu}$ in the Lagrangian there are other g's, the field itself. I kept worrying something was going to happen from that. In the Yang-Mills theory this is not so, that's not the cause of the trouble, and so on. That's one advantage — it limits the number of possibilities. And the second great advantage was that the Yang-Mills theory is enormously easier to compute with than the gravity theory, and therefore I continued most of my investigations on the Yang-Mills theory, with the idea, if I ever cure that one, I'll turn around and cure the other. Because I can demonstrate one thing; line for line it's a translation like music transcribed to a different score; everything has its analogue precisely, so it is a very good example to work with. Incidentally, to give you some idea of the difference in order to calculate this diagram Fig. 4b the Yang-Mills case took me about a day; to calculate the diagram in the case of gravitation I tried again and again and was never able to do it; and it was finally put on a computing machine—I don't mean the arithmetic, I mean the algebra of all the terms coming in, just the algebra; I did the integrals myself later, but the algebra of the thing was done on a machine by John Matthews, so I couldn't have done it by hand. In fact, I think it's historically interesting that it's the first problem in algebra that I know of that was done on a machine that has not been done by hand.

Well, what then, now you have the difficulty; how do you cure it? Well I tried the following idea: I assumed the tree theorem to be true, and used it in reverse. If every closed ring diagram can be expressed as trees, and if trees produce no trouble and can be computed, then all you have to do is to say that the closed loop diagram is the sum of the corresponding tree diagrams, that it should be. Finally in each tree diagram for which a graviton line has been opened, take only real transverse graviton to represent that term. This then serves as the definition of how to calculate closed-loop diagrams; the old rules, involving a propagator $1/k^2+i\varepsilon$ *etc.* being superseded. The advantage of this is, first, that it will be gauge invariant, second, it will be unitary, because unitarity is a relation between a closed diagram and an open one, and is one of the class of relations I was talking about, so there's no difficulty. And third, it's completely unique as to what the answer is; there's no arbitrary fiddling around with different gauges and so forth, in the inside ring as there was before. So that's the plan.

Now, the plan requires, however, one more point. It's true that we proved here that every ring diagram can be broken up into a whole lot of trees; but, a given tree is not gauge invariant. For instance the tree diagram of Fig. 2A is not. Each one of the four diagrams of Fig. 2 is not gauge-invariant, nor is any combination of them except the sum of all four. So the thing is the following. Suppose I take all the processes, all of them that belong together in a given order; for example, all the diagrams of fourth order, of which Fig. 4 illustrates three; I break the whole mess into trees, lots of trees. Then I must gather

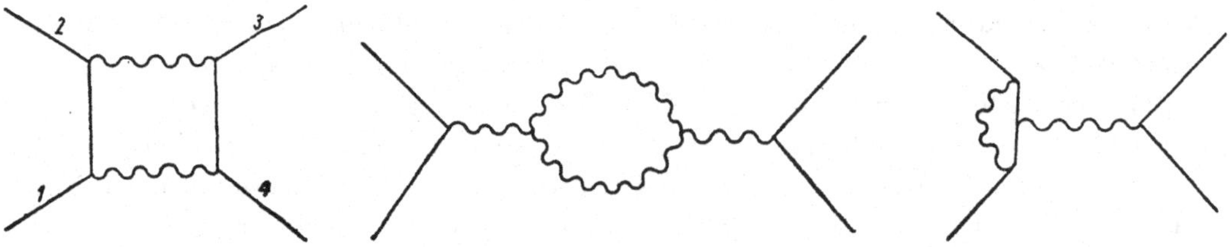

Fig. 5

the trees into baskets again, so that each basket contains the total of all of the diagrams of some specific process (for example the four diagrams of Fig. 2), you see, not just some particular tree diagram but the complet set for some process. The business of gathering the tree diagrams together in bunches representing all diagrams for complet processes is important, for only such a complet set is gauge invariant. The question is: Will any odd tree diagrams be left out or can they all be gathered into processes? The question is: Can we express the closed ring diagrams for some process into a sum over various other processes of tree diagrams for these processes?

Well, in the case with one ring only, I am sure it can be done, I proved it can be done and I have done it and it's all fine. And therefore the problem with one ring is fundamentally solved; because we say, you express it in terms of open parts, you find the processes that they correspond to, compute each process and add them together.

You might be interested in what the rule is for one ring; it's the sum of several pieces: first it is the sum of all the processes which you get in the lower order, in which you scatter one extra particle from the system. For instance, in Fig. 4 we have the rings for two particles scattering. There is no external graviton but there are two internal ones; now we compute in the same order a new problem in which there are two particles scattering, but while that's happening another particle, for example a graviton scatters forward. Some of the diagrams for this are illustrated in Fig. 5. State f the same state as g; so another graviton comes in and is scattered forward. In other words we do the forward scattering of an extra graviton. In addition, from breaking matter lines we have terms for the forward scattering of an extra positron, plus the forward scattering of an extra electron, and so on; one adds the forward scattering of every possible extra particle together. That is the first contribution. But when you break up the trees, you also sometimes break two lines, and then you get diagrams like Fig. 6 with two extra particles scattering (here a graviton and electron) so it turns out you must now subtract all the diagrams with two extra particles of all kinds scattering. Then add all diagrams with 3 extra particles scattering and so on. It's a nice rule, its's quite beautiful; it took me quite a while to find; I have other proofs for orther cases that are easy to understand.

Now, the next thing that anybody would ask which is a natural, interesting thing to ask, is this. Is it possible to go back and to find the rule by which you could have integrated the closed rings directly? In other words, change the rule for integrating the closed rings, so that when you integrate them in a more natural fashion, with the new method, it will

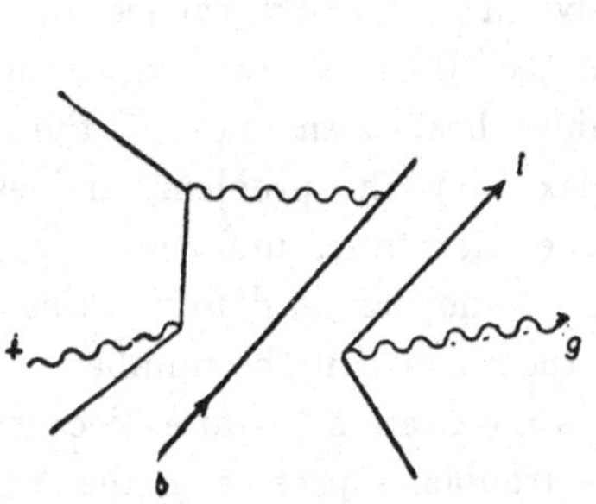

Fig. 6

give the same answer as this unique, absolute, definite thing of the trees. It's not necessary to do this, because, of course, I've defined everything; but it's of great interest to do this, because maybe I'll understand what I did wrong before. So I investigated that in detail. It turns out there are two changes that have to be made — it's a little hard to explain in terms of the gravitation of which I'll only tell about one. Well, I'll try to explain the other, but it might cause some confusion. Because I have to explain in general what I'm doing when I do a ring. Most what it corresponds to is this: first you subtract from the Lagrangian this

$$\int \sqrt{g}\, \bar{H}^{\mu\nu}{}_{;\nu} H^{\sigma}_{\mu;\sigma}\, d\tau.$$

In that way the equation of motion that results is non-singular any more. Let me write what it really is so that there's no trouble. You say to me what is this, there's a g in it and an H in it? Yes. In doing a ring, there's a field variation over which you're integrating, which I call H; and there's a g — which is the representative of all the outside disturbances which can be summarized as being an effective external field g. And so you add to the complicated Lagrangian that you get in the ordinary way an extra term, which makes it no longer singular. That's the first thing; I found it out by trial and error before, when I made it gauge invariant. But then secondly, you must subtract from the answer, the result that you get by imagining that in the ring which involves only a graviton going around, instead you calculate with a different particle going around, an artificial, dopey particle is coupled to it. It's a vector particle, artificially coupled to the external field, so designed as to correct the error in this one. The forms are evidently invariant, as far as your g-space is concerned; these are like tensors in the g world; and therefore it's clear that my answers are gauge invariant or coordinate transformable, and all that's necessary. But are also quantum-mechanically satisfactory in the sense that they are unitary.

Now, the next question is, what happens when there are two or more loops? Since I only got this completely straightened out a week before I came here, I haven't had time to inwestigate the case of 2 or more loops to my own satisfaction. The preliminary investigations that I have made do not indicate that it's going to be possible so easily gather the things into the right barrels. It's surprising, I can't understand it; when you gather the trees into processes, there seems to be some loose trees, extra trees. I don't understand them at the moment, and I therefore do not claim that this method of quantization can be obviously and evidently carried on to the next order. In short, therefore, we are still not sure, of the radiative corrections to the radiative corrections to the Lamb shift, the uncertainty lies in energies of the order of magnitude of 10^{-255} rydbergs. I can therefore relax from the problem, and say: for all practical purposes everything is all right. In the meantime, unfortunately, although I could retire from the field and leave you experts who are used to working in gravitation to worry about this matter, I can't retire on the claim that the number is so small and that the thing is now really irrational, if it was not irrational before. Because, unfortunately, I also discovered in the process that the trouble is present in the Yang-Mills theory; and secondly I have incidentally discovered a tree-ring connection which is of very great interest and importance in the meson theories and so on. And so I'm stuck to have to continue this investigation, and of course you all appreciate that this is the secret reason for doing any work, no matter how absurd and irrational and academic it looks; we all realize that no matter how small a thing is, if it has physical interest and is thought about carefully enough, you're bound to think of something that's good for something else.

DISCUSSION

Møller: May I, as a non-expert, ask you a very simple and perhaps foolish question. Is this theory really Einstein's theory of gravitation in the sense that if you would have here many gravitons the equations would go over into the usual field equations of Einstein?

Feynman: Absolutely.

Møller: You are quite sure about it?

Feynman: Yes, in fact when I work out the fields and I don't say in what order I'm working, I have to do it in an abstract manner which includes any number of gravitons; and then the formulas are definitely related to the general theory's formulas; and the invariance is the same; things like this that you see labelled as loops are very typical quantum-mechanical things; but even here you see a tendency to write things with the right derivatives, gauge invariant and everything. No, there's no question that the thing is the Einsteinian theory. The classical limit of this theory that I'm working on now is a non-linear theory exactly the same as the Einsteinian equations. One thing is to prove it by equations; the other is to check it by calculations. I have mathematically proven to myself so many things that aren't true. I'm lousy at proving things — I always make a mistake. I don't notice when I'm doing a path integral over an infinite number of variables that the Lagrangian does not depend upon one of them, the integral is infinite and I've got a ratio of two infinities and I could get a different answer. And I don't notice in the morass of things that something, a little limit or sign, goes wrong. So I always have to check with calculations; and I'm very poor at calculations — I always get the wrong answer. So it's a lot of work in these things. But I've done two things. I checked it by the mathmatics, that the forms of the mathematical equations are the same; and then I checked it by doing a considerable number of problems in quantum mechanics, such as the rate of radiation from a double star held together by quantum-mechanical force, in several orders and so on, and it gives the same answer in the limit as the corresponding classical problem. Or the gravitational radiation when two stars — excuse me, two particles — go by each other, to any order you want (not for stars, then they have to be particles of specified properties; because obviously the rate of radiation of the gravity depends on the give of the starstides are produced). If you do a real problem with real physical things in in then I'm sure we have the right method that belongs to the gravity theory. There's no question about that. It can't take care of the cosmological problem, in which you have matter out to infinity, or that the space is curved at infinity. It could be done I'm sure, but I haven't investigated it. I used as a background a flat one way out at infinity.

Møller: But you say you are not sure it is renormalizable.

Feynman: I'm not sure, no.

Møller: In the limit of large number of gravitons this would not matter?

Feynman: Well, no; you see, there is still a classical electrodynamics; and it's not got to do with the renormalizability of quantum electrodynamics. The infinities come in different places. It's not a related problem.

Rosen: I'm not sure of this, not being one of the experts; but I have the impression that because of the non-linearity of the Einstein equations there exists a difficulty of the following kind. If the linear equations have a solution in the form of an infinite plane monochromatic wave, there does not seem to correspond to that a more exact solution; because you get piling up of energies in space and the solution then diverges at infinity. Could that have any bearing on the accuracy of this kind of calculation?

Feynman: No, I take that into account by a series of corrections. A single graviton is not the same thing as an infinite gravitational wave, because there's a limited energy in it. There's only one $\hbar\omega$.

Rosen: But you're using a momentum expansion which involves infinite waves.

Feynman: Yes, there are corrections. You see what happens if one calculates the corrections. If you have here a graviton coming in this way, then there are corrections for such a ring as this and so on. And these produce first, a divergence as usual; but second, a term in the logarithm of q^2; which means that if this thing is absolutely a free plane

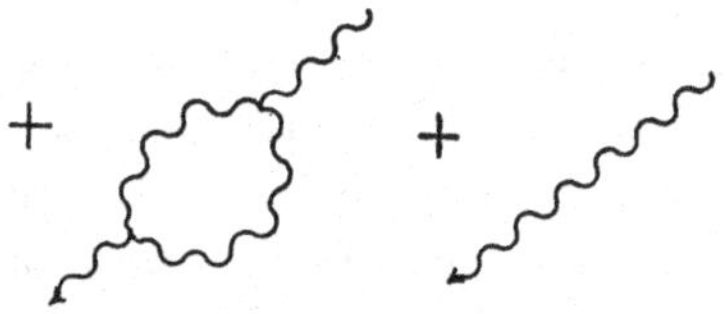

wave, there's no meaning to the correction. So it must be understood in this way, that the thing was emitted some time far in the past, and is going to be absorbed some time in the future; and has not absolutely been going on forever. Then there's a very small coefficient in front of the logarithm and then for any reasonable q^2, like the diameter of the universe or something, I can still get a sensible answer; this is the shadow of the phenomenon you're talking about, that the corrections to the propagation of a graviton, dependent on the logarithm of the momentum squared carried by the graviton and which would be infinite if it were really a zero momentum graviton exactly. And so a free graviton just like that does not quite exist. And this is the correction for that. Strictly we would have to work with wave packets, but they can be of very large extent compared to the wave length of the gravitons.

Anderson: I'd like to ask if you get the same difficulty in the electromagnetic case that you did in the Yang-Mills and gravitational cases?

Feynman: No, sir, you do not. Gauge invariance of diagrams such as Fig. 2 (there is no $2D$) is satisfied whether b is a free wave or not. That is because photons are not the source of photons; they are uncharged.

Anderson: The other thing I would like to suggest is that in putting of things into baskets, you might be able to get easily by always only starting out with vacuum diagrams and opening those successively.

Feynman: I tried that and it didn't go successfully.

Ivanenko: If I understood you correctly, you had used in the initial presentation the transmutation of two particles into gravitons. Yes?

Feynman: It was one of the examples.

Ivanenko: Yes. This process was considered, perhaps in a preliminary manner, by ourselves and by Prof. Weber and Brill. I ask you two questions. Do you possess the effective cross-section? Can you indicate the effects for which high-energy processes play an important role?

Feynman: I never went to energies more than one billion-billion BeV. And then the cross-sections of any of these processes are infinitesimal.

Ivanenko: They increase very, very sharply with energy. Yes, because the radiation is quadrupole, so it increases sharply in contrast to the electromagnetic transmutation of an electron-positron pair.

Feynman: It increases very sharply indeed. On the other hand, it starts out so low that one has to go pretty far to get anywhere. And the distance that you have to go is involved in this thing — the thing that's the analogue of $e^2/\hbar c$ in electricity, which is 1/137 is non-existent in gravitation; it depends on the problem; this is so because of the dimensions of G. So if E is the energy of some process, then if you take $GE^2/\hbar c$ you get an equivalent to this $e^2/\hbar c$. It may be less than that, but at least it can't be any bigger than this. So in order to make this thing to be of the order of 1%, in which case the rate is similar to the rate of photon annihilation, at ordinary energies, we need the GE^2 to be of the order of hc, and as has been pointed out many times, that's an energy of the order 10^{-5} grams, which is 10^{18} BeV. You can figure out the answer right away; just take the energy that you are interested in, square, multiply by G and divide by hc; if that becomes something, then you're getting somewhere. You still might not get somewhere, because the cross-section might not go up that fast, but at least it can't get up any worse than that. So I think that in order to get an appreciable effect, you've got to go to ridiculous energies. So you either have a ridiculously small effect or a ridiculous energy.

Weber: I have a cross-section which may be a partial answer to Ivanenko's question. Could I write it on the board? We have carried out a canonical quantization, which is not as fancy as the one you have just heard about; but considering the interaction of photons and gravitons; and it turns out that even in the linear approximation that one has the possibility of the graviton production by scattering of photons in a Coulomb field. And the scattering cross-section for this case turns out to be $8\pi^2$ times the constant of gravitation times the energy of the scatterer times the thickness of the scatterer in the direction of propagation of the photon through it divided by c^4. This assumes that all of the dimensions of the scatterer are large in comparison with the wave length of the photon. We obtained this result by quantization, and noticed that it didn't have Planck's constant in it, so we turned around and calculated it classically. Now, if one puts numbers in this, one finds that the scattering cross-section of a galaxy due to a uniform magnetic field through it is 10^{28} cm^2, a much larger number than the object that you talked about. This represents a conversion of photons into gravitons of about 1 part in 10^{16}. This is of course too small to measure. Also, we considered the possibility of using this cross-section for a laboratory experiment in which one had a scatterer consisting, say of a million gauss magnetic field over something like a cubic meter. This turned out to be entirely impossible, a result in total contradiction to what has appeared in the Russian literature. In fact, the theory of fluctuations shows that for a laboratory experiment involving the production of gravitons by scattering of photons in a Coulomb field, the scattered power has to be greater than twice the square root of kT times the photon power divided by the averaging time of the experiment. I believe that the incorrect results that have appeared in the literature have been due to the statement that ΔP has to be greater than kT over τ; dimensionally these things are the same, but order of magnitude-wise this kind of experiment for the scatterer of which I spoke requires something like 10^{50} watts. Maybe I can say something about this afternoon; I don't want to take any more time.

De Witt: I should like to ask Prof. Feynman the following questions. First, to give us a careful statement of the tree theorem; and then outline, if he can to a brief extent, the nature of the proof of the theorem for the one-loop case, whish I understand does work. And then, to also show in a little bit more detail the structure and nature of the fictitious particle needed if you want to renormalize everything directly with the loops. And if you like, do it for the Yang-Mills, if things are prettier that way.

Feynman: I usually don't find that to go into the mathematical details of proofs in a large company is a very effective way to do anything; so, although that's the question that you asked me — I'd be glad to do it — I could instead of that give a more physical explanation of why there is such a theorem; how I thought of the theorem in the first place, and things of this nature; although I do have a proof — I'm not trying to cover up.

De Witt: May we have a statement of the theorem first?

Feynman: That I do not have. I only have it for one loop, and for one loop the careful statement of the theorem is... — look, let me do it my way. First — let me tell you how I thought of this crazy thing. I was invited to Brussels to give a talk on electrodynamics — the 50th anniversary of the 1911 Solvay Conference on radiation. And I said I'd make believe I'm coming back, and I'm telling an imaginary audience of Einstein, Lorentz and so on what the answer was. In other words, there are going to be intelligent guys, and I'll tell them the answer. So I tried to explain quantum electrodynamics in a very elementary way, and started out to explain the self-energy, like the hydrogen Lamb shift. How can you explain the hydrogen Lamb shift easily? It turns out you can't at all — they didn't even know there was an atomic nucleus. But, never mind. I thought of the following. I would explain to Lorentz that his idea that he mentioned in the conference, that classically the electromagnetic field could be represented by a lot of oscillators was correct. And that Planck's idea that the oscillators are quantized was correct, and that Lorentz's suggestion, which is also in that thing, that Planck should quantize the oscillators that the field is equivalent to, was right. And it was really amusing to discover that all that was in 1911. And that the paper in which Planck concludes that the energy of each oscillator was not $n\hbar\omega$ but $(n+1/2)\hbar\omega$ which was also in that, was also right; and that this produced a difficulty, because each of the harmonic oscillators of Lorentz in each of the modes had a frequency of $\hbar\omega/2$ which is an infinite amount of energy, because there are an infinite number of modes. And that that's a serious problem in quantum electrodynamics and the first one we have to remove. And the method we use to remove it is to simply redefine the energy so that we start from a different zero, because, of course, absolute energy doesn't mean anything. (In this gravitational context, absolute energy does mean something, but it's one of the technical points I can't discuss, which did require a certain skill to get rid of, in making a gravity theory; but never mind.) Now look — I make a little hole in the box and I let in a little bit of hydrogen gas from a reservoir; such a small amount of hydrogen gas, that the density is low enough that the index of refraction in space differs from one by an amount proportional to A, the number of atoms. With the index being somewhat changed, the frequency of all the normal modes is altered. Each normal mode has the same wavelenght as before, because it must fit into the box; but the frequencies are all altered. And therefore the $\hbar\omega$'s should all be shifted a trifle, because of the shift of index, and therefore there's a slight shift of the energy. Although we subtract $\hbar\omega/2$ for the vacuum, there's a correction when we put the gas in; and this correction is proportional to the number of atoms, and can be associated with an energy for each atom. If you say, yes, but you had that energy already when you had the gas in back in the reservoir, I say, but let us only compare the difference in energy between the $2S$ and $2P$ state. When we change the excitation of the hydrogen gas from $2S$ to $2P$ then it changes its index without removing anything; and the energy difference that is needed to change the energy from $2S$ to the $2P$ for all these atoms is not only the energy that you calculate with disregard of the zero point energy; but the fact is that the zero point energy is changed very slightly. And this very slight difference should be the Lamb effect. So I thought, it's a nice argument; the only question is, is it true. In the first place it's interesting, because as you well know the index differs from one by an amount which is proportional to the forward scattering for γ rays of momentum k and therefore that shift in energy is essentially the sum over all momentum states of the forward scattering for γ rays of momentum k. So I looked at the forward scattering and compared it with the right formula for the Lamb shift, and it was not true, of course; it's too simple an argument. But then I said, wait, I forgot something. Dirac, explained to us that there are negative energy states for the electron but that the whole sea of negative energy states is filled. And, of course, if I put the hydrogen atoms in here all those electrons in negative energy states are also ascattering off the hydrogen atoms; and therefore their states are all shifted; and therefore the energy levels of all those are shifted a tiny bit. And therefore there's shift in the energy due to those. And so there must be an additional term which is the forward scattering of positrons, which is the same as scattering of negative energy electrons. Actually, for the symmetry of things it is better to take half the case where you make the positrons the holes and the other half where you make the electrons the holes; so it should be 1/2 forward scattering by electrons, 1/2 scattering by positrons and scattering by γ rays — the sum of all those forward scattering amplitudes ought to equal the self-energy of the hydrogen atom. And thats' right. And it's simple, and it's very peculiar. The reason it's peculiar is that these forward scatterings are real processes. At last I had discovered a formula I had always wanted, which is a formula for energy differences (which are defined in terms of virtual fields) in terms of actual measurable quantities, no matter how difficult the experiment may be — I mean I have to be able to scatter these things. Many times in studying the energy difference due to electricity (I suppose) between the proton and the neutron, I had hoped for a theorem which would go something like this — this energy difference between proton and neutron must be equal to the following sum of a bunch of cross-sections for a number of processes, but all real physical processes, I don't care how hard they are to measure. So this is the beginning of such a formula. It's rather surprising.

716

It's not the same as the usual formula — it's equal to it but it's not the same. I have no formulation of the laws of quantum gravidynamics; I have a proposal on how to make the calculations. When I make the proposal on how to do the closed loops, the obvious proposal does not work; it gives non-unitarity and stuff like that. So the obvious proposal is no good; it works O.K. for trees; so how am I going to define the answer for would correspond to a ring? The one I happen to have chosen is the following: I take the ring in general for any meson theory, one closed ring can be written as equivalent to a whole lot of processes each one of which is trees. I then define, as my belief as to what the ring ought to be in the grand theory, that it's going to be also equal to the corresponding physical set of trees. When I said this is equal to this. I didn't worry about gauge or anything else; what I means was, if these weren't gravitons but photons or any other neutral object — it doesn't make any difference what they are — this theorem is right. So I suppose it's right also for real gravitons, and I suppose also that what's being scattered is only transverse and is only a real free graviton with $q^2 = 0$. Therefore, I say let this ring equal this set of trees. Every one of these terms can be completely computed — it's a tree. And it's gauge invariant; that is, if I added an extra potential on the whole thing, another outside disturbance of a type which is nothing but a coordinate transformation — in short a pure gradient wave — to the whole diagram then it comes on to all of these processes; but it makes no effect on any of them, and therefore makes no effect on the sum; and therefore I know my definition of this ring is gauge-invariant. Second, unitarity is a property of the breaking of this diagram; the imaginary part of this equals something; if you take the imaginary part of this side, it's already broken up, in fact, and you can prove immediately that it's the correct unitarity rule. Therefore it's going to be unitarity and so on and so on. And so I therefore define gravity with one ring in this way. Now what prevents me from doing it with two rings? The lack of a complet statement of what two rings is equal to in terms of processes; that is I can open the ring all right; but I can't put the pieces — the broken diagrams — back together again into complete sets that each one is a complete physical process. In other words some of them correspond to the scattering of a graviton, but leaving out some diagrams. But the scattering of a graviton leaving out diagrams is no longer gauge invariant, I mean, not evidently gauge invariant, and so the power of the whole thing collapses. I don't know what to do with it. So that's the situation; that's why it is crucial to the particular plan. There's always, of course, another way out. And that's the following (and that's what I tried to describe at the end of the talk — maybe I talked too fast): After all now I've defined what this results is equal to — by definition not that you should do a loop some way and get this, but that a loop is equal to this by defition, and I'm not going to do a loop any other way. But, of course, from a practical point of view or from the point of view purely of interest, the question is, can you come back now and calculate the ring directly by some particular mathematical shenanigans, and get the same answer as you get by adding the trees. And I found the way to do that. I have another way, in other words, to do the ring integral directly. I have to subtract something from a vector particle going around the instead of a graviton to get the answer right. So I known the rule, and I know why the rule is, and I have a proof of the rule for one loop. I have two ways of extending. I can either break this two loop diagram open and get it back into the processes, like I did with the one ring — where so far I'm stuck. Or,

I can take the rule which I found here and try to guess the generalization for any number of rings. Also stuck. But I've only had a week, gentlemen; I've only been able to straighten out the difficulty of a single ring a week ago when I got everything cleaned up. It's more than a week — I had to take a lot of time checking and checking; but I was only finished checking to make sure of everything for this conference. And of course you're always asking me about the thing I haven't had time to make sure about yet, and I'm sorry; I worked hard to be sure of something, and now you ask me about those things I haven't had time. I hoped that I would be able to get it. I still have a few irons to try; I'm not completely stuck—maybe.

DeWitt: Because of the interest of the tricky extra particle that you mentioned at the end, and its possible connection, perhaps, with some work of Dr Białynicki-Birula, have you got far enough on that so that you could repeat it with just a little more detail? The structure of it and what sort of an equation it satisfies, and what is its propagator? These are technical points, but they have an interest.

Feynman: Give me ten minutes. And let me show how the analysis of these tree diagrams, loop diagrams and all this other stuff is done mathematical way. Now I will show you that I too can write equations that nobody can understand. Before I do that I should like to say that there are a few properties that this result has that are interesting. First of all in the Yang-Mills case there also exists a theory which violates the original idea of symmetry of the isotopic spin (from which was originally invented) by the simple assumption that the particle has a mass. That means to add to the Lagrangian a term $-\mu^2 a_\mu a^\mu$ where a_μ is an isotopic vector. You add this to the Lagrangian. This destroys the gauge invariance of the theory — it's just like electrodynamics with a mass, it's no longer gauge-invariant, it's just a dirty theory. Knowing that there is no such field with zero mass people say: „let's put the mass term on". Now when you put a mass term on it is no longer gauge invariant. But then it is also no longer singular. The Lagrangian is no longer singular for the same reason that it is not invariant. And therefore everything can be solved precisely. The propagator instead of being $\delta_{\mu\nu}$ between two currents is

$$\frac{\delta_{\mu\nu}-q_\mu q_\nu/\mu^2}{q^2-\mu^2}, \tag{10}$$

where q_μ is the momentum of propagating particle. The factor $1/(q^2-\mu^2)$ is typical for mass μ but the part $-q_\mu q_\nu/\mu^2$ is an important term which can be taken to be zero in electrodynamics but it is not obvious whether it can be taken to be zero in the case of Yang-Mills theory. In fact it has been proved it cannot be taken to be zero; this propagator is used between two currents. I am using the Yang-Mills example instead of the gravity example. I really want only the case $\mu^2 = 0$, and am asking whether I can get there by first calculating finite μ^2, then taking the limit $\mu^2 = 0$.

Now, with $\mu^2 \neq 0$ this is a definite propagator and there are no ambiguities at the closed rings, the closed loops. I have no freedom, I must compute this propagator. I mean there is no reason for trouble, and there is no trouble. There is no gauge invariance either.

And of course I checked. I broke the rings and I computed by the broken ring theorem method a closed loop problem of fair complexity (which in fact was the interaction of two electrons). I computed it by the open ring method and by the closed ring method, and of course it agreed, there is no reason that it shouldn't. It turned out that for tree diagrams you don't have to worry about this $q_\mu q_\nu/\mu^2$ term, you can drop it — but not for the closed ring — only for tree. Therefore the tree diagrams have the definite limit as μ^2 goes to zero. And yet I have the closed ring diagram which is equal to the tree diagram when the mass is anything but zero, and therefore it ought to be true that the limit as μ^2 goes to zero of the ring is equal to the case when $\mu = 0$. It sounds like a great idea why don't you define the desired $\mu^2 = 0$ theory that way? Answer: You can't put μ^2 equal zero in the form (10). You can't do it because of the $q_\mu q_\nu/\mu^2$. So it was necessary next to see if there is a way to re-express the ring diagrams, for the case with $\mu^2 \neq 0$, in a new form with a propagator different from (10), that didn't have a μ^2 in it, in such a form that you can take the limits as μ^2 goes to zero. Then that would be a new way to do the μ equal zero case; and that's the way I found the formula. I'll try to explain how to find that theory.

We start with a definite theory, the Yang-Mills theory with a mass (the reason I do that is that there's no ambiguity about what I am trying to do) and later on I take the mass to zero, then the theory works something like this. You have the Lagrangian $\mathcal{L}(A, \varphi)$ which involves the vector potential of this field and the fields φ representing the matter with which this object is interacting for zero mass, to which, for finite mass we add the term $\mu^2 A_\mu A_\mu$. This is the Lagrangian that has to be integrated and the idea is that you integrate this over all fields A and φ; and that is the answer for the amplitude of the problem

$$X = \int e^{\int \mathcal{L}(A,\varphi)d\tau + \mu^2 A_\mu A_\mu d\tau} DAD\varphi. \tag{11}$$

But wait, what about the initial and final conditions? You have certain particles coming in and going out. To simplify things (this is not essential) I'll just study the case that corresponds only to gravitons in and out. I'll call them gravitons and mesons even though they are vector particles. The question is first, what is the right answer if you have gravitons represented by plane waves, A_1, A_2, A_3 ... going in (positive frequency in A_1) or out (negative frequency). You make the following field up. Let A_{asym} be defined as α times the wave function A_1 that represents the first graviton coming in a plane wave, plus β times A_2 plus γ times A_3 and so on.

$$A_{\text{asym}} = \alpha A_1 + \beta A_2 + \gamma^{A_3} \ldots, \quad A \to A_{\text{asym}}. \tag{12}$$

Then you calculate this integral (11) subject to the condition that A approaches A_{asym} at infinity. The result of this is of course a function of α, β, γ ... and so on. Then what you want for X is just the term first order in α, β, γ ... That means just one of each these gravitons coming in and out. That's the right formula for a regular theory, for meson theory, You calculate the integral subject to the asymptotic condition, when you imagine all these waves, but you take the first order perturbation with respect to each one of the incoming waves. You never let the same photon operate twice; a photon operating twice is not a photon, it is a classical wave. So you take the derivative of this with respect to α, β, γ and so on, then setting them all equal to zero. That's problem. (In general there's φ asymptotic too.)

Now the way I happened to do this is the following: Let us call A_0 the A which satisfies the classical eqatiuons of motion, which in this particular case will be

$$\left.\frac{\partial \mathcal{L}}{\partial A}\right|_{A^\circ} + \mu^2 A^\circ = 0 \tag{13}$$

I solve this subject to the condition that A_0 equals A_{asym}. In other words, I find what is the maximum or minimum — whatever it is — of the action in (11), subject to the asymptotic condition. That's the beginning of analysing this.

The next thing is to make the simple substitution $A = A_0 + B$ and put it back in equation (11). Then if you take $\mathcal{L}$ od A_0+B (if B is negligible you get $\mathcal{L}$ of A_0 and so forth) so you get something like this

$$e^{i[\mathcal{L}(A_0)+\mu^2 A_0 A_0]} \int e^{[\mathcal{L}(A^0+B)-\mathcal{L}(A^0)]+\mu^2 BB+2\mu^2 AB}\, DB. \tag{14}$$

The integral is over all B, and B must go to zero asymptotically. This business can be expanded in powers of B.

$$\mathcal{L}(A+B)-\mathcal{L}(A)+\mu^2 BB+2\mu^2 AB = \text{Quad}\,(B)+\text{Cubic}\,(B)+\ldots+\mu^2 BB. \tag{15}$$

The zeroth power B is evidently zero. The first power of B is also zero because A_0 minimized the original thing. So this starts out quadratic in B plus cubic in B plus *etc.*, that's what this is here. These quadratic forms Quad (B) and so on of course depend on A_0, the cubic form involves A_0 in some complex, maybe very complicated, locked-up mess, but as far as B is concerned it is second power and higher powers.

Now I would like to point something out. First — it turns out if you analyze it, that the contribution of the first factor here alone (if you had forgotten the intergal and called it one) is exactly the contribution of all trees to the problem. So that's like the classical theories related to trees. Next, if you drop the term cubic in B in the exponent completely and just integrated the result over DB, that corresponds to the contribution from one ring, or from two isolated rings, or three isolated rings, but not interlocked rings. If you start to include the cubic term is has to come in a second power to do anything, because of the evenness and oddness of function. And as soon as it comes in second power, the cubic term, having three of these things come together twice, makes a terrible thing like ∞ which is a double ring. So you don't get to a double ring until you bring a cubic term down to the second order. So if I disregard that and just work with this second order term Quad (B) $+\mu^2 BB$, I'm studying the contribution from one ring. If I study this I am working from the trees. And now you see I have in my hands an expression for the contribution of a ring correct in all orders no matter how many lines come in. I also have expressions for the contributions from trees and so on. I can compare them in different mathematical circumstances, and it's on this basis that I have been able to prove everything I have been able to prove relating one ring to trees.

Now, let me explain how the theorem was obtained that takes the case for the mass and for a ring. Now we have to discuss a ring, which is a formula like this

$$X = \int e^{(\text{Quadr}(B)+\mu^2 B^2)}\, DB. \tag{15}$$

720

The quadratic form involves A_0 so the answer depends on A_0 — it's some complicated functional of A_0. Anyway I won't say that all the time, I'll just remember that. We have to integrate over all B. And the difficulty is — not difficulty, but the point is — that this quadratic form in B is singular, because it came from the piece of the action that has an invariance and this invariance keeps chasing us along. And there are certain transformations of B which leave this Quard B part unchanged in first order. That transformation in the Yang-Mills theory is

$$\vec{B}'_\mu = \vec{B}_\mu + \vec{\nabla}\alpha + (\vec{\alpha}\times\vec{A}) = \vec{B}_\mu + \alpha_{;\mu}\,. \tag{16}$$

where the vectors are in isotopic spin space and α is considered as first order. This transformation leaves the quadratic form invariant so the Quad (B) thing by itself is singular. But it doesn't make any difference, because of the addition of the $\mu^2 BB$. If $\mu^2 \neq 0$, there is no problem, but if $\mu^2 \to 0$, I'd be in trouble.

I discovered that if I make this change (16) in the actual Lagrangian and carry everything up to second order it is exact, in fact because it's only second order. If I do it with the exact change, the thing isn't invariant, it is only invariant to first order in α. But if I make the substitution exactly, then I get a certain addition to the Lagrangian, in other words the Lagrangian of B' (this includes the μ^2, the Lagrangian plus the μ^2 term in B) is the Lagrangian plus the μ^2 term in B plus something like this

$$\mu^2 B_\mu \cdot a_{,\mu} + \frac{1}{2}\mu^2 a_{,\mu}\, a_{;\mu}$$

I have to explain that the semicolon is analogous to the semicolon in gravity. The semicolon derivative $X_{;\mu}$ means the ordinary derivative of X minus A cross X and that's the analogue of the Christoffel symbols. Anyway, I find out what happens to L when I make this transformation. Now comes the idea, the trick, the nonsense: you start with the following thing; you, say, suppose instead of writing the original terms down, instead of writing the original Lagrangian I were to write the following:

$$\int e^{\mathcal{L}(B) + \frac{1}{2}(B_{\mu,\mu} - \alpha_{;\mu\mu} + \mu^2\alpha)^2}\, \mathcal{D}\alpha\mathcal{D}B.$$

Now I say that the integral over α is some constant or other. So all I have done is to multiply my original integral by $\mathcal{L}$ of B (by $\mathcal{L}$ of B I mean the whole thing, I mean this whole thing is going to be $\mathcal{L}$ of B). If I can claim that when I integrate α I get something which is independent of B, which is not self-evident. If I integrate over all α it does not look as if it is independent of B — but after a moment's consideration you see that it is. Because if I can solve a certain equation, which is $\alpha^{\mu}_{;\mu} - \mu^2\alpha = B^{\mu}_{\mu}$, I can shift the value of α by that amount, and then this term would disappear. In other words if I can solve this, and call this solution α_0 and change α to α_0, then the B would cancel and it would only be α' here. I did it a little abstractly which is a little easier to explain, therefore, this term that I've added can be thought of as an integral of the following nature: Integral of some B, plus an operator acting on α (this complicated operator is the second derivative and so on) squared $\mathcal{D}\alpha$. And then by that substitution I've just mentioned, this becomes equal to 1/2 the operator on A

times α' squared $\mathcal{D}\alpha$, which is equal to the integral e to the one half of α times A, the operator A, times the operator A times α integrated over primed α. Now when you integrate a quadratic form, which is a quadratic with an operator like this you get one over the square root of the determinant of the operator. So this thing is one over the square root of the determinant of the operator AA. The determinant of the operator A times A is square of the determinant of A. So this is one over the determinant of the operator A, or better it is one over square root of the determinant of the operator A squared, you'll see in a minute why I like to write it in this way. In other words, when I've written this thing down I've written the answer that I want. Let's call X the unknown answer that I want. Then this is equal to X divided by this determinant's square root squared. Now comes the trick — I now make the change from B to B'. We notice that B changed to B' is simply... oh!, this is wrong, that's what's wrong, it should be just this. Now I've got it. The change from B' to B is to add something to B. Therefore to the differential of B it adds nothing, it's just shifting the B to a new value. So I make the transformation from B to B' everywhere. So then I have $d\alpha$ and dB, and now I have a new thing up here where I make use of the formula for $\mathcal{L}$ of B':

$$\mathcal{L}(B') = \mathcal{L}(B) + \mu^2 B_\mu \alpha_\mu + \frac{1}{2}\mu^2 \alpha_{,\mu}\alpha_{;\mu}$$

You see there is a certain cross term generated here and another cross term coming from expanding this out and the net result, with a little algebra here, is that becomes $\mathcal{L}$ of B, but the quadratic term doesn't cancel out and is left; there's one half of $B_{\mu,\mu}$ squared; that's from this term; the cross term here cancels the cross term in there; and then we have only the quadratic — I mean the α terms

$$\int e^{\mathcal{L}(B)+\frac{1}{2}(B_{\mu,\mu})^2}\mathcal{D}B e^{\frac{\mu^2}{2}(\alpha_{,\mu}\alpha_{;\mu}+\mu^2\alpha^2)}\mathcal{D}\alpha.$$

And the problem is now to do this integral on α; well, another miraculous thing happens. I have the operator A, but that this down thing is $\alpha A \alpha$, and therefore its result is just determinant once; or the square of this integral is equal to this determinant, or something like that. Therefore, when you get all the factors right, X, the unknown, is equal to

$$X = \left[\int e^{\mathcal{L}(B)+\frac{1}{2}(B_{\mu,\mu})^2}\mathcal{D}B\right] : \left[\int e^{\frac{\mu^2}{2}(\alpha_{,\mu}\alpha_{;\mu}+\mu^2\alpha^2)}\mathcal{D}\alpha\right].$$

Sachs: I want to ask a question about long-range hopes. Perhaps for irrational reasons people are particularly interested in those parts of the theory where is a possibility of real qualitative differences: what do the coordinates or topology mean in a quantized theory, and this kind of junk. Now I wonder if you think that this perturbation theory can eventually be jazzed up to cover also this kind of questions?

Feynman: The present theory is not a theory as it is incomplete. I do not give a rule on how to do all problems. I expect of course that if I spend more time on figuring out how to untangle the pretzels I shall be able to make it into such a theory. So let's suppose I did. Now you can ask the question would the completed job, assuming it exists, be of any interest to esoteric question about the quantization of gravity. Of course it would be, because it would be the expression of the quantum theory; there is today no expression of the quantum theory which is consistent. You say: but it's perturbation theory. But it isn't. I worked on the thing analyzing it in the series of increasing accuracy, but that's only, obviously, when I am doing problems and checking, or doing things like I just did. But even there I haven't said how many times the vector potential A_0 is attacking the diagram, there is no limit to what order of external lines are involved in the calculation of A_0, for example. And so if I get my general theorem for all orders, I'll have some kind of a formulation. The fact is, that in such things as electrodynamics and other theories, it has not been possible to figure out the consequences of the quantum field theory in the case of strong interactions, because of technical difficulties which are not technical difficulties just of the gravitation theory, but exist all over the quantum field theory. I do not expect that the gravitational problems will be any easier in that region than they are in any other field theory, so I can say very little there. But at least one should certainly formulate the theory that you're trying to calculate first, and then find out what the consequences are, before trying to do it the other way round. So I think that you'll be frustrated by the difficulties that do appear whenever any theory diverges. On other hand, if you ask about the physical significance of the quantization of geometry, in other words about the philosophy behind it; what happens to the metric, and all such questions, those I believe will be answerable, yes. I think you would be able to figure out the physics of it afterwards, but I won't to think about that until I have it completely formulated, I don't want to start to work out the anser to something unless I know what the equation is I am trying to analyze. But I dont' have the doubt that you will be able to do something, because after all you are describing the phenomena that you would expect, and if you describe the phenomena then you expect you can then find some kind of framework in which to talk to help to understand the phenomena.

FEYNMAN DIAGRAMS FOR THE YANG-MILLS FIELD

L. D. FADDEEV and V. N. POPOV
Mathematical Institute, Leningrad, USSR

Received 1 June 1967

Feynman and De Witt showed, that the rules must be changed for the calculation of contributions from diagrams with closed loops in the theory of gauge invariant fields. They suggested also a specific recipe for the case of one loop. In this letter we propose a simple method for calculation of the contribution from arbitrary diagrams. The method of Feynman functional integration is used.

It is known, that one can associate the field of the Yang-Mills type with an arbitrary simple group G [1-3]. It is appropriate to describe this field by means of the matricies $B_\mu(x)$ with values in the Lie algebra of this group.

The gauge group consits of the transformations

$$B_\mu \to B_\mu^\Omega = \Omega B_\mu \Omega^{-1} + \epsilon^{-1}\partial_\mu \Omega\Omega^{-1}$$

where $\Omega(x)$ is an arbitrary function with values in the group G.

The Lagrange function

$$\mathcal{L}(x) = -\tfrac{1}{4}\mathrm{Sp}G_{\mu\nu}G^{\mu\nu};\quad G_{\mu\nu} = \partial_\nu B_\mu - \partial_\mu B_\nu + \epsilon[B_\mu, B_\nu]$$

is invariant with respect to these transformations. It is clear, that

$$\mathcal{L} = \mathcal{L}_0 + \mathcal{L}_1$$

where $\mathcal{L}_0$ is a quadratic form, and $\mathcal{L}_1$ is the sum of trilinear and quartic forms in B. In the quantization of the Feynman type $\mathcal{L}_1$ generates vertices with three and four external lines and $\mathcal{L}_0$ is to define the propagator function. However the form $\mathcal{L}_0$ is singular and the longitudinal part of the propagator can not be found unambiguously. This ambiguity does not influence the physical results in quantum electrodynamics. It seems that Feynman [4] was the first to show that the matter is not so simple in the cases of Yang-Mills and gravitational fields. Namely the contribution of the closed loop diagrams depends essentially on the longitudinal part of the propagator and spoils the transversality and unitarity properties of scattering amplitudes. Feynman himself described the necessary change of rules for calculation the contribution from diagrams with one closed loop. A more detailed derivation of the new rules was given by De Witt [5]. However it seems that nobody gave the generalization of these rules for arbitrary diagrams.

The formal considerations below are to give a simple explanation of the described difficulties and a quite workable recipe to circumvent them.

We know from Feynman [6] that every element of the S-matrix can be written down as the functional integral

$$\langle \mathrm{in}|\mathrm{out}\rangle \sim \int \exp\{iS[B]\}\prod_x \mathrm{d}B(x)$$

up to an (infinite) normalizing factor. Here $S[B] = \int \mathcal{L}(x)\mathrm{d}x$ is the action functional and one is to integrate over all fields $B(x)$ with the asymptote at $t = x_0 \to \pm\infty$ prescribed by in-and out-states. The diagrams appear naturally in the perturbative calculation of this integral.

In the case of gauge invariant theory it is necessary to transform this integral a little. In fact, we can say, using the natural geometrical language, that the integrand is constant on the "orbits" $B_\mu \to B_\mu^\Omega$ of the gauge group in the manifold of all fields $B_\mu(x)$. It follows that the integral itself is proportional to the volume of this orbit which can be expressed as the integral $\int \prod_x \mathrm{d}\Omega(x)$ over all matrices $\Omega(x)$. This integral should be factorized before using the perturbation theory.

There exist several methods for this purpose. The idea of one of them is to integrate over the orbits and some transversal surface. It is appropriate to choose for the latter the "plane" $\partial_\mu B^\mu = 0$. Then the integral reduces to the following

$$\int \exp\{iS[B]\}\ \Delta[B] \prod_x \delta\left(\partial_\mu B^\mu(x)\right) \mathrm{d}B(x) \int \prod_x \mathrm{d}\Omega(x)$$

where the factor $\prod_x \delta\left(\partial_\mu B^\mu\right)$ symbolizes that we integrate over transverse fields and $\Delta[B]$ is to be chosen such that the condition

$$\Delta[B] \int \prod_x \delta\left(\partial_\mu (B^\mu)^\Omega\right) \mathrm{d}\Omega = \mathrm{const}$$

holds. It is the nontriviality of $\Delta[B]$ which distinguishes the theories of Yang-Mills and gravitational fields from quantum electrodynamics.

We must know $\Delta[B]$ only for transverse fields and in this case all contribution to the last integral is given by the neighbourhood of the unit element of the group. After appropriate linearization we come to the condition

$$\Delta[B] \int \prod_x \delta\left(\Box u - \epsilon[B^\mu, \partial_\mu u]\right) \mathrm{d}u(x) = \mathrm{const}$$

where $\Box$ is the D'Alembert operator and $u(x)$ are functions with values in the Lie algebra of the group G.

Formally $\Delta[B]$ is equal to the determinant of the operator

$$Au = \Box u - \epsilon[B^\mu, \partial_\mu u] \equiv A_0 u - \epsilon V(B) u$$

After extracting the trivial infinite factor $\det A_0$ we obtain the following expression for $\ln\Delta[B]$

$$\ln\Delta[B] = \ln(\det A/\det A_0) = \mathrm{Sp}\ln(1 - \epsilon A_0^{-1} V(B))$$

Developing the right hand side in a power series in ϵ we have the following expressions for the coefficients

$$\int \mathrm{d}x_1 \ldots \int \mathrm{d}x_n\ \mathrm{Sp}\left(B^{\mu_1}(x_1) \ldots B^{\mu_n}(x_n)\right) \times$$
$$\times\ \partial_{\mu_1} G(x_1 - x_2) \ldots \partial_{\mu_n} G(x_n - x_1)$$

where $G(x)$ is a Green function of the D'Alembert operator. This expression corresponds to the closed loop with the scalar particle propagating along it and interacting with the transverse vector particles according to the law $\sim \epsilon\ \mathrm{Sp}(\varphi[B^\mu \partial_\mu \varphi])$.

There results the diagram technique with the following features:

1. The pure transversal Green function is to be used as a propagator for the vector particles (Landau gauge).

2. It is necessary to take into account the new vertex with two scalar and one vector external line in addition to the ordinary vertices with three and four lines.

Concrete calculations with these changes in the rules give transverse and unitary expressions for the scattering amplitudes.

It must be stressed that the Landau gauge is essential for the new rules. It is connected with the chosen method of extracting the fact or $\int \prod_x \mathrm{d}\Omega(x)$.

An other method leads to the expression

$$\int \exp\{iS[B] - \tfrac{1}{2}i \int \mathrm{Sp}(\partial_\mu B^\mu)^2\, \mathrm{d}x\}\ \varphi[B] \prod_x \mathrm{d}B$$

where the factor $\varphi[B]$ must be found from the condition

$$\varphi[B] \int \exp\{-\tfrac{1}{2}i \int \mathrm{Sp}\left(\partial_\mu (B^\mu)^\Omega\right)^2 \mathrm{d}x\} \prod_x \mathrm{d}\Omega = \mathrm{const}$$

This integral gives the perturbation series with Feynman propagator, but the calculation of $\varphi[B]$ is more cumbersome than that of $\Delta[B]$.

We conclude with the comment that one can proceed in an analogous way with gravitation theory. The analog for $\Delta[B]$ is the determinant of the Beltrami - Laplace operator in a harmonic coordinate system.

References

1. C.N.Yang and R.L.Mills, Phys. Rev. 96 (1954) 191.
2. R.Utiyama, Phys. Rev. 101 (1956) 1597.
3. S.L.Glashow and M.Gell-Mann, Ann. of Phys. 15 (1961) 437.
4. R.P.Feynman, Acta Physica Polonica, 24 (1963) 697.
5. B.S.De Witt, Relativity, groups and topology. (Blackie and Son Ltd 1964) pp 587-820.
6. R.P.Feynman, Phys. Rev. 80 (1950) 440.

7.A.1

Nuclear Physics B35 (1971) 167–188. North-Holland Publishing Company

RENORMALIZABLE LAGRANGIANS FOR MASSIVE YANG-MILLS FIELDS

G.'t HOOFT

Institute for Theoretical Physics, University of Utrecht

Received 13 July 1971

Abstract: Renormalizable models are constructed in which local gauge invariance is broken spontaneously. Feynman rules and Ward identities can be found by means of a path integral method, and they can be checked by algebra. In one of these models, which is studied in more detail, local SU(2) is broken in such a way that local U(1) remains as a symmetry. A renormalizable and unitary theory results, with photons, charged massive vector particles, and additional neutral scalar particles. It has three independent parameters.

Another model has local SU(2)⊗U(1) as a symmetry and may serve as a renormalizable theory for ρ-mesons and photons.

In such models electromagnetic mass-differences are finite and can be calculated in perturbation theory.

1. INTRODUCTION

In a preceding article [1], henceforth referred to as I, it has been shown that, owing to their large symmetry, mass-less Yang-Mills fields may be renormalized, provided that a certain set of Ward identities is not violated by renormalization effects. With this we mean that anomalies like those of the axial current Ward identities in nucleon-nucleon interactions [2–4], which are due to an unallowed shift of integration variables in the "formal" proof, must not occur. In I it is proved that such anomalies are absent in diagrams with one closed loop, if there are no parity-changing transformations in the local gauge group. We do know an extension of this proof for diagrams with an arbitrary number of closed loops, but it is rather involved and we shall not present it here.

Thus, our prescription for the renormalization procedure is consistent, so the ultraviolet problem for mass-less Yang-Mills fields has been solved. A much more complicated problem is formed by the infrared divergencies of the system. Weinberg [5] has pointed out that, contrary to the quantum electrodynamical case, this problem cannot merely be solved by some closer contemplation of the measuring process. The disaster is such that the perturbation expansion breaks down in the infrared region, so we have no rigorous field theory to describe what happens.

However, although the Lagrangian is invariant under local gauge transformations, the physical solutions we are interested in may provide us with a certain preference gauge, in which these solutions take a simple form. If this is the case, then the local gauge invariance is hidden, and it is very well possible that all Yang-Mills bosons become massive vector particles [6]. We do not know whether such a thing can happen with mass-less Yang-Mills fields alone, but it surely can happen in other models, of which we present some.

In all these models additional scalar fields are introduced, which are representations of the local gauge group. If, in some gauge, these fields have a non-zero vacuum expectation value, then they may fix the gauge, either completely, or partly. In the latter case, invariance under transformations of a local subgroup of the original invariance group remains evident, and some of the Yang-Mills bosons remain mass-less.

The transition from a "symmetric" to a "non-symmetric" representation is done in a way analogous to the treatment of the σ-model by Lee and Gervais [7, 8]. The difference is of course that we have a local invariance, and we have no symmetry breaking term in the Lagrangian.

Our result is a large set of different models with massive, charged or neutral, spin one bosons, photons, and massive scalar particles. Due to the local symmetry our models are renormalizable, causal, and unitary. They all contain a small number of independent physical parameters.

A nice feature is that in certain models the electromagnetic mass-differences are finite and can be expressed in terms of the other parameters.

In sect. 2 we give a short review of the results in the preceding paper (I) on massless Yang-Mills fields. A general procedure appears to exist for deriving Feynman rules for models with a local gauge invariance. One statement must be made on our use of path integrals here: we only apply path integral techniques in order to get some idea of what the Feynman rules and Ward-identities might be. Consistency and unitarity of the renormalized expressions must always be checked later on. This has been done for the models described in this paper.

In sect. 3 we consider SU(2) gauge fields and an additional scalar isospin one boson. We show how the vacuum expectation value of this boson field can become non-zero due to dynamical effects, and how two of the Yang-Mills bosons become massive, oppositely charged, vector particles, while the third becomes an ordinary photon. Of the original scalar fields one component survives in the form of a neutral spinless particle. Interaction and gauge are formulated in such a way that the theory remains renormalizable. In sect. 4 a renormalization scheme is presented, but for a more elaborate description of the renormalization procedures we refer to I. In sect. 5 we prove that the model of sect. 3 is unitary and it is easily seen that the proof applies also to the other models.

In sect. 6 we describe an example where local invariance seems broken, while global invariance remains evident. All Yang-Mills particles get equal mass, and the model resembles very much the massive Yang-Mills field studied by other authors

[9, 10] except for the presence of one extra neutral scalar boson with arbitrary mass. The model can be used to describe ρ-mesons as elementary particles.

In sect. 7 it is shown that our "ρ-meson model" can be enriched with electromagnetic interactions without destroying renormalizability or unitarity. $\rho^0 - \gamma$ mixture leads to phenomena like vector-dominance.

In the appendix we formulate the Feynman rules for the various models.

2. RESUME MASSLESS YANG-MILLS FIELDS

The massless Yang-Mills field has been discussed in I. The Lagrangian is

$$\mathcal{L}_{\text{YM}} = -\tfrac{1}{4} G^a_{\mu\nu} G^a_{\mu\nu} + \mathcal{L}^c(\partial_\mu W^a_\mu) , \tag{2.1}$$

with

$$G^a_{\mu\nu} \equiv \partial_\mu W^a_\nu - \partial_\nu W^a_\mu + g f_{abc} W^b_\mu W^c_\nu , \tag{2.2}$$

where W^a_μ are the Yang-Mills field components and f_{abc} are the structure constants of the underlying gauge group. g is a coupling constant.

$\mathcal{L}^c$ is an extra term, only depending on the divergence of the field W^a_μ, and it may be chosen in an arbitrary way, thus fixing the gauge.

From (2.1) the Feynman rules may be constructed by ordinary Feynman path integral methods: the procedure is clarified in the appendix. But, because of the gauge non-invariance of $\mathcal{L}^c$, an extra complex ghost particle field φ^a must be introduced, described by the Lagrangian

$$\mathcal{L}_\varphi = -\partial_\mu \varphi^{*a} (D_\mu \varphi)^a , \tag{2.3}$$

where D_μ is the covariant derivative, defined as

$$(D_\mu X)^a \equiv \partial_\mu X^a + g f_{abc} W^b_\mu X^c . \tag{2.4}$$

Furthermore, an extra factor -1 must be inserted in the amplitude for each closed loop of φ's.

The φ particles (and antiparticles) do not occur in the intermediate states in the unitarity condition, because they cancel the helicity-0 states in the W-field.

The Lagrangian (2.3) is related to the behaviour under local gauge transformations of the gauge non-invariant part $\mathcal{L}^c$ of the Lagrangian (2.1):

$$\mathcal{L}^c = \mathcal{L}^c(\partial_\mu W^a_\mu) .$$

Under an infinitesimal gauge transformation generated by $\Lambda^a(x)$, the quantity $\partial_\mu W^a_\mu$

transforms as:

$$\partial_\mu W_\mu^{a'} = \partial_\mu W_\mu^a - g^{-1}\partial_\mu (D_\mu \Lambda)^a \, . \tag{2.5}$$

Because the fields W_μ^a occur explicitly in the covariant derivative D_μ in eq. (2.5), a non-trivial Jacobian factor is needed in the gauge dependent expressions for the amplitude, which is precisely the φ-particle contribution. In appendix A of I it is shown how to derive the Lagrangian (2.3) from eq. (2.5).

The choice

$$\mathcal{L}^c = -\alpha(\partial_\mu W_\mu^a - C^a)^2 \, ; \qquad \alpha \to \infty \tag{2.6}$$

leads to the Landau gauge for $C = 0$, and from the fact that the amplitudes must be independent of $C^a(x)$, one can derive Ward identities. The most appropriate choice however is

$$\mathcal{L}^c = -\tfrac{1}{2}(\partial_\mu W_\mu^a(x) - J^a(x))^2 \, . \tag{2.7}$$

For $J = 0$ we have the Feynman gauge, in which the propagators are:

$$\frac{\delta_{\mu\nu}}{k^2 - i\epsilon} \tag{2.8}$$

and again one can derive Ward identities.

These Ward identities supply a unique prescription for the subtraction constants in a renormalization procedure, and from them unitarity of the system follows.

If we introduce other fields which are representations of the gauge group, then all derivatives in their Lagrangian parts must be replaced by covariant derivatives, thus ensuring local gauge invariance and unitarity.

3. SELECTION OF A PREFERENCE GAUGE BY INTRODUCING AN ISOSPIN 1 SCALAR FIELD

Consider the case that the local gauge group is SU(2), and suppose that an isospin one scalar field or current $X^a(x)$ exists which has (in a certain gauge) a non-zero vacuum expectation value [6]. Then this isovector is apt to select a preference gauge, which may be taken to be the gauge in which

$$X^1(x) = X^2(x) = 0 \tag{3.1}$$

for all x. However, a gauge transformation of the original system to this "X-field gauge" would in general involve a non-polynomial Jacobian factor (cf. I), thus destroying renormalizability of our model. So, in general we shall abandon the gauge (3.1), but its formal possibility indicates clearly that the components X^1 and X^2 are unobservable, and X^3 acts as a "schizon": our symmetry *seems* to be broken.

In our renormalizable model, X is simply a boson field, and we fix the gauge by adding some functional $\mathcal{L}^c$ to the Lagrangian as in sect. 2. In the symmetric representation the Lagrangian is

$$\mathcal{L} = \mathcal{L}_{\text{YM}} - \tfrac{1}{2}(D_\mu X)^2 - \tfrac{1}{2}\mu^2 X^2 - \tfrac{1}{8}\lambda (X^2)^2 \,. \tag{3.2}$$

This Lagrangian corresponds to a renormalizable theory. The last term is necessary for fixing the counterterm in divergent graphs with four X-lines. Thus we have three independent parameters g, μ and λ.

In order to get some insight in what might happen let us consider the tree-approximation, that is, we disregard all graphs with closed loops. In this approximation all fields may be considered as being classical, and the vacuum corresponds to the equilibrium state, where all fields are constant and the total energy has a minimum. (If we specify the gauge, then this energy can be written as an integral over space pf an Hamiltonian density $\mathcal{H}(x)$.) In order for this vacuum to exist, the parameter λ must be positive, but μ^2 may have a negative value. In the latter case we expect that the X-field is non-zero in the equilibrium state: for slowly varying X^a, and $W_\mu^a \simeq 0$, we have

$$\mathcal{H} \simeq \int_V \mathrm{d}x (\tfrac{1}{2}\mu^2 X^2 + \tfrac{1}{8}\lambda (X^2)^2) \,.$$

This has a minimum for

$$X^a(x) = e^a \sqrt{\frac{-2\mu^2}{\lambda}} \,, \tag{3.3}$$

with e^a an arbitrary vector with modulus unity. By a global gauge transformation this vector e^a can always be pointed in the z-direction.

If we do quantum mechanics things do not change drastically. Eq. (3.3) must be replaced by

$$\langle 0 | X^a(x) | 0 \rangle = F \begin{pmatrix} 0 \\ 0 \\ 1 \end{pmatrix} \,,$$
$$F = \sqrt{\frac{-2\mu^2}{\lambda}} + O(g) \,, \tag{3.4}$$

(the parameter λ is of order g^2).

We now proceed as Lee did for the σ-model [7]. We write

$$X^a(x) \equiv F\begin{pmatrix}0\\0\\1\end{pmatrix} + A_a(x)\,, \tag{3.5}$$

with

$$\langle 0|A_a(x)|0\rangle \equiv 0\,. \tag{3.6}$$

The Lagrangian (3.2) then becomes

$$\begin{aligned}\mathcal{L} = \mathcal{L}_{\mathrm{YM}} &- \tfrac{1}{2}(D_\mu A)^2 - \tfrac{1}{2}\lambda F^2 A_3^2 - \tfrac{1}{2}g^2F^2(W_\mu^{1\,2} + W_\mu^{2\,2}) - \tfrac{1}{2}\lambda F A^2 A_3 - \tfrac{1}{8}\lambda(A^2)^2\\ &- g^2FA_3(W_\mu^{1\,2} + W_\mu^{2\,2}) + g^2FW_\mu^3(A_1W_\mu^1 + A_2W_\mu^2) - \beta(\tfrac{1}{2}A^2 + FA_3)\\ &+ gF(A_1\partial_\mu W_\mu^2 - A_2\partial_\mu W_\mu^1)\,,\end{aligned} \tag{3.7}$$

where

$$\beta \equiv \mu^2 + \tfrac{1}{2}\lambda F^2\,.$$

Note, that the "tadpole condition" (3.6) implies that $\beta = 0$ in first order of g and λ, in accordance with eq. (3.4).

We deliberately have not yet specified the local gauge. We have seen that the gauge (3.1) is no good, because it renders the theory unrenormalizable. One could very well proceed like in sect. 2 and choose the local gauge by adding

$$\mathcal{L}^c = \mathcal{L}^c(\partial_\mu W_\mu) = -\tfrac{1}{2}(\partial_\mu W_\mu^a - J_a(x))^2\,,$$

but the resulting Feynman rules are rather complicated and a massless ghost remains. It is more convenient to choose:

$$\begin{aligned}\mathcal{L}^c = &-\tfrac{1}{2}(\partial_\mu W_\mu^3 - J_3(x))^2 - \tfrac{1}{2}(\partial_\mu W_\mu^1 - gFA_2 - J_1(x))^2\\ &-\tfrac{1}{2}(\partial_\mu W_\mu^2 + gFA_1 - J_2(x))^2\,.\end{aligned} \tag{3.8}$$

In here the functions $J_a(x)$ will be put equal to zero, but the fact that all physical amplitudes are independent of them enables us to formulate Ward-identities.

The fields in eq. (3.8) transform as follows under infinitesimal local gauge transformations:

$$\partial_\mu W_\mu^{3'} = \partial_\mu W_\mu^3 - g^{-1}\partial_\mu (D_\mu \Lambda)^3 ;$$

$$(\partial_\mu W_\mu^1 - gFA_2)' = \partial_\mu W_\mu^1 - gFA_2 - g^{-1}\partial_\mu (D_\mu \Lambda)^1 + gF^2\Lambda^1 - gF\epsilon_{2bc}\Lambda^b A_c ; \quad (3.9)$$

$$(\partial_\mu W_\mu^2 + gFA_1)' = \partial_\mu W_\mu^2 + gFA_1 - g^{-1}\partial_\mu (D_\mu \Lambda)^2 + gF^2\Lambda^2 + gF\epsilon_{1bc}\Lambda^b A_c .$$

With the same procedure as in sect. 2 we derive the φ-ghost Lagrangian‡:

$$\mathcal{L}_\varphi = -\partial_\mu \varphi^{*a}(D_\mu\varphi)^a - g^2F^2(\varphi^{*1}\varphi^1 + \varphi^{*2}\varphi^2)$$
$$+ g^2F(\varphi^{*1}\epsilon_{2bc}\varphi^b A_c - \varphi^{*2}\epsilon_{1bc}\varphi^b A_c) . \quad (3.10)$$

Because of our choice (3.8) for $\mathcal{L}^c$ the last term in (3.7) is cancelled; likewise the term

$$+\tfrac{1}{2}(\partial_\mu W_\mu^a)^2 \quad \text{in} \quad -\tfrac{1}{4}G_{\mu\nu}^a G_{\mu\nu}^a .$$

Let us finally replace the three parameters by

$$M \equiv gF > 0 ,$$
$$\alpha \equiv \lambda/g^2 > 0 , \quad (3.11)$$

and g.

The resulting theory has two massive, charged vector particles $W_\mu^1 \pm i\, W_\mu^2$, with the propagator

$$\frac{\delta_{\mu\nu}}{k^2 + M^2 - i\epsilon} ; \quad (3.12)$$

a massless photon W_μ^3, with the propagator

$$\frac{\delta_{\mu\nu}}{k^2 - i\epsilon} \quad (3.13)$$

and a neutral, scalar particle A_3, with mass $M\sqrt{\alpha}$.

There are two different types of ghosts:

‡ Note that this expression is not Hermitian; the φ-ghost restores unitarity. Because of these features Feynman rules must be derived by path integral methods. The heuristic Feynman rules are here the correct ones, as is shown in the appendix.

First: the complex φ-ghost with the oriented propagators: $(k^2+M^2-i\epsilon)^{-1}$ for $\varphi^{1,2}$ and $(k^2-i\epsilon)^{-1}$ for φ^3.

A minus sign must be inserted for each closed loop of φ-ghosts.

Second: the real $A^{1,2}$ ghosts, with no minus sign, and mass M^2 (resulting from a contribution of $\mathcal{L}^c$, eq. (3.8)).

These, and all other Feynman rules, except for the above mentioned minus sign, may be derived with ordinary Feynman path-integral techniques, from the Lagrangian

$$\mathcal{L} = -\tfrac{1}{4}G^a_{\mu\nu}G^a_{\mu\nu} - \tfrac{1}{2}(\partial_\mu W^a_\mu)^2 + \mathcal{L}_\varphi - \tfrac{1}{2}(D_\mu A)^2 - \tfrac{1}{2}M^2(W^{1\,2}_\mu + W^{2\,2}_\mu)$$

$$-\frac{\alpha M^2}{2}A_3^2 - \frac{M^2}{2}(A_1^2+A_2^2) - \frac{\alpha}{2}gMA_a^2A_3 - \frac{\alpha g^2}{8}(A_a^2)^2 - gMA_3(W^{1\,2}_\mu + W^{2\,2}_\mu)$$

$$+ gMW^3_\mu(A_1W^1_\mu + A_2W^2_\mu) - \beta(\tfrac{1}{2}A^2 + \frac{M}{g}A_3) + J_3(x)\partial_\mu W^3_\mu + J_1(x)$$

$$\times\ [\partial_\mu W^1_\mu - MA_2] + J_2(x)\,[\partial_\mu W^2_\mu + MA_1] - \tfrac{1}{2}J_a^2(x) \tag{3.14}$$

with

$$\mathcal{L}_\varphi = -\partial_\mu\varphi^{*a}(D_\mu\varphi)^a - M^2(|\varphi^1|^2 + |\varphi^2|^2) + gM(\varphi^{*1}\epsilon_{2bc} - \varphi^{*2}\epsilon_{1bc})\varphi^bA_c\,. \tag{3.15}$$

The functions $J_a(x)$ are arbitrary, which enables us to formulate Ward identities. They may be chosen to be zero. The constant β must be adjusted in such a way, that the tadpole condition (3.6) holds.

The complex φ-particles with their unphysical "Fermi statistics", and the $A^\pm$ particles, with positive definite metric, must all be considered as ghosts. The most compelling reason for this is the unitarity condition; In sect. 5 we derive that these particles cancel the unphysical polarisation directions of the W-fields in the intermediate states, resulting from the anomalous propagators (3.12) and (3.13).

4. RENORMALIZATION

The expression (3.8) for $\mathcal{L}^c$, has especially been chosen in order to acquire the simple, quadratically convergent propagators (3.12) and (3.13), and to arrive at Feynman rules which are those of a renormalizable theory.

However, renormalization requires the introduction of a cut-off procedure, and in general this modifies the theory such that the symmetry, and therefore also unitarity, get lost. Thus, the cut-off procedure must be chosen in accordance with our symmetry requirement. To this purpose we can use the observation that in all orders of g the physical amplitudes must be independent of the source function $J_a(x)$. The Feynman rules for the contribution of their Fourier transform $J_a(k)$ are given in

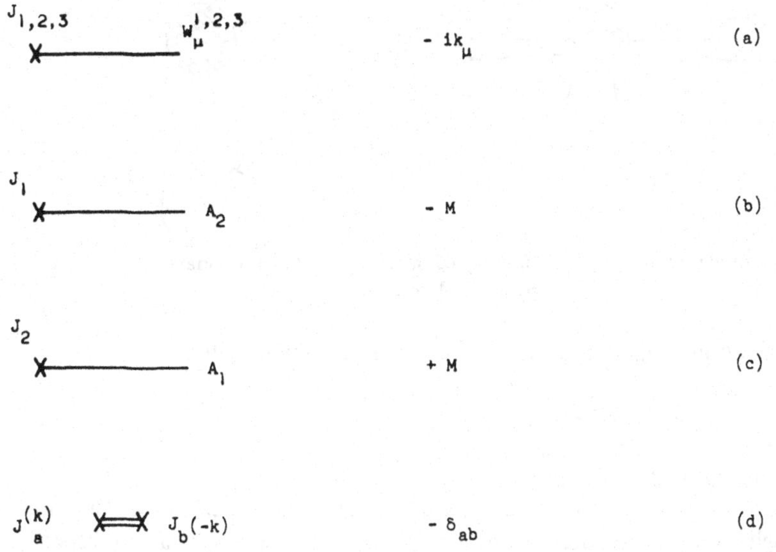

Fig. 1. Feynman rules for the contribution of the source function J_a to the amplitude.

fig. 1 (compare eq. (3.14)). (Note that (d) is cancelled by contributions of bare W and A particles.)

A graphical notation for the Ward identities is shown in fig. 2. The number of "J-lines" must be non-zero. A combinatorial proof of these Ward identities can be given in the same way as in the case of the massless Yang-Mills fields.

Further, also in this model, a variation upon these identities can be found for the case that one of the external W-lines on mass shell has a non-physical polarization direction (fig. 3). The identity in fig. 3 can be proven either by combinatorics, or by using a formulation in terms of path integrals: one must consider an infinitesi-

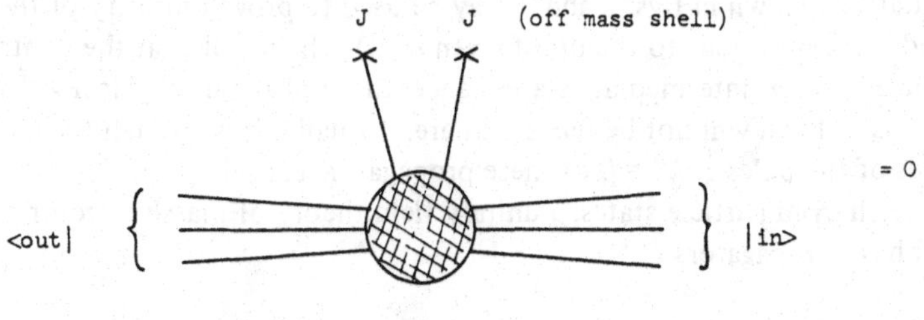

Fig. 2. Example of Ward identity expressing the fact that physical amplitudes are independent of the J-source.

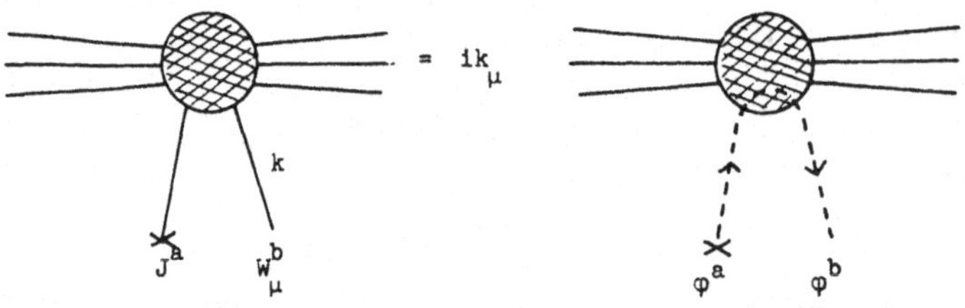

Fig. 3. Ward identity for the case that one of the external W-lines on mass shell has a non-physical polarization direction.

mal gauge transformation generated by $\Lambda^a(x)$ with the property

$$\partial_\mu D_\mu \Lambda(x) = J(x)\,, \tag{4.1}$$

($J(x)$ being also infinitesimal).

One of the simplest ways to use these Ward identities to calculate the renormalized higher order corrections to the amplitudes, is to apply subtracted dispersion relations. The behaviour of the amplitudes for the momenta going to infinity is prescribed by the condition that renormalizability must not be destroyed at higher orders: hence, the amplitude for a diagram with N outgoing boson lines must behave at infinity like $(k)^{4-N}L(k)$, where (k) stands for the momenta of the outgoing lines, and L is a logarithmic factor. Thus the number of subtractions has been determined, whereas the Ward identities give a large restriction on the possible values of the subtraction constants. The procedure sketched here can be proven to be consistent as soon as some gauge covariant set of regulators is found. Such a set can indeed be formulated for diagrams with one closed loop by the introduction of a fifth dimension in Minkowsky space (cf. I). By introducing more dimensions one can give a consistency proof for all orders, but we shall not present it here.

5. UNITARITY

The equations shown in figs. 2 and 3 may be used to prove unitarity of the model. The procedure is analogous to the proof given in (I). The proof that the contributions of the φ^3 and $\bar{\varphi}^3$ intermediate states cancel those of the unphysical W^3_μ-states is the same as in I and will not be repeated here. Actually, it is enough to show that the residues of the poles at $k^2 = 0$ of these propagators cancel.

As to the charged particle states, a unitary field theory of massive vector particles would have propagators

$$\frac{\delta_{\mu\nu} + \dfrac{k_\mu k_\nu}{M^2}}{k^2 + M^2 - i\epsilon} \tag{5.1}$$

instead of (3.12). Hence, the $W^{1,2}$ propagators have anomalous parts

$$-\frac{k_\mu k_\nu}{M^2(k^2+M^2-i\epsilon)}. \tag{5.2}$$

Now, indeed, we see that as a consequence of the equations in figs. 2 and 3, the residues of the poles at $k^2=-M^2$ of the unphysical propagators cancel. In fig. 4 it is shown which combination of these propagators has to be considered in order to prove this cancellation. (Note the explicitly written minus sign for the φ-loops; the integration over k has not yet been carried out.) For more details of this proof we refer to the treatment of the analogous case in massless Yang-Mills fields, given in (I). It is because of this phenomenon that we can consider the anomalous part of the W-field, and the A_1, A_2 and φ-particles, as unphysical. They can be left out of the intermediate states without invalidating the unitarity equation.

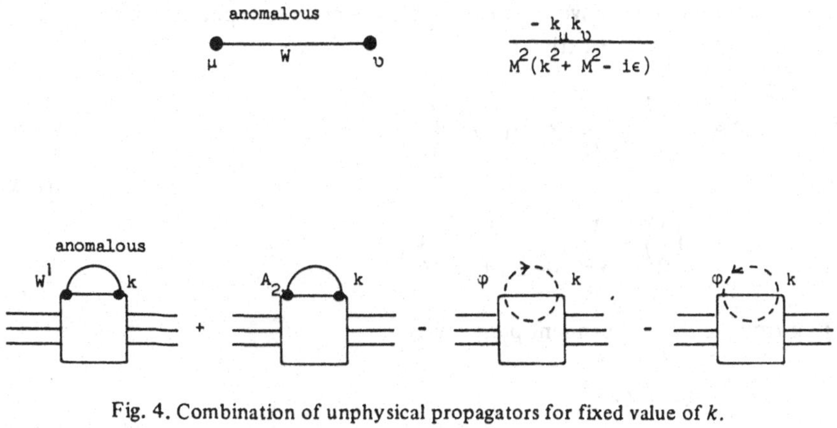

Fig. 4. Combination of unphysical propagators for fixed value of k.

6. ISOSPIN-$\frac{1}{2}$ SCALAR FIELD

Our most important conclusion from the foregoing is that a basic principle like gauge invariance can lead to renormalizable, unitary theories with massive, charged, vector particles. Many different models may be constructed this way. We like to mention in more detail a very interesting case: local SU(2)-gauge invariance with an isospin-$\frac{1}{2}$ "symmetry breaking" field

$$K(x) \equiv \begin{pmatrix} K_1' + iK_1'' \\ K_2' + iK_2'' \end{pmatrix}.$$

Let the Lagrangian in the symmetric representation be

$$\mathcal{L} = \mathcal{L}_{\mathrm{YM}} - (D_\mu K)^* D_\mu K - \mu^2 K^* K - \tfrac{1}{2}\lambda (K^* K)^2 \,. \tag{6.1}$$

The covariant derivative of an isospin-$\frac{1}{2}$ field is:

$$D_\mu K \equiv \partial_\mu K - \tfrac{1}{2} i g \tau^a W_\mu^a K \,. \tag{6.2}$$

For negative values of μ^2, the field K is expected to have a non-zero vacuum expectation value, which by a suitable global rotation in isospin space can be taken to be:

$$\langle 0 | K(x) | 0 \rangle \equiv F \begin{pmatrix} 1 \\ 0 \end{pmatrix} \,. \tag{6.3}$$

Now, it appears to be convenient to express the complex spinor K in terms of a real isospin singlet and a real triplet:

$$\begin{aligned} K &\equiv F \begin{pmatrix} 1 \\ 0 \end{pmatrix} + \frac{1}{\sqrt{2}} (Z + i \psi_a \tau^a) \begin{pmatrix} 1 \\ 0 \end{pmatrix} , \\ &= F \begin{pmatrix} 1 \\ 0 \end{pmatrix} + \frac{1}{\sqrt{2}} \begin{pmatrix} Z + i\psi_3 \\ -\psi_2 + i\psi_1 \end{pmatrix} , \end{aligned} \tag{6.4}$$

and to introduce the independent parameters

$$\begin{aligned} M^2 &\equiv \tfrac{1}{4} g^2 F^2 \,, \\ \alpha &= \lambda / g^2 \,, \end{aligned} \tag{6.5}$$

and g.

In this representation, the Lagrangian is:

$$\begin{aligned} \mathcal{L} = & -\tfrac{1}{4} G_{\mu\nu}^a G_{\mu\nu}^a + \mathcal{L}^c - \tfrac{1}{2} M^2 W_\mu^2 - \tfrac{1}{2} (\partial_\mu Z)^2 - \tfrac{1}{2} (\partial_\mu \psi) D_\mu \psi - \tfrac{1}{2} 4 \alpha M^2 Z^2 \\ & + \tfrac{1}{2} g W_\mu^a (Z \partial_\mu \psi_a - \psi_a \partial_\mu Z) - \tfrac{1}{8} g^2 W_\mu^2 (\psi^2 + Z^2) - \tfrac{1}{2} g M W^2 Z - \alpha M g Z (\psi^2 + Z^2) \\ & - \tfrac{1}{8} \alpha g^2 (\psi^2 + Z^2)^2 - \beta \left[\tfrac{1}{2} (Z^2 + \psi^2) + \frac{2M}{g} Z \right] - M \psi_a \partial_\mu W_\mu^a \,, \end{aligned} \tag{6.6}$$

where the parameter $\beta \equiv \mu^2 + \lambda F^2$ must be chosen in such a way that all tadpoles cancel. It is of order g^2.

$\mathcal{L}^c$ is chosen to be

$$\mathcal{L}^c = -\tfrac{1}{2}(\partial_\mu W^a_\mu - M\psi_a - J_a)^2 \,, \tag{6.7}$$

so that again the Feynman propagator (3.12) for the W-field emerges, and the last term of eq. (6.6) cancels.

By studying the behaviour of $\mathcal{L}^c$ under gauge transformations we derive the Lagrangian for the ghost field φ (compare sects. 2 and 3)

$$\mathcal{L}_\varphi = -\partial_\mu \varphi^* D_\mu \varphi - M^2 \varphi^* \varphi - \tfrac{1}{2} M g \varphi^* \varphi Z + \tfrac{1}{2} M g \epsilon_{abc} \varphi^{*a} \varphi^b \psi^c \,. \tag{6.8}$$

The Feynman rules for the source function contributions are now those of fig. 5.

Now we observe that the Lagrangian (6.6) as well as the Ward identities remain invariant under global isospin transformations, if the ψ fields are considered as a triplet and the Z as a singlet. Only local gauge invariance has been broken. Here the ψ fields act as additional ghosts, and all three isospin components of the W^a_μ-fields have become massive. The Z is an additional physical particle.

Many authors [9, 10] have considered the massive Yang-Mills theory described by the Lagrangian

$$\mathcal{L}_{\mathrm{YM}} - \tfrac{1}{2} M^2 W^2_\mu \,. \tag{6.9}$$

That model appears to be non renormalizable, although many of the divergencies can be seen to cancel by the introduction of two kinds of ghost fields [11, 12]. In our model also two ghosts appear: the complex φ-field, with Fermi statistics, and the ψ-field. But now we see that the introduction of one physical isospin-zero particle Z can render the model renormalizable. Its mass is a new independent parameter. For large values of this mass we get something very similar to the old model (6.9).

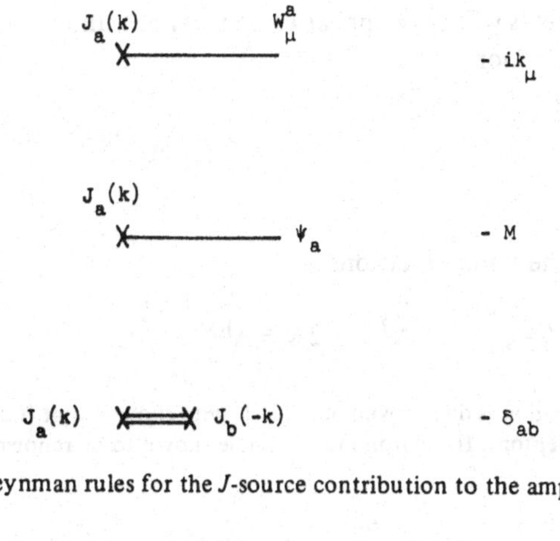

Fig. 5. Feynman rules for the J-source contribution to the amplitude.

7. ISOSPIN AND ELECTROMAGNETISM; VECTOR DOMINANCE

In the previous model, electromagnetism can be introduced in an elegant way†. Consider first the symmetric representation (6.1). Let us assume the presence of a "hyperelectromagnetic" field, $\widetilde{A}_\mu$, which does not break isospin. Let in (6.1) only the K particle have a "hypercharge" q. The Lagrangian is then:

$$\mathcal{L} = \mathcal{L}_{\mathrm{YM}} - \tfrac{1}{4}\widetilde{F}_{\mu\nu}\widetilde{F}_{\mu\nu} - (\widetilde{D}_\mu K)^* \widetilde{D}_\mu K - \mu^2 K^* K - \tfrac{1}{2}\lambda(K^*K)^2 ,$$

with

$$\begin{aligned} \widetilde{F}_{\mu\nu} &\equiv \partial_\mu \widetilde{A}_\nu - \partial_\nu \widetilde{A}_\mu , \\ \widetilde{D}_\mu K &\equiv D_\mu K + iq\widetilde{A}_\mu K . \end{aligned} \tag{7.1}$$

The gauge group in this model is SU(2) ⊗ U(1). Let us consider an infinitesimal gauge transformation:

$$\begin{aligned} K' &= (1 - \tfrac{1}{2} i \Lambda^a \tau^a - i\widetilde{\Lambda}) K , \\ W'^a_\mu &= W^a_\mu - g^{-1}(D_\mu \Lambda)^a , \\ \widetilde{A}'_\mu &= \widetilde{A}_\mu + q^{-1}\partial_\mu \widetilde{\Lambda} , \end{aligned} \tag{7.2}$$

where $\Lambda^a(x)$, $\widetilde{\Lambda}(x)$ are generators of an infinitesimal gauge transformation.

Now if the K field has a non-zero vacuum expectation value:

$$\langle 0 | K(x) | 0 \rangle = F \begin{pmatrix} 1 \\ 0 \end{pmatrix} , \tag{7.3}$$

then the physical fields will only appear to be invariant under those transformations (7.2) that leave the spinor

$$\begin{pmatrix} 1 \\ 0 \end{pmatrix}$$

invariant; that are the transformations with

$$\Lambda^1 = \Lambda^2 = 0 ; \qquad \Lambda^3 = -2\widetilde{\Lambda} \equiv \Lambda^{\mathrm{EM}} . \tag{7.4}$$

† The model of this section is due to Weinberg [13], who showed that it can describe weak interactions between leptons. His lepton model can be shown to be renormalizable.

Let us call such transformations electromagnetic gauge transformations. If we define

$$
\begin{aligned}
&W_\mu^{1,2} \equiv \rho_\mu^{1,2} && \text{(a)} \\
&W_\mu^3 \equiv \rho_\mu^3 + \frac{2q}{g}\tilde{A}_\mu && \text{(b)} \\
&K \equiv F\begin{pmatrix}1\\0\end{pmatrix} + \frac{1}{\sqrt{2}}\begin{pmatrix}Z + i\psi_3\\ -\psi_2 + i\psi_1\end{pmatrix} && \text{(c)}
\end{aligned}
\tag{7.5}
$$

then these quantities transform under electromagnetic gauge transformations like:

$$
\begin{aligned}
&Z' = Z\,, \\
&\psi_3' = \psi_3\,, \\
&\psi_1' \pm i\psi_2' = e^{\pm i\Lambda^{\mathrm{EM}}}(\psi_1 \pm i\psi_2)\,, \\
&\rho_\mu^{3'} = \rho_\mu^3\,, \\
&\rho_\mu^{1'} \pm i\rho_\mu^{2'} = e^{\pm i\Lambda^{\mathrm{EM}}}(\rho_\mu^1 \pm i\rho_\mu^2)\,, \\
&\tilde{A}_\mu' = \tilde{A}_\mu - \frac{1}{2q}\partial_\mu\Lambda^{\mathrm{EM}}\,.
\end{aligned}
\tag{7.6}
$$

Finally, we make the substitutions:

$$
\tilde{A}_\mu \equiv A_\mu\left(1 + \frac{4q^2}{g^2}\right)^{-\frac{1}{2}}; \qquad e \equiv 2q\left(1 + \frac{4q^2}{g^2}\right)^{-\frac{1}{2}}. \tag{7.7}
$$

In terms of these variables the Lagrangian (7.1) becomes:

$$
\begin{aligned}
\mathcal{L} = &-\tfrac{1}{4}\rho^a_{\mu\nu}\rho^a_{\mu\nu} - \tfrac{1}{4}F_{\mu\nu}F_{\mu\nu} - \frac{e}{2g}F_{\mu\nu}\rho^3_{\mu\nu} + \mathcal{L}^c - \tfrac{1}{2}M^2\rho_\mu^2 - \tfrac{1}{2}(\partial_\mu Z)^2 - \tfrac{1}{2}4\alpha M^2 Z^2 \\
&- \tfrac{1}{2}(\partial_\mu^{\mathrm{EM}}\psi)D_\mu^{\mathrm{EM}}\psi + \tfrac{1}{2}g\rho_\mu^a(Z\partial_\mu^{\mathrm{EM}}\psi_a - \psi_a\partial_\mu Z) - \tfrac{1}{2}gM\rho^2 Z - \tfrac{1}{8}g^2\rho^2(\psi^2 + Z^2) \\
&- \alpha MgZ(\psi^2 + Z^2) - \tfrac{1}{8}\alpha g^2(\psi^2 + Z^2)^2 - \beta\left(\tfrac{1}{2}(Z^2 + \psi^2) + \frac{2M}{g}Z\right) - M\psi_a\partial_\mu^{\mathrm{EM}}\rho_\mu^a\,,
\end{aligned}
\tag{7.8}
$$

where

$$\begin{aligned}
&\partial_\mu^{\mathrm{EM}}\psi_3 \equiv \partial_\mu\psi_3 ,\\
&\partial_\mu^{\mathrm{EM}}(\psi_1 \pm i\psi_2) \equiv (\partial_\mu \pm ieA_\mu)(\psi_1 \pm i\psi_2) ,\\
&(D_\mu^{\mathrm{EM}}\psi)_a \equiv \partial_\mu^{\mathrm{EM}}\psi_a + g\epsilon_{abc}\rho_\mu^b\psi_c ,\\
&\rho_{\mu\nu}^a \equiv \partial_\mu^{\mathrm{EM}}\rho_\nu^a - \partial_\nu^{\mathrm{EM}}\rho_\mu^a + g\epsilon_{abc}\rho_\mu^b\rho_\nu^c .
\end{aligned} \tag{7.9}$$

Let us choose

$$\mathcal{L}^c = -\tfrac{1}{2}(\partial_\mu^{\mathrm{EM}}\rho_\mu^a - M\psi_a - J_a)^2 - \tfrac{1}{2}(\partial_\mu A_\mu - J^{\mathrm{EM}})^2 , \tag{7.10}$$

so that the last term in (7.8) is cancelled, while the photon, and the ρ-particle have the Feynman propagators (3.12) and (3.13) resp.

The contributions of the φ-ghosts is described by

$$\mathcal{L}_\varphi = -\partial_\mu^{\mathrm{EM}}\varphi^* D_\mu^{\mathrm{EM}}\varphi - M^2\varphi^*\varphi - \tfrac{1}{2}Mg\varphi^*\varphi Z + \tfrac{1}{2}Mg\epsilon_{abc}\varphi^{*a}\varphi^b\psi^c , \tag{7.11}$$

again with the additional minus sign for each closed loop of φ's.

So, we have arrived at a renormalizable model containing photons, ρ-mesons and neutral Z-particles. There are four independent parameters: g, M, α and e. The parameter β is dictated by the tadpole condition:

$$\langle 0|Z(x)|0\rangle = 0 . \tag{7.12}$$

Note the third term in (7.8), which leads to phenomena like vector dominance. It is a consequence of the translation (7.5b).

8. CALCULATION OF ELECTROMAGNETIC MASS DIFFERENCES

One of the main virtues of the models presented here is, that there are no ambiguities due to infinities, and the number of independent parameters is small. It is interesting to calculate some "electromagnetic" mass differences. For instance, in the model of sect. 3 one may introduce an isospin one fermion†

$$\mathcal{L}_N = -\bar{N}(m + \gamma_\mu D_\mu)N . \tag{8.1}$$

† It appears that counterterms of the form $\bar{N}^a\epsilon_{abc}X^bN^c$, or, in the asymmetric representation, $\bar{N}^a\epsilon_{abc}A^bN^c$, are not needed for renormalization.

A direct computation of the difference in mass of the $N^{\pm}$ and N^{0} seems to lead to ambiguous results because the integrals for the self-energy corrections diverge. But, one of the Ward identities states that the graph of fig. 6a equals zero, if the charged particle N_2, and the neutral N_3 are both on mass shell. Let us consider all terms of this graph up to third order in g (fig. 6b).

The first term only contributes if $m^{\pm} \neq m^{0}$. Now the mass difference $m^{\pm} - m^{0}$ will be of order g^2, so q can be taken of order g^2. Hence the last four diagrams of fig. 6b will not contribute in the third order of g.

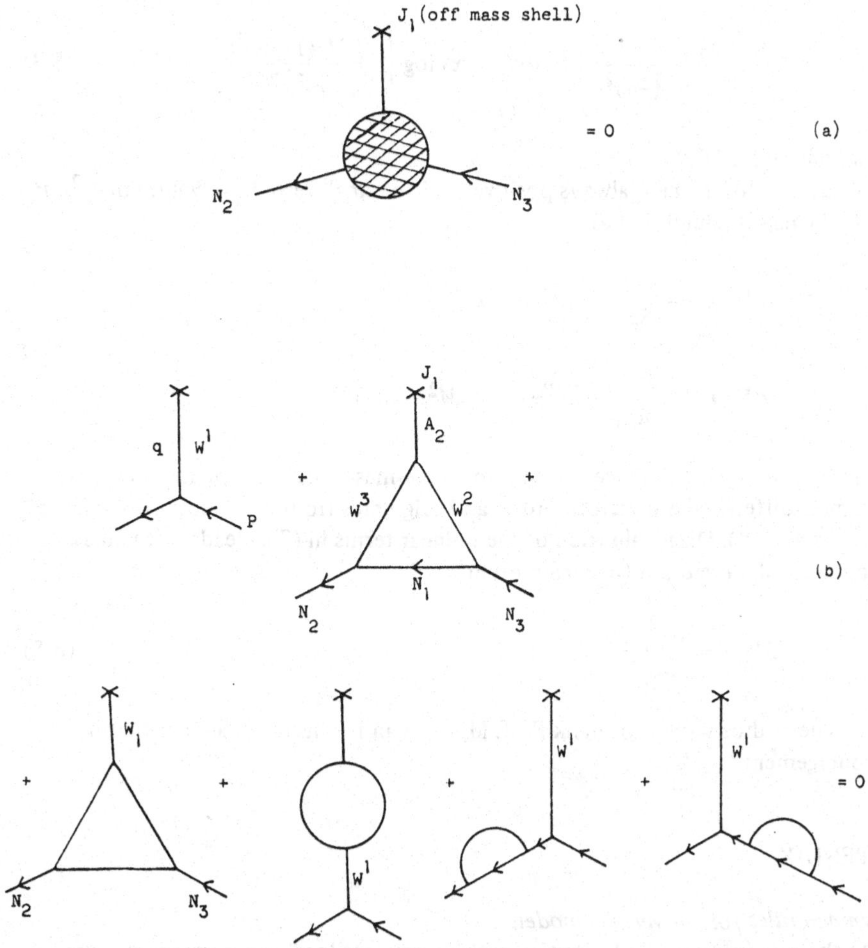

Fig. 6. a. Ward identity for the model of sect. 3, augmented with isospin-one fermions. b. All contributions up to order g^3.

Thus, the Ward identity reads

$$g(m_{\pm} - m_{0}) - M.B = 0 \tag{8.2}$$

where B is the second graph of fig. 6b.

The momentum transfer q may now be taken to be zero, and both external N-lines have momentum p with $p^2 = -m^2$. The integral in B converges, and the result is:

$$\frac{m^{\pm} - m^{o}}{m} = \frac{2g^2}{(4\pi)^2} \int_0^1 dx(1+x) \log \left[1 + \frac{M^2(1-x)}{m^2 x^2}\right] \tag{8.3}$$

(in second order of g).

This mass-difference is always positive, and for small resp. large values of M^2/m^2 eq. (8.3) may be simplified to

$$m^{\pm} - m^{o} = \frac{g^2 M}{8\pi} \qquad (M^2 \ll m^2) ;$$

$$m^{\pm} - m^{o} = \frac{3g^2 m}{(4\pi)^2} \log \frac{M^2}{m^2} \qquad (M^2 \gg m^2) . \tag{8.4}$$

A negative mass-difference is found for the ρ-meson itself in the model of sect. 7. This mass-difference is of zeroth order, and originates from the "vector dominance" term in eq. (7.8). Diagonalisation of the bilinear terms in (7.8) leads to the mass formula for the ρ-mesons (in zeroth order):

$$M_o^2 = \frac{M^{\pm 2}}{1 - e^2/g^2} . \tag{8.5}$$

The author wishes to thank Prof. M.Veltman for his invaluable criticism and encouragement.

APPENDIX

Feynman rules for the various models

In the preceding article I a system with a complete local gauge invariance was quantized using a path integral technique. It was shown how to make the integrand gauge non-invariant without changing the total amplitude. The same procedure has been applied here. The term which breakes the gauge invariance is always called $\mathcal{L}^c$. As a consequence of this procedure, the Feynman rules must always be derived from the final Lagrangians by performing the path integral, and not by canonical quantization. This means that the propagators are always the inverse of the matrices in the bilinear terms of the Lagrangian, and the vertices are the coefficients in front of the remaining terms in the Lagrangian, regardless whether time derivatives occur or not. We mention here some of these Feynman rules.

Model of section 3. (compare (3.14) and (3.15)).
physical particles:

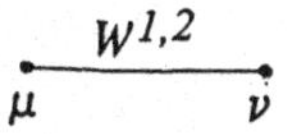

$$\frac{\delta_{\mu\nu}}{k^2+M^2-i\epsilon} \tag{A.1}$$

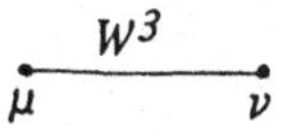

$$\frac{\delta_{\mu\nu}}{k^2-i\epsilon} \tag{A.2}$$

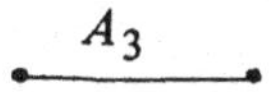

$$\frac{1}{k^2+\alpha M^2-i\epsilon} \tag{A.3}$$

ghosts:

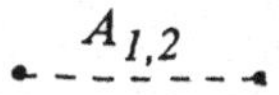

$$\frac{1}{k^2+M^2-i\epsilon} \tag{A.4}$$

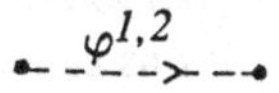

$$\frac{1}{k^2+M^2-i\epsilon} \tag{A.5}$$

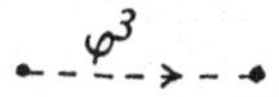

$$\frac{1}{k^2-i\epsilon} \tag{A.6}$$

for each closed loop of φ lines: -1 (A.7)

some of the vertices:

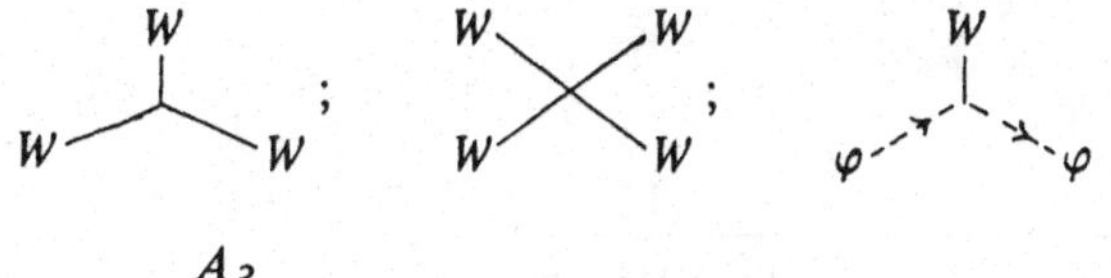

as in massless Yang-Mills fields.

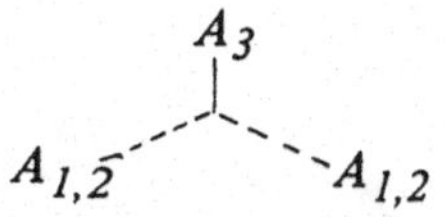

$$-\alpha g M \tag{A.8}$$

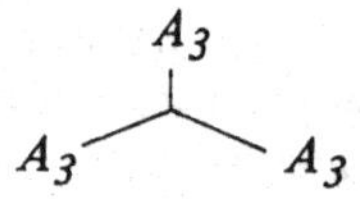

$-3\alpha g M$ (A.9)

etc.

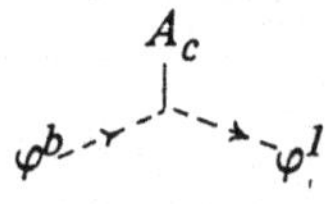

$gM\epsilon_{2bc}$ (A.10)

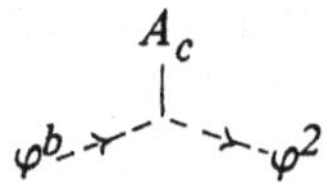

$-gM\epsilon_{1bc}$ (A.11)

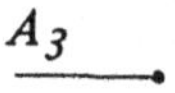

$(k = 0)$ $-\dfrac{\beta M}{g}$ (A.12)

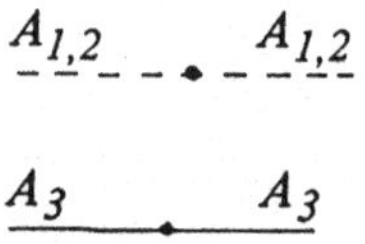

$-\beta$ (A.13)

The vertices (A.12) and (A.13) must be added to higher order tadpole- and self-energy diagrams, with β chosen such, that all tadpole contributions cancel.

Model of section 6. (compare (6.6), (6.7) and (6.8)).
physical particles:

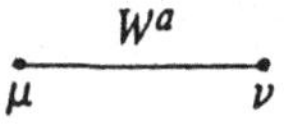

$$\frac{\delta_{\mu\nu}}{k^2 + M^2 - i\epsilon} \qquad (A.14)$$

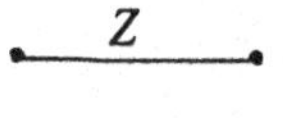

$$\frac{1}{k^2 + 4\alpha M^2 - i\epsilon} \qquad (A.15)$$

ghosts:

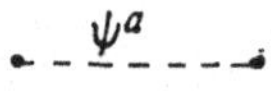

$$\frac{1}{k^2 + M^2 - i\epsilon} \qquad (A.16)$$

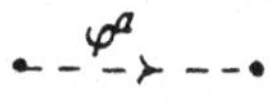

$$\frac{1}{k^2 + M^2 - i\epsilon} \qquad (A.17)$$

some of the vertices:

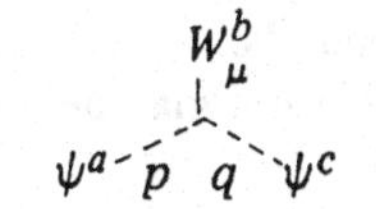

$$-\tfrac{1}{2} i g \epsilon_{abc}(p-q)_\mu \tag{A.18}$$

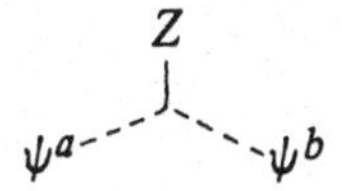

$$-2\alpha M g\, \delta_{ab} \tag{A.19}$$

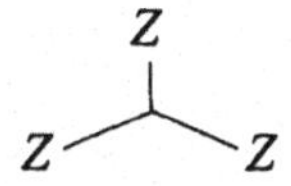

$$-6\alpha M g \tag{A.20}$$

etc.

Model of section 7. (compare (7.8), (7.9), (7.10) and (7.11)).
Rules as in preceding model, but with additional photon lines:

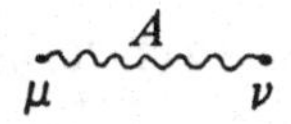

$$\frac{\delta_{\mu\nu}}{k^2 - i\epsilon} \tag{A.21}$$

and vertices:

$$-\frac{e}{g}(k^2\delta_{\mu\nu} - k_\mu k_\nu) \tag{A.22}$$

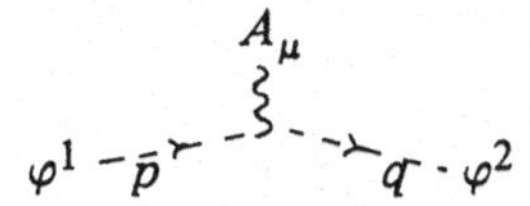

$$e(p-q)_\mu \tag{A.23}$$

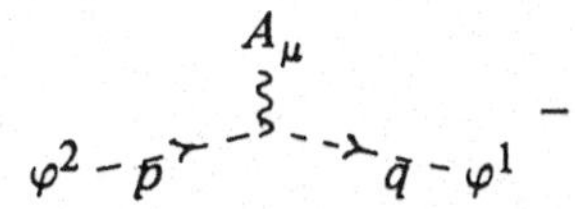

$$-e(p-q)_\mu \tag{A.24}$$

etc.

The J-source can now emit a photon and a ρ-meson simultaneously:

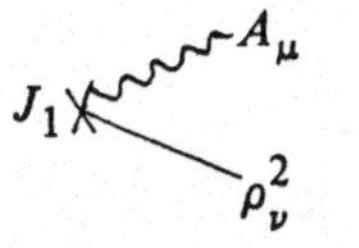

$$-e\,\delta_{\mu\nu} \tag{A.25}$$

etc.

The vertex (A.22) is a consequence of the fact that we did not diagonalize the bilinear terms in (7.8) completely. The diagonalized propagators have a rather complicated form. For small e it is easier to leave (A.22) as it stands.

If necessary, vertices like (A.25) for the J-source contribution may be avoided by choosing another expression for $\mathcal{L}^c$ (eq. (7.10)).

REFERENCES

[1] G.'t Hooft, Nucl. Phys. B33 (1971) 173.
[2] J.S.Bell and R.Jackiw, Nuovo Cimento 60A (1969) 47.
[3] S.L.Adler, Phys. Rev. 177 (1969) 2426.
[4] I.S.Gerstein and R.Jackiw, Phys. Rev. 181 (1969) 1955.
[5] S.Weinberg, Phys. Rev. 140 (1965) B516.
[6] T.W.B.Kibble, Phys. Rev. 155 (1967) 1554.
[7] B.W.Lee, Nucl. Phys. B9 (1969) 649.
[8] J.L.Gervais and B.W.Lee, Nucl. Phys. B12 (1969) 627.
[9] M.Veltman, Nucl. Phys. B7 (1968) 637; Nucl. Phys. B21 (1970) 288.
[10] D.G.Boulware, Ann. of Phys. 56 (1970) 140.
[11] A.A.Slavnow, L.D.Faddeyev, Massless and massive Yang-Mills field, Moscow preprint 1970.
[12] E.S.Fradkin, U.Esposito and S.Termini, Rivista del Nuovo Cimento 2 (1970) 498.
[13] S.Weinberg, Phys. Rev. Letters 19 (1967) 1264.

Nuclear Physics B139 (1978) 1–19

QUANTIZATION OF NON-ABELIAN GAUGE THEORIES

V.N. GRIBOV
Leningrad Nuclear Physics Institute, Gatchina, Leningrad 188350, USSR

Received 13 January 1978

It is shown that the fixing of the divergence of the potential in non-Abelian theories does not fix its gauge. The ambiguity in the definition of the potential leads to the fact that, when integrating over the fields in the functional integral, it is apparently enough for us to restrict ourselves to the potentials for which the Faddeev-Popov determinants are positive. This limitation on the integration range over the potentials cancels the infrared singularity of perturbation theory and results in a linear increase of the charge interaction at large distances.

1. Introduction

The quantization problem for non-Abelian gauge theories within the framework of perturbation theory was solved by Feynman [1], DeWitt [2] and Faddeev and Popov [3]. A subsequent analysis of perturbation theory in such theories (Politzer [4], Gross and Wilczek [5], Khriplovich [6]) has shown that they possess a remarkable property called asymptotic freedom. This property resides in the fact that zero-point field oscillations increase the effective charge not in the high-momentum region as in QED [7], but in the low-momentum region, i.e. at large distances between the charges. This gave hope that such theories may incorporate the phenomenon of color confinement which is fundamental to present day ideas concerning the structure of hadrons.

Answering the question as to whether color confinement occurs in non-Abelian theories proved to be a very difficult problem since the non-Abelian fields possessing charges ("color") strongly interact in the large-wavelength region.

Strong interaction between vacuum fluctuations in the region of large wavelengths means that at these wavelengths a significant role is played by field oscillations with large amplitudes, for which the substantially non-linear character of non-Abelian theories is decisive. Thus, the problem of color confinement is closely connected with that of the quantization of large non-linear oscillations. In this paper we show that in the region of large field amplitudes the method of quantization by Faddeev and Popov is to be improved.

As will be demonstrated, it is very likely that this improvement reduces simply to an additional limitation on the integration range in the functional space of non-Abelian fields, which consists in integrating only over the fields for which the Faddeev-Popov determinant is positive. This additional limitation is not significant for high-frequency oscillations, but substantially reduces the effective oscillation amplitudes in the low-frequency region. This in turn results in the fact that the "effective" charge interaction does not tend to infinity at finite distances as occurs in perturbation theory, but goes to infinity at infinitely large distances between charges, if at all.

2. Non-uniqueness of gauge conditions

The difficulties in the quantization of gauge fields are caused by the fact that the gauge field Lagrangian

$$\mathcal{L} = -\frac{1}{4g^2}\,\mathrm{Sp}\,F_{\mu\nu}F_{\mu\nu}\,, \tag{1}$$

$$F_{\mu\nu} = \partial_\mu A_\nu - \partial_\nu A_\mu + [A_\mu A_\nu]\,, \tag{2}$$

where A_μ are antihermitian matrices, $\mathrm{Sp}\,A_\mu = 0$, being invariant with respect to the transformation

$$A_\mu = S^+ A'_\mu S + S^+ \partial_\mu S\,, \qquad S^+ = S^{-1}, \tag{3}$$

contains non-physical variables which must be eliminated before quantization. A conventional method of relativistic invariant quantization [3] is as follows. Let us consider a functional integral

$$W = \int e^{-\int \mathcal{L} d^4x} \prod dA'_\mu \tag{4}$$

in Euclidean space-time and imagine the functional space A_μ in the form shown in

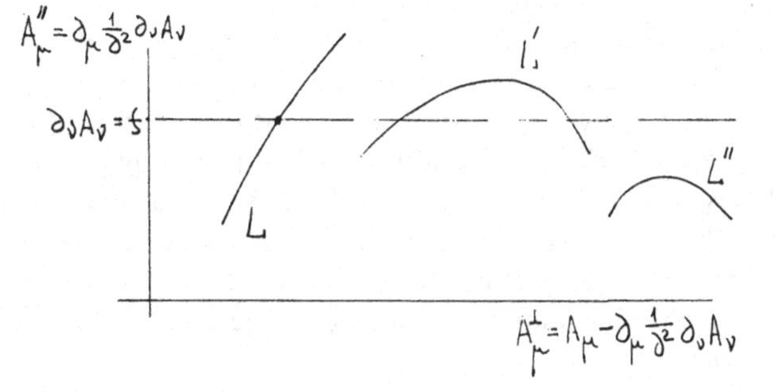

Fig. 1

fig. 1, where the transverse and longitudinal components of the field A_μ are plotted along the horizontal and vertical axes, respectively. Then for fixed A_μ, eq. (3) defines the line L (as a function of S) on which $\mathcal{L}$ is constant. The Faddeev and Popov idea is that, instead of integrating over A'_μ, one should integrate over matrices S and fields A_μ which have a certain divergence $f = \partial_\mu A_\mu$. Then W is written in the form

$$W = \int e^{-\int \mathcal{L} d^4x} \prod \int dA' \frac{1}{\Delta(A')}$$

$$\times \int dS \cdot S^+ \delta[f - S^+ \{\partial_\mu A'_\mu + [\nabla_\mu(A')S\partial_\mu S^+]\} S] , \tag{5}$$

where

$$\Delta(A) = \int dS \cdot S^+ \delta[f - S^+\{\partial_\mu A'_\mu + [\nabla_\mu(A')S\partial_\mu S^+]\}S] , \tag{6}$$

$$\nabla_\mu(A') = \partial_\mu + A'_\mu .$$

Since the variation with respect to S of the expression under the sign of the δ function is $\partial_\mu[\nabla_\mu(A), S^+\partial S]$ then

$$\frac{1}{\Delta(A')} = \|\tilde{\Box}(A)\| , \tag{7}$$

where the operator $\tilde{\Box}(A)$ is defined by the equation

$$\tilde{\Box}(A)\psi = \partial^2\psi + \partial_\mu[A_\mu \psi] \equiv \partial_\mu[\nabla_\mu(A)\psi] . \tag{8}$$

Replacing in (4) the variables

$$A'_\mu = SA_\mu S^+ + S\partial_\mu S^+ , \tag{9}$$

we obtain

$$W = \int e^{-\int \mathcal{L} d^4x} \delta(f - \partial_\mu A_\mu)\, dA\, \|\tilde{\Box}(A)\|\, dS \cdot S^+ . \tag{10}$$

Since (10) does not depend upon f, we may integrate over f with any weight function, $\exp\{(1/2\alpha g^2)\, \mathrm{Sp} \int f^2 dx\}$ being commonly used for this purpose. In so doing, with the integration over S omitted, W is obtained in the form

$$W = \int \exp\left(-\int \mathcal{L} d^4x + \frac{1}{2\alpha g^2} \mathrm{Sp} \int (\partial_\mu A_\mu)^2 d^4x\right) \|\tilde{\Box}(A)\|\, dA . \tag{11}$$

This conclusion is correct under the essential condition that, given a field A'_μ, one can always find a unique field A_μ with a prescribed divergence f, i.e. there are neither situations where curve (3) crosses the line $\partial_\mu A_\mu = f$ several times (curve L′) nor where it does not cross it at all (curve L″). We do not know any examples of situations of the type L″, where one cannot find a field A_μ, with a certain divergence, which is gauge-equivalent to a given field A'_μ. However, a situation of the type where many gauge-equivalent fields A_μ with a given divergence correspond to a given field A'_μ is typical in non-Abelian theories. Indeed, in order for two gauge-equivalent fields $A_{1\mu}$ and $A_{2\mu}$ with the same divergence to exist, there should be a unitary matrix S connecting $A_{1\mu}$ and $A_{2\mu}$,

$$A_{2\mu} = S^+ A_{1\mu} S + S^+ \partial_\mu S \,, \tag{12}$$

and satisfying the equation

$$\partial_\mu S^+ [\nabla_\mu(A_1), S] = 0 \,, \tag{13}$$

or obtained from it through the substitution of $A_{2\mu}$ for $A_{1\mu}$ and S^+ for S. In an Abelian theory, where $S = e^{i\varphi}$ is a unit matrix, eq. (13) reduces to the Laplace equation

$$\partial^2 \varphi = 0 \,, \tag{14}$$

and to eliminate non-uniqueness it is sufficient for us to confine ourselves to the fields which vanish at infinity. In a non-Abelian case, the non-linear equation (13) cannot have growing solutions and hence even for $A_{1\mu} = 0$ it has solutions for S leading to a decreasing $A_{2\mu}$. In the appendix we consider examples of the solution to eq. (13) for $A_{1\mu} = 0$, from which it will be evident that a set of these solutions, i.e. of the transverse potentials equivalent to the vacuum, are in order of magnitude similar to a set of solutions to the Laplace equation, which grow at infinity, but that all of these, though corresponding to such S that do not tend to unity at infinity, result in the potentials $A_{2\mu}$ decreasing as $1/r$.

In the appendix we shall also show that, with values of $A_{1\mu}$ large enough, (13) has solutions for S which tend to unity at $r \to \infty$ and result in rapidly decreasing $A_{2\mu}$. Under these circumstances, to calculate correctly the functional integral in a non-Abelian theory, we must either replace eq. (11) by the expression

$$W = \int e^{-\int \tilde{\mathcal{L}} d^4x} \frac{\|\Box(A)\|}{1 + N(A)} dA \,, \tag{15}$$

where N is the number of fields gauge-equivalent to a given field A and having the same divergence, or restrict the integration range in the functional space so as to have no repetitions.

An intermediate case, when both things are required, is possible. For instance, when integrating only over A_μ vanishing at infinity faster than $1/r$, we eliminate the fields gauge-equivalent to "small" fields, but for large enough A_μ the gauge-equivalent fields will remain and hence $N(A)$ in (15) will be needed. In this case, the problem of calculating $N(A)$ reduces to the analysis of solutions of eq. (13) tending to unity under $r \to \infty$ which depend on the character and the magnitude of the field A_μ. This problem seems to be almost hopeless, but we shall demonstrate below that there exists a possibility of a sufficiently universal solution leading to physically interesting results.

3. A limitation on the integration range in the functional space

In order to gain some insight into the nature of non-uniqueness in the functional space A_μ, let us see for what fields A_μ there exist gauge-equivalent fields close to the former and having the same divergence, i.e. what are the conditions for solving eq. (13) with S close to unity. Substituting into (13)

$$S = 1 + \alpha\,, \qquad \alpha^+ = -\alpha\,, \tag{16}$$

we get

$$\widetilde{\Box}(A)\alpha = \partial_\mu [\nabla_\mu(A), \alpha] = 0\,. \tag{17}$$

Since $\widetilde{\Box}(A)$ is the operator whose determinant enters into the functional integral, and eq. (17) is simply an equation for the eigenfunction of this operator with a zero eigenvalue, we draw the conclusion that the field A_μ can only have a close equivalent field when the Faddeev-Popov determinant for this field turns into zero, or (which is the same) if the field is such that the Faddeev-Popov ghost has a zero-mass bound state. Clearly, if the field A_μ is sufficiently small in the sense that the product of the width of the region where A_μ differs from zero with its amplitude over the same region is small, then there are no bound states in such a field, i.e. the equation

$$-\widetilde{\Box}(A)\psi = \epsilon\psi \tag{18}$$

is solvable for positive ϵ only. For a sufficiently large magnitude and a particular sign of the field there appears a solution with $\epsilon = 0$, which becomes one with a negative ϵ as the field increases further. For a particular still greater magnitude of the field, the level with a zero ϵ reappears, etc. Thus, one can imagine the fields for which eq. (17) is solvable as dividing the functional space into regions over each of which eq. (18) has a given number of eigenvalues, i.e. there exist a given number of bound states for the Faddeev-Popov ghosts. Fig. 2 shows this division of the field space into the regions $C_0, C_1, ..., C_n$, over which the ghosts have 0, 1, 2, ..., n bound states, by the lines $\ell_1, \ell_2, ..., \ell_n$ on which the ghosts have zero-mass levels.

Hence, if we imagine the space of the fields A_μ in this way, it may be asked whether two near equivalent fields that can exist close to the line, say, ℓ_1, are located on different sides of this line, i.e. one field within the region C_0, another in C_1, or may be arbitrarily situated. We shall demonstrate below that, indeed, if there are only two near equivalent fields, they will always lie on different sides of the corresponding curve ℓ_1. Moreover, we shall show that for any field located within the region C_1 close to the curve ℓ_1 there is an equivalent field within the region C_0 close to the same curve.

If we could prove that not only for small neighbourhoods close to the curves ℓ_n, but also for any field in the region C_n there is an equivalent field in the region C_{n-1}, we would prove that instead of integrating over the entire space of the fields A_μ, it would be sufficient for us to confine ourselves to the region C_0, i.e. to integrating only with respect to the fields A_μ which create no bound states for the ghosts (up

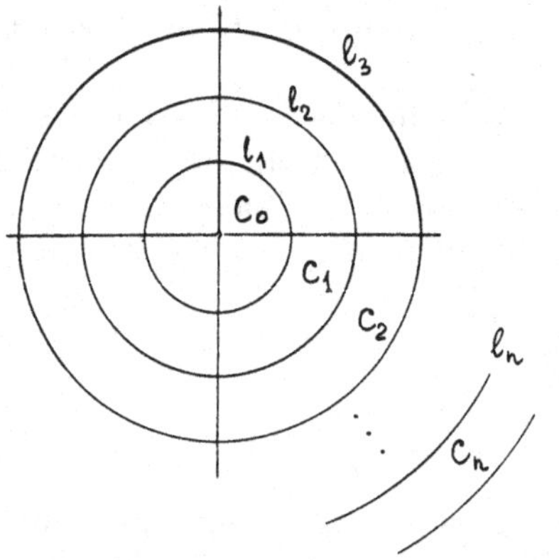

Fig. 2

to the first zero of the Faddeev-Popov determinant). However, even at the level of the things we can prove, there is a significant statement that for the functional integral, integration can be cut off on the boundary of the region C_0, and if there exist fields A_μ not equivalent to those over the region C_0, they are separated by a finite region from the boundary ℓ_1 and are in no way connected with the region of small fields A_μ lying within C_0 for which perturbation theory holds.

We shall assume below (until the contrary is established) that these additions are either non-existent or insignificant and that the integral (15) is determined over the region C_0. Generally speaking, we must retain $N(A)$ in (15) because we have not proved that there are no equivalent fields over the region inside C_0. We shall return to this subject below.

4. Proof of the field equivalence over the regions C_0 and C_1 close to their boundary

We shall first of all show that if the field A_μ is close to ℓ_1, then there is always a similar field equivalent to the former, i.e. for such a field eq. (13) always has a solution with S little different from unity and tending to unity at $r \to \infty$. The condition for $S \to 1$ as $r \to \infty$ is required because a solution with $S \not\to 1$ yields equivalent fields greatly different from the initial field. As shown in the appendix, these fields are located within the region C_∞. We write the field A_μ in the form

$$A_\mu = C_\mu + a_\mu \,, \tag{19}$$

where C_μ lies on ℓ_1, i.e. there exists φ_0 decreasing at infinity and satisfying the equation

$$\partial_\mu [\nabla_\mu(C), \varphi_0] = 0 \,, \tag{20}$$

and a_μ is small compared to C_μ in the sense defined below.

Substituting the matrix S in the form exp α into (13) and confining ourselves to the terms quadratic in α, we obtain

$$\partial_\mu [\nabla_\mu(A), \alpha] - \tfrac{1}{2} \partial_\mu [\alpha[\nabla_\mu, \alpha]] = 0 . \tag{21}$$

The solution to (21) may be sought for in the form

$$\alpha = \varphi_0 + \tilde{\alpha}, \qquad \tilde{\alpha} << \varphi_0 . \tag{22}$$

Substituting (19), (22) into (21) and taking into account (20), we get

$$\partial_\mu [\nabla_\mu(C), \tilde{\alpha}] = -\partial_\mu [a_\mu, \varphi_0] + \tfrac{1}{2} \partial_\mu [\varphi_0 [\nabla_\mu(C), \varphi_0]] . \tag{23}$$

Since the right-hand side of (23) vanishes at infinity, for a vanishing solution it is sufficient for the r.h.s. to be orthogonal with respect to φ_0, i.e.

$$\mathrm{Sp} \int \mathrm{d}^4x \, \{\varphi_0 \partial_\mu [a_\mu, \varphi_0] - \tfrac{1}{2} \varphi_0 \partial_\mu [\varphi_0 [\nabla_\mu(C), \varphi_0]]\} = 0 . \tag{24}$$

This equation defines the normalization of φ_0, i.e. the difference between S and unity.

Hence, we have found S, and thus may now find the field $A'_\mu = C_\mu + a'_\mu$, equivalent to the field $A_\mu = C_\mu + a_\mu$,

$$a'_\mu = a_\mu + [\nabla_\mu(C), \varphi_0] , \tag{25}$$

and clear up the question of whether A'_μ and A_μ lie on both sides of ℓ_1, i.e. whether it is true that there exists a bound ghost state in one of the fields A_μ, A'_μ and that there is no such state in the other. For this purpose, it is sufficient to calculate the shift of the level from zero due to the fields a_μ and a'_μ:

$$\epsilon(a) = -\frac{1}{N^2} \mathrm{Sp} \int \mathrm{d}^4x \, \varphi_0 \partial_\mu [a_\mu, \varphi_0] ,$$

$$\epsilon(a') = -\frac{1}{N^2} \mathrm{Sp} \int \mathrm{d}^4x \, \{\varphi_0 \partial_\mu [a_\mu, \varphi_0] + \varphi_0 \partial_\mu [[\nabla_\mu(C), \varphi_0] \varphi_0]\} ,$$

$$N^2 = \mathrm{Sp} \int \varphi_0^2 \, \mathrm{d}^4x . \tag{26}$$

Using (24), which defines the normalization φ_0, we obtain

$$\epsilon(a) = -\epsilon(a') , \tag{27}$$

which was to be proved.

Non-strictness of the derivation due to the fact that we ignored the continuous character of the spectrum at $\epsilon > 0$ can be easily avoided, and we shall not dwell on that. The derivation can just as readily be repeated for the fields close to any ℓ_n, imposing the orthogonality condition of the eigenfunction φ_0 on the eigenfunctions of other bound states together with eq. (24).

In essence, one can gain some insight into the cause of the field equivalence over

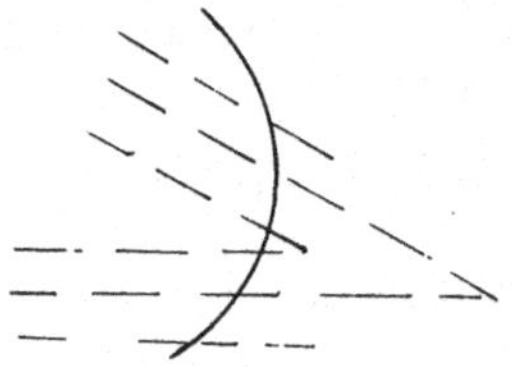

Fig. 3

the regions C_0 and C_1 when considering the effective Lagrangian in (15),

$$\tilde{\mathcal{L}} = -\frac{1}{4g^2}\,\mathrm{Sp}\,F_{\mu\nu}F_{\mu\nu} - \frac{1}{2\alpha g^2}\,\mathrm{Sp}(\partial_\mu A_\mu)^2\,, \tag{28}$$

close to ℓ_1. The property of the fields C_μ lying on the line ℓ_1, is that there exists a solution to eq. (20). Consider now the fields of the form

$$A_\mu = C_\mu + [\nabla_\mu(C), \varphi_0]\,. \tag{29}$$

Then the Lagrangian $\tilde{\mathcal{L}}$, to second-order in φ_0, takes the form

$$\tilde{\mathcal{L}}(A) = \tilde{\mathcal{L}}(C) - \tfrac{1}{4}\,\mathrm{Sp}\{[F_{\mu\nu}(C), \varphi_0]\,[F_{\mu\nu}(C), \varphi_0] + 2F_{\mu\nu}(C)\,[[\nabla_\mu, \varphi_0]\,[\nabla_\nu, \varphi_0]]\}\,, \tag{30}$$

i.e. $\tilde{\mathcal{L}}$ has an extremum along $[\nabla_\mu(C), \varphi_0]$ and does not change upon replacing φ_0 by $-\varphi_0$. From this it follows that the fields $A^{\pm}_\mu = C_\mu \pm [\nabla_\mu(C), \varphi_0]$ are equivalent. Clearly $\epsilon(A^+_\mu) = -\epsilon(A^-_\mu)$.

Since the direction $[\nabla_\mu(C), \varphi_0]$ is generally not orthogonal to the curve ℓ_1, the distribution of the equivalent fields can correspond to that shown in fig. 3, where the equivalent fields are located on dashed lines in the opposite directions to ℓ_1. If structural lines can intersect as shown in fig. 3, this will result in the existence of at least pairs of equivalent fields within the region C_0. As will be demonstrated in the appendix, such intersections actually occur, and in this sense the equivalent fields do exist in C_0, but in the examples considered, these equivalent fields turn out to be mirror-reflected ones that are always equivalent. Hence, the field doubling thus obtained is independent of the field magnitude and insignificant in the functional integral.

5. Gauge non-uniqueness and limitation on the integration range over the fields in physical space-time

So far we have discussed the functional integration over non-Abelian fields defined in four-dimensional Euclidean space. This somewhat simplifies the mathematics, but makes it more difficult to understand the physical content of the theory and leaves a feeling of dissatisfaction related to the need for analytical continuation of the results.

Certainly, the general statement that the integration should only be performed over non-equivalent fields is independent of the nature of the space, and formula (15) holds. The difference is in the real form of eq. (13) defining the number of equivalent fields. Now, this equation is a hyperbolic one having non-zero solutions even in an Abelian case ($S = \exp \varphi; \varphi$ is an arbitrary solution to a wave equation). One of the ways for eliminating this non-uniqueness is a change-over to Euclidean space. In normal space-time this non-uniqueness does not show itself because, according to Feynman, we integrate over fields A_μ which have only positive frequencies at $t \to -\infty$ and negative frequencies at $t \to +\infty$. In this case, if we want to have two equivalent A_μ and A'_μ under the same boundary conditions, then $A'_\mu - A_\mu$ as $t \to -\infty$ and $+\infty$ should contain only negative and positive frequencies, respectively. This indicates that equivalent trajectories will only occur through those solutions of eq. (13) which have only negative and positive frequencies as $t \to -\infty$ and $+\infty$, respectively. Clearly, these conditions play the same part as those at infinity in the Euclidean case, and the linear equations will have no solution at $A_\mu = 0$ because of frequency conservation. Such solutions will exist for non-linear equations or for sufficiently large fields A_μ. For instance, eq. (18) at $\epsilon = 0$ is one for the ghost wave function in the external field A_μ. If the field A_μ is situated on the line ℓ_1, such an equation has a solution under the boundary condition specified above, and defines the ghost transition from the state with negative energy to that with positive energy. Since the ghosts are quantized in the same way as fermions, the process is apparently interpreted as the classic formation of ghost pairs in the external field. In a similar manner it can be said that solutions of eq. (13) result in the fields A'_μ which differ from A_μ in pairs of the gauge quanta produced.

The restriction of the integration in the functional integral to the region C_0 implies the restriction to the fields in which no classical ghost formation occurs because the formation of ghosts merely redefines the fields A_μ.

6. The effect of the field magnitude restriction on the zero-point oscillations and interaction in the low-momenta region

In this section we shall try to analyze how a limitation on the integration range over the field in the functional integral affects the physical properties of non-Abelian theories.

We shall proceed from eq. (15) for the action, disregarding the possibility for the equivalent fields to exist in C_0:

$$W = \int e^{-\int \tilde{\mathcal{L}} d^4x} \, \|\tilde{\Box}(A)\| \mathcal{V}(\Box) \, dA \,, \tag{31}$$

where $\mathcal{V}(\Box)$ indicates that the integration is performed only over the region C_0.

First of all, let us see whether the restriction $\mathcal{V}(\Box)$ is significant from the standpoint of what we know from the perturbation theory analysis. For this purpose, consider the Green function of the Faddeev-Popov ghost

$$G(k) = -\frac{1}{W}\int e^{-\int \tilde{\mathcal{L}} d^4x} \langle k | \frac{1}{\tilde{\Box}(A)} | k \rangle \|\Box\| \mathcal{V}(\Box)\, dA \,. \tag{32}$$

It is well known that, if we calculate $G(k)$ in perturbation theory, i.e., perform the integration over A in (32), omitting $\mathcal{V}(\Box)$ and expanding over the coupling constant, we get

$$G(k) = \frac{1}{k^2} \frac{1}{\left(1 - \frac{11g^2C_2}{48\pi^2} \ln \frac{\Lambda^2}{k^2}\right)^{(3/22)(3/2-\alpha/2)}} \,, \tag{33}$$

where Λ is the ultraviolet cutoff, α is the gauge parameter in $\tilde{\mathcal{L}}$. From this it is obvious that $G(k)$ becomes large at $\alpha < 3$ and physical k^2 (in the Euclidean space) such that

$$1 - \frac{11g^2C_2}{48\pi^2} \ln \frac{\Lambda^2}{k^2} \sim g^2 \,,$$

where (33) still holds. From the standpoint of (32), $G(k)$ can be large only due to the integration range for the fields where $\tilde{\Box}$ is small, i.e. close to the lines ℓ_n.

It is interesting that transverse fields (low α) act on the ghosts as attractive fields and longitudinal fields as repulsive ones. Since the influence of longitudinal fields cancels in the calculations of gauge-invariant quantities, we may say that we study the contribution to the functional integral close to the curves ℓ_n when calculating $G(k)$ near the "infrared pole", and hence $\mathcal{V}(\tilde{\Box})$ is definitely significant at momenta below or of the order of the "infrared pole" position, whereas at large k we are within C_0 (low A), where $\mathcal{V}(\tilde{\Box})$ is insignificant and perturbation theory works.

Furthermore, $\mathcal{V}(\tilde{\Box})$ makes it impossible for a singularity of $G(k)$ to exist at finite k^2 because, with k^2 below the singularity position, $G(k)$ would either reverse its sign or become complex. Both things would indicate that $\tilde{\Box}$ has ceased to be a positively defined quantity, i.e. we have left the region C_0 when integrating over A_μ. The only possibility that now remains is that $k^2G(k)$ has a singularity at $k^2 = 0$. Such a possibility would indicate that at $k^2 = 0$ we feel the fields on the line ℓ_1.

Up to now, all attempts at finding the mechanism for removal of the "infrared pole" have not been successful. Higher corrections [8–10] and instantons [11,12] only bring it nearer. If no other causes are found, $\mathcal{V}(\tilde{\Box})$ will be the cause. The fact that there are no other causes for the interaction cutoff is equivalent to the statement that without $\mathcal{V}(\tilde{\Box})$ zero fluctuations of the fields tend to leave the region C_0. Hence it appears quite natural that the fields closest to the boundary of the region C_0, i.e. connected with the singularity of $k^2G(k)$ at $k^2 = 0$, will correspond to the real vacuum if $\mathcal{V}(\tilde{\Box})$ is taken into account.

For checking the above by a specific calculation, one must write $\mathcal{V}(\tilde{\Box})$ in a constructive way, but unfortunately we have not yet succeeded in doing this. All we could have done was to write this criterion to second order in perturbation theory and then calculate the functional integral taking no account of the interaction except for $\mathcal{V}(\tilde{\Box})$. In this case it turns out that there appears a characteristic scale κ^2 defined by the condition $g^2 \ln \Lambda^2/\kappa^2 \sim 1$, so that at $k^2 >> \kappa^2$ the gluon and ghost Green functions remain free. The gluon Green function $D(k)$ has complex singularities and is non-singular at $k^2 \to 0$. The ghost Green function under $k^2 \to 0$ is $G(k) \sim C/k^4$.

If it were not for the roughness of the calculations and difficulties with complex singularities of $D(k)$, this would be the right thing for the colour confinement theory.

Let us show the way this is obtained. We write $G_{aa}(k, A) = -\langle k, a| 1/\tilde{\Box} |k, a\rangle$ in the form of an expansion in perturbation theory (where a is the isotopic index)

$G_{aa}(k, A) =$  $+ \cdots .$ (34)

The first-order term gives no contribution to the diagonal element. The second-order term is

$$= V \frac{4}{k^4} \int \frac{d^4 q}{(2\pi)^4} \frac{A^a_\mu(q) A^a_\nu(-q) k_\mu (k-q)_\nu}{(k-q)^2} \equiv \frac{V}{k^2} \sigma(k, A) . \tag{35}$$

$A^a_\mu(q)$ is the Fourier component of the potential A_μ, V the volume of the system, $\sigma(k, A)$ defines positions of the poles $G(k, A)$, if any, to a second Born approximation since

$$G(k, A) \approx \frac{1}{k^2} \frac{V}{1 - \sigma(k, A)} . \tag{36}$$

In this case we assume, of course, that k is conserved in a typical field of zero-point fluctuations ($\langle k | 1/\Box | k' \rangle |_{k'=k}$ is proportional to the volume of the system which is replaced by $\delta(k - k')$ after averaging). The no-pole condition at a given k is $\sigma(k, A) < 1$. For simplicity, we choose a transverse gauge ($\alpha = 0$). On averaging over the gluon polarization directions λ, we have

$$\sigma(k, A) = 4 \int \frac{d^4 q}{(2\pi)^4} \frac{|A^{a,\lambda}(q)|^2}{(k-q)^2} \left(1 - \frac{(kq)^2}{k^2 q^2}\right) . \tag{37}$$

If $|A^{a,\lambda}(q)|^2$ over the main range of integration with respect to q decreases monotonically with q^2, as will prove to be the case in what follows, then $\sigma(k, A)$ decreases as k^2 increases and hence as a no-level condition use can be made of

$$\sigma(0, A) = 3 \int \frac{d^4 q}{(2\pi)^4} \frac{|A^{a,\lambda}(q)|^2}{q^2} < 1 . \tag{38}$$

Taking (38) as a condition for $\mathcal{V}(\widetilde{\Box})$, replacing $\mathcal{L}$ by $\mathcal{L}_0$ in (31) and omitting $\|\Box\|$, instead of (31) we obtain a functional integral which is easy to calculate, if $\mathcal{V}(1-\sigma(k,A))$ is written in the form

$$\mathcal{V}(1-\sigma(0,A)) = \int \frac{d\beta}{2\pi i\beta}\, e^{\beta(1-\sigma(0,A))}\,, \tag{39}$$

$$W = \int \frac{d\beta}{2\pi i\beta}\, e^{\beta} \int \prod dA^{a,\lambda}$$

$$\times \exp\left\{-\frac{1}{g^2}\sum q^2 |A^{a,\lambda}(q)|^2 - \frac{\beta}{V}\sum \frac{|A^{a,\lambda}|^2}{q^2}\right\}\,, \tag{40}$$

where V is the volume of the system. Calculating the integral over A, we get

$$W \sim \int \frac{d\beta}{2\pi i\beta}\, e^{\beta} \prod_q \frac{1}{(q^2 + \beta g^2/Vq^2)^{3n/2}}\,, \tag{41}$$

n being the number of isotopic states. The integral over β can be obtained by the steepest-descent method, with the saddle-point value β_0 determined by

$$\frac{3n}{2}\frac{g^2}{V}\sum_q \frac{1}{q^4 + \beta_0 g^2/V} - 1 + \frac{1}{\beta_0} = 0\,. \tag{42}$$

Setting $\beta_0 g^2/V = \kappa^4$, with $V \to \infty$ we get

$$\tfrac{3}{2} n g^2 \int \frac{d^4q}{(2\pi)^4}\frac{1}{q^4+\kappa^4} = 1 \tag{43}$$

or

$$\frac{3n}{32\pi} g^2 \ln\frac{\Lambda^2}{\kappa^2} = 1\,. \tag{44}$$

If the saddle-point value β_0 is known, we can return to the functional integral (40), substituting $\beta = \beta_0$ in it and omitting the integration over β, so as to obtain an effective functional integral for calculating the correlation functions of the fields A. In this case, W is

$$W = \int dA \exp -\frac{1}{g^2}\sum_{k,\lambda,a}\left(k^2 + \frac{\kappa^4}{k^2}\right)|A^{\lambda,a}(k)|^2\,. \tag{45}$$

Consequently, the gluon and ghost Green functions are

$$D^{ab}_{\mu\nu}(k) = A^a_\mu(k)\, A^b_\nu(k) = \delta_{ab}\left(\delta_{\mu\nu} - \frac{k_\mu k_\nu}{k^2}\right)\frac{g^2 k^2}{k^4+\kappa^4}\,, \tag{46}$$

$$G(k) = \frac{1}{k^2(1-\sigma(k))} = \left\{1 - \frac{4g^2}{k^2}\int \frac{d^4q}{(2\pi)^4}\frac{k_\mu k_\nu D^{aa}_{\mu\nu}(q^2)}{(k-q)^2}\right\}^{-1}$$

$$= \frac{1}{g^2 n k^2}\left\{\int \frac{d^4 q}{(2\pi)^4}\, \frac{k^2 - 2kq}{q^4 + \kappa^4}\, \frac{1}{(k-q)^2}\left(1 - \frac{(kq)^2}{k^2 q^2}\right)\right\}^{-1}, \tag{47}$$

respectively, due to (43).

As $k^2 \to 0$

$$G(k) \approx \frac{32\pi\kappa^2}{g^2 n k^4}, \tag{48}$$

in accordance with the above. The fact that the significant range of integration in the functional integral turns out to coincide with the boundary of the region ℓ_1, is evident without calculating $G(k)$ because, when calculating the saddle-point value β_0, the last term in (42) has no effect at $V \to \infty$ and hence $\mathcal{V}(1-\sigma)$ is equivalent to $\delta(1-\sigma)$. We would obtain the same result when calculating with the function $(1-\sigma)\mathcal{V}(1-\sigma)$, which is equivalent to an attempt at taking into account the effect of the determinant in (31).

7. Coulomb gauge

In sect. 6 we discussed the effect of limiting the integration over the fields on the properties of vacuum fluctuations in the invariant Euclidean formulation of the theory. In so doing, we adduced arguments for singularity of the ghost Green function as $k^2 \to 0$ (for example, $1/k^4$). This certainly is an indication of a substantial long-range effect in the theory that may result in colour confinement, but the ghost Green function in an arbitrary gauge is not connected directly with the Coulomb interaction at large distances. Hence, in this section we shall rewrite the foregoing analysis for the Coulomb gauge [13] where the Green function of the ghost determines directly the Coulomb interaction. We shall show that the situation which involves a restriction on the integration range over fields and a cutoff of the infrared singularity found in perturbation theory is exactly the same as in invariant gauges. The arguments for singularity of the ghost Green function hold here as well. In this case, however, a singularity of the ghost Green function as $k^2 \to 0$ of the type $1/k^4$ is indicative of a linear increase in the Coulomb interaction with distance.

The most natural way of formulating the Coulomb gauge is the Hamiltonian form which shows up vividly the unitarity of the theory because of the lack of ghosts. To this end, the functional integral W incorporates the fields which satisfy the three-dimensional transversality condition

$$\frac{\partial A_i}{\partial x_i} = 0, \tag{49}$$

and momenta π_i which are canonically conjugated with them and stand for the transverse part of the electric field

$$\pi_i = E_i^{\perp} = \left(\frac{\partial A_i}{\partial t} - [\nabla_i A_0]\right)^{\perp}. \tag{50}$$

The integral over A_0 is calculated for fixed A_i and cancels the Faddeev-Popov determinant.

As a result, the functional integral takes the form

$$W = \int \exp\left[\frac{i}{g^2}\int d^4x\,\{\pi_i \dot{A}_i - \mathcal{H}(\pi_i, A_i)\}\right] dA\, d\pi\,, \tag{51}$$

$$\mathcal{H}(\pi_i, A_i) = -\tfrac{1}{2}\{\pi_i^2 + H^2(A_i) + \partial_i\varphi\partial_i\varphi\}\,,$$

$$H^2 = \tfrac{1}{2}F_{ij}E_{ij}\,, \tag{52}$$

$$\tilde{\Delta}(A)\varphi = -\rho, \quad \tilde{\Delta}(A)\varphi \equiv [\nabla_i(A), \partial_i\varphi], \quad \rho = [A_i, \pi_i]\,. \tag{53}$$

In this case, the integration should be performed over gauge-inequivalent transverse fields, as for invariant gauges in (51).

The number of fields equivalent to a given field A_i is determined by the number of solutions of the equation similar to eq. (13)

$$[\nabla_i(A), S\partial_i S^+] = 0\,. \tag{54}$$

The condition for the existence of two equivalent fields is the existence of a solution of the equation

$$[\nabla_i(A), \partial_i\varphi] = 0\,, \tag{55}$$

a zero eigenvalue of the operator $\tilde{\Delta}(A)$ defining the Coulomb potential according to eq. (53). Repeating the arguments given above for the four-dimensional case (see also the appendix), we draw a conclusion that the integration range in (51) should be restricted to the region C_0 where the operator $\tilde{\Delta}(A)$ has no eigenvalues.

This region coincides with the one where the Coulomb energy density

$$V_{\text{Coul}} = \tfrac{1}{2}\rho\frac{1}{\tilde{\Delta}}\partial^2\frac{1}{\tilde{\Delta}}\rho \tag{56}$$

does not go to infinity anywhere except for its boundary. In the Coulomb gauge, instead of the functional integral we can use the Schrödinger equation

$$\frac{1}{g^2}\int d^3x\,\mathcal{H}\left(-ig\frac{\partial}{\partial A_i}, A_i\right)\psi(A_i) = E\psi(A_i)\,, \tag{57}$$

bearing in mind that $\mathcal{H}$ is determined only for the region C_0.

Let us discuss now how a limitation by the fields within C_0 affects the spectrum and zero oscillations defined by eq. (57). As before, for the no-level condition in the field A_i we set

$$\tfrac{4}{3}\int\frac{d^3k}{(2\pi)^3}\,\frac{\sum\limits_{\lambda,q}|A^{\lambda,q}(k)|^2}{k^2} < 1\,, \tag{58}$$

obtained in a similar manner from the Green function. Omitting in $\mathcal{H}(\pi, A)$ all terms except for $\mathcal{H}_0 = -\frac{1}{2}(\pi_i^2 - A_i\partial^2 A_i)$, we obtain, instead of (57), an equation for an oscillator system

$$\sum_{k,\lambda,q} \{\pi_{\lambda,q}^2(k) + k^2 a_{\lambda,q}^2(k)\}\,\psi(a) = E\psi(a)\,, \tag{59}$$

provided that

$$\frac{4g^2}{3V}\sum_{k,\lambda,q}\frac{|a_{\lambda,q}(k)|^2}{k^2} < 1\,. \tag{60}$$

Taking no account of (60), for free oscillations $\overline{|a_{\lambda,q}(k)|^2} \sim 1/k$ and hence the left-hand side of (60) is infinite. This means that free oscillations correspond to the fields far outside the region C_0. Eqs. (59), (60) can be solved approximately by the variational method assuming that

$$\psi = \prod_{k,\lambda,q} f(a_{\lambda,q}(k))\,. \tag{61}$$

Calculating the energy minimum for the system with the fixed value of the left-hand side in (60), we get an oscillator equation for f with $k^2 + \kappa^4/k^2$ instead of k^2, where κ^4 is the variational parameter. In this case, the ground-state energy and average squares of oscillation amplitudes will be

$$E = \tfrac{1}{2}\sum_{k,\lambda,q}\sqrt{k^2 + \frac{\kappa^4}{k^2}}\,,$$

$$\overline{|a_{\lambda,q}(k)|^2} = \frac{k}{\sqrt{k^4 + \kappa^4}}\,. \tag{62}$$

The energy is at a minimum with $\kappa = 0$, but as this takes place, the left-hand side in (60) equals infinity. Therefore, κ is determined from the condition

$$\tfrac{8}{3}g^2\int\frac{d^3k}{(2\pi)^3}\,\frac{1}{k\sqrt{k^4 + \kappa^4}} \approx \frac{2g^2}{3\pi^2}\ln\frac{\Lambda^2}{\kappa^2} = 1\,. \tag{63}$$

The ghost Green function is

$$G(k) = \langle k|\frac{1}{\tilde{\Delta}}|k\rangle$$

$$= \frac{1}{k^2 4g^2 n}\left\{\int\frac{d^3k'}{(2\pi)^3}\cdot\frac{(k^2 - 2kk')(1-(kk')^2/k^2k'^2)}{k'(k'-k)^2\sqrt{k'^4 + \kappa^4}}\right\}^{-1}; \tag{64}$$

as $k^2 \to 0$

$$G(k) = \frac{6}{5g^2 n k^4 \ln(\kappa^2/k^2)}\,. \tag{65}$$

Thus, this crude calculation shows that again there is a characteristic scale beyond which zero-point oscillations become small and the Coulomb potential increases linearly with distance.

In conclusion, it should be noted that since zero-point oscillations of the fields in vacuum turn out to be on the boundary of the region C_0, we have no right to ignore the Coulomb energy which may go to infinity on this boundary. However, considering that the number of zero oscillation modes pushing the system outside the region C_0 is infinitely large and that the boundary equation comprises one condition (60), we have good cause to think that the system will remain close to the boundary despite the Coulomb interaction. In this case, the condition (60) will be satisfied all the same, which is indicative of a decrease in the zero oscillation amplitudes for momenta below a particular value and of a linear increase in the Coulomb potential with distance.

Finally, I wish to express my sincere gratitude to A.A. Belavin, A.M. Polyakov, L.N. Lipatov and Yu.L. Dokshitser for numerous most helpful discussions.

Appendix

In this appendix we consider the properties of gauge-equivalent fields with equal divergence, giving the simplest examples. Let us begin with the case of the three-dimensional space (Coulomb gauge) and the group SU(2). We consider "spherically symmetric" fields A_i, i.e., the fields dependent on one unit vector $n_i = x_i/r$ ($r = \sqrt{x_i^2}$). The general expression for such a field has the form

$$A_i(x) = f_1(r)\frac{\partial \hat{n}}{\partial x_i} + f_2(r)\hat{n}\frac{\partial \hat{n}}{\partial x_i} + f_3(r)\hat{n}n_i\,, \tag{A.1}$$

$\hat{n} = in_a\sigma_a$, σ_a, are Pauli matrices, $\hat{n}^2 = -1$. Under the spherically symmetric gauge transformation of the form

$$S = \exp\{\tfrac{1}{2}\alpha(r)\hat{n}\} = \cos(\tfrac{1}{2}\alpha) + \hat{n}\sin(\tfrac{1}{2}\alpha) \tag{A.2}$$

$A_i(x)$ goes into $\widetilde{A}_i(x) = S^+A_iS + S^+\partial_iS$ so that

$$\tilde{f}_1 = f_1\cos\alpha + (f_2 + \tfrac{1}{2})\sin\alpha\,,$$

$$\tilde{f}_2 + \tfrac{1}{2} = -f_1\sin\alpha + (f_2 + \tfrac{1}{2})\cos\alpha\,,$$

$$\tilde{f}_3 = f_3' + \tfrac{1}{2}\alpha'\,, \tag{A.3}$$

$$\partial A_i/\partial x_i = \hat{n}[f_3' + (2/r)f_3 - 2f_1/r^2]\,. \tag{A.4}$$

The condition

$$\partial\widetilde{A}_i/\partial x_i = \partial A_i/\partial x_i \tag{A.5}$$

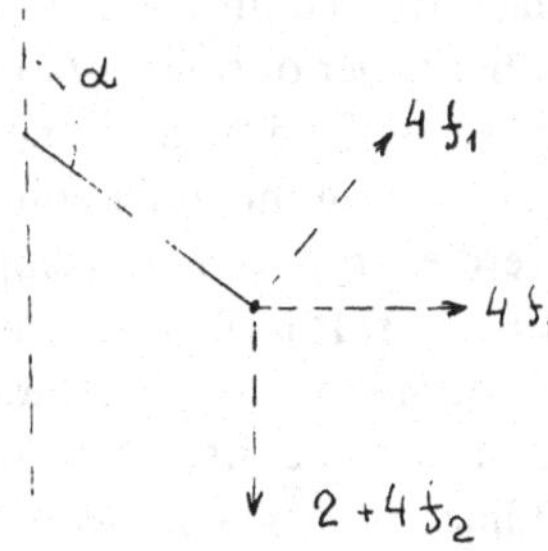

Fig. 4

is equivalent to the equation

$$\alpha'' + (2/r)\alpha' - (4/r^2)\{(f_2 + \tfrac{1}{2}) \sin\alpha + f_1(\cos\alpha - 1)\} = 0 \,. \tag{A.6}$$

If we introduce the variable $\tau = \ln r$, eq. (A.6) reduces to the equation for a pendulum with damping in the field of the vertical force $4f + 2$, horizontal force $4f_1$, and the force perpendicular to the pendulum, $-4f_1$

$$\ddot{\alpha} + \dot{\alpha} - (2 + 4f_2)\sin\alpha + 4f_1(1 - \cos\alpha) = 0 \,. \tag{A.7}$$

From this analogy the general properties of solutions to the equations of the equivalence conditions (A.5), (A.6) are readily seen.

If the forces f_1 and f_2 are equal to zero as $r \to 0$ ($\tau \to -\infty$; otherwise the field A_1 is singular as $r \to 0$) and tend to zero as $r \to \infty$ ($\tau \to +\infty$), then for a solution to exist at finite τ, as $\tau \to -\infty$, the pendulum should be in the position of unstable equilibrium, $\alpha = 0$. In such an event, if its initial velocity as $\tau \to -\infty$ is not specifically selected, upon executing a number of oscillations in the field, the pendulum starts damping and once only the vertical force remains, it comes to stable equilibrium. Such a solution corresponds to $S \to \hat{n}$, as $r \to \infty$, and the equivalent field

$$A_i = -\hat{n}\partial\hat{n}/\partial x_i \sim 1/r \,. \tag{A.8}$$

decreases slowly at infinity.

However, exceptional cases are possible. If sufficiently large, these forces can under specifically selected initial conditions restore the pendulum to its unstable equilibrium position. In this case, we obtain the equivalent field $\widetilde{A}_i$ which decreases fairly rapidly at infinity. We consider several versions of such a possibility. Let the forces and initial conditions be such that throughout the whole "time" $-\infty < \tau < +\infty$, $|\alpha(\tau)| \ll 1$. Then, to a zero approximation the equation

$$\ddot{\alpha} + \dot{\alpha} - 2\alpha(1 + 2f_2) = 0 \tag{A.9}$$

should be satisfied. This equation is simply eq. (20) for the three-dimensional case. In order for eq. (A.9) to hold, $-\int f_2 \, d\tau$ should have a particular and sufficiently large value. The field with a corresponding f_2 lies on the curve ℓ_1 independently of the values for f_1 and f_3. Taking into account the quadratic term in (A.7) enables us to get a solution with a somewhat larger or smaller f_2 (within or outside the region C_0) and to determine the amplitude of α as demonstrated in the text.

If $f_1 = 0$, there are no second-order terms and a solution exists only for f_2 larger than the value required for a zero energy level. The amplitude of α is determined by taking into account third-order terms. It is obvious that in this case there are two solutions that differ from one another by the sign of α. It can easily be shown that they are in the region C_0. This situation corresponds to the phenomenon of intersection of the equivalent field lines discussed in the text and illustrated in fig. 3. The field for which $f_1 = 0$ lies on the intersection of such lines. The occurrence of two equivalent fields in C_0 in this example does not point to the necessity of introducing $N(A)$ in the region C_0 because it shows the symmetry of the problem with respect to reflection. It is easily shown that the fields f_1, f_3, f_2 and $-f_1, -f_3, -f_2$ are gauge-equivalent and have equal divergence. Hence the functional integral should be divided by two, independently of the magnitudes of the fields f_1, f_3, f_2, which do not matter. Accordingly, eq. (A.6) always has two solutions $\alpha_1 = \alpha(f_1, f_2)$ and $\alpha_2 = -\alpha(-f_1, f_2)$. The second solution in the field $-f_1, -f_3, f_2, \alpha_2(-f_1, f_2)$, leads to an equivalent, reflected field obtained from f_1, f_2, f_3 using the solution $\alpha_1(f_1, f_2)$. With $f_1 = 0$, both reflected fields are obtained from one initial field. Two other solutions $\alpha_1(-f_1, f_2)$ and $\alpha_2(f_1, f_3)$ are outside C_0 anyway, for small f_1 and f_2 close to ℓ_1.

Another possibility for the pendulum to regain its unstable equilibrium position is to execute one complete revolution (or more). This possibility may exist even with $f_2 = 0$, i.e., deep inside the region C_0. Such solutions exist at large enough f_1, but the equivalent fields corresponding to them differ greatly from the initial field and are likely to lie outside C_0. For example, with $f_2 = 0, \tilde{f}_2$ is of the order of unity according to (A.3) because α in a complete revolution changes from 0 to 2π.

Hence, the fields inside C_0 have their equivalents of two types, i.e. the fields which possess asymptotics at ∞ of the type (A.8) ($\alpha \to \pi$) and lying within the region C_∞ (it is easy to see that with $f_2 = 1$ as $r \to \infty$, (A.9) has an infinite number of solutions) and the fields situated in C_n with a finite n.

Finally, let us discuss the question as to whether for a particular field A_i an equivalent field A_i' with a specified difference $\hat{n}\Delta f$ in their divergence can always be found. The equation for a corresponding $\alpha(r)$ will differ from (A.7) in the external force $2\Delta f e^{2\tau}$ on the right-hand side perpendicular to the pendulum. In this case, it is likely that there exists, "almost" without exception, a solution with α tending to $2n\pi$ as $r \to \infty$ because, as we have seen, if f_1 and f_2 are large, the solution comes into play through choosing the initial conditions; should f_1, f_2 and Δf be small, we have an inhomogeneous linear equation for which the choice is made in a trivial way.

We now turn to the four-dimensional space. In this case, it is convenient to deal with the group O(4) from which SU(2) is trivially separated.

Instead of $i\tau_i$ as antihermitian matrices for infinitesimal transformations in the group O(4), one may choose $\sigma_{\mu\nu} = \frac{1}{2}(\gamma_\mu\gamma_\nu - \gamma_\nu\gamma_\mu)$. For constructing a scalar, we need an antisymmetric tensor, i.e. at least two vectors are required. This indicates that the field cannot be spherically symmetric. It can be axially symmetric if we choose as antisymmetric tensor

$$F_{\mu\nu} = \frac{n_\mu l_\nu - n_\nu l_\mu}{\sqrt{1-(n_\alpha l_\alpha)^2}} , \tag{A.10}$$

where $n_\mu = x_\mu/\sqrt{x^2}$ and l_μ is a constant unit vector. The gauge transformation matrix between such axially symmetric fields can be written as

$$S = \exp\{\tfrac{1}{2}\beta(r, n_l)\hat{\psi}\} = \cos\tfrac{1}{2}\beta + \hat{\psi}\sin\tfrac{1}{2}\beta ,$$
$$\hat{\psi} = \tfrac{1}{2}\sigma_{\mu\nu}F_{\mu\nu} , \qquad n_l = l_\mu n_\mu , \qquad r = \sqrt{x_\mu^2} . \tag{A.11}$$

The field A_μ which preserves its shape under this transformation has the form

$$A_\mu = f_1 \partial_\mu \psi + f_2 \psi \partial_\mu \psi + \psi \partial_\mu f_3 . \tag{A.12}$$

The transformation formulae between $\tilde{f}_i$ and f_i coincide with (A.3), if α is replaced by β.

The equivalence condition is

$$\partial^2\beta - \frac{4}{r^2(1-n_l)^2}\,[(f_2+\tfrac{1}{2})\sin\beta - f_1(1-\cos\beta)] = 0 . \tag{A.13}$$

With $f_1 = f_2 = 0$, there is a solution similar to (A.6) which is dependent on one variable $\rho^2 = r^2(1-n_l^2)$ and has the same asymptotics. Despite two variables, which make this equation more cumbersome, its structure is much the same as that of (A.6), and we do not see any reasons why the structure of its solutions should differ markedly from (A.6).

References

[1] R.P. Feynman, Acta Phys. Pol. 24 (1963) 262.
[2] B. DeWitt, Phys. Rev. 160 (1967) 113; 162 (1967) 1195, 1293.
[3] L.D. Faddeev and V.N. Popov, Phys. Lett. 25B (1967) 30.
[4] H.D. Politzer, Phys. Rev. Lett. 30 (1973) 1346.
[5] D.J. Gross and P. Wilczek, Phys. Rev. Lett. 30 (1973) 1343.
[6] I.B. Khriplovich, ZhETF (USSR) 10 (1969) 409.
[7] L.D. Landau, A.A. Abrikosov and I.M. Khalatnikov, DAN 95 (1954) 497, 773, 1117.
[8] A.A. Belavin and A.A. Migdal, ZhETF Pisma 19 (1974) 317.
[9] D.R.T. Jones, Nucl. Phys. B75 (1974) 530.
[10] W. Caswell, Phys. Rev. Lett. 33 (1974) 244.
[11] A.A. Belavin, A.M. Polyakov, Yu. Tyupkin and A.S. Schwartz, Phys. Lett. 59B (1975) 85.
[12] C.G. Callan, R. Dashen and D.J. Gross, Princeton Univ. preprint C00-2220-115 (August 1977).
[13] V.N. Gribov, Materials for the 12th LNPI Winter School, 1977, Vol. 1, p. 147.

GAUGE FIELD MODELS

C. BECCHI
University of Genova
Italy

A. ROUET
Max-Planck Institut für
Physik und Astrophysik
München

R. STORA
Centre de Physique Théorique
C.N.R.S. - Marseille

Lectures given by C. BECCHI

INTRODUCTION

The renormalized versions of the Schwinger Quantum Action Principle [1] (Q.A.P.) have been among the most important subjects of the first week of this School [2,3]. It turns out that the naive identities, which can be deduced on a very formal ground, expressing in terms of the Green functions the quantum equivalent of the Noether theorem are affected with quantum correction depending on the particular renormalization rules. In some "clever" renormalization scheme [3], (e.g. in those which are based on the dimensional regularization of the Feynman integrals) these quantum corrections seem to satisfy a not yet well defined criterion of minimality. In a more general framework they have known power counting properties [2,4]. This weak form of the Q.A.P. proved in the framework of the Bogoliubov-Parasiuk-Hepp-Zimmermann [5] (B.P.H.Z.) renormalization scheme has been stated in Stora's lectures [6]. Exploiting the B.P.H.Z. version of the Q.A.P. Stora has been able to prove rather general results on the renormalization of lagrangian theories with broken symmetries [7].

The purpose of these lectures is to continue Stora's

G. Velo and A. S. Wightman (eds.), Renormalization Theory, 269–297.

analysis discussing with the same technique a more sophisticated class of lagrangian models, that is the nonabelian (Yang-Mills) gauge field models (G.F.M.). This is the only known class of renormalizable models involving a self interacting system of vector mesons. Hence it is of particular interest in the construction of renormalizable models based on vector intermediate bosons for the weak and electromagnetic interactions [8].

In the G.F.M. in which the gauge group is semisimple the gauge photons must be massless except if the gauge symmetry is spontaneously broken according to the Higgs-Kibble (H.K.) mechanism [9]. In order to avoid infrared difficulties we shall discard models in which massless fields are involved, limiting ourselves to the analysis of theories in which the gauge symmetry is completely broken according to the H.K. mechanism.

It turns out clearly from the study of quantum electrodynamics, (Q.E.D.), the oldest and best known gauge field model, that a gauge invariant lagrangian model cannot be directly quantized apparently preserving the local structure of the theory and leading to a renormalizable perturbation theory. (Here by renormalizable we mean satisfying the necessary power counting prescriptions.) The classical way out of this difficulty implies the introduction of a certain number of degrees of freedom which are not physically interpretable (e.g. the longitudinal, scalar, photons and the Faddeev-Popov [10] ($\Phi.\Pi.$) ghosts). This procedure completely changes the renormalization problem. We have a model depending on a finite number of parameters and we have to show that the original physical properties of the theory are not affected by the introduction of the new "unphysical" degrees of freedom. More precisely we have to prove that the restriction of the scattering operator of the theory to a certain "physical" subspace of the Fock space is unitary in the perturbative sense and independent from the properties of the unphysical particles.

In much the same way as in Q.E.D. this is the consequence of a set of Ward-Takahashi identities, the Slavnov identities [11], which, owing to the Q.A.P., express the invariance of the Faddeev-Popov lagrangian of the G.F.M. under a set of nonlinear field transformations (Slavnov transformations) [12] explicitely involving the $\Phi.\Pi.$ ghost fields. Hence our first goal will be to prove that one can fulfill the Slavnov identities to all orders of the renormalized perturbation theory.

Then we shall discuss in some details the gauge independence of the physical scattering operator and we shall outline the connection between its unitarity and the Slavnov symmetry. The complete combinatorial proof of unitarity will not be examined here since it is discussed in details in the referred literature [12,13]. We shall restrict ourselves to models involving semisimple gauge groups. This will greatly simplify the analysis of the possible quantum corrections to the Q.A.P. which in technical terms will be reduced to the study of the cohomology group of the Lie algebra characterizing the gauge theory [7,13]. For what concerns the basic tools and definitions (in particular the functional method) that we shall extensively use in the following we refer to the first section of Stora's lectures.

In section two we discuss at the classical level the algebraic properties of the SU(2) Higgs-Kibble-Englert-Brout-Faddeev-Popov lagrangian and we exhibit its invariance under Slavnov transformations.

In section three we study the renormalization of the Slavnov identity in the G.F.M. involving semisimple gauge groups.

Section four deals with unitarity and gauge independence of the physical S operator in the SU(2) H.K. model.

2. THE CLASSICAL SU(2) MODEL

In this section we shall examine in some details the classical structure of the SU(2) Higgs-Kibble-Englert-Brout model and the problems which are encountered in its quantization.

Let $\{\varphi_a\} = \{\pi_\alpha, \sigma\}$ with $a = 1, \ldots 4$, and $\alpha = 1,2,3$ be a multiplet of scalar fields and $a_{\mu\alpha}$, $\alpha = 1,2,3$, a multiplet of Yang-Mills photons.
An infinitesimal gauge transformation of parameter $\delta\omega_\alpha(x)$ is defined by

$$\delta\pi_\alpha(x) = \frac{e}{2}\left[\varepsilon_\alpha{}^{\beta\gamma}\,\delta\omega_\beta\,\pi_\gamma + (\sigma + F)\,\delta\omega_\alpha\right](x) \equiv \int dy\, \frac{\delta\pi_\alpha(x)}{\delta\omega_\beta(y)}\,\delta\omega_\beta(y)$$

$$\delta\sigma(x) = -\frac{e}{2}\left[\pi^\alpha\,\delta\omega_\alpha\right](x) \equiv \int dy\, \frac{\delta\sigma(x)}{\delta\omega_\alpha(y)}\,\delta\omega_\alpha(y) \qquad (1)$$

$$\delta a^\mu_\alpha(x) = \left[\partial^\mu\,\delta\omega_\alpha + \varepsilon_\alpha{}^{\beta\gamma}\,\delta\omega_\beta\, a^\mu_\gamma\right](x) \equiv \int dy\, \frac{\delta a^\mu_\alpha(x)}{\delta\omega_\beta(y)}\,\delta\omega_\beta(y)$$

The parameter F plays the role of the σ field vacuum expectation value. A lagrangian invariant under such transformations is constructed in terms of the antisymmetric covariant field tensor :

$$G_\alpha^{\mu\nu} = \partial^\mu a_\alpha^\nu - \partial^\nu a_\alpha^\mu - e\, \varepsilon_\alpha{}^{\beta\gamma} a_\beta^\mu a_\gamma^\nu \tag{2}$$

and of the covariant derivatives :

$$(D_\mu \varphi)_a = \left\{ \partial_\mu \pi_\alpha - \frac{e}{2}\left[\varepsilon_\alpha{}^{\beta\gamma} a_\beta^\mu \pi_\gamma - (\sigma+F) a_\alpha^\mu\right], \partial^\mu \sigma + \frac{e}{2} \pi^\alpha a_\alpha^\mu \right\} \tag{3}$$

it has the form :

$$\mathcal{L}_{inv}(\varphi_a, a_\alpha^\mu) = -\frac{1}{4}(G_{\mu\nu}^\alpha)^2 + \frac{1}{2}(D_\mu \varphi_a)^2 - H([\varphi+F]^2) \tag{4}$$

Here H is a polynomial in $\pi^\alpha \pi_\alpha + (\sigma+F)^2 \equiv [\varphi+F]^2$ invariant under the transformation (1) and satisfying the condition :

$$\partial_{\varphi_a} H([\varphi+F]^2)\Big|_{\varphi=0} = 0 \tag{5}$$

expressing that the classical vacuum corresponds to the configuration $\varphi_a = 0$. From the invariance property of H and from Eq.(5) it turns out that the mass matrix of the scalar fields :

$$\partial_{\varphi_a} \partial_{\varphi_b} H([\varphi+F]^2)\Big|_{\varphi=0} \tag{6}$$

has three null eigenvalues corresponding to the π_α modes and a positive eigenvalue corresponding to the σ field. This result expresses the content of the Goldstone theorem [14].

To discuss the quantization of this model we have to examine the wave operator corresponding to the free part (that bilinear in the fields) of the lagrangian. In particular the free lagrangian of the photon - π channel is :

$$\mathcal{L}^{o}_{inv} = -\frac{1}{4}\left(\partial^{\mu} a^{\nu}_{\alpha} - \partial^{\nu} a^{\mu}_{\alpha}\right)^2 + \frac{e^2}{8} F^2 (a^{\mu}_{\alpha})^2 + + \frac{1}{2}(\partial_{\mu}\pi_{\alpha})^2 + \frac{eF}{2}\partial_{\mu}\pi^{\alpha} a^{\mu}_{\alpha} \tag{7}$$

The Fourier transformed wave operator factorizes in two submatrices. The first associated with the transverse modes of the photon:

$$K^{T\,\alpha\beta}_{\mu\nu}(p) = -\delta^{\alpha\beta}\left(g^{\mu\nu} - \frac{p^{\mu}p^{\nu}}{p^2}\right)\left(p^2 - \frac{e^2F^2}{4}\right) \tag{8}$$

exhibits the Higgs phenomenon [9] (the photons are massive). The second, corresponding to the coupled longitudinal photon - π channel, is :

$$\left\| K^{\alpha\beta}_{a_L,\pi}(p)\right\| = \delta^{\alpha\beta}\begin{vmatrix} e^2F^2 & -ip\frac{eF}{2} \\ ip\frac{eF}{2} & p^2 \end{vmatrix} \tag{9}$$

The determinant of this matrix is identically zero. It follows that our invariant theory is not directly quantizable.

In the cases in which the theory is spontaneously broken one can eliminate the degrees of freedom with degenerate wave operator by means of a field dependent gauge transformations [15]. This reduces the degrees of freedom of the model to a minimal number, the transverse modes of the photons and the σ field, to which "physical" particles are associated. However the resulting lagrangian is not renormalizable by power counting.

Another solution is based on the Faddeev-Popov lagrangian [10]. This is the natural extension to the case of non abelian G.F.M. of the usual Q.E.D. lagrangian. It involves, beyond the scalar and vector fields, two conjugate multiplets of scalar fields $\{c_{\alpha}\}, \{\bar{c}_{\alpha}\}$ quantized according to the Fermi statistics in a Fock space carrying indefinite metric . The lagrangian is :

$$\mathcal{L}(\varphi_a, a^{\mu}_{\alpha}, c_{\alpha}, \bar{c}_{\alpha}) = \mathcal{L}_{inv}(\varphi_a, a^{\mu}_{\alpha}) - \frac{1}{\kappa}\left[\frac{(g^{\alpha})^2}{2} - c_{\alpha}(m\bar{c})^{\alpha}\right] \tag{10}$$

with

$$g_\alpha = \partial_\mu a^\mu_\alpha + \rho \pi_\alpha \tag{11}$$

and

$$(m\bar{c})_\alpha(x) = \int dy \frac{\delta g_\alpha(x)}{\delta\omega_\beta(y)} \bar{c}_\beta(y) \equiv$$
$$\equiv \left\{ \Box \bar{c}_\alpha + e\,\varepsilon_\alpha{}^{\beta\gamma} \partial_\mu (\bar{c}_\beta a^\mu_\gamma) + \frac{\rho e}{2}\left[\varepsilon_\alpha{}^{\beta\gamma}\bar{c}_\beta \pi_\gamma + (\sigma+F)\bar{c}_\alpha\right]\right\}(x) \tag{12}$$

This new lagrangian can be straightforwardly quantized in a Fock space with indefinite metric ; the corresponding perturbation theory is renormalizable by power counting. Indeed now :

$$\det \left\| K^{\alpha\beta}_{a_L,\pi}(p^2)\right\| \propto \left(p^2 - \frac{\rho e F}{2}\right)^2 \text{ and } K_{c\bar{c}}(p^2) = \frac{1}{\kappa}\left(p^2 - \frac{\rho e F}{2}\right) \tag{13}$$

However choosing the $\Phi.\Pi.$ lagrangian we have increased the number of the unphysical degrees of freedom of our model. We are then left with the problem of proving that the restriction of the S operator to the above defined physical subspace of the free field Fock space is unitary in the perturbative sense and independent from the parameters specifying the $\Phi.\Pi.$ term of the lagrangian. In the tree approximation this is true if the matrix elements of the gauge operator $g_\alpha(x)$ within physical states vanish. This generalizes the well known Q.E.D. supplementary condition. We are now going to show that this follows from the Slavnov identity, which at the classical level expresses the invariance of the lagrangian under the following set of infinitesimal transformations (Slavnov transformations)

$$\begin{aligned}
\delta\pi_\alpha(x) &= \delta\lambda \int dy \frac{\delta\pi_\alpha(x)}{\delta\omega_\beta(y)} \bar{c}_\beta(y) \equiv \delta\lambda\, P_\alpha(x) \\
\delta\sigma(x) &= \delta\lambda \int dy \frac{\delta\sigma(x)}{\delta\omega_\beta(y)} \bar{c}_\beta(y) \equiv \delta\lambda\, P_0(x) \\
\delta a^\mu_\alpha(x) &= \delta\lambda \int dy \frac{\delta a^\mu_\alpha(x)}{\delta\omega_\beta(y)} \bar{c}_\beta(y) \equiv \delta\lambda\, P^\mu_\alpha(x) \\
\delta\bar{c}_\alpha(x) &= \delta\lambda \frac{e}{2} \varepsilon_\alpha{}^{\beta\gamma}(\bar{c}_\beta\bar{c}_\gamma)(x) \equiv \delta\lambda\, \overline{P}_\alpha(x) \\
\delta c_\alpha(x) &= \delta\lambda\, g_\alpha(x)
\end{aligned} \tag{14}$$

$\delta\lambda$ is a space-time independent infinitesimal parameter commuting with $\{\varphi_a\}$ and $\{a_{\mu\alpha}\}$ but anticommuting with the $\Phi.\Pi.$ ghost fields, and for two transformations labelled by $\delta\lambda_1$ and $\delta\lambda_2$, $\delta\lambda_1$ and $\delta\lambda_2$ anticommute.

The Slavnov transformation laws can be cast together introducing the functional differential operator :

$$\mathcal{s}=\int dx\left[P_\alpha\frac{\delta}{\delta\pi_\alpha}+P_0\frac{\delta}{\delta\sigma}+P_\alpha^{\mu}\frac{\delta}{\delta a_\alpha^{\mu}}+\bar{P}_\alpha\frac{\delta}{\delta\bar{c}_\alpha}+g_\alpha\frac{\delta}{\delta c_\alpha}\right](x) \tag{15}$$

Then the transformation law for the field $\{\psi\}=\{\varphi_a, a_\mu^\alpha, c_\alpha, \bar{c}_\alpha\}$ is :

$$\delta\psi=\delta\lambda\,\mathcal{s}\,\psi \tag{16}$$

The invariance of $\mathcal{L}$ under the transformations (14) can be checked immediately taking into account the composition law of the gauge transformations :

$$\left[\frac{\delta}{\delta\omega_\alpha(x)},\frac{\delta}{\delta\omega_\beta(y)}\right]=e\,\varepsilon_\gamma^{\alpha\beta}\,\delta(x-y)\frac{\delta}{\delta\omega_\gamma(x)} \tag{17}$$

Let us now introduce the system of external fields :

$$\{\eta\}=\{\gamma_\alpha,\gamma_0,\gamma_\alpha^\mu,\zeta_\alpha\} \tag{18}$$

whose couplings are defined by the new lagrangian ;

$$\mathcal{L}^{(\eta)}(\psi,\eta)=\mathcal{L}(\psi)+\gamma_\alpha P^\alpha+\gamma_0 P^0+\gamma_\alpha^\mu P_\mu^\alpha+\zeta_\alpha\bar{P}^\alpha \tag{19}$$

Since, owing to the group structure underlying the Slavnov transformations, the polynomials $\{P_\alpha, P_\alpha^\mu, P_0, \bar{P}_\alpha\}$ are invariant, also $\mathcal{L}^{(\eta)}$ is invariant. This fact can be translated by means of the Q.A.P. into a system of identities relating the Green functions of the theory whenever the possible quantum corrections are absent, hence, at least in the tree approximation. These identities are easily expressed in terms of the connected Green functional Z_c [6]. Upon introducing the sources $\{J\}=\{J_\alpha, J_0, j_\alpha^\mu, \xi_\alpha, \bar{\xi}_\alpha\}$ for the fields $\{\psi\}=\{\pi_\alpha,\sigma,a_\alpha^\mu,c_\alpha,\bar{c}_\alpha\}$ respectively yields:

$$0=\mathcal{S}Z_c(J,\eta)\equiv \tag{20}$$

$$\equiv\int dx\left[J_\alpha\frac{\delta}{\delta\gamma_\alpha}+J_0\frac{\delta}{\delta\gamma_0}+j_\alpha^\mu\frac{\delta}{\delta\gamma_\alpha^\mu}-\zeta_\alpha\frac{\delta}{\delta\xi_\alpha}-\bar{\xi}_\alpha\left(\partial_\mu\frac{\delta}{\delta j_{\mu\alpha}}+\rho\frac{\delta}{\delta J_\alpha}\right)\right](x)Z_c(J,\eta)$$

The corresponding identity for the vertex functional :

$$\Gamma(\psi,\eta) = Z_c(J,\eta) - \int dx\, \big(J,(\psi+F)\big)(x)\Big|_{\psi+F=\frac{\delta Z_c}{\delta J}} \tag{21}$$

is

$$\int dx\Big[\frac{\delta}{\delta\pi_\alpha}\Gamma\frac{\delta}{\delta\gamma_\alpha}\Gamma + \frac{\delta}{\delta\sigma}\Gamma\frac{\delta}{\delta\gamma_0}\Gamma + \frac{\delta}{\delta a^\alpha_\mu}\Gamma\frac{\delta}{\delta\gamma^\mu_\alpha}\Gamma + \\ + \frac{\delta}{\delta\bar{c}_\alpha}\Gamma\frac{\delta}{\delta\zeta^\alpha}\Gamma + g_\alpha\frac{\delta}{\delta c_\alpha}\Gamma\Big](\psi,\eta)(x) = 0 \tag{22}$$

Let us now recall that we have defined physical degrees of freedom of our model the transverse components of the photons and the σ field. Comparing with Eq.(14) we see that the Slavnov transformed of the physical fields do not contain any term linear in c . In terms of Feynman diagrams this property can be translated into the following : if ψ_{phys} is any physical field, an amplitude containing $s\,\psi_{phys}$ as a vertex is one particle irreducible with respect to cuts between this vertex and the gauge function g^α , since from Eq.(10) and Eq.(14) we see that :

$$s\, g_\alpha(x) = (m\bar{c})_\alpha(x) = \kappa\,\xi_\alpha(x) \tag{23}$$

Hence this property of the physical fields is not altered if we combine them with g_α .

From the L.S.Z. reduction formulae it turns out that we can write the restriction of the S operator to the physical space in the form [6] :

$$S_{phys} = :\exp\int dx\,dy\Big\{\sigma^{in}(x)K_\sigma(x,y)\frac{\delta}{\delta J_0(y)} + a^{T\,in}_{\mu\alpha}(x)K^{\alpha\beta}_{\mu\nu}(x,y)\frac{\delta}{\delta J^\beta_\nu(y)}\Big\}: \\ \exp\frac{i}{\hbar}Z_c(J,\eta)\Big|_{J=\eta=0} \equiv \Sigma_{phys}\,Z(J,\eta)\Big|_{J=\eta=0} \tag{24}$$

where σ^{in}, K_σ , $a^{T,in}_{\mu,\alpha}$, $K^{\mu\nu}_{\alpha\beta}$ are the canonically quantized asymptotic fields and the corresponding wave operators. Now the previously mentioned property of the physical fields implies that the operator s defined in Eq.(15) and Σ_{phys} commute, since Σ_{phys} amputates on the mass shell the physical legs. It follows that the field g^α is decoupled from the physical states :

$$\frac{\hbar}{i}\frac{\delta}{\delta\xi_\alpha(0)}\Sigma_{phys}\, s\, Z_c(J,\eta)\Big|_{J=\eta=0} = \frac{\hbar}{i}\left(\partial_\mu\frac{\delta}{\delta J^\alpha_\mu(0)}+\rho\frac{\delta}{\delta J_\alpha(0)}\right)\Sigma_{phys}Z_c(J,\eta)\Big|_{J=\eta=0} = 0 \qquad (25)$$

This is the desired supplementary condition.

This fulfillement of the Slavnov identity can be considered as the fundamental property of the G.F.M.. However to base our renormalization program on the Slavnov identity we have to verify that our lagrangian is uniquely determined (up to the addition of a divergence and a multiplicative renormalization of the fields) by the condition of Slavnov invariance, the dimensional constraints ensuring the renormalizability of the theory being understood. This is easily verified if the gauge group is semisimple. In the non semisimple case there may arise new terms. For instance in Q.E.D. in the Stueckelberg gauge the mass term:

$$\int dx\left[\frac{a_\mu^2}{2}+c\bar{c}\right](x) \qquad (26)$$

is Slavnov invariant.

If we assign dimension two to the external fields and we introduce the $\Phi.\Pi.$charge $Q^{\Phi\Pi}$:

$$Q^{\Phi\Pi}\zeta_\alpha = 2\zeta_\alpha$$

$$Q^{\Phi\Pi}c_\alpha = c_\alpha\,,\quad Q^{\Phi\Pi}\gamma_\alpha=\gamma_\alpha\,,\quad Q^{\Phi\Pi}\gamma_0=\gamma_0,\quad Q^{\Phi\Pi}\gamma^\alpha_\mu=\gamma^\alpha_\mu\,,$$

$$Q^{\Phi\Pi}\pi_\alpha = Q^{\Phi\Pi}\sigma = Q^{\Phi\Pi}a^\alpha_\mu = 0 \qquad (27)$$

$$Q^{\Phi\Pi}\bar{c}_\alpha = -\bar{c}_\alpha$$

the most general dimension four lagrangian carrying null charge is :

$$\bar{\mathcal{L}}^{(\eta)}(\psi,\eta) = \bar{\mathcal{L}}(\psi)+\gamma_\alpha\mathcal{P}^\alpha+\gamma_0\mathcal{P}^0+\gamma^\mu_\alpha\mathcal{P}^\alpha_\mu+\zeta_\alpha\bar{\mathcal{P}}^\alpha \qquad (28)$$

where the polynomials $\{\mathcal{P}^\alpha, \mathcal{P}^0, \mathcal{P}^\alpha_\mu, \bar{\mathcal{P}}^\alpha\}$ have dimension non exceeding two. The Slavnov invariance of the action is expressed in the functional differential form :

$$\int dx\left[\mathcal{P}_\alpha \frac{\delta}{\delta \pi_\alpha} + \mathcal{P}_0 \frac{\delta}{\delta \sigma} + \mathcal{P}^\mu_\alpha \frac{\delta}{\delta a^\mu_\alpha} + \bar{\mathcal{P}}^\alpha \frac{\delta}{\delta \bar{c}_\alpha} + g_\alpha \frac{\delta}{\delta c_\alpha}\right](x) \int \bar{\mathcal{L}}^{(\eta)}(y)\, dy$$
$$= \bar{s} \int dy\ \bar{\mathcal{L}}^{(\eta)}(y) = 0 \tag{29}$$

Looking in particular at the coefficients of the external fields we get :

$$\bar{s}\,\mathcal{P}_\alpha = \bar{s}\,\mathcal{P}_0 = \bar{s}\,\mathcal{P}^\mu_\alpha = \bar{s}\,\bar{\mathcal{P}}_\alpha = 0 \tag{30}$$

from which the structure of the Slavnov transformations (Eq.(14)) can be reconstructed. Indeed, for example, the general form of $\bar{\mathcal{P}}_\alpha$ is :

$$\bar{\mathcal{P}}_\alpha = \frac{1}{2}\,\Gamma_\alpha^{\beta\gamma}\,\bar{c}_\beta\,\bar{c}_\gamma \tag{31}$$

and Eq.(30)) writes :

$$\bar{s}\,\bar{\mathcal{P}}_\alpha = \frac{1}{2}\,\Gamma_\alpha^{\beta\lambda}\,\Gamma_\lambda^{\gamma\delta}\,\bar{c}_\beta\,\bar{c}_\gamma\,\bar{c}_\delta = 0 \tag{32}$$

which is nothing but the Jacobi identity :

$$\Gamma_\alpha^{\beta\lambda}\,\Gamma_\lambda^{\gamma\delta} + \Gamma_\alpha^{\gamma\lambda}\,\Gamma_\lambda^{\delta\beta} + \Gamma_\alpha^{\delta\lambda}\,\Gamma_\lambda^{\beta\gamma} = 0 \tag{33}$$

whose solutions are stable under perturbations [13]. That is if a set of structure constants differs from another one by perturbation, to the first order in the perturbation the two sets are equivalent.

Now, concerning the external field independent part of the Lagrangian, if $\bar{\mathcal{L}}(\psi)$ is Slavnov invariant a fortiori :

$$s^2 \int dx\ \bar{\mathcal{L}}(\psi)(x) = \int dx\,(\partial_\mu P^\mu_\alpha + \rho P_\alpha)(x) \frac{\delta}{\delta c_\alpha(x)} \int dy\ \bar{\mathcal{L}}(\psi)(y)$$
$$= \int dx\left[(m\bar{c})_\alpha \frac{\delta}{\delta c_\alpha}\right](x) \int dy\ \bar{\mathcal{L}}(\psi)(y) = 0 \tag{34}$$

Making the dependence on the $\Phi.\Pi.$ fields explicit, $\bar{\mathcal{L}}(\psi)$ assumes the form :

$$\bar{\mathcal{L}}(\psi) = \bar{\mathcal{L}}_{inv}(\varphi_a, a^\mu_\alpha) + \Delta\mathcal{L}(\varphi_a, a^\mu_\alpha) + $$
$$+ L^{\alpha\beta\gamma\delta} c_\alpha c_\beta \bar{c}_\gamma \bar{c}_\delta + c_\alpha (K\bar{c})^\alpha \tag{35}$$

where $\bar{\mathcal{L}}_{\text{INV}}$ is invariant under gauge transformations and the corresponding action is Slavnov invariant. By Eq.(34) it follows that :

$$L^{\alpha\beta\gamma\delta} = 0 \tag{36}$$

and that :

$$\int dx\, (m\bar{c}_\alpha)(x)\; (K\bar{c})^\alpha(x) = 0 \tag{37}$$

The general solution of Eq.(37) is, if the gauge group is semisimple :

$$(K\bar{c})_\alpha = \Gamma_{\alpha\alpha'}\,(m\bar{c})^{\alpha'} \tag{38}$$

where $\Gamma_{\alpha\alpha'}$ is a symmetrical matrix. Going back to Eq.(29) yields :

$$\bar{\mathcal{L}}(\psi) = \bar{\mathcal{L}}_{\text{INV}}(\varphi_a, a^\mu_\alpha) + c_\alpha \Gamma^{\alpha\alpha'}(m\bar{c})_{\alpha'} - \frac{1}{2}\, g_\alpha \Gamma^{\alpha\alpha'} g_{\alpha'} \tag{39}$$

which is identical with Eq.(12) modulo a redefinition of g^α and c^α .

3. RENORMALIZATION OF THE SLAVNOV IDENTITY

In this section we discuss the existence of quantum extensions of G.F.M. with semisimple gauge group, satisfying the appropriate Slavnov identities.

In order to study the general case we shall use a slightly different notation labelling the quantized fields by :

$$\{\psi\} = \{\phi_i, c_\alpha, \bar{c}_\alpha\} \tag{40}$$

with $\{\phi_i\} = \{\varphi_a, a^\mu_\alpha\}$. Here $\{c_\alpha, \bar{c}_\alpha\}$ are the $\Phi.\Pi.$ fields, $\{a^\mu_\alpha\}$ the gauge photons, and $\{\varphi_a\}$ a system of matter fields. We also introduce the sources:

$$\{J\} = \{J_i, \bar{\xi}_\alpha, \xi_\alpha\} \tag{41}$$

and the external fields :

$$\{\eta\} = \{\gamma_i, \zeta_\alpha\} \tag{42}$$

The $\Phi.\Pi.$ charge and dimension assignements are the natural extensions of those in the previous section.

The gauge function g^α is chosen linear in the fields:

$$g^\alpha(x) = g_i^\alpha \, \phi_i(x) \qquad \text{with} \qquad g^\alpha_{\beta,\mu} = g^\alpha_\beta \, \partial_\mu \tag{43}$$

A quantized extension of a G.F.M. is defined by means of an effective lagrangian :

$$\mathcal{L}_{eff}(\psi,\eta) = \mathcal{L}_{eff}(\psi) + \gamma^i P_i + \zeta^\alpha \bar{P}_\alpha \tag{44}$$

to which the Zimmermann [5] subtraction index four is assigned.

The lagrangian $\mathcal{L}_{eff}(\psi,\eta)$ is a formal power series in $\hbar$:

$$\mathcal{L}_{eff}(\psi,\eta) = \sum_{0}^{\infty}{}_n \, \hbar^n \, \overset{(n)}{\mathcal{L}}_{eff}(\psi,\eta) \tag{45}$$

Hence in particular :

$$\overset{(0)}{\bar{P}}_\alpha(\psi) = \frac{e}{2} \, \overset{(0)}{f}{}_\alpha^{\beta\gamma} \, \bar{c}_\beta \bar{c}_\gamma \tag{46}$$

and

$$\overset{(0)}{P}_i(\psi) = e\left(\overset{(0)\alpha}{t}_{ij} \phi_j + \overset{(0)\alpha}{q}\right) \qquad \text{with : } \overset{(0)\alpha}{q}_a = \overset{(0)\alpha}{t}_{ab} q_b \tag{47}$$

$$\text{and} \quad \overset{(0)\alpha}{q}_{\beta\mu} = \frac{1}{e} \, \delta^\alpha_\beta \, \partial_\mu$$

The coefficients $\overset{(0)}{f}{}_\alpha^{\beta\gamma}$ and $\overset{(0)\alpha}{t}_{ij}$ are the structure constants and a representation of the Lie algebra associated with the gauge group.

The purpose of this section is to show that the Slavnov identity :

$$\delta Z_c(J,\eta) \equiv \int dx \left[J_i \frac{\delta}{\delta \gamma_i} - \xi_\alpha \frac{\delta}{\delta \xi_\alpha} - \bar{\xi}_\alpha g_i^\alpha \frac{\delta}{\delta J_i} \right](x) Z_c(J,\eta) = 0 \quad (48)$$

can be solved in terms of the coefficients of $\mathcal{L}_{eff}(\psi,\eta)$. These coefficients are themselves formal power series in $\hbar$.

We shall proceed in two steps. First we shall discuss the necessary condition :

$$\delta^2 Z_c(J,\eta) \equiv -\int dx \left(\bar{\xi}_\alpha g_i^\alpha \frac{\delta}{\delta \gamma_i} \right)(x)\, Z_c(J,\eta) = 0 \quad (49)$$

Then we shall go back to Eq.(48).

After a Legendre transformation Eq.(49) becomes :

$$\int dx \left(\frac{\delta}{\delta C_\alpha} \Gamma\, g_\alpha^i \frac{\delta}{\delta \gamma_i} \Gamma \right)(x) = 0 \quad (50)$$

which is solved by :

$$\frac{\delta}{\delta C_\alpha(x)} \Gamma(\psi,\eta) - \kappa\, g_i^\alpha \frac{\delta}{\delta \gamma_i(x)} \Gamma(\psi,\eta) \equiv D_\alpha(x) \Gamma(\psi,\eta) = 0 \quad (51)$$

Now for a generic choice of the effective lagrangian the Q.A.P. yields :

$$D_\alpha(x)\, \Gamma(\psi,\eta) = \left(\Delta_\alpha(x) . \Gamma \right)_3 (\psi,\eta) \quad (52)$$

where the right-hand side means the insertion of the vertex $N_3(\Delta_\alpha(x))$ into the appropriate Green function. The vertex $\Delta_\alpha(x)$ is a polynomial in the fields and their derivatives whose coefficients are formal power series in $\hbar$:

$$\Delta_\alpha(x) = \sum_{\nu}^{\infty}{}_{n}\, \hbar^n \Delta_\alpha^{(n)}(x) \qquad \text{with } \nu \geq 1 \quad (53)$$

To prove that Eq.(51) can be solved in terms of $\mathcal{L}_{eff}$ is equivalent to show that the equation :

$$\Delta_\alpha^{(\nu)} = 0 \quad (54)$$

can be solved in terms of $\mathcal{L}_{eff}^{(\nu)}$ for any $\nu \geq 1$. Now we know that :

$$\Delta_\alpha^{(\nu)}(x) = D_\alpha(x)\int dy\, \mathcal{L}_{eff}^{(\nu)}(\varphi,\eta)(y) + \mathcal{Q}_\alpha^{(\nu)}(x,\mathcal{L}_{eff}) \quad (55)$$

where $\mathcal{Q}_\alpha^{(\nu)}(x,\mathcal{L}_{eff})$ sums up the quantum correction to the naive version of the Q.A.P. and consequently depends on $\mathcal{L}_{eff}^{(\kappa)}$ only for κ smaller than ν. Clearly Eq.(54) can be solved if:

$$\mathcal{Q}_\alpha^{(\nu)}(x,\mathcal{L}_{eff}) = D_\alpha(x)\,\hat{\mathcal{Q}}(\mathcal{L}_{eff}) \quad (56)$$

for some local functional $\hat{\mathcal{Q}}(\mathcal{L}_{eff})$ of dimension four. From a simple analysis it turns out that Eq.(56) is fulfilled if $\mathcal{Q}_\alpha^{(\nu)}(x,\mathcal{L}_{eff})$ and consequently $\Delta_\alpha^{(\nu)}(x)$ do not contain terms of the form :

$$\Gamma^{\widetilde{\alpha\beta}\,\gamma\delta}\,(c_\beta\,\bar{c}_\gamma\,\bar{c}_\delta)(x) \quad (57)$$

where the symbol $\widetilde{\alpha\beta}$ means symmetry under the exchange of the two indices. Now, taking into account the anticommuting character of the operator $D_\alpha(x)$, we have

$$\underbrace{\left[D_\alpha(x),D_\beta(y)\right]_+}_{=0}\Gamma(\varphi,\eta)=\left\{D_\alpha(x)\left(\Delta_\beta(y).\Gamma\right)_3+D_\beta(y)\left(\Delta_\alpha(x).\Gamma\right)_3\right\}(\varphi,\eta) \quad (58)$$

Since picking out of Eq.(52) the term of order $\hbar^\nu$ yields :

$$\left(\Delta_\beta(y)\,\Gamma\right)_3^{(\nu)}(\varphi,\eta) = \Delta_\beta^{(\nu)}(y) \quad (59)$$

we have to the same order the consistency condition :

$$D_\alpha(x)\,\Delta_\beta^{(\nu)}(y) + D_\beta(y)\,\Delta_\alpha^{(\nu)}(x) = 0 \quad (60)$$

from which the absence from $\Delta_\alpha^{(\nu)}(x)$ of terms of the form (57) follows immediately.

This is a particularly simple example of our general algebraic method. The possible quantum anomalies to the Q.A.P. are constrained, in addition to power counting, by a system of consistency conditions which to the first non vanishing order in $\hbar$ assume the naive (classical) form (as in Eq.(60)). We shall apply this method to the Slavnov identity and show that if the gauge group is semisimple, the consistency conditions are sufficient to ensure that by a suitable choice of the effective lagrangian all the possible anomalies can be eliminated in the absence of the Adler-Bardeen anomaly [16]. This is the full exploitation of the idea, due to Wess and Zumino [17], of defining the Adler-Bardeen anomaly by means of a system of consistency conditions.

We now go back to the Slavnov identity. Assuming from now on that the $\Phi.\Pi.$ part of the effective lagrangian is determined in such a way as to satisfy Eq.(49) we get from the Q.A.P.:

$$\mathcal{S} Z_c(J,\eta) = (\Delta . Z_c)_s (J,\eta) \tag{61}$$

where the vertex :

$$\Delta = \int dx\, \Delta(x) \tag{62}$$

has dimension lower than six and :

$$\Delta(x) = \left[- \mathcal{S}^{(P)} \mathcal{L}_{eff}(\varphi,\eta)(x) + \hbar\, Q(x, \mathcal{L}_{eff}, \hbar) \right] \tag{63}$$

where :

$$\mathcal{S}^{(P)} = \int dx \left[P_i \frac{\delta}{\delta \Phi_i} + \bar{P}_\alpha \frac{\delta}{\delta \bar{C}_\alpha} + g_\alpha \frac{\delta}{\delta C_\alpha} \right](x) \tag{64}$$

Furthermore there exists an integer $\nu \geq 1$ such that :

$$\Delta(x) = \sum_{\nu}^{\infty} {}_n\, \hbar^n \Delta^{(n)}(x) \tag{65}$$

and from Eq.(63):

$$\begin{aligned} \Delta^{(\nu)}(x) = - \left(\mathcal{S}^{(P^{(\nu)})} \mathcal{L}^{(0)}_{eff}(\varphi,\eta) + \mathcal{S}_0 \mathcal{L}^{(\nu)}_{eff}(\varphi,\eta) \right)(x) \, + \\ + R^{(\nu)}(\mathcal{L}_{eff}, x) \end{aligned} \tag{66}$$

where

$$\mathcal{S}_0 = \mathcal{S}^{(P)} \Big|_{\substack{P_i = P_i^{(0)} \\ \bar{P}_\alpha = \bar{P}_\alpha^{(0)}}} \quad \text{and} \quad \mathcal{S}^{(P^{(\nu)})} = \mathcal{S}^{(P)} \Big|_{\substack{P_i = P_i^{(\nu)} \\ \bar{P}_\alpha = \bar{P}_\alpha^{(\nu)} \\ g^\alpha = 0}} \tag{67}$$

The remainder $R^{(\nu)}$ depends on the coefficients of $\mathcal{L}_{eff}(\varphi,\eta)$ up to order $\nu - 1$. Making explicit the dependence on the external fields we have :

$$\overset{(\nu)}{\Delta}(x) = \overset{(\nu)}{\Delta_0}(x) + \left[\gamma^i \overset{(\nu)}{\Delta_i}\right](x) + \left[\zeta^\alpha \overset{(\nu)}{\Delta_\alpha}\right](x) \tag{68}$$

with :

$$\overset{(\nu)}{\Delta_0}(x) = -\left[\mathcal{S}^{(P^{(\nu)})} \mathcal{L}^{(0)}_{eff}(\varphi) + \mathcal{S}_0 \mathcal{L}^{(\nu)}_{eff}(\varphi\right] + \overset{(\nu)}{R_0}(x, \mathcal{L}_{eff})$$

$$\overset{(\nu)}{\Delta_i}(x) = \mathcal{S}^{(P^{(\nu)})} P_i^{(0)}(x) + \mathcal{S}_0 P_i^{(\nu)}(x) + \overset{(\nu)}{R_i}(x, \mathcal{L}_{eff}) \tag{69}$$

$$\overset{(\nu)}{\Delta_\alpha}(x) = -\left[\mathcal{S}^{(P^{(\nu)})} \bar{P}^{(0)}_\alpha + \mathcal{S}_0 \bar{P}^{(\nu)}_\alpha\right](x) + \overset{(\nu)}{R_\alpha}(x, \mathcal{L}_{eff})$$

This is the structure of the possible anomalies to the Slavnov identity. However if Eq.(49) is satisfied, $\Delta(x)$ is constrained by a system of consistency conditions which we are now going to derive. First we introduce a new external anticommuting field $\beta(x)$ carrying $\Phi.\Pi.$ charge $Q^{\Phi\Pi} = 1$ and coupled to the vertex $N_5[\Delta(x)]$. In the theory defined by the new lagrangian :

$$\mathcal{L}_{eff}(\varphi, \eta, \beta) = \left[\mathcal{L}_{eff}(\varphi, \eta) + (\beta\Delta)\right](x) \tag{70}$$
$$= \mathcal{L}_{eff}(\varphi) + \beta\Delta_0(\varphi) + \gamma^i(P_i - \beta\Delta_i) + \zeta^\alpha(\bar{P}_\alpha + \beta\Delta_\alpha)$$

the Slavnov identity writes :

$$\mathcal{S} Z_c(J, \eta, \beta) = (\Delta . Z_c)_5 (J, \eta, \beta) + (\overset{(\beta)}{\Delta} . Z_c)_5 (J, \eta, \beta)$$
$$= \int dx \frac{\delta}{\delta\beta(x)} Z_c(J, \eta, \beta) + (\overset{(\beta)}{\Delta} . Z_c)_5 (J, \eta, \beta) \tag{71}$$

In the right-hand side of Eq.(71) the insertion $\Delta^{(\beta)}$ lumps together the quantum corrections involving β -couplings. To order ν these new terms can only arise from the naive variation of $\Delta(x)$ since the β -dependent radiative corrections appear to an order in $\hbar$ strictly higher than ν (that of the new coupling). Hence we can write :

$$\Delta^{(\beta)} = \int dx\, \beta(x) \left\{\left[-\mathcal{S}^{(\Delta)} \mathcal{L}^{(0)}_{eff}(\varphi) + \gamma^i P_i^{(0)} + \zeta^\alpha \bar{P}^{(0)}_\alpha + \mathcal{S}_0 \Delta\right] + \right.$$
$$\left. + \hbar^{\nu+1} Q^{(\beta)}\right\}(x) \tag{72}$$

with :

$$\mathcal{S}^{(\Delta)} = \int dx \left[\Delta_i \frac{\delta}{\delta\phi_i} - \Delta_\alpha \frac{\delta}{\delta\bar{c}_\alpha}\right](x) \tag{73}$$

Now we can compute :

$$\delta^2 Z_c(J,\eta) = \delta^2 Z_c(J,\eta,\beta)\Big|_{\beta=0} = \delta\int dx \frac{\delta}{\delta\beta(x)} Z_c(J,\eta,\beta)\Big|_{\beta=0}$$

$$= -\int dx \frac{\delta}{\delta\beta(x)} \delta Z_c(J,\eta,\beta)\Big|_{\beta=0} = -\int dx\left(\left[\frac{\delta}{\delta\beta(x)}\Delta^{(\beta)}\right]\cdot Z_c\right)_6 (J,\eta) = 0 \tag{74}$$

since :

$$\int dx\, dy \frac{\delta^2}{\delta\beta(x)\,\delta\beta(y)} = 0 \tag{75}$$

vanishes owing to the anticommuting character of the external field

To order ν Eq.(74) yields the desired system of consistency conditions for Δ :

$$\begin{aligned} &\delta^{(\Delta^{(\nu)})}\bar{P}^{(0)}_\alpha + \delta_0 \Delta^{(\nu)}_\alpha = 0 \\ &\delta^{(\Delta^{(\nu)})} P^{(0)}_i - \delta_0 \Delta^{(\nu)}_i = 0 \\ &\int dx \left[\delta^{(\Delta^{(\nu)})} \mathcal{L}^{(0)}_{eff}(\psi) + \delta_0 \Delta^{(\nu)}_0(\psi)\right](x) = 0 \end{aligned} \tag{76}$$

The first equation ensures that the system :

$$\Delta^{(\nu)}_\alpha = \Delta^{(\nu)}_i = 0 \tag{77}$$

can be solved in terms of $\bar{P}^{(\nu)}_\alpha$ and $P^{(\nu)}_i$. Indeed by a completely algebraic (cohomological) method which is fully exhibited in Appendix A and B of reference [13], it is possible to show that the first two consistency equations yield :

$$\begin{aligned} \Delta^{(\nu)}_\alpha &= -\left(\delta^{(\Pi^{(\nu)})}\bar{P}^{(0)}_\alpha + \delta_0 \Pi^{(\nu)}_\alpha\right) \\ \Delta^{(\nu)}_i &= \delta^{(\Pi^{(\nu)})} P^{(0)}_i + \delta_0 \Pi^{(\nu)}_i \end{aligned} \tag{78}$$

with :

$$\delta^{(\Pi^{(\nu)})} = \int dx \left(\Pi^{(\nu)}_i \frac{\delta}{\delta\phi_i} + \Pi^{(\nu)}_\alpha \frac{\delta}{\delta \bar{c}_\alpha}\right)(x) \tag{79}$$

for some $\Pi^{(\nu)}_i$ and $\Pi^{(\nu)}_\alpha$ linear in $\Delta^{(\nu)}_i$ and $\Delta^{(\nu)}_\alpha$.

Comparing Eq.(79) with Eq.(69) we get :

$$R_\alpha^{(\nu)}(x,\mathcal{L}_{eff}) = \mathcal{S}^{(P^{(\nu)})}\bar{P}_\alpha^{(0)} + \mathcal{S}_0 \mathcal{P}_\alpha^{(\nu)}$$
$$R_i^{(\nu)}(x,\mathcal{L}_{eff}) = -\left(\mathcal{S}^{(P^{(\nu)})}P_i^{(0)} + \mathcal{S}_0 \mathcal{P}_i^{(\nu)}\right) \tag{80}$$

for some $\mathcal{P}_i^{(\nu)}$ and $\mathcal{P}_\alpha^{(\nu)}$ depending on the coefficients of $\mathcal{L}_{eff}(\psi,\eta)$ up to order $\nu-1$. Then the system (77) is solved by the choice :

$$\bar{P}_\alpha^{(\nu)} = \mathcal{P}_\alpha^{(\nu)} \quad \text{and} \quad P_i^{(\nu)} = \mathcal{P}_i^{(\nu)} \tag{81}$$

We are then left with the external field independent part of the anomaly $\Delta_0^{(\nu)}(x)$ subject to the condition :

$$\mathcal{S}_0 \int dx\, \Delta_0^{(\nu)}(x) = 0 \tag{82}$$

and we want to show that the equation :

$$\int dx\, \Delta_0^{(\nu)}(x) = 0 \tag{83}$$

can be solved in terms of $\mathcal{L}_{eff}^{(\nu)}(\psi)$. Comparing with Eq.(69) we see that this is ensured if we can show that :

$$\int dx\, \Delta_0^{(\nu)}(x) = \mathcal{S}_0 \int dx\, \hat{\Delta}_0^{(\nu)}(x) \tag{84}$$

for some $\hat{\Delta}_0^{(\nu)}$ linear in Δ_0. Making explicit the dependence on the $\Phi.\Pi.$ fields we can decompose $\Delta_0^{(\nu)}$ into the form :

$$\Delta_0^{(\nu)}(x) = (\Delta^\alpha \bar{c}_\alpha)(x) + \int dy\, dz\, \left(c_\alpha \Delta^{\alpha,\beta\gamma}(y,z)\, \bar{c}_\beta(y)\bar{c}_\gamma(z)\right)(x) + \tag{85}$$
$$+ \Delta^{\alpha\beta,\gamma\delta\eta}(c_\alpha c_\beta \bar{c}_\gamma \bar{c}_\delta \bar{c}_\eta)(x)$$

To exploit the constraint (82) we first write :

$$\mathcal{S}_0^2 \int dx\, \Delta_0^{(\nu)}(x) = \int dy \left[(\mathcal{S}_0^2 c_\beta)\frac{\delta}{\delta c_\beta}\right](y) \int dx\, \Delta_0^{(\nu)}(x) = 0 \tag{86}$$

from which

$$\Delta^{\alpha\beta,\gamma\delta\eta} = 0 \tag{87}$$

and

$$\int dy\,dz\,\left(C_\alpha \Delta^{\alpha,\beta\gamma}(y,z)\,\bar{C}_\beta(y)\,\bar{C}_\gamma(z)\right)(x) = \Delta^{\widetilde{\alpha\beta}\gamma}\left(C_\alpha (s_0^2 C_\beta)\bar{C}_\gamma\right)(x) \tag{88}$$

follow immediately. Here $\Delta^{\widetilde{\alpha\beta}\gamma}$ is a tensor symmetrical under the permutation of the first two indices. Then applying Eq.(82) directly to $\Delta_0^{(\nu)}$ we get :

$$\Delta^{\widetilde{\alpha\beta}\gamma} = 0$$

and

$$s_0 \int dx\,(\Delta^\alpha \bar{C}_\alpha)(x) = 0 =$$

$$= \frac{1}{2}\int dx\,dy\,\bar{C}_\alpha(x)\,\bar{C}_\beta(y)\left\{\frac{\delta}{\delta\omega_\alpha(x)}\Delta_\beta(y) - \frac{\delta}{\delta\omega_\beta(y)}\Delta_\alpha(x) - f^{\alpha\beta}{}_\gamma\,\delta(x-y)\,\Delta^\gamma(x)\right\} \tag{89}$$

One can show that the general solution of Eq.(89), which is nothing but the Wess-Zumino [17] consistency condition, or, in other words, the first cohomology equation for the gauge Lie algebra, is, if the gauge group is semisimple :

$$\Delta^\alpha(x) = \frac{\delta}{\delta\omega_\alpha(x)}\hat{\Delta}_0 + h^\alpha(x) \tag{90}$$

for some dimension four functional $\hat{\Delta}_0$. Here $h^\alpha(x)$ is the Adler-Bardeen [16] anomaly which has been discussed in Stora's lecture concerning the current algebra Ward identities [6]. Such an anomaly can only arise if the tree lagrangian contains $\varepsilon_{\mu\nu\rho\sigma}$ or γ_5 symbols and if the Lie algebra admits a non trivial invariant completely symmetrical $D^{\alpha\beta\gamma}$ tensor. If the Adler-Bardeen anomaly is absent we get :

$$\Delta_0^{(\nu)} = \int dx\,(\Delta^\alpha \bar{C}_\alpha)(x) = s_0\,\hat{\Delta}_0 \tag{91}$$

which completes the proof.

4. GAUGE INVARIANCE

Given a gauge model renormalizable in such a way as to preserve the Slavnov identity to all orders of perturbation theory, there remains to show that one can interpret it in physical terms in spite of the presence of many ghost fields. First one should introduce the physical parameters (masses, coupling constants) into the theory, connecting them, through a system of normalization conditions, to the parameters which are left arbitrary in the lagrangian. Then one has to specify a physical subspace within which the S operator is unitary and independent from the properties (masses,...) of the ghost particles.

We shall discuss here in some details the gauge independence of the physical S operator of the SU(2) H.K.E.B. model whose classical limit has been discussed in the first section.

The quantized fields are

$$\{\psi\} = \{\sigma, \pi_\alpha, a^\mu_\alpha, c_\alpha, \bar{c}_\alpha\}$$

The classical lagrangian, given in Eq.(12), is invariant under a group of global (non space dependent) rotations transforming the π^α, a^α_μ, c^α, $\bar{c}^\alpha$ fields as vectors and leaving the σ field invariant. Under this restriction the Slavnov transformations, whose classical form is given in Eq.(14), depends on five parameters. The dependence on three of them, namely ρ, e, and F is explicit in Eq.(14). There are two more hidden parameters which can be introduced by the substitution :

$$\pi_\alpha \to \chi\,\pi_\alpha \qquad \rho \to \rho/\chi$$
$$\bar{c}_\alpha \to \tau\,\bar{c}_\alpha \qquad e \to e/\tau \tag{92}$$

The most general lagrangian invariant under Slavnov transformations :

$$\mathcal{L}(\psi) = -\frac{Z_a}{4}\left(G^a_{\mu\nu}\right)^2 + \frac{Z_1}{2}\left(D_\mu\varphi_a\right)^2 + \frac{\mu^2}{2}(\varphi+F)^2 - \frac{\lambda^2}{4!}\left((\varphi+F)^2\right)^2$$
$$-\left[\frac{\mu^2F^2}{2} - \frac{\lambda^2}{4!}F^4\right] - \frac{1}{\kappa}\left[\frac{(g^\alpha)^2}{2} - c^\alpha(m\bar{c})_\alpha\right] \tag{93}$$

(compare with Eq.(4) and Eq.(12)), depends on five parameters, namely : Z_a, Z_1, μ, λ, κ. These however have to be adjusted so that the coefficient of the term linear in

σ vanishes :

$$\mu^2 + \frac{F^2\lambda^2}{3!} = 0 \tag{94}$$

The theory thus depends on ten parameters, four specifying the transformation laws, one related with the σ field vacuum expectation value, five specifying the external field independent part of the lagrangian. One can alternatively specify the position of the poles in the transverse photon, σ and c-$\bar{c}$ propagators (m, M, $m_{\phi\pi}$), the residues at these poles (Z_a^{-1}, Z_1^{-1}, κ) and a coupling constant ε, by the normalization conditions :

$$\left(g^{\mu\nu} - \frac{p^\mu p^\rho}{p^2}\right) \frac{\delta}{\delta \tilde{a}^\rho_\alpha(p)} \frac{\delta}{\delta a_{\nu\beta}(0)} \Gamma \Big|_{\psi=\eta=0} \equiv \Gamma^{\mu\nu,\alpha\beta}_{a_T a_T}(p) =$$

$$= - Z_a\, \delta^{\alpha\beta}\left(g^{\mu\nu} - \frac{p^\mu p^\nu}{p^2}\right)\left[p^2 - m^2 + O((p^2-m^2)^2)\right].$$

$$\frac{\delta}{\delta\tilde{\sigma}(p)} \frac{\delta}{\delta\sigma(0)} \Gamma \Big|_{\psi=\eta=0} = \Gamma_{\sigma\sigma}(p) = Z_1 (p^2 - M^2) + O((p^2-M^2)^2) \tag{95}$$

$$\Gamma^{\mu\nu,\alpha\beta}_{a_T a_T \sigma}(m^2, m^2, M^2) = \delta^{\alpha\beta} g^{\mu\nu} \varepsilon$$

$$\frac{\delta}{\delta\tilde{\bar{c}}^\alpha(p)} \frac{\delta}{\delta \bar{c}^\beta(0)} \Gamma \Big|_{\psi=\eta=0} = \Gamma^{\alpha\beta}_{c\bar{c}}(p) = \frac{\delta_{\alpha\beta}}{\kappa} (p^2 - m^2_{\phi\pi}) + O((p^2 - m^2_{\phi\pi})^2)$$

These normalization conditions together with Eq.(94) which is equivalent to :

$$\frac{\delta}{\delta J_0(x)} Z_c(J,\eta)\Big|_{J=\eta=0} = 0 \tag{96}$$

fix the values of z_a, z_1, μ, λ, κ, ρ, e, F, leaving free the unphysical parameters τ and χ. It is easy to show that, as a consequence of the Slavnov invariance, the masses associated with the coupled longitudinal photon-π channel are pairwise degenerated with those of the c-$\bar{c}$ channel. Hence, owing to the global isotopic spin symmetry, all the ghost masses are degenerate and fixed by the normalization conditions (95).

According to the analysis of section 3 it is possible to find an effective lagrangian fulfilling the Slavnow identity (20) and the normalization conditions (95) and (96) to all orders of perturbation theory. To prove in this theory the gauge invariance of the physical S operator we have to show that:

$$\partial_\lambda \Sigma_{phys} Z(J,\eta)\Big|_{J=\eta=0} = \partial_\lambda S_{phys} = 0 \tag{97}$$

(compare with Eq.(24)), where λ is one of the parameters $\mathcal{K}$, $m_{\phi\pi}$.

Here we shall sketch out the proof of Eq.(97) referring for the details to the existing literature [13]. If "a" is a parameter of our lagrangian we define the invariant derivative with respect to a :

$$D_a = \partial_a + \frac{\partial_a \rho}{\rho}\int dx \left[J_\alpha \frac{\delta}{\delta J_\alpha} + \gamma_\alpha \frac{\delta}{\delta \gamma_\alpha} \right](x) \equiv \partial_a + \frac{\partial_a \rho}{\rho}\Lambda \tag{98}$$

It is easy to verify that :

$$\mathcal{S} D_a Z(J,\eta) = 0 \tag{99}$$

and that :

$$\Sigma_{phys} \partial_a Z(J,\eta)\Big|_{J=\eta=0} = \Sigma_{phys} D_a Z(J,\eta)\Big|_{J=\eta=0} \tag{100}$$

Since the Lowenstein action principles yields ;

$$D_a Z(J,\eta) = \frac{i}{\hbar}\left(\Delta_a^{S} \cdot Z\right)_4 (J,\eta) \tag{101}$$

for some insertion (D.V.O.) $\Delta_a^{(S)}$ of dimension four, we can write Eq.(99) and Eq.(100) in the form :

$$\mathcal{S}\left(\Delta_a^{(S)} \cdot Z\right)_4 (J,\eta) = 0 \tag{102}$$

and

$$\Sigma_{phys} \partial_a Z(J,\eta)\Big|_{J=\eta=0} = \frac{i}{\hbar} \Sigma_{phys} \left(\Delta_a^{(S)} \cdot Z\right)_4 (J,\eta)\Big|_{J=\eta=0} \tag{103}$$

In the following we shall call symmetrical an insertion satisfying Eq.(102). We shall also call an insertion Δ non physical if :

$$\sum_{phys} (\Delta . Z)_4 (J,\eta)\Big|_{J=\eta=0} = 0 \tag{104}$$

Our aim is to prove that the symmetrical insertion $\Delta_\lambda^{(s)}$ (for $\lambda = \mathcal{K},\ m_{\phi\pi}$) is non physical. (Eq.(100)). Now the set of the symmetrical insertions of our model is a linear space of dimension nine, since, for a prescribed Slavnov operator $\mathcal{S}$ (ρ fixed) our theory depends on nine parameters. Also, one can construct explicitly (as in reference [13] section 3-C) in terms of functional differential operators four independent non physical insertions :

$$\Delta_i^{(o,s)} \qquad i = 1, \dots 4 \tag{105}$$

The remaining five independent symmetrical insertions :

$$\Delta_i^{(s)} \qquad i = 1, \dots . 5 \tag{106}$$

are such that the matrix with columns :

$$\begin{aligned}
\Delta_{i,1}^{(s)} &= (\Delta_i^{(s)} . \Gamma)_{a_T a_T} (m) \\
\Delta_{i,2}^{(s)} &= \frac{\partial}{\partial p^2} \left(\Delta_i^{(s)} . \Gamma \right)_{a_T a_T} (p^2) \Big|_{p^2 = m^2} \\
\Delta_{i,3}^{(s)} &= (\Delta_i^{(s)} . \Gamma)_{\sigma\sigma} (M^2) \\
\Delta_{i,4}^{(s)} &= \frac{\partial}{\partial p^2} \left(\Delta_i^{(s)} . \Gamma \right)_{\sigma\sigma} (p^2) \Big|_{p^2 = M^2} \\
\Delta_{i,5}^{(s)} &= (\Delta_i^{(s)} . \Gamma)_{a_T a_T \sigma} (m^2, m^2, M^2)
\end{aligned} \tag{107}$$

has non vanishing determinant :

$$\det \left\| \Delta_{i,j}^{(s)} \right\| \neq 0 \tag{108}$$

This can be easily proved by constructing explicitly the five insertions in the tree approximation. Now the insertions (105) and (106) are a basis of the linear space of symmetrical insertions. Hence in particular :

$$\Delta_\lambda^{(s)} = \sum_1^4 {}_i\, c_{\lambda,i}\, \Delta_i^{(o,s)} + \sum_1^5 {}_j\, d_{\lambda,j}\, \Delta_j^{(s)} \tag{109}$$

and :

$$\partial_\lambda \Gamma(\psi,\eta) = -\frac{\partial_\lambda \rho}{\rho} \Lambda \Gamma(\psi,\eta) + \frac{i}{\hbar}\left(\Delta_\lambda^{(s)} \cdot \Gamma\right)(\psi,\eta) \tag{110}$$

The independence from λ of the physical parameters m , M , z_a , z_1 , $\mathcal{E}$ is expressed by the system :

$$\begin{aligned}
&(\partial_\lambda \Gamma)_{a_T a_T}(m^2) = 0 \\
&(\partial_\lambda \partial_{p^2} \Gamma)_{a_T a_T}(m^2) = 0 \\
&(\partial_\lambda \Gamma)_{\sigma\sigma}(M^2) = 0 \\
&(\partial_\lambda \partial_{p^2} \Gamma)_{\sigma\sigma}(M^2) = 0 \\
&(\partial_\lambda \Gamma)_{a_T a_T \sigma}(m^2\, m^2\, M^2) = 0
\end{aligned} \tag{111}$$

which, taking into account Eq.(110) is equivalent to :

$$\sum_1^5 {}_j\, d_{\lambda,j}\, \Delta_{j,i}^{(s)} = 0 \qquad i = 1,\dots 5 \tag{112}$$

which completes the proof.

Let us now give an outline of the connection between the Slavnov symmetry and the unitarity of the restriction of the S operator to the physical space [12,13].

We shall first introduce a new coordinate frame for the unphysical degrees of freedom of our model. Namely we shall replace the longitudinal photon and π fields by their independent linear combinations g^α (the gauge function) and :

$$\bar{g}_\alpha = \frac{\rho \pi_\alpha - \partial_\mu a^\mu_\alpha}{2\rho\, \Gamma_\gamma (m^2_{\phi\pi})} \tag{114}$$

where :

$$\Gamma_\gamma (p^2) = \frac{\delta}{\delta \tilde{\bar{c}}_\alpha (p)} \frac{\delta}{\delta \gamma^\alpha (0)} \Gamma(\psi, \eta) \Big|_{\psi=\eta=0} \tag{115}$$

Owing to the Slavnov symmetry the g-$\bar{g}$ propagator has a simple pole at $p^2 = m^2_{\phi\pi}$ with the same residue as the c-$\bar{c}$ propagator. Is is also easy to verify that the Fourier transformed g-g propagator is independent from the momentum.

We shall also put ourselves in the restricted t'Hooft gauge defined by the normalization condition :

$$i p_\mu \frac{\delta}{\delta \tilde{a}_{\mu\alpha}(p)} \frac{\delta}{\delta \pi_\beta (0)} \Gamma(\psi, \eta) \Big|_{\substack{\psi=\eta=0 \\ p^2 = m^2_{\phi\pi}}} = 0 \tag{116}$$

The main advantage of the restricted t'Hooft gauge is the absence of double poles in the propagators of the ghost fields. This greatly simplifies the analysis of the ghost contribution to the unitarity sum.

Introducing the complete system of asymptotic fields

$$\{\psi^{in}\} = \{\sigma^{in}, a^{T\,in}_{\mu\alpha}, g^{in}_\alpha, \bar{g}^{in}_\alpha, c^{in}_\alpha, \bar{c}^{in}_\alpha\}$$

and the corresponding wave operators K the scattering operator of our model is given by the L.S.Z. reduction formulae [6], namely :

$$\begin{aligned} S &= \Sigma\, Z(J, \eta)\big|_{J=\eta=0} \\ &= :\exp \int dx\, dy\, (\psi^{in}(x)\, K_{xy} \frac{\delta}{\delta J(y)}): Z(J, \eta)\big|_{J=\eta=0} \end{aligned} \tag{117}$$

Now the Slavnov identity can be translated into a symmetry property for S . Indeed introducing the functional differential operator :

$$\bar{\mathcal{S}} = \int dx \left[J^{\alpha}_{\bar{g}} \frac{\delta}{\delta \xi_{\alpha}} - \xi^{\alpha} \frac{\delta}{\delta J^{\alpha}_{g}} \right](x) \tag{118}$$

which corresponds to the field transformations :

$$\begin{aligned} \delta c^{\alpha} &= \delta\lambda \bar{\mathcal{S}} c^{\alpha} = \delta\lambda\, g^{\alpha} \\ \delta \bar{g}^{\alpha} &= \delta\lambda \bar{\mathcal{S}} \bar{g}^{\alpha} = \delta\lambda\, \bar{c}^{\alpha} \\ \delta \bar{c}^{\alpha} &= \delta g^{\alpha} = \delta\sigma = \delta a^{T}_{\mu\alpha} = 0 \end{aligned} \tag{119}$$

the Slavnov identity for the amputated Green functions on the mass shell simplifies to

$$\sum \bar{\mathcal{S}}\, Z(J, \eta)\Big|_{J=\eta=0} = 0 \tag{120}$$

The essential meaning of this equation is that on the mass shell the g-$\bar{g}$ and c-$\bar{c}$ pairs are equally coupled. Since, owing to the anticommuting character of the $\Phi.\Pi.$ fields, their contribution to the unitarity sum is opposite in sign, it turns out that the total contribution to the physical unitarity of states involving g-$\bar{g}$ and (or) c-$\bar{c}$ pairs vanish. This is the whole contribution coming from the unphysical states since the g field is decoupled from the physical states (Eq.(25)) and the g-g propagator is momentum independent.

5. CONCLUSION

The gauge field models are characterized by the fulfilment of Slavnov identities. If the gauge group is semisimple and all the fields are massive an algebraic analysis of the possible quantum corrections to the action principle shows that, indeed, the Slavnov identities can be fulfilled in the absence of the Adler-Bardeen anomaly.

We have also discussed in the SU(2) Higgs-Kibble model the connection between the Slavnov identity and the unitarity and gauge invariance of the physical S operator.

The analysis of gauge independent local operators, although not touched here, should also be based on their Slavnov invariance [12].

Concerning the possible extensions of our results, Lowenstein will show at the end of his lectures how the infrared problems connected with the gauge field models can be overcome at least in the extreme case (pure Yang-Mills models) in which all the particules are massless [2,18] . The extension to G.F.M. with non semisimple gauge group (i.e. for example the Weinberg models) can be worked out supplementing the algebraic analysis of the anomalies by a power counting analysis similar to that used in reference [7] to deal with the invariant-abelian anomalies in the lagrangian models with broken symmetry.

REFERENCES

[1] J. SCHWINGER
Phys. Rev. 82, 918 (1951).

C.S. LAM
Nuovo Cimento 38, 1754 (1965).

[2] J. LOWENSTEIN
Lectures given at this School.

[3] E. SPEER
Lectures given at this School.

P. BREITENLOHNER
Lectures given at this School.

P. BREITENLOHNER, D. MAISON
Dimensional Renormalization and the Action Principle
MPI-PAE/PTh25/74 (May 1975).

[4] Y.M.P. LAM
Phys. Rev. D6, 2145 (1972).

[5] W. ZIMMERMANN
Ann. Phys. 77, 536-570 (1973).

J. LOWENSTEIN
Lectures given at this School.

[6] R. STORA
Lectures given at this School.

[7] C. BECCHI, A. ROUET, R. STORA
Renormalizable Theories with Symmetry Breaking
Marseille Preprint 75/P.734 (June 1975).

[8] A complete bibliography about the theory of gauge fields can be found for instance in :

E.S. ABERS, B.W. LEE
Phys. Reports, 9C n°1 (1973).

M. VELTMAN
Invited talk presented at the International Symposium on Electron and Photons at High Energies, Bonn, 27-31 August 1973.

J. ZINN-JUSTIN
Lectures given at the International Summer Institute

for Theoretical Physics, Bonn 1974.

[9] P. HIGGS
Phys. Letters 12, 132 (1964) ;
Phys. Rev. 145, 1156 (1966).

T.W.B. KIBBLE
Phys. Rev. 155, 1554 (1967).

F. ENGLERT, R. BROUT
Phys. Rev. Letters, 13, 321 (1964).

[10] L.D. FADDEEV, V.N. POPOV
Phys. Letters, 25B, 29 (1957).

[11] A.A. SLAVNOV
Teor. i Mat. Fiz. 10, 153 (1972).

J.C. TAYLOR
Nucl. Phys. B33, 436 (1971).

[12] C. BECCHI, A. ROUET, R. STORA
Phys. Letters 52B, 344 (1974).

C. BECCHI, A. ROUET, R. STORA
Commun. math. Phys. 42, 127 (1975).

[13] C. BECCHI, A. ROUET, R. STORA
Renormalization of Gauge Theories,
Marseille Preprint 75/P.723 (April 1975).

[14] J. GOLSTONE
Nuovo Cimento 19, 154 (1961).

[15] B. ZUMINO
in Lectures on Elementary Particles and Quantum Field Theory,
Brandeis University Summer Institute in Theoretical Physics ; vol.2 (the MIT Press, Cambridge Mass.)

[16] W.A. BARDEEN
Phys. Rev. 184, 1848 (1969).

[17] J. WESS, B. ZUMINO
Phys. Letters 37B, 95 (1971).

VOLUME 30, NUMBER 26 PHYSICAL REVIEW LETTERS 25 JUNE 1973

Ultraviolet Behavior of Non-Abelian Gauge Theories*

David J. Gross† and Frank Wilczek
Joseph Henry Laboratories, Princeton University, Princeton, New Jersey 08540
(Received 27 April 1973)

It is shown that a wide class of non-Abelian gauge theories have, up to calculable logarithmic corrections, free-field–theory asymptotic behavior. It is suggested that Bjorken scaling may be obtained from strong-interaction dynamics based on non-Abelian gauge symmetry.

Non-Abelian gauge theories have received much attention recently as a means of constructing unified and renormalizable theories of the weak and electromagnetic interactions.[1] In this note we report on an investigation of the ultraviolet (UV) asymptotic behavior of such theories. We have found that they possess the remarkable feature, perhaps unique among renormalizable theories, of asymptotically approaching free-field theory. Such asymptotically free theories will exhibit, for matrix elements of currents between on-mass-shell states, Bjorken scaling. We therefore suggest that one should look to a non-Abelian gauge theory of the strong interactions to provide the explanation for Bjorken scaling, which has so far eluded field-theoretic understanding.

The UV behavior of renormalizable field theories can be discussed using the renormalization-group equations,[2,3] which for a theory involving one field (say $g\varphi^4$) are

$$[m\partial/\partial m + \beta(g)\,\partial/\partial g - n\gamma(g)]\,\Gamma_{\text{asy}}^{(n)}(g; P_1, \ldots, P_n) = 0. \tag{1}$$

$\Gamma_{\text{asy}}^{(n)}$ is the asymptotic part of the one-particle–irreducible renormalized n-particle Green's function, $\beta(g)$ and $\gamma(g)$ are finite functions of the renormalized coupling constant g, and m is either the renormalized mass or, in the case of massless particles, the Euclidean momentum at which the theory is renormalized.[4] If we set $P_i = \lambda q_i^0$, where q_i^0 are (nonexceptional) Euclidean momenta, then (1) determines the λ dependence of $\Gamma^{(n)}$:

$$\Gamma^{(n)}(g; P_i) = \lambda^D \Gamma^{(n)}(\bar{g}(g,t); q_i) \exp\left[-n\int_0^t \gamma(\bar{g}(g,t'))\,dt'\right], \tag{2}$$

where $t = \ln\lambda$, D is the dimension (in mass units) of $\Gamma^{(n)}$, and $\bar{g}$, the invariant coupling constant, is the solution of

$$d\bar{g}/dt = \beta(\bar{g}), \quad \bar{g}(g,0) = g. \tag{3}$$

The UV behavior of $\Gamma^{(n)}$ $(\lambda \to +\infty)$ is determined by the large-t behavior of $\bar{g}$ which in turn is controlled by the zeros of β: $\beta(g_f) = 0$. These fixed points of the renormalization-group equations are said to be UV stable [infrared (IR) stable] if $\bar{g} \to g_f$ as $t \to +\infty$ $(-\infty)$ for $\bar{g}(0)$ near g_f. If the physical coupling constant is in the domain of attraction of a UV-stable fixed point, then

$$\Gamma^{(n)}(g; P_i) \underset{\lambda\to\infty}{\approx} \lambda^{D - n\gamma(g_f)} \Gamma^{(n)}(g_f; q_i) \exp\left\{-n\int_0^\infty [\gamma(\bar{g}(g,t)) - \gamma(g_f)]\,dt\right\}, \tag{4}$$

so that $\gamma(g_f)$ is the anomalous dimension of the field. As Wilson has stressed, the UV behavior is determined by the theory at the fixed point ($g = g_f$).[5]

In general, the dimensions of operators at a fixed point are not canonical, i.e., $\gamma(g_f) \neq 0$. If we wish to explain Bjorken scaling, we must assume the existence of a tower of operators with canonical dimensions. Recently, it has been argued for all but gauge theories, that this can only occur if the fixed point is at the origin, $g_f = 0$, so that the theory is asymptotically free.[6,7] In that case the anomalous dimensions of all operators vanish, one obtains naive scaling up to finite and calculable powers of $\ln\lambda$, and the structure of operator products at short distances is that of free-field theory.[7] Therefore, the existence of such a fixed point, for a theory of the strong interactions, might explain Bjorken scaling and the success of naive light-cone or parton-model relations. Unfortunately, it appears that the fixed point at the origin, which is common to all theories, is not UV stable.[8,9] The only exception would seem to be non-Abelian gauge theories, which hitherto have not been explored in this regard.

Let us consider a Yang-Mills theory given by the Lagrangian

$$\mathcal{L} = -\tfrac{1}{4}F_{\mu\nu}{}^{a}F_{a}{}^{\mu\nu},$$
$$F_{\mu\nu}{}^{a} = \partial_\mu A_\nu{}^a - \partial_\nu A_\mu{}^a - gC_{abc}A_\mu{}^b A_\nu{}^c, \qquad (5)$$

where the C_{abc} are the structure constants of some (semisimple) Lie group G. Since the theory is massless, the renormalization is performed at an (arbitrary) Euclidean point. For example, the wave-function renormalization constant $Z_3(g, \Lambda/m)$ will be defined in terms of the unrenormalized vector-meson propagator $D_{\mu\nu}{}^{ab}$ (in the Landau gauge),

$$D_{\mu\nu}{}^{ab}(P)|_{P^2=-m^2} = \left(g_{\mu\nu} + \frac{P_\mu P_\nu}{m^2}\right)\frac{iZ_3}{m^2}\delta_{ab}. \qquad (6)$$

(For a thorough discussion of the renormalization see the work of Lee and Zinn-Justin.[10]) The renormalization-group equations for this theory are easily derived.[11] In the Landau gauge they are identical with (1). In order to investigate the stability of the origin, it is sufficient to calculate β to lowest order in perturbation theory. To this order we have

$$\beta(g) = \left.\frac{\partial g}{\partial \ln m}\right|_{\Lambda, g_0} = -g\frac{\partial}{\partial \ln\Lambda}\left(\frac{Z_3{}^{3/2}}{Z_1}\right), \qquad (7)$$

where Λ is a UV cutoff, and Z_1 the charge-renormalization constant. In Abelian gauge theories $Z_3 = Z_1 = 1 - g^2C\ln\Lambda$ $(C>0)$, as a consequence of gauge invariance and the Källén-Lehman representation, and thus $\beta(g) \cong g^3$ which leads to IR stability at $g=0$. Non-Abelian theories have no such requirement; Z_3 and Z_1 are gauge dependent and can be greater than 1. Thus $\beta(g)$, which must be gauge independent in lowest order, could have any sign at $g=0$. We have calculated Z_1 and Z_3 for the above Lagrangian, and we find that[12]

$$\beta_V = -(g^3/16\pi^2)\tfrac{11}{3}C_2(G) + O(g^5), \qquad (8)$$

where $C_2(G)$ is the quadratic Casimir operator of the adjoint representation of the group G: $\sum_{b,c} c_{abc} \times c_{dbc} = C_2(G)\delta_{ad}$ [e.g., $C_2(\mathrm{SU}(N)) = N$]. The solution of (3) is then $\bar{g}^2(t) = g^2/(1-2\beta_V g^{-1}t)$, and $\bar{g} \to 0$ as $t\to\infty$ as long as the physical coupling constant g is in the domain of attraction of the origin.[13]

We have thus established that for all non-Abelian gauge theories based on semisimple Lie groups the origin is UV stable. It is easy to incorporate fermions into such a theory without destroying the UV stability. The fermion interaction is given by $L_F = \bar{\psi}(i\gamma\cdot\partial - g\gamma\cdot B^a M^a)\psi +$ mass terms, where M^a are the matrices of some representation R of the gauge group G. The only effect of the fermions is to change the value of $\beta(g)$ by the amount[11]

$$\beta_F(R) = (g^3/16\pi^2)\tfrac{4}{3}T(R), \qquad (9)$$

where $\mathrm{Tr}(M^aM^b) = T(R)\delta_{ab}$, $T(R) = C_2(R)d(R)/r$, $d(R)$ is the dimension of the representation R, and r is the order of the group, i.e., the number of generators, and $C_2(R)$ is the quadratic Casimir operator of the representation. Although the fermions tend to destabilize the origin, there is room to spare. For example, in the case of SU(3): $\beta_V = -11$, whereas $\beta_F(\underline{3}) = \tfrac{2}{3}$, $\beta_F(\underline{8}) = 4$, etc., so that one could accomodate as many as sixteen triplets. One can therefore construct many asymptotically free theories with fermions. The vector mesons, however, will remain massless until the gauge symmetry is spontaneously broken. One might hope that this would be a consequence of the dynamics,[14] but at the present the only known way of achieving this is to introduce scalar Higgs mesons, whose nonvanishing vacuum expectation values break the symmetry.

The introduction of scalar mesons has a very destabilizing effect on the UV stability of the origin. Their contribution to $\beta(g)$ is small; a scalar meson transforming under a complex (real) representation R of the gauge group adds to β a term equal to $\tfrac{1}{4}$ $(\tfrac{1}{8})$ of Eq. (9). The problem with scalar mesons is that they necessarily have their own quartic couplings, and one must deal with a new coupling constant. Consider the Lagrangian for the coupling of scalars belonging to a representation R of G:

$$\mathcal{L} = \tfrac{1}{2}[(\partial_\mu - igB_\mu{}^aM^a)\vec{\varphi}]^2 - \lambda(\vec{\varphi}\cdot\vec{\varphi})^2 + V(\vec{\varphi}). \qquad (10)$$

where $V(\varphi)$ contains cubic, quadratic, and linear terms in φ (which have no effect on the UV behavior of the theory) plus, perhaps, additional quartic terms invariant under G. The renormalization-group equations have an additional term, $\beta_\lambda(g,\lambda)\,\partial/\partial\lambda$, and one must investigate the UV stability of the origin $(g=\lambda=0)$ with respect to both g and λ [if there are other quartic invariants in $V(\varphi)$ there will be additional coupling constants to consider]. The structure of the renormalization-group equation for g is unchanged to lowest order, whereas for the coupling constant $\Gamma \equiv \lambda/g^2$ we have[11]

$$d\bar{\Gamma}(\Gamma, t, g^2)/dt = \bar{g}^2[A\bar{\Gamma}^2 + B\bar{\Gamma} + C] \qquad (11)$$

(where we have neglected terms of order g^4, $g^4\Gamma$,

VOLUME 30, NUMBER 26 PHYSICAL REVIEW LETTERS 25 JUNE 1973

$g^4\Gamma^2$, and $g^4\Gamma^3$). In the absence of vector mesons ($g=0$) this equation is UV unstable at $\lambda=0$, since A is strictly positive and λ must be positive.[15] The vector mesons contribute to B and C and tend to stabilize the origin. If the right-hand side of (11) has positive zeros ($C>0$, $B<0$, and $B^2>4AC$), then for Γ less than the larger zero of (11) we will have that $\lambda\to+0$ as $t\to\infty$. We have investigated the structure of these equations for a large class of gauge theories and representations of the scalar mesons. We have found many examples of theories which contain scalar mesons and are UV stable.[11] These include (a) SU(N) if the scalar mesons belong to the adjoint representation for $N\geqslant 6$; (b) SU(N) $\otimes$SU(N) if the scalars belong to the $(N,\bar{N})$ representation for $N\geqslant 5$; (c) SU(N) with the scalars transforming as a symmetric tensor for $N\geqslant 9$; and many others. In all of these models it is necessary for the theory to contain a large number of fermions in order to make β_g small; otherwise $\bar{g}$ approaches zero too rapidly for the vector mesons to stabilize the scalar couplings.

Unfortunately, in none of these models can the gauge symmetry be totally broken by the Higgs mechanism. The requirement that the interactions of the scalar mesons be renormalizable so severely constrains the form of Lagrangian that the ground state invariably is invariant under some non-Abelian subgroup of the gauge group. If one tries to overcome this by larger representations for the scalar mesons, UV instability inevitably occurs.

It thus appears to be very difficult to retain UV stability and break the gauge symmetry by explicitly introducing Higgs mesons. Since the Higgs mesons are so restrictive, we would prefer to believe that spontaneous symmetry breaking would arise dynamically.[14] This is suggested by the IR instability of the theories, which assures us that perturbation theory is not trustworthy with respect to the stability of the symmetric theory nor to its particle content.

With this hope in mind one can construct many interesting models of the strong interactions. One particularly appealing model is based on three triplets[16] of fermions, with Gell-Mann's SU(3)$\otimes$SU(3) as a global symmetry and an SU(3) "color" gauge group to provide the strong interactions. That is, the generators of the strong-interaction gauge group commute with ordinary SU(3)$\otimes$SU(3) currents and mix quarks with the same isospin and hypercharge but different "color." In such a model the vector mesons are neutral, and the structure of the operator product expansion of electromagnetic or weak currents is (assuming that the strong coupling constant is in the domain of attraction of the origin!) essentially that of the free quark model (up to calculable logarithmic corrections).[11]

Finally, we note that theories of the weak and electromagnetic interactions, built on semisimple Lie groups,[17] will be asymptotically free if we again ignore the complications due to the Higgs particles. This suggests that the program of Baker, Johnson, Willey, and Adler[18] to calculate the fine-structure constant as the value of the UV-stable fixed point in quantum electrodynamics might fail for such theories.

*Research supported by the U.S. Air Force Office of Scientific Research under Contract No. F-44620-71-C-0180.

†Alfred P. Sloan Foundation Research Fellow.

[1]S. Weinberg, Phys. Rev. Lett. 19, 1264 (1967). For an extensive review as well as a list of references, see B. W. Lee, in Proceedings of the Sixteenth International Conference on High Energy Physics, National Accelerator Laboratory, Batavia, Illinois, 1972 (to be published).

[2]M. Gell-Mann and F. E. Low, Phys. Rev. 95, 1300 (1954).

[3]C. G. Callan, Phys. Rev. D 2, 1541 (1970); K. Symanzik, Commun. Math. Phys. 18, 227 (1970).

[4]The basic assumption underlying the derivation and utilization of the renormalization group equations is that the large Euclidean momentum behavior of the theory is the same as the sum, to all orders, of the leading powers in perturbation theory.

[5]K. Wilson, Phys. Rev. D 3, 1818 (1971).

[6]G. Parisi, to be published.

[7]C. G. Callan and D. J. Gross, to be published.

[8]A. Zee, to be published.

[9]S. Coleman and D. J. Gross, to be published.

[10]B. W. Lee and J. Zinn-Justin, Phys. Rev. D 5, 3121 (1972).

[11]Full details will be given in a forthcoming publication: D. J. Gross and F. Wilczek, to be published.

[12]After completion of this calculation we were informed of an independent calculation of β for gauge theories coupled to fermions by H. D. Politzer [private communication, and following Letter, Phys. Rev. Lett. 30, 1346 (1973)].

[13]K. Wilson has suggested that the coupling constants of the strong interactions are determined to be IR-stable fixed points. For nongauge theories the IR stability of the origin in four-dimensional field theories implies that theories so constructed are trivial, at least in a domain about the origin. Our results suggest that non-Abelian gauge theories might possess IR-stable fixed points at nonvanishing values of the coupling constants.

[14]Y. Nambu and G. Jona-Lasino, Phys. Rev. 122, 345 (1961); S. Coleman and E. Weinberg, Phys. Rev. D 7, 1888 (1973).
[15]K. Symanzik (to be published) has recently suggested that one consider a $\lambda\varphi^4$ theory with a negative λ to achieve UV stability at $\lambda=0$. However, one can show, using the renormalization-group equations, that in such theory the ground-state energy is unbounded from below (S. Coleman, private communication).
[16]W. A. Bardeen, H. Fritzsch, and M. Gell-Mann, CERN Report No. CERN-TH-1538, 1972 (to be published).
[17]H. Georgi and S. L. Glashow, Phys. Rev. Lett. 28, 1494 (1972); S. Weinberg, Phys. Rev. D 5, 1962 (1972).
[18]For a review of this program, see S. L. Adler, in Proceedings of the Sixteenth International Conference on High Energy Physics, National Accelerator Laboratory, Batavia, Illinois, 1972 (to be published).

Reliable Perturbative Results for Strong Interactions?*

H. David Politzer
Jefferson Physical Laboratories, Harvard University, Cambridge, Massachusetts 02138
(Received 3 May 1973)

An explicit calculation shows perturbation theory to be arbitrarily good for the deep Euclidean Green's functions of any Yang-Mills theory and of many Yang-Mills theories with fermions. Under the hypothesis that spontaneous symmetry breakdown is of dynamical origin, these symmetric Green's functions are the asymptotic forms of the physically significant spontaneously broken solution, whose coupling could be strong.

Renormalization-group techniques hold great promise for studying short-distance and strong-coupling problems in field theory.[1,2] Symanzik[2] has emphasized the role that perturbation theory might play in approximating the otherwise unknown functions that occur in these discussions. But specific models in four dimensions that had been investigated yielded (in this context) disappointing results.[3] This note reports an intriguing contrary finding for any generalized Yang-Mills theory and theories including a wide class of fermion representations. For these one-coupling-constant theories (or generalizations involving product groups) the coefficient function in the Callan-Symanzik equations commonly called $\beta(g)$ is negative near $g=0$.

The constrast with quantum electrodynamics (QED) might be illuminating. Renormalization of QED must be carried out at off-mass-shell points because of infrared divergences. For small e^2, we expect perturbation theory to be good in some neighborhood of the normalization point. But what about the inevitable logarithms of momenta that grow as we approach the mass shell or as some momenta go to infinity? In QED, the mass-shell divergences do not occur in observable predictions, when we take due account of the experimental situation. The renormalization-group technique[4] provides a somewhat opaque analysis of this situation. Loosely speaking,[5] the effective coupling of soft photons goes to zero, compensating for the fact that there are more and more of them. But the large-p^2 divergence represents a real breakdown of perturbation theory. It is commonly said that for momenta such that $e^2\ln(p^2/m^2)\sim 1$, higher orders become comparable, and hence a calculation to any finite order is meaningless in this domain. The renormalization group technique shows that the effective coupling grows with momenta.

The behavior in the two momentum regimes is reversed in a Yang-Mills theory. The effective coupling goes to zero for large momenta, but as p^2's approach zero, higher-order corrections become comparable. Thus perturbation theory tells *nothing* about the mass-shell structure of the symmetric theory. Even for arbitrarily small g^2, there is no sense in which the interacting theory is a small perturbation on a free multiplet of massless vector mesons. The truly catastrophic infrared problem makes a symmetric particle interpretation impossible. Thus, though one can well approximate asymptotic Green's functions, to what particle states do they refer?

Consider theories defined by the Lagrangian

$$\mathcal{L} = -\tfrac{1}{4}F_{\mu\nu}{}^{a}F^{a\mu\nu} + i\bar{\psi}_i\gamma\cdot D_{ij}\psi_j, \qquad (1)$$

where

$$F_{\mu\nu}{}^{a} = \partial_\mu A_\nu{}^{a} - \partial_\nu A_\mu{}^{a} + gf^{abc}A_\mu{}^{b}A_\nu{}^{c},$$

and

$$D_{ij}^{\mu} = \partial^{\mu}\delta_{ij} - igA^{a\mu}T_{ij}^{a},$$

the f^{abc} are the group structure constants, and the T^a are representation matrices corresponding to the fermion multiplet. (One may be interested in models with massless fermions because of their group structure or because they have the same asymptotic forms[6] as massive theories.) The normalizations of the conventionally defined irreducible vertices for n mesons and n' fermions, $\Gamma^{n,n'}$, must refer to some mass M. The renormalization-group equation reads

$$\left(M\frac{\partial}{\partial M} + \beta(g)\frac{\partial}{\partial g} + n\gamma_A(g) + n'\gamma_\psi(g)\right)\Gamma^{n,n'} = 0. \qquad (2)$$

Putting it in this form makes use of the first available simplification, proper choice of gauge. Equation (2) describes how finite renormalizations accompanied by a change in g and a rescaling of the fields leave the $\Gamma^{n,n'}$ unchanged. Consider gauges defined by α in the zeroth-order propagator

$$\Delta_{\mu\nu}(p^2) = \frac{-g_{\mu\nu} + p_\mu p_\nu/p^2}{p^2} + \alpha\frac{p_\mu p_\nu}{p^4}.$$

The generalized Ward identities[7] imply that there are no higher-order corrections to the longitudinal part. But if the fields are rescaled as in Eq. (2), α must be changed to leave Γ^2 invariant. Hence α should occur in Eq. (2) much as g does, and one would have to study the $\Gamma^{n,n'}$ for arbitrary α to determine the coefficient functions perturbatively. But for $\alpha = 0$ initially, it remains zero under finite renormalizations; so it suffices to study the theory in a Landau gauge.

To first order, the meson inverse propagator is

$$\Gamma_{\mu\nu}^{2ab}(p,-p) = \delta^{ab}(-g_{\mu\nu}p^2 + p_\mu p_\nu)[1 + (\tfrac{13}{3}c_1 - \tfrac{8}{3}c_2)(g/4\pi)^2\ln(-p^2/M^2)], \qquad (3)$$

where

$$f_{acd}f_{bcd} = 2c_1\delta_{ab}, \quad \mathrm{tr}(T^aT^b) = 2c_2\delta_{ab},$$

and $c_1 > 0$ and $c_2 \geq 0$. [For SU(2), $c_1 = 1$, c_2(isodoublet) $= \frac{1}{4}$, and c_2(isotriplet) $= 1$.] To first order (only), the fermion self-energy is proportional to the self-energy in massless QED, which vanishes in the Landau gauge. Similarly, the contribution to the fermion-vector three-point vertex correction proportional to the first-order QED correction needs no subtractions and contains no reference to M. Calculation of the remaining correction, involving the meson self-coupling, yields

$$\Gamma_{\mu ij}^{1,2a}(0,p,-p) = gT_{ij}^{a}\gamma_\mu[1 - \tfrac{3}{2}c_1(g/4\pi)^2\ln(-p^2/M^2)]. \qquad (4)$$

Applying Eq. (2) to these functions at their normalization points yields

$$\gamma_\psi(g) = 0 + O(g^4), \quad \gamma_A(g) = (\tfrac{13}{3}c_1 - \tfrac{8}{3}c_2)(g/4\pi)^2 + O(g^4), \quad \beta(g) = -(\tfrac{22}{3}c_1 - \tfrac{8}{3}c_2)g(g/4\pi)^2 + O(g^5).$$

It is also apparent, by inspecting the graphs, that to this order the coupling constants of product groups do not enter into each other's β functions.

For the case where there are no fermions, the coefficient functions can be obtained by setting $c_2 = 0$. (Even though the fermion-vector vertex, which had been used implicitly to define g, is no longer present, it can be simulated by introducing two multiplets of spinor fields with the same group transformations but opposite statistics. The physical effects of internal fermions are canceled by the ghosts —spinor fields with Bose statistics.) Alternatively, one can study the corrections to the three-meson vertex. Define F by

$$\Gamma_{\lambda\mu\nu}^{3abc}(p,-p,0) = f^{abc}(p_\lambda g_{\mu\nu} + p_\mu g_{\nu\lambda} - 2p_\nu g_{\lambda\mu})gF(p^2/M^2, g^2). \qquad (5)$$

The normalization condition is $F(-1, g^2) = 1$ (up to a phase convention.) To first order

$$F = 1 + \tfrac{17}{6}c_1(g/4\pi)^2\ln(-p^2/M^2) \qquad (6)$$

which yields the same β as described above.

The renormalization-group "improved" perturbation theory[4,5] extends results valid near the normalization point by effectively moving that point. The improved vertex functions are constructed from the straightforward perturbative ones, involving a momentum-scale-dependent effective coupling $g'(g,t)$, where $t = \frac{1}{2}\ln(s/M^2)$ and s sets the scale, e.g., $s = \sum(-p_i^2)$. $g'(g,t)$ is defined by

$$\partial g'/\partial t = \beta(g'), \qquad g'(g,0) = g. \qquad (7)$$

VOLUME 30, NUMBER 26 PHYSICAL REVIEW LETTERS 25 JUNE 1973

For the approximate β's derived above, $\beta = -bg^3$,

$$g'^2 \approx g^2/(1+2bg^2t). \tag{8}$$

Thus for a pure meson theory or for theories including not too many fermions (in the sense that $c_2 < \frac{11}{4}c_1$), g' goes to zero for asymptotic momenta, i.e., $t \to \infty$. The $\Gamma^{n,n'}$ show a well-defined slow approach to quasifree field values.

It is worth remembering that successive orders of perturbation theory give the behavior of β for infinitesimal g and, strictly speaking, say nothing about finite g. Making a polynomial fit to a perturbative result for β is pure conjecture.

Hypothesizing that β stays negative (at least into the domain of strong coupling constant) relates all theories defined by Eq. (1) [with g less than the first zero of $\beta(g)$] to the model with g arbitrarily small by a change in mass scale. They all share the same asymptotic Green's functions, differing only by how large is asymptotic.

To utilize this result, we make the following hypothesis: The gauge symmetry breaks down spontaneously as a result of the dynamics. Consequently, the fields obtain (in general massive) particle interpretation—the Higgs phenomenon. As yet, nothing is known about the particle spectrum, the low-energy dynamics, or particles describable only by composite fields. But the Callan-Symanzik analysis says that the asymptotic Green's functions for the "dressed" fundamental fermion and vector fields are the symmetric functions discussed above.[8]

[An alternative is to introduce fundamental scalar fields, in terms of which the group transformation properties of the vacuum can be studied.[9] But these theories are not in general ultraviolet stable in terms of the additional coupling constants that must be introduced. Particular models which are ultraviolet stable as well as spontaneously asymmetric have been found.[10] But gauge theories of fermions (only) have aesthetic attractions, including the possibility of a dynamical determination of the dimensionless coupling constant.[9]]

Hypotheses of this type go back to the work of Nambu and collaborators.[11] In the renormalizable massless theories including scalars that have been studied,[9] infrared instability is a necessary condition for spontaneous symmetry breakdown.[12] The model of Nambu and Jona-Lasinio can be treated by the methods of Coleman and Weinberg.[9] The model is defined by

$$\mathcal{L} = i\bar{\psi}\gamma\cdot\partial\psi + g_0[(\bar{\psi}\psi)^2 - (\bar{\psi}\gamma_5\psi)^2] \tag{9}$$

and the stipulation that the momentum integrals are cut off at some Euclidean mass squared Λ^2. Define a scalar

$$\varphi(x) = g_0\bar{\psi}(x)\psi(x)$$

and an analogous pseudoscalar, which one can do because of the cutoff. A study of the Green's functions in the one-loop approximation yields all the original results. But the existence of the vacuum-degenerate solution requires the dimensionless parameter characterizing the theory to satisfy $g_0\Lambda^2 > 2\pi^2$. But this is the condition that the one-loop correction to fermion-fermion scattering be at least as important as the tree approximation.

The situation is similar in the renormalizable models. $\lambda\varphi^4$ is stable for small λ because the one-loop corrections are small. But in massless scalar QED, photon-loop corrections of order e^4 can dominate over the lowest order φ-φ scattering (order λ) for both λ and e arbitrarily small. The requirement is just that $\lambda < e^4$. In this light, the problem with the Nambu–Jona-Lasinio model is not its nonrenormalizability but that in the domain of large $g_0\Lambda^2$, where spontaneous breakdown is alleged to occur, higher loop corrections are likely to dominate. (In the framework of the original solution, more complex infinite chains and self-energy graphs dominate over the ones studied.) In theories defined by Eq. (1), composite scalar densities can also be defined and studied in perturbation theory. But the condition that the one-loop approximation imply vacuum degeneracy requires that the expansion parameter be large, rendering the application of perturbation theory suspect.

The author thanks Sidney Coleman and Erick Weinberg, who have offered insights and advice freely, and the latter especially for his help in the computations.

---

*Work supported in part by the U. S. Air Force Office of Scientific Research under Contract No. F44620-70-C-0030.

[1]Of central importance is the work reviewed in K. Wilson and J. Kogut, "The Renormalization Group and the ϵ Expansion" (to be published); K. Johnson and M. Baker, "Some Speculations on High Energy Quantum Electrodynamics" (to be published); S. Adler, Phys. Rev. D 5, 3021 (1972).

[2]K. Symanzik, DESY Report No. DESY 72/73, 1972

(to be published), and references therein.

[3]A. Zee, "Study of the Renormalization Group for Small Coupling Constants" (to be published).

[4]N. N. Bogoliubov and D. V. Shirkov, *Introduction to the Theory of Quantized Fields* (Interscience, New York, 1959).

[5]Definitions of the relevant quantities will be given, but for the general theory and derivations see Refs. 1, 2, and 4; S. Coleman, in the Proceedings of the 1971 International Summer School "Ettore Majorana" (Academic, New York, to be published); S. Coleman and E. Weinberg, Phys. Rev. D 7, 1888 (1973), whose conventions we follow.

[6]Asymptotic refers to a particular set of Euclidean momenta as they are collectively scaled upward.

[7]E. g., B. W. Lee and J. Zinn-Justin, Phys. Rev. D 5, 3121 (1972); G. 't Hooft, Nucl. Phys. B33, 173 (1971), which also include details of Feynman rules, regularization, etc.

[8]Configurations where the symmetric theory has infrared singularities not present in the massive case are discussed in detail by Symanzik (Ref. 2).

[9]Coleman and Weinberg, Ref. 5.

[10]D. Gross and F. Wilczek, preceding Letter [Phys. Rev. Lett. 30, 1343 (1973)].

[11]Y. Nambu and G. Jona-Lasinio, Phys. Rev. 122, 345 (1961).

[12]$\lambda\varphi^4$ theory with $\lambda < 0$ is ultraviolet stable (Ref. 2) and hence infrared unstable but cannot be physically interpreted in perturbation theory. Using the computations of Ref. 9, for $\lambda < 0$ "improved" perturbation theory is arbitrarily good for large field strengths. In particular, the potential whose minimum determines the vacuum decreases without bound for large field.

60

Quantised Singularities in the Electromagnetic Field.

By P. A. M. Dirac, F.R.S., St. John's College, Cambridge.

(Received May 29, 1931.)

§ 1. *Introduction.*

The steady progress of physics requires for its theoretical formulation a mathematics that gets continually more advanced. This is only natural and to be expected. What, however, was not expected by the scientific workers of the last century was the particular form that the line of advancement of the mathematics would take, namely, it was expected that the mathematics would get more and more complicated, but would rest on a permanent basis of axioms and definitions, while actually the modern physical developments have required a mathematics that continually shifts its foundations and gets more abstract. Non-euclidean geometry and non-commutative algebra, which were at one time considered to be purely fictions of the mind and pastimes for logical thinkers, have now been found to be very necessary for the description of general facts of the physical world. It seems likely that this process of increasing abstraction will continue in the future and that advance in physics is to be associated with a continual modification and generalisation of the axioms at the base of the mathematics rather than with a logical development of any one mathematical scheme on a fixed foundation.

There are at present fundamental problems in theoretical physics awaiting solution, *e.g.*, the relativistic formulation of quantum mechanics and the nature of atomic nuclei (to be followed by more difficult ones such as the problem of life), the solution of which problems will presumably require a more drastic revision of our fundamental concepts than any that have gone before. Quite likely these changes will be so great that it will be beyond the power of human intelligence to get the necessary new ideas by direct attempts to formulate the experimental data in mathematical terms. The theoretical worker in the future will therefore have to proceed in a more indirect way. The most powerful method of advance that can be suggested at present is to employ all the resources of pure mathematics in attempts to perfect and generalise the mathematical formalism that forms the existing basis of theoretical physics, and *after* each success in this direction, to try to interpret the new mathematical features in terms of physical entities (by a process like Eddington's Principle of Identification).

A recent paper by the author* may possibly be regarded as a small step according to this general scheme of advance. The mathematical formalism at that time involved a serious difficulty through its prediction of negative kinetic energy values for an electron. It was proposed to get over this difficulty, making use of Pauli's Exclusion Principle which does not allow more than one electron in any state, by saying that in the physical world almost all the negative-energy states are already occupied, so that our ordinary electrons of positive energy cannot fall into them. The question then arises as to the physical interpretation of the negative-energy states, which on this view really exist. We should expect the uniformly filled distribution of negative-energy states to be completely unobservable to us, but an unoccupied one of these states, being something exceptional, should make its presence felt as a kind of hole. It was shown that one of these holes would appear to us as a particle with a positive energy and a positive charge and it was suggested that this particle should be identified with a proton. Subsequent investigations, however, have shown that this particle necessarily has the same mass as an electron† and also that, if it collides with an electron, the two will have a chance of annihilating one another much too great to be consistent with the known stability of matter.‡

It thus appears that we must abandon the identification of the holes with protons and must find some other interpretation for them. Following Oppenheimer,§ we can assume that in the world as we know it, *all*, and not merely nearly all, of the negative-energy states for electrons are occupied. A hole, if there were one, would be a new kind of particle, unknown to experimental physics, having the same mass and opposite charge to an electron. We may call such a particle an anti-electron. We should not expect to find any of them in nature, on account of their rapid rate of recombination with electrons, but if they could be produced experimentally in high vacuum they would be quite stable and amenable to observation. An encounter between two hard γ-rays (of energy at least half a million volts) could lead to the creation simultaneously of an electron and anti-electron, the probability of occurrence of this process being of the same order of magnitude as that of the collision of the two γ-rays on the assumption that they are spheres of the same size as classical

* 'Proc. Roy. Soc.,' A, vol. 126, p. 360 (1930).

† H. Weyl, 'Gruppentheorie und Quantenmechanik,' 2nd ed. p. 234 (1931).

‡ I. Tamm, 'Z. Physik,' vol. 62, p. 545 (1930); J. R. Oppenheimer, 'Phys. Rev.,' vol. 35, p. 939 (1930); P. Dirac, 'Proc. Camb. Philos. Soc.,' vol. 26, p. 361 (1930).

§ J. R. Oppenheimer, 'Phys. Rev.,' vol. 35, p. 562 (1930).

electrons. This probability is negligible, however, with the intensities of γ-rays at present available.

The protons on the above view are quite unconnected with electrons. Presumably the protons will have their own negative-energy states, all of which normally are occupied, an unoccupied one appearing as an anti-proton. Theory at present is quite unable to suggest a reason why there should be any differences between electrons and protons.

The object of the present paper is to put forward a new idea which is in many respects comparable with this one about negative energies. It will be concerned essentially, not with electrons and protons, but with the reason for the existence of a smallest electric charge. This smallest charge is known to exist experimentally and to have the value e given approximately by*

$$hc/e^2 = 137. \tag{1}$$

The theory of this paper, while it looks at first as though it will give a theoretical value for e, is found when worked out to give a connection between the smallest electric charge and the smallest magnetic pole. It shows, in fact, a symmetry between electricity and magnetism quite foreign to current views. It does not, however, force a complete symmetry, analogous to the fact that the symmetry between electrons and protons is not forced when we adopt Oppenheimer's interpretation. Without this symmetry, the ratio on the left-hand side of (1) remains, from the theoretical standpoint, completely undetermined and if we insert the experimental value 137 in our theory, it introduces quantitative differences between electricity and magnetism so large that one can understand why their qualitative similarities have not been discovered experimentally up to the present.

§ 2. *Non-integrable Phases for Wave Functions.*

We consider a particle whose motion is represented by a wave function ψ, which is a function of x, y, z and t. The precise form of the wave equation and whether it is relativistic or not, are not important for the present theory. We express ψ in the form

$$\psi = Ae^{i\gamma}, \tag{2}$$

where A and γ are real functions of x, y, z and t, denoting the amplitude and phase of the wave function. For a given state of motion of the particle, ψ will be determined except for an arbitrary constant numerical coefficient, which must be of modulus unity if we impose the condition that ψ shall be normalised.

* h means Planck's constant divided by 2π.

The indeterminacy in ψ then consists in the possible addition of an arbitrary constant to the phase γ. Thus the value of γ at a particular point has no physical meaning and only the difference between the values of γ at two different points is of any importance.

This immediately suggests a generalisation of the formalism. We may assume that γ has no definite value at a particular point, but only a definite difference in values for any two points. We may go further and assume that this difference is not definite unless the two points are neighbouring. For two distant points there will then be a definite phase difference only relative to some curve joining them and different curves will in general give different phase differences. The total change in phase when one goes round a closed curve need not vanish.

Let us examine the conditions necessary for this non-integrability of phase not to give rise to ambiguity in the applications of the theory. If we multiply ψ by its conjugate complex ϕ we get the density function, which has a direct physical meaning. This density is independent of the phase of the wave function, so that no trouble will be caused in this connection by any indeterminacy of phase. There are other more general kinds of applications, however, which must also be considered. If we take two different wave functions ψ_m and ψ_n, we may have to make use of the product $\phi_m \psi_n$. The integral

$$\int \phi_m \psi_n \, dx \, dy \, dz$$

is a number, the square of whose modulus has a physical meaning, namely, the probability of agreement of the two states. In order that the integral may have a definite modulus the integrand, although it need not have a definite phase at each point, must have a definite phase difference between any two points, whether neighbouring or not. Thus the change in phase in $\phi_m \psi_n$ round a closed curve must vanish. This requires that the change in phase in ψ_n round a closed curve shall be equal and opposite to that in ϕ_m and hence the same as that in ψ_m. We thus get the general result :— *The change in phase of a wave function round any closed curve must be the same for all the wave functions.*

It can easily be seen that this condition, when extended so as to give the same uncertainty of phase for transformation functions and matrices representing observables (referring to representations in which x, y and z are diagonal) as for wave functions, is sufficient to insure that the non-integrability of phase gives rise to no ambiguity in all applications of the theory. Whenever a ψ_n appears, if it is not multiplied into a ϕ_m, it will at any rate be multiplied into something of a similar nature to a ϕ_m, which will result in the uncertainty of phase cancelling out, except for a constant which does not matter. For example, if ψ_n is to be transformed to another representation in which, say, the observables ξ are diagonal, it must be multiplied by the transformation function $(\xi \,|\, xyzt)$ and integrated with respect to x, y and z. This transformation function will have the same uncertainty of phase as a ϕ, so that the transformed wave function will have its phase determinate, except for a constant independent of ξ. Again, if we multiply ψ_n by a matrix $(x'y'z't \,|\, \alpha \,|\, x''y''z''t)$, representing an observable α, the uncertainty in the phase as concerns the column [specified by x'', y'', z'', t] will cancel the uncertainty in ψ_n and the uncertainty as concerns the row will survive and give the necessary uncertainty in the new wave function $\alpha\psi_n$. The superposition principle for wave functions will be discussed a little later and when this point is settled it will complete the proof that all the general operations of quantum mechanics can be carried through exactly as though there were no uncertainty in the phase at all.

The above result that the change in phase round a closed curve must be the same for all wave functions means that this change in phase must be something determined by the dynamical system itself (and perhaps also partly by the representation) and must be independent of which state of the system is considered. As our dynamical system is merely a simple particle, it appears that the non-integrability of phase must be connected with the field of force in which the particle moves.

For the mathematical treatment of the question we express ψ, more generally than (2), as a product

$$\psi = \psi_1 \, e^{i\beta}, \tag{3}$$

where ψ_1 is any ordinary wave function (*i.e.*, one with a definite phase at each point) whose modulus is everywhere equal to the modulus of ψ. The uncertainty of phase is thus put in the factor $e^{i\beta}$. This requires that β shall not be a function of x, y, z, t having a definite value at each point, but β must have definite derivatives

$$\kappa_x = \frac{\partial\beta}{\partial x}, \quad \kappa_y = \frac{\partial\beta}{\partial y}, \quad \kappa_z = \frac{\partial\beta}{\partial z}, \quad \kappa_0 = \frac{\partial\beta}{\partial t},$$

at each point, which do not in general satisfy the conditions of integrability $\partial\kappa_x/\partial y = \partial\kappa_y/\partial x$, etc. The change in phase round a closed curve will now be, by Stokes' theorem,

$$\int (\boldsymbol{\kappa}, \, d\mathbf{s}) = \int (\text{curl}\,\boldsymbol{\kappa}, \, d\mathbf{S}), \tag{4}$$

where **ds** (a 4-vector) is an element of arc of the closed curve and **dS** (a 6-vector) is an element of a two-dimensional surface whose boundary is the closed curve. The factor ψ_1 does not enter at all into this change in phase.

It now becomes clear that the non-integrability of phase is quite consistent with the principle of superposition, or, stated more explicitly, that if we take two wave functions ψ_m and ψ_n both having the same change in phase round any closed curve, any linear combination of them $c_m\psi_m + c_n\psi_n$ must also have this same change in phase round every closed curve. This is because ψ_m and ψ_n will both be expressible in the form (3) with the same factor $e^{i\beta}$ (*i.e.*, the same κ's) but different ψ_1's, so that the linear combination will be expressible in this form with the same $e^{i\beta}$ again, and this $e^{i\beta}$ determines the change in phase round any closed curve. We may use the same factor $e^{i\beta}$ in (3) for dealing with all the wave functions of the system, but we are not obliged to do so, since only curl $\boldsymbol{\kappa}$ is fixed and we may use κ's differing from one another by the gradient of a scalar for treating the different wave functions.

From (3) we obtain

$$-ih\frac{\partial}{\partial x}\psi = e^{i\beta}\left(-ih\frac{\partial}{\partial x} + h\kappa_x\right)\psi_1, \qquad (5)$$

with similar relations for the y, z and t derivatives. It follows that if ψ satisfies any wave equation, involving the momentum and energy operators **p** and W, ψ_1 will satisfy the corresponding wave equation in which **p** and W have been replaced by $\mathbf{p} + h\boldsymbol{\kappa}$ and $\mathrm{W} - h\kappa_0$ respectively.

Let us assume that ψ satisfies the usual wave equation for a free particle in the absence of any field. Then ψ_1 will satisfy the usual wave equation for a particle with charge $-e$ moving in an electromagnetic field whose potentials are

$$\mathbf{A} = hc/e\,.\,\boldsymbol{\kappa}, \qquad \mathrm{A}_0 = -\,h/e\,.\,\kappa_0. \qquad (6)$$

Thus, since ψ_1 is just an ordinary wave function with a definite phase, our theory reverts to the usual one for the motion of an electron in an electromagnetic field. This gives a physical meaning to our non-integrability of phase. We see that we must have the wave function ψ always satisfying the same wave equation, whether there is a field or not, and the whole effect of the field when there is one is in making the phase non-integrable.

The components of the 6-vector curl $\boldsymbol{\kappa}$ appearing in (4) are, apart from numerical coefficients, equal to the components of the electric and magnetic fields **E** and **H**. They are, written in three-dimensional vector-notation,

$$\operatorname{curl}\boldsymbol{\kappa} = \frac{e}{hc}\mathbf{H}, \qquad \operatorname{grad}\kappa_0 - \frac{\partial\boldsymbol{\kappa}}{\partial t} = \frac{e}{h}\mathbf{E}. \qquad (7)$$

The connection between non-integrability of phase and the electromagnetic field given in this section is not new, being essentially just Weyl's Principle of Gauge Invariance in its modern form.* It is also contained in the work of Iwanenko and Fock,† who consider a more general kind of non-integrability based on a general theory of parallel displacement of half-vectors. The present treatment is given in order to emphasise that non-integrable phases are perfectly compatible with all the general principles of quantum mechanics and do not in any way restrict their physical interpretation.

§ 3. *Nodal Singularities.*

We have seen in the preceding section how the non-integrable derivatives κ of the phase of the wave function receive a natural interpretation in terms of the potentials of the electromagnetic field, as the result of which our theory becomes mathematically equivalent to the usual one for the motion of an electron in an electromagnetic field and gives us nothing new. There is, however, one further fact which must now be taken into account, namely, that a phase is always undetermined to the extent of an arbitrary integral multiple of 2π. This requires a reconsideration of the connection between the κ's and the potentials and leads to a new physical phenomenon.

The condition for an unambiguous physical interpretation of the theory was that the change in phase round a closed curve should be the same for all wave functions. This change was then interpreted, by equations (4) and (7), as equal to (apart from numerical factors) the total flux through the closed curve of the 6-vector **E**, **H** describing the electromagnetic field. Evidently these conditions must now be relaxed. The change in phase round a closed curve may be different for different wave functions by arbitrary multiples of 2π and is thus not sufficiently definite to be interpreted immediately in terms of the electromagnetic field.

To examine this question, let us consider first a very small closed curve. Now the wave equation requires the wave function to be continuous (except in very special circumstances which can be disregarded here) and hence the change in phase round a small closed curve must be small. Thus this change cannot now be different by multiples of 2π for different wave functions. It must have one definite value and may therefore be interpreted without

* H. Weyl, 'Z. Physik,' vol. 56, p. 330 (1929).

† D. Iwanenko and V. Fock, 'C. R.,' vol. 188, p. 1470 (1929); V. Fock, 'Z. Physik,' vol. 57, p. 261 (1929). The more general kind of non-integrability considered by these authors does not seem to have any physical application.

ambiguity in terms of the flux of the 6-vector **E**, **H** through the small closed curve, which flux must also be small.

There is an exceptional case, however, occurring when the wave function vanishes, since then its phase does not have a meaning. As the wave function is complex, its vanishing will require two conditions, so that in general the points at which it vanishes will lie along a line.* We call such a line a *nodal line.* If we now take a wave function having a nodal line passing through our small closed curve, considerations of continuity will no longer enable us to infer that the change in phase round the small closed curve must be small. All we shall be able to say is that the change in phase will be close to $2\pi n$ where n is some integer, positive or negative. This integer will be a characteristic of the nodal line. Its sign will be associated with a direction encircling the nodal line, which in turn may be associated with a direction along the nodal line.

The difference between the change in phase round the small closed curve and the nearest $2\pi n$ must now be the same as the change in phase round the closed curve for a wave function with no nodal line through it. It is therefore this difference that must be interpreted in terms of the flux of the 6-vector **E**, **H** through the closed curve. For a closed curve in three-dimensional space, only magnetic flux will come into play and hence we obtain for the change in phase round the small closed curve

$$2\pi n + e/hc \, . \int (\mathbf{H}, d\mathbf{S}).$$

We can now treat a large closed curve by dividing it up into a network of small closed curves lying in a surface whose boundary is the large closed curve. The total change in phase round the large closed curve will equal the sum of all the changes round the small closed curves and will therefore be

$$2\pi \Sigma n + e/hc \, . \int (\mathbf{H}, d\mathbf{S}), \tag{8}$$

the integration being taken over the surface and the summation over all nodal lines that pass through it, the proper sign being given to each term in the sum. This expression consists of two parts, a part $e/hc \, . \int (\mathbf{H}, d\mathbf{S})$ which must be the same for all wave functions and a part $2\pi \Sigma n$ which may be different for different wave functions.

* We are here considering, for simplicity in explanation, that the wave function is in three dimensions. The passage to four dimensions makes no essential change in the theory. The nodal lines then become two-dimensional nodal surfaces, which can be encircled by curves in the same way as lines are in three dimensions.

Expression (8) applied to any surface is equal to the change in phase round the boundary of the surface. Hence expression (8) applied to a closed surface must vanish. It follows that Σn, summed for all nodal lines crossing a closed surface, must be the same for all wave functions and must equal $-e/2\pi hc$ times the total magnetic flux crossing the surface.

If Σn does not vanish, some nodal lines must have end points inside the closed surface, since a nodal line without such end point must cross the surface twice (at least) and will contribute equal and opposite amounts to Σn at the two points of crossing. The value of Σn for the closed surface will thus equal the sum of the values of n for all nodal lines having end points inside the surface. This sum must be the same for all wave functions. Since this result applies to *any* closed surface, it follows that *the end points of nodal lines must be the same for all wave functions. These end points are then points of singularity in the electromagnetic field.* The total flux of magnetic field crossing a small closed surface surrounding one of these points is

$$4\pi\mu = 2\pi nhc/e,$$

where n is the characteristic of the nodal line that ends there, or the sum of the characteristics of all nodal lines ending there when there is more than one. Thus at the end point there will be a magnetic pole of strength

$$\mu = \tfrac{1}{2}nhc/e.$$

Our theory thus allows isolated magnetic poles, but the strength of such poles must be quantised, the quantum μ_0 being connected with the electronic charge e by

$$hc/e\mu_0 = 2. \tag{9}$$

This equation is to be compared with (1). The theory also requires a quantisation of electric charge, since any charged particle moving in the field of a pole of strength μ_0 must have for its charge some integral multiple (positive or negative) of e, in order that wave functions describing the motion may exist.

§ 4. *Electron in Field of One-Quantum Pole.*

The wave functions discussed in the preceding section, having nodal lines ending on magnetic poles, are quite proper and amenable to analytic treatment by methods parallel to the usual ones of quantum mechanics. It will perhaps help the reader to realise this if a simple example is discussed more explicitly.

Let us consider the motion of an electron in the magnetic field of a one-quantum pole when there is no electric field present. We take polar co-ordinates r, θ, ϕ with the magnetic pole as origin. Every wave function must now have a nodal line radiating out from the origin.

We express our wave function ψ in the form (3), where β is some non-integrable phase having derivatives κ that are connected with the known electromagnetic field by equations (6). It will not, however, be possible to obtain κ's satisfying these equations all round the magnetic pole. There must be some singular line radiating out from the pole along which these equations are not satisfied, but this line may be chosen arbitrarily. We may choose it to be the same as the nodal line for the wave function under consideration, which would result in ψ_1 being continuous. This choice, however, would mean different κ's for different wave functions (the difference between any two being, of course, the four-dimensional gradient of a scalar, except on the singular lines). This would perhaps be inconvenient and is not really necessary. We may express all our wave functions in the form (3) with the same $e^{i\beta}$, and then those wave functions whose nodal lines do not coincide with the singular line for the κ's will correspond to ψ_1's having a certain kind of discontinuity on this singular line, namely, a discontinuity just cancelling with the discontinuity in $e^{i\beta}$ here to give a continuous product.

The magnetic field **H**, lies along the radial direction and is of magnitude μ_0/r^2, which by (9) equals $\frac{1}{2}hc/er^2$. Hence, from equations (7), curl $\boldsymbol{\kappa}$ is radial and of magnitude $1/2r^2$. It may now easily be verified that a solution of the whole of equations (7) is

$$\kappa_0 = 0, \qquad \kappa_r = \kappa_\theta = 0, \qquad \kappa_\phi = 1/2r \,.\, \tan \tfrac{1}{2}\theta, \tag{10}$$

where κ_r, κ_θ, κ_ϕ are the components of $\boldsymbol{\kappa}$ referred to the polar co-ordinates. This solution is valid at all points except along the line $\theta = \pi$, where κ_ϕ becomes infinite in such a way that $\int (\boldsymbol{\kappa}, \mathbf{ds})$ round a small curve encircling this line is 2π. We may refer all our wave functions to this set of κ's.

Let us consider a stationary state of the electron with energy W. Written non-relativistically, the wave equation is

$$-h^2/2m \,.\, \nabla^2\psi = \mathrm{W}\psi.$$

If we apply the rule expressed by equation (5), we get as the wave equation for ψ_1

$$-h^2/2m \,.\, \{\nabla^2 + i\,(\boldsymbol{\kappa}, \boldsymbol{\nabla}) + i\,(\boldsymbol{\nabla}, \boldsymbol{\kappa}) - \kappa^2\}\,\psi_1 = \mathrm{W}\psi_1. \tag{11}$$

The values (10) for the κ's give

$$(\kappa, \nabla) = (\nabla, \kappa) = \kappa_\phi \frac{1}{r \sin \theta} \frac{\partial}{\partial \phi} = \frac{1}{4r^2} \sec^2 \tfrac{1}{2}\theta \frac{\partial}{\partial \phi}$$

$$\kappa^2 = \kappa_\phi{}^2 = \frac{1}{4r^2} \tan^2 \tfrac{1}{2}\theta,$$

so that equation (11) becomes

$$-\frac{h^2}{2m} \left\{ \nabla^2 + \frac{i}{2r^2} \sec^2 \tfrac{1}{2}\theta \frac{\partial}{\partial \phi} - \frac{1}{4r^2} \tan^2 \tfrac{1}{2}\theta \right\} \psi_1 = \mathrm{W} \psi_1.$$

We now suppose ψ_1 to be of the form of a function f of r only multiplied by a function S of θ and ϕ only, *i.e.*,

$$\psi_1 = f(r) \, \mathrm{S}(\theta\phi).$$

This requires

$$\left\{ \frac{d^2}{dr^2} + \frac{2}{r} \frac{d}{dr} - \frac{\lambda}{r^2} \right\} f = -\frac{2m\mathrm{W}}{h^2} f, \tag{12}$$

$$\left\{ \frac{1}{\sin \theta} \frac{\partial}{\partial \theta} \sin \theta \frac{\partial}{\partial \theta} + \frac{1}{\sin^2 \theta} \frac{\partial^2}{\partial \phi^2} + \tfrac{1}{2} i \sec^2 \tfrac{1}{2}\theta \frac{\partial}{\partial \phi} - \tfrac{1}{4} \tan^2 \tfrac{1}{2}\theta \right\} \mathrm{S} = -\lambda \mathrm{S}, \tag{13}$$

where λ is a number.

From equation (12) it is evident that there can be no stable states for which the electron is bound to the magnetic pole, because the operator on the left-hand side contains no constant with the dimensions of a length. This result is what one would expect from analogy with the classical theory. Equation (13) determines the dependence of the wave function on angle. It may be considered as a generalisation of the ordinary equation for spherical harmonics.

The lowest eigenvalue of (13) is $\lambda = \frac{1}{2}$, corresponding to which there are two independent wave functions

$$\mathrm{S}_a = \cos \tfrac{1}{2}\theta, \quad \mathrm{S}_b = \sin \tfrac{1}{2}\theta \, e^{i\phi},$$

as may easily be verified by direct substitution. The nodal line for S_a is $\theta = \pi$, that for S_b is $\theta = 0$. It should be observed that S_a is continuous everywhere, while S_b is discontinuous for $\theta = \pi$, its phase changing by 2π when one goes round a small curve encircling the line $\theta = \pi$. This is just what is necessary in order that both S_a and S_b, when multiplied by the $e^{i\beta}$ factor, may give continuous wave functions ψ. The two ψ's that we get in this way are both on the same footing and the difference in behaviour of S_a and S_b is due to our having chosen κ's with a singularity at $\theta = \pi$.

The general eigenvalue of (13) is $\lambda = n^2 + 2n + \frac{1}{2}$. The general solution of this wave equation has been worked out by I. Tamm.*

* Appearing probably in ' Z. Physik.'

§ 5. *Conclusion.*

Elementary classical theory allows us to formulate equations of motion for an electron in the field produced by an arbitrary distribution of electric charges and magnetic poles. If we wish to put the equations of motion in the Hamiltonian form, however, we have to introduce the electromagnetic potentials, and this is possible only when there are no isolated magnetic poles. Quantum mechanics, as it is usually established, is derived from the Hamiltonian form of the classical theory and therefore is applicable only when there are no isolated magnetic poles.

The object of the present paper is to show that quantum mechanics does not really preclude the existence of isolated magnetic poles. On the contrary, the present formalism of quantum mechanics, when developed naturally without the imposition of arbitrary restrictions, leads inevitably to wave equations whose only physical interpretation is the motion of an electron in the field of a single pole. This new development requires *no change whatever* in the formalism when expressed in terms of abstract symbols denoting states and observables, but is merely a generalisation of the possibilities of representation of these abstract symbols by wave functions and matrices. Under these circumstances one would be surprised if Nature had made no use of it.

The theory leads to a connection, namely, equation (9), between the quantum of magnetic pole and the electronic charge. It is rather disappointing to find this reciprocity between electricity and magnetism, instead of a purely electronic quantum condition, such as (1). However, there appears to be no possibility of modifying the theory, as it contains no arbitrary features, so presumably the explanation of (1) will require some entirely new idea.

The theoretical reciprocity between electricity and magnetism is perfect. Instead of discussing the motion of an electron in the field of a fixed magnetic pole, as we did in § 4, we could equally well consider the motion of a pole in the field of fixed charge. This would require the introduction of the electromagnetic potentials B satisfying

$$\mathrm{E} = \operatorname{curl} \mathrm{B}, \quad \mathrm{H} = \frac{1}{c}\frac{\partial \mathrm{B}}{\partial t} + \operatorname{grad} \mathrm{B}_0,$$

to be used instead of the A's in equations (6). The theory would now run quite parallel and would lead to the same condition (9) connecting the smallest pole with the smallest charge.

There remains to be discussed the question of why isolated magnetic poles are not observed. The experimental result (1) shows that there must be some cause of dissimilarity between electricity and magnetism (possibly connected with the cause of dissimilarity between electrons and protons) as the result of which we have, not $\mu_0 = e$, but $\mu_0 = 137/2 \, . \, e$. This means that the attractive force between two one-quantum poles of opposite sign is $(137/2)^2 = 4692\frac{1}{4}$ times that between electron and proton. This very large force may perhaps account for why poles of opposite sign have never yet been separated.

Nuclear Physics B79 (1974) 276–284. North-Holland Publishing Company

MAGNETIC MONOPOLES IN UNIFIED GAUGE THEORIES

G. 't HOOFT
CERN, Geneva

Received 31 May 1974

Abstract: It is shown that in all those gauge theories in which the electromagnetic group U(1) is taken to be a subgroup of a larger group with a compact covering group, like SU(2) or SU(3), genuine magnetic monopoles can be created as regular solutions of the field equations. Their mass is calculable and of order 137 M_W, where M_W is a typical vector boson mass.

1. Introduction

The present investigation is inspired by the work of Nielsen et al. [1], who found that quantized magnetic flux lines, in a superconductor, behave very much like the Nambu string [2]. Their solution consists of a kernel in the form of a thin tube which contains most of the flux lines and the energy; all physical fields decrease exponentially outside this kernel. Outside the kernel we do have a transverse vector potential A, but there it is rotation-free: if we put the kernel along the z axis, then

$$A(x) \propto (y, -x, 0)/(x^2 + y^2) . \tag{1.1}$$

$A(x)$ can be obtained by means of a gauge transformation $\Omega(\varphi)$ from the vacuum. Here φ is the angle about the z axis:

$$\Omega(0) = \Omega(2\pi) = 1 .$$

It is obvious that such a string cannot break since we cannot have an end point: it is impossible to replace a rotation over 2π continuously by $\Omega(\varphi) \to 1$. Or: magnetic monopoles do not occur in the system. Also it is easy to see that these strings are oriented: two strings with opposite direction can annihilate; if they have the same direction they may only join to form an even tighter string.

Now, let us suppose that the electromagnetism in the superconductor is in fact described by a unified gauge theory, in which the electromagnetic group U(1) is a subgroup of, say, SO(3). In such a non-Abelian theory one can only imagine non-oriented strings, because a rotation over 4π can be continuously shifted towards a fixed Ω. What happened with our original strings? The answer is simple: in an SO(3) gauge theory magnetic monopoles with twice the flux quantum (i.e., the

Schwinger [3,4] value), occur. Two of the original strings, oriented in the same direction, can now annihilate by formation of a monopole pair [5].

From now on we shall dispose of the original superconductor with its quantized flux lines. We consider free monopoles in the physical vacuum. That these monopoles are possible, as regular solutions of the field equations, can be understood in the following way. Imagine a sphere, with a magnetic flux Φ entering at one spot (see fig. 1). Immediately around that spot, on the contour C_0 in fig. 1, we must have a magnetic potential field A, with $\oint(A \cdot dx) = \Phi$. It can be obtained from the vacuum by applying a gauge transformation Λ:

$$A = \nabla \Lambda. \tag{1.2}$$

This Λ is multivalued. Now we require that all fields, which transform according to

$$\psi \to \psi \, e^{ni\Lambda} , \tag{1.3}$$

to remain single valued, so Φ must be an integer times 2π: we then have a complete gauge rotation along the contour in fig. 1.

In an Abelian gauge theory we must necessarily have some other spot on the sphere where the flux lines come out, because the rotation over $2k\pi$ cannot continuously change into a constant while we lower the contour C over the sphere. In a non-Abelian theory with compact covering group, however, for instance the group O(3), a rotation over 4π may be shifted towards a constant, without singularity: we may have a vacuum all around the sphere. In other theories, even rotations over 2π

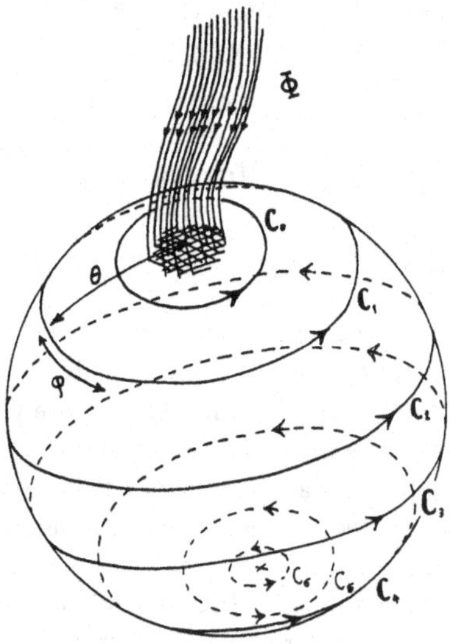

Fig. 1. The contour C on the sphere around the monopole. We deplace it from C_0 to C_1, etc., until it shrinks at the bottom of the sphere. We require that there be no singularity at that point.

may be shifted towards a constant. This is why a magnetic monopole with twice or sometimes once the flux quantum is allowed in a non-Abelian theory, if the electromagnetic group U(1) is a subgroup of a gauge group with compact covering group. There is no singularity anywhere in the sphere, nor is there the need for a Dirac string.

This is how we were led to consider solutions of the following type to the classical field equations in a non-Abelian Higgs–Kibble system: a small kernel occurs in the origin of three dimensional space. Outside that kernel a non-vanishing vector potential exists (and other non-physical fields) which can be obtained from the vacuum* by means of a gauge transformation $\Omega(\theta,\varphi)$. At one side of the sphere ($\cos\theta \to 1$) we have a rotation over 4π, which goes to unity at the other side of the sphere ($\cos\theta \to -1$). For such a rotation one can, for instance, take the following SU(2) matrix:

$$\Omega(\theta,\varphi) = \cos\tfrac{1}{2}\theta\begin{pmatrix} e^{i\varphi} & 0 \\ 0 & e^{-i\varphi}\end{pmatrix} + \sin\tfrac{1}{2}\theta\begin{pmatrix} 0 & i \\ i & 0\end{pmatrix}. \tag{1.4}$$

Now consider one rotation of the angle φ over 2π. At $\theta = 0$, this Ω rotates over 4π (the spinor rotates over 2π). At $\theta = \pi$, this Ω is a constant. One easily checks that

$$\Omega\Omega^\dagger = 1\,. \tag{1.5}$$

In the usual gauge theories one normally chooses the gauge in which the Higgs field is a vector in a fixed direction, say, along the positive z axis, in isospin space. Now, however, we take as a gauge condition that the Higgs field is $\Omega(\theta,\varphi)$ times this vector. As we shall see in sect. 2, this leads to a new boundary condition at infinity, to which corresponds a non-trivial solution of the field equations: a stable particle is sitting at the origin. It will be shown to be a magnetic monopole. If we want to be conservative and only permit the normal boundary condition at infinity, with Higgs fields pointing in the z direction, then still monopole-antimonopole pairs, arbitrarily far apart, are legitimate solutions of the field quations.

2. The model

We must have a model with a compact covering group. That, unfortunately, excludes the popular SU(2) × U(1) model of Weinberg and Salam [6]. There are two classes of possibilities.

(i) In models of the type described by Georgi and Glashow [7], based on SO(3), we can construct monopoles with a mass of the order of 137 M_W, where M_W is the

* As we shall see this vacuum will still contain a radial magnetic field. This is because the incoming field in fig. 1 will be spread over the whole sphere.

mass of the familiar intermediate vector boson. In the Georgi–Glashow model, $M_W < 53\ \text{GeV}/c^2$.

(ii) The Weinberg–Salam model can still be a good phenomenological description of processes with energies around hundreds of GeV, but may need extension to a larger gauge group at still higher energies. Weinberg [8] proposed SU(3) × SU(3) which would then be compact. Then the monopole mass would be 137 times the mass of one of the superheavy vector bosons.

We choose the first possibility for our sample calculations, because it is the simplest one. We take as our Lagrangian:

$$\mathcal{L} = -\tfrac{1}{4}G^a_{\mu\nu}G^a_{\mu\nu} - \tfrac{1}{2}D_\mu Q_a D_\mu Q_a - \tfrac{1}{2}\mu^2 Q_a^2 - \tfrac{1}{8}\lambda(Q_a^2)^2 , \tag{2.1}$$

where

$$G^a_{\mu\nu} = \partial_\mu W^a_\nu - \partial_\nu W^a_\mu + e\,\epsilon_{abc} W^b_\mu W^c_\nu ,$$

$$D_\mu Q_a = \partial_\mu Q_a + e\,\epsilon_{abc} W^b_\mu Q_c . \tag{2.2}$$

W^a_μ and Q_a are a triplet of vector fields and scalar fields, respectively.

We choose the parameter μ^2 to be negative so that the field Q gets a non-zero vacuum expectation value [6,7,9]:

$$\langle Q_a \rangle^2 = F^2 , \qquad \mu^2 = -\tfrac{1}{2}\lambda F^2 . \tag{2.3}$$

Two components of the vector field will acquire a mass:

$$M_{W_{1,2}} = eF , \tag{2.4}$$

whereas the third component describes the surviving Abelian electromagnetic interactions. The Higgs particle has a mass:

$$M_H = \sqrt{\lambda}\, F . \tag{2.5}$$

We are interested in a solution where the Higgs field is not rotated everywhere towards the positive z direction. If we apply the transformation Ω of eq. (1.5) to the isospin-one vector $F(0,0,1)$ we get

$$F(\sin\theta\,\cos\varphi,\ \sin\theta\,\sin\varphi,\ \cos\theta). \tag{2.6}$$

We shall take this isovector as our boundary condition for the Higgs field at spacelike infinity. As one can easily verify, it implies that the Higgs field must have at least one zero. This zero we take as the origin of our coordinate system.

We now ask for a solution of the field equations that is time-independent and spherically symmetric, apart from the obvious angle dependence. Introducing the vector

$$r_a = (x,y,z) , \qquad r_a^2 = r^2 , \tag{2.7}$$

we can write

$$Q_a(x,t) = r_a Q(r)\,, \qquad W_\mu^a(x,t) = \epsilon_{\mu ab} r_b W(r)\,, \tag{2.8}$$

where $\epsilon_{\mu ab}$ is the usual ϵ symbol if $\mu = 1, 2, 3$, and $\epsilon_{4ab} = 0$.

In terms of these variables the Lagrangian becomes

$$L = \int \mathcal{L}\,\mathrm{d}^3x = 4\pi \int_0^\infty r^2\,\mathrm{d}r \Big[-r^2\left(\frac{\mathrm{d}W}{\mathrm{d}r}\right)^2 - 4rW\frac{\mathrm{d}W}{\mathrm{d}r} - 6W^2 - 2er^2W^3$$

$$- \tfrac{1}{2}e^2r^4W^4 - \tfrac{1}{2}r^2\left(\frac{\mathrm{d}Q}{\mathrm{d}r}\right)^2 - rQ\frac{\mathrm{d}Q}{\mathrm{d}r} - \tfrac{3}{2}Q^2 - 2er^2WQ^2 - e^2r^4W^2Q^2$$

$$+ \tfrac{1}{4}\lambda F^2r^2Q^2 - \tfrac{1}{8}\lambda r^4Q^4 - \tfrac{1}{8}\lambda F^4\Big]\,, \tag{2.9}$$

where the constant has been added to give the vacuum a vanishing action integral. The field equations are obtained by requiring L to be stationary under small variations of the functions $W(r)$ and $Q(r)$. The energy of the system is then given by

$$E = -L\,, \tag{2.10}$$

since our system is stationary.

Before calculating this energy, let us concentrate on the boundary condition at $r \to \infty$. From the preceding arguments we already know that we must insist on

$$Q(r) \to F/r\,. \tag{2.11}$$

The field W must behave smoothly, as some negative power of r:

$$W(r) \to ar^{-n}\,. \tag{2.12}$$

From (2.9) we find the Lagrange equation

$$\frac{\mathrm{d}}{\mathrm{d}r}\left(2r^4\frac{\mathrm{d}W}{\mathrm{d}r} + 4r^3W\right)$$

$$= r^2\left[4r\frac{\mathrm{d}W}{\mathrm{d}r} + 12W + 6er^2W^2 + 2e^2r^4W^3 + 2er^2Q^2 + 2e^2r^4WQ^2\right]. \tag{2.13}$$

So, substituting (2.11) and (2.12),

$$(3-n)(4-2n)ar^{2-n} \xrightarrow[r\to\infty]{} -4nar^{2-n} + 12ar^{2-n} + 6ea^2r^{4-2n} + 2e^2a^3r^{6-3n}$$

$$+ 2eF^2r^2 + 2e^2aF^2r^{4-n}\,. \tag{2.14}$$

The only solution is

$$n = 2\,, \qquad a = -1/e\,. \tag{2.15}$$

So, far from the origin, the fields are

$$W_\mu^a(x,t) \to -\epsilon_{\mu ab} r_b/er^2 , \qquad Q_a(x,t) \to Fr_a/r . \tag{2.16}$$

Now most of these fields are not physical. To find the physically observable fields, in particular the electromagnetic ones, $F_{\mu\nu}$, we must first give a gauge invariant definition, which will yield the usual definition in the gauge where the Higgs field lies along the z direction everywhere. We propose:

$$F_{\mu\nu} = \frac{1}{|Q|} Q_a G_{\mu\nu}^a - \frac{1}{e|Q|^3} \epsilon_{abc} Q_a (D_\mu Q_b)(D_\nu Q_c) , \tag{2.17}$$

because, if after a gauge rotation, $Q_a = |Q|(0,0,1)$ everywhere within some region, then we have there

$$F_{\mu\nu} = \partial_\mu W_\nu^3 - \partial_\nu W_\mu^3 ,$$

as one can easily check. (Observe that the definition (2.17) satisfies the usual Maxwell equations, except where $Q^a = 0$; this is one other way of understanding the possibility of monopoles in this theory.) From (2.16), we get (see the definitions (2.2)):

$$Q_a G_{\mu\nu}^a = -\frac{F}{er^3} \epsilon_{\mu\nu a} r_a , \tag{2.18}$$

$$D_\mu Q_a = \partial_\mu Q_a + e\epsilon_{abc} W_\mu^b Q_c = 0 . \tag{2.19}$$

Hence

$$F_{\mu\nu} = -\frac{1}{er^3} \epsilon_{\mu\nu a} r_a . \tag{2.20}$$

Again, the ϵ symbol has been defined to be zero as soon as one of its indices has the value 4. So, there is a radial magnetic field

$$B_a = r_a/er^3 , \tag{2.21}$$

with a total flux

$$4\pi/e .$$

Hence, our solution is a magnetic monopole, as we expected. It satisfies Schwinger's condition

$$eg = 1 \tag{2.22}$$

(in units where $\hbar = 1$). In sect. 4, however, we show that in certain cases only Dirac's condition

$$eg = \tfrac{1}{2} n , \qquad n \text{ integer} ,$$

is satisfied.

3. The mass of the monopole

Let us introduce dimensionless parameters:

$$w = W/F^2 e\,, \qquad q = Q/F^2 e\,,$$

$$x = eFr\,, \qquad \beta = \lambda/e^2 = M_{\mathrm{H}}^2/M_{\mathrm{W}}^2\,. \tag{3.1}$$

From (2.9) and (2.10) we find that the energy E of the system is the minimal value of

$$\frac{4\pi M_{\mathrm{W}}}{e^2}\int_0^\infty x^2\,\mathrm{d}x\left[x^2\left(\frac{\mathrm{d}w}{\mathrm{d}x}\right)^2 + 4xw\frac{\mathrm{d}w}{\mathrm{d}x} + 6w^2 + 2x^2w^3 + \tfrac{1}{2}x^4w^4 + \tfrac{1}{2}x^2\left(\frac{\mathrm{d}q}{\mathrm{d}x}\right)^2\right.$$

$$\left. + xq\frac{\mathrm{d}q}{\mathrm{d}x} + \tfrac{3}{2}q^2 + 2x^2wq^2 + x^4w^2q^2 - \tfrac{1}{4}\beta x^2q^2 + \tfrac{1}{8}\beta x^4q^4 + \tfrac{1}{8}\beta\right]. \tag{3.2}$$

The quantity between the brackets is dimensionless and the extremum can be found by inserting trial functions and adjusting their parameters.

We found that the mass of the monopole (which is equal to the energy E since the monopole is at rest) is

$$M_{\mathrm{m}} = \frac{4\pi}{e^2}M_{\mathrm{W}}C(\beta)\,, \tag{3.3}$$

where $C(\beta)$ is nearly independent of the parameter β. It varies from 1.1 for $\beta = 0.1$ to 1.44 for $\beta = 10$ *.

Only in the Georgi–Glashow model (for which we did this calculation) is the parameter M_{W} in eq. (3.3) really the mass of the conventional intermediate vector boson. In other models it will in general be the mass of that boson which corresponds to the gauge transformations of the compact covering group: some of the superheavies in Weinberg's SU(3) × SU(3) for instance.

4. Conclusions

The relation between charge quantization and the possible existence of magnetic monopoles has been speculated on for a long time [10] and it has been observed that the gauge theories with compact gauge groups provide for the necessary charge quantization [11]. On the other hand, solutions of the field equations with abnormally rotated boundary conditions for the Higgs fields have also been considered before [1,12]. Nevertheless, it had escaped to our notion until now that magnetic monopoles occur among the solutions in those theories, and that their properties are predictable and calculable.

* These values may be slightly too high, as a consequence of our approximation procedure.

Our way of formulating the theory of magnetic monopoles avoids the introduction of Dirac's string [3], We expect no fundamental problems in calculating quantum corrections to the solution although they might be complicated to carry out.

The prediction is the most striking for the Georgi–Glashow model, although even in that model the mass is so high that that might explain the negative experimental evidence so far. If Weinberg's SU(2) × U(1) model wins the race for the presently observed weak interactions, then we shall have to wait for its extension to a compact gauge model, and the predicted monopole mass will be again much higher. Finally, one important observation. In the Georgi–Glashow model, one may introduce isospin $\frac{1}{2}$ representations of the group SU(2) describing particles with charges $\pm\frac{1}{2}e$. In that case our monopoles do not obey Schwinger's condition, but only Dirac's condition

$$qg = \tfrac{1}{2} ,$$

where q is the charge quantum and g the magnetic pole quantum, in spite of the fact that we have a completely quantized theory. Evidently, Schwinger's arguments do not hold for this theory [13]. We do have, in our model

$$\Delta q g = 1 ,$$

where Δq is the charge-difference between members of a multiplet, but this is certainly not a general phenomenon. In Weinberg's SU(3) × SU(3) the monopole quantum is the Dirac one and in models where the leptons form an SU(3) × SU(3) octet [14] the monopole quantum is three times the Dirac value (note the possibility of fractionally charged quarks in that case).

We thank H. Strubbe for help with a computer calculation of the coefficient $C(\beta)$, and B. Zumino and D. Gross for interesting discussions.

References

[1] H.B. Nielsen and P. Olesen, Niels Bohr Institute preprint, Copenhagen (May 1973);
B. Zumino, Lectures given at the 1973 Nato Summer Institute in Capri, CERN preprint TH. 1779 (1973).
[2] Y. Nambu, Proc. Int. Conf. on symmetries and quark models, Detroit, 1969 (Gordon and Breach, New York, 1970) p. 269;
L. Susskind, Nuovo Cimento 69A (1970) 457;
J.L. Gervais and B. Sakita, Phys. Rev. Letters 30 (1973) 716.
[3] P.A.M. Dirac, Proc. Roy. Soc. A133 (1934) 60; Phys. Rev. 74 (1948) 817.
[4] J. Schwinger, Phys. Rev. 144 (1966) 1087.
[5] G. Parisi, Columbia University preprint CO-2271-29.
[6] S. Weinberg, Phys. Rev. Letters 19 (1967) 1264.
[7] M. Georgi and S.L. Glashow, Phys. Rev. Letters 28 (1972) 1494.
[8] S. Weinberg, Phys. Rev. D5 (1972) 1962.

[9] F. Englert and R. Brout, Phys. Rev. Letters 13 (1964) 321;
P.W. Higgs, Phys. Letters 12 (1964) 132; Phys. Rev. Letters 13 (1964) 508; Phys. Rev. 145 (1966) 1156;
G.S. Guralnick, C.R. Hagen and T.W.B. Kibble, Phys. Rev. Letters 13 (1964) 585.
[10] D.M. Stevens, Magnetic monopoles: an updated bibliography, Virginia Poly. Inst. and State University preprint VPI-EPP-73-5 (October 1973).
[11] C.N. Yang, Phys. Rev. D1 (1970) 2360.
[12] A. Neveu and R. Dashen, Private communication.
[13] B. Zumino, Strong and weak interactions, 1966 Int. School of Physics, Erice, ed. A. Zichichi (Acad. Press, New York and London) p. 711.
[14] A. Salam and J.C. Pati, University of Maryland preprint (November 1972).

Particle spectrum in quantum field theory

A. M. Polyakov

L. D. Landau Institute of Theoretical Physics
(Submitted July 5, 1974)
ZhETF Pis. Red. **20**, No. 6, 430–433 (September 20, 1974)

We wish to call attention in this article to a hitherto uninvestigated part of the spectrum of ordinary Hamiltonians used in quantum field theory. As the first example we consider the model of a self-interacting scalar field in two-dimensional space-time, with Hamiltonian

$$H = \frac{1}{2}\int_{-\infty}^{+\infty} dx\left\{\pi^2 + \left(\frac{d\phi}{dx}\right)^2 - \mu^2\phi^2 + {}^{\lambda}\!/_2\,\phi^4\right\}$$
$$\pi(x) = -i\,{}^{\delta}\!/_{\delta\phi(x)}\quad. \tag{1}$$

In this model, the vacuum is filled with the Bose condensate $\bar{\phi}^2 = \mu^2/\lambda$. Over the vacuum there is a single-particle state with mass $\sqrt{2}\mu$, which represents small oscillations of the constant condensate. However, constant $\bar{\phi}$ is not the only stable equilibrium state. There is another extremal of the potential energy in (1), determined from the equation

$$\phi_c'' + \mu^2\phi_c - \lambda\phi_c^3 = 0 \tag{2}$$

with boundary conditions $\phi_c^2(\pm\infty) = \mu^2/\lambda$, which mean that the vacuum is perturbed only within a finite volume. The solution of (2) is

$$\phi_c(x) = \frac{\mu}{\sqrt{\lambda}}\,\mathrm{th}\,\frac{\mu x}{\sqrt{2}}\;. \tag{3}$$

To calculate the spectrum of the vibrational energy levels near the considered equilibrium point, we write

$$\phi(x) = \phi_c(x) + \phi(x)$$

and neglect the terms ϕ^3 and ϕ^4 in the Hamiltonian (the latter is valid if $\lambda \ll \mu^2$). Diagonalizing the obtained quadratic Hamiltonian, we obtain the mass spectrum

$$M_n = \frac{2\sqrt{2}}{3}\frac{\mu^3}{\lambda} + \frac{\sqrt{3}+2}{2\sqrt{2}}\mu + \frac{\mu}{\sqrt{2}}\sqrt{n(4-n)}$$
$$n = 0,\,1,\,2\,. \tag{4}$$

In formula (4), the first term is the potential energy at the equilibrium point, the second is the zero-point oscillation energy, and the third is the excitation energy. Thus, in this model there are three types of particles with anomalously large masses. We call these objects "extremons."

In the generalized formalization, the results consists in the fact that each stationary regular solution of the classical equations of motion corresponds in quantum field theory with weak coupling to its own set of extremons, the masses of which can in principle be calculated. We shall show that extremons exist in three-dimensional models. We consider the theory of the Higgs isovector field $\phi_a(x)$, $a = 1, 2, 3$,

$$H = \int\left\{\frac{1}{2}\pi_a^2 + \frac{1}{2}(\nabla\phi_a)^2 - \frac{1}{2}\mu^2\phi_a^2 + \frac{\lambda}{4}(\phi_a^2)^2\right\}d^3x\,. \tag{5}$$

The equation for the extremal

$$\nabla^2\phi_a + \mu^2\phi_a - \lambda\sum_b\phi_b^2\quad\phi_a = 0$$

has a solution

$$\phi_a = x_a u(r)r^{-1}\,,$$

where u is subject to the equation

$$u'' + \frac{2}{r}u' + \left(\mu^2 - \frac{2}{r^2}\right)u - \lambda u^3 = 0$$

$$u(\infty) = \mu/\sqrt{\lambda}\,;\qquad u(r) = \mathrm{const}\ r\,.$$

We call this the "hedgehog" solution, inasmuch as the isovector at a given point of space is directed along the radius vector. The solitary hedgehog is not an extremon, since its energy diverges linearly at large distances, owing to the inhomogeneity of the distribution of the directions of the field ϕ_a. There are two ways of overcoming this difficulty. The first is to connect to the hedgehog a Yang-Mills field, i.e., to make the substitution

$$\nabla_\mu\phi_a \to \nabla_\mu\phi_a + g\epsilon_{abc}A_\mu^b\phi_c$$

and to add the Yang-Mills Hamiltonian to (5). By virtue of the gauge symmetry of the second kind, the inhomogeneity of the directions then becomes physically unrealizable and makes no contribution to the energy, which therefore turns out to be finite. The solution of the classical equations is

$$\phi_a(x) = x_a u(r)r^{-1},$$
$$A_\mu^a(x) = \epsilon_{\mu ab}x_b\left(a(r) - \frac{1}{gr^2}\right), \tag{6}$$

where u and a satisfy the equations

$$\begin{cases} u'' + \dfrac{2}{r}u' + (\mu^2 - 2g^2a^2(r))\,u - \lambda u^3 = 0 \\ a'' + \dfrac{4}{r}a' - \dfrac{3}{r^2}a - g^2r^2a^3 - g^2u^2a = 0 \end{cases}.$$

The mass of the resultant extremon is of the order of

$$M \sim \frac{\mu^2}{\lambda}m_V^{-1} \sim m_V/g^2$$

(where $m_V = g^2u^2(\infty)$ is the mass of the vector bosons).

As is well known, the model under consideration has one massless vecton and two massive vectons. If the first of them is identified with the photon, then the hedgehog has by virtue of (6) a magnetic charge.[1)]

It is easy to construct hedgehogs in which all the components of the gauge field are massive and concentrated within the region $1/m_V$. To this end it suffices to consider an isotensor or isospinor Higgs field. In the former case the solution should be sought in the form

$$\phi_{ab} = r^{-2}(x_a x_b - \frac{1}{3}\delta_{ab} r^2) u(r)$$

and in the latter case it has a more complicated form and will be described elsewhere.

Another possibility for the construction of an extremon is formation of a hedgehog-antihedgehog pair. It is easy to show that the pair energy is $E = AR$, where R is the distance between the pair components. To stabilize this state it is necessary to consider levels with zero angular momentum L

$$E_{eff} = AR + B\frac{L^2}{R^2}. \quad (7)$$

The mass spectrum is given by

$$M^2(L) = \text{const}\ L^{4/3}. \quad (8)$$

A rigorous justification of (7) calls for the solution of the equation for the extremal with allowance for the fixed angular momentum. We have derived such an equation, but are unable to solve it exactly. Therefore the validity of (8) depends on the hypothesis concerning the character of the hedgehog rotation. In particular, if it is assumed that the region of space contained between the hedgehogs takes part in the rotation, then we obtain in place of (8) the formula

$$M^2(L) = \text{const}\ L.$$

There is no doubt, however, that a formula of the type (8) for the energy should lead to growing Regge trajectories.

Thus, in the interpretation of the elementary-particle spectrum on the basis of field theory one should bear in mind the many new possibilities that are afforded by taking the extremon states into account.

Ideas very close to those described above were developed in[1–4] but it seems to me that both the analysis method and the results obtained above are new to some degree.

I am grateful to V.L. Berezinskiĭ, L.B. Okun', and L.D. Faddeev for useful discussions.

[1)]This circumstance was pointed out to me by L. B. Okun'. After completing the work, I obtained a preprint by t'Hooft, which contains a similar result.

[1]H. Nielsen and P. Olesen, Nucl. Phys. B61, 45 (1973).
[2]Ya. B. Zel'dovich, I. Yu. Kobzarev, and L. B. Okun', IPM (Inst. Appl. Math.) Preprint, 1974.
[3]T. D. Lee and G. C. Wick, Columbia University Preprint, COO-2271-20 (1974).
[4]L. D. Faddeev, Phys. Lett., in press.

Nuclear Physics B61 (1973) 45–61. North-Holland Publishing Company

VORTEX-LINE MODELS FOR DUAL STRINGS

H.B. NIELSEN and P. OLESEN
The Niels Bohr Institute, University of Copenhagen, DK-2100 Copenhagen Ø, Denmark

Received 6 June 1973

Abstract: We call attention to the possibility of constructing field theories for the dual string. As an example, we show that a Higgs type of Lagrangian allows for vortex-line solutions, in analogy with the vortex lines in a type II superconductor. These vortex lines can approximately be identified with the Nambu string. In the strong coupling limit we speculate that the vortex lines make up all low energy phenomena. It turns out that this strong coupling limit is "super quantum mechanical" in the sense that the typical action of the theory is very small in comparison with Planck's constant.

1. Introduction and motivation

The many good experimental and theoretical features of the Veneziano model seem to suggest that an underlying string structure [1–4] of hadronic matter is a likely possibility. On the other hand it is not unreasonable to expect that some field theory describes relativistic physics (presumably including strong interactions and certainly electromagnetic interactions). For example, crossing symmetry which is crucial for the Veneziano model, is based on ideas taken from field theory. It is therefore of interest to see if one can cook up a local field theory which gives string structures behaving like dual strings. The spectrum of the Veneziano model would then be brought into contact with a local field theory.

It is the purpose of the present article to point out that it is easy to build up classical field theories allowing for vortex lines (or similar string-like structures) with the property of having equations of motion identical with those of the Nambu dual string [2]. The string description (and thus also the dual string equations of motion) are obtained only in the approximation where the radius of curvatures of the string is much larger than the width of the string. The width can be computed in terms of the parameters of the specific field theory model that one uses.

There are several motivations for being interested in field theories for dual strings:

(i) We have good reasons to believe that both field theory (of a kind which is so far not known) and dual strings (with some yet unknown degrees of freedom) are in fact realized in nature. It is therefore likely that nature has decided to merge some field theory with some dual string structure.

(ii) One may hope that by building some field theory for the dual string one is able to secure that the good properties of the former are inherited by the latter. Thus, for example one might hope that by choosing a field theory with positive definite Hamiltonian one might avoid tachyons in the corresponding dual string model. As we shall see by an example, this hope is apparently not supported when quantum mechanics is taken into account, at least not if this is done in the sloppy way of this paper. But we can still hope that a more careful treatment of quantization can resolve this difficulty.

(iii) One may hope to understand the features of dual strong models better by considering the corresponding field theories, and it would be of great interest to translate the requirement of a critical dimension ($d = 26$ in the conventional model, $d = 10$ in the Neveu-Schwarz model) into a field theoretic language. In particular, the question is as to whether the unphysical amount of dimensions needed in dual models [4] may be related to internal symmetries of the hadrons. Also, the possibility of understanding what happens when a couple of strings collide seems to have a chance in field theory.

(iv) We may use field theory to get ideas for how to modify the existing dual (string) models to get perhaps some day the right model. The trouble with generalizing dual models is that they are so tight, whereas field theory allows many possibilities.

As the main example of a field theory with a dual string structure we consider the theory of an Abelian gauge field coupled to a charged scalar field. This model has been used by Higgs to illustrate the Higgs mechanism. The relevance to dual models of an Abelian gauge field was first pointed out by us in ref. [5]. The Higgs model may also be considered as a relativistic generalization of the Ginzburg-Landau phenomenological field theory of superconductivity (see ref. [6]). In the Ginzburg-Landau case one knows the existence of a vortex-line solution, and it is exactly this fact which allows us to connect the Higgs-type of Lagrangian with the dual string, provided we identify the vortex-line with the dual string.

It turns out (see end of sect. 4) that the limit in which the Higgs Lagrangian gives the dual string solution is a sort of super-quantum mechanical limit. Although this may be bad from the point of view of quantization, it is still interesting that in order to understand hadronic structure in the sense of dual strings one has (in some sense) to consider the quantum of action h to be very large. That is to say, the typical action in the theory is much smaller than the fundamental quantum of action given by Planck's constant h.

In sect. 2 we consider the Higgs Lagrangian in the static case, and it turns out that this Lagrangian becomes identical to the Ginzburg-Landau free energy in the theory of type II superconductors. We recapitulate the most relevant features of this theory, and we discuss how the vortex solution comes out. In sect. 3 we show by a very general argument how the vortex solution leads to a Nambu Lagrangian [1]. This argument is very independent of the specific features of the Lagrangian. In sect. 4 we deal with the problem of getting the width of the vortex line sufficiently small in order that we can get a sufficiently good approximation to the dual string. It is shown that we have to consider the strong coupling limit. In sect. 5 we give a very brief and preliminary discussion of the problem of quantization. In an appendix we deal with the case of a Yang-Mills Lagrangian.

2. An example of a static vortex solution of the Ginzburg-Landau type

In this section we shall consider a special example of a field theory with a vortex solution. We start from the Lagrangian

$$\mathcal{L} = -\tfrac{1}{4}F_{\mu\nu}F^{\mu\nu} + \tfrac{1}{2}|(\partial_\mu + ieA_\mu)\phi|^2 + c_2|\phi|^2 - c_4|\phi|^4 , \tag{2.1}$$

and the equations of motion are

$$(\partial_\mu + ieA_\mu)^2\phi = -2c_2\phi + 4c_4\phi^2\phi^* , \tag{2.2}$$

$$\partial^\nu F_{\mu\nu} \equiv j_\mu = \tfrac{1}{2}ie(\phi^*\partial_\mu\phi - \phi\partial_\mu\phi^*) + e^2A_\mu\phi^*\phi . \tag{2.3}$$

We are now looking for a solution of the vortex type‡. In this case the field $F_{\mu\nu}$ has a simple meaning: the field F_{12} measures the number of vortex lines (going in the 3-direction) which pass a unit square in the (12)-plane. This interpretation is identical to the one proposed by us in ref. [5]. We want to identify the vortex line with a dual string, and it is thus necessary that the flux is quantized. To see that this is the case, we use that the flux is given by

$$\Phi = \int F_{\mu\nu}\mathrm{d}\sigma^{\mu\nu} = \oint A_\mu(x)\,\mathrm{d}x^\mu , \tag{2.4}$$

where $\mathrm{d}\sigma_{\mu\nu}$ is a two-dimensional surface element in Minkowski-space. Writing

$$\phi = |\phi|\,\mathrm{e}^{i\chi(x)} , \tag{2.5}$$

we get from eq. (2.3)

$$A_\mu = \frac{1}{e^2}\frac{j_\mu}{|\phi|^2} - \frac{1}{e}\partial_\mu\chi . \tag{2.6}$$

Next let us perform the integration in (2.4) over a closed loop without any current. Then

$$\Phi = \oint A_\mu(x)\,\mathrm{d}x^\mu = -\frac{1}{e}\oint \partial_\mu\chi(x)\,\mathrm{d}x^\mu . \tag{2.7}$$

‡ Most of the results obtained in this section are known in the theory of type II superconductors (see e.g. ref. [6]). The main new result is the identification of the Ginzburg-Landau theory with the static solution of the Higgs type of Lagrangian (2.1).

The line integral over the gradient of the phase of ϕ does not necessarily vanish. The only general requirement on the phase is that ϕ is single valued, i.e., χ varies by $2\pi n$ (n = integer) when we make a complete turn around a closed loop. Thus,

$$\Phi = n\Phi_0 , \qquad \Phi_0 = -\frac{2\pi}{e} . \tag{2.8}$$

Thus, the flux of vortex lines is quantized, $-2\pi/e$ being the quantum.

We still have to show that the equations of motion (2.2) and (2.3) allow a string-like solution. Let us consider the static case, with a gauge choice $A_0 = 0$. We look for a cylindrically symmetric solution, with axis along the z-direction. We write

$$A(r) = \frac{r \times e_z}{r} |A(r)| , \tag{2.9}$$

where e_z is a unit vector along the z-direction. The flux is given by

$$\Phi(r) = 2\pi r |A(r)| , \tag{2.10}$$

so that

$$|H| = \frac{1}{2\pi r} \frac{\mathrm{d}}{\mathrm{d}r} \Phi(r) = \frac{1}{r} \frac{\mathrm{d}}{\mathrm{d}r} (r|A|) . \tag{2.11}$$

With cylindrical symmetry around the z-axis the equations of motion (2.2) and (2.3) give

$$-\frac{1}{r}\frac{\mathrm{d}}{\mathrm{d}r}\left(r\frac{\mathrm{d}}{\mathrm{d}r}|\phi|\right) + \left[\left(\frac{1}{r} - e|A|\right)^2 - 2c_2 + 4c_4|\phi|^2\right] |\phi| = 0 , \tag{2.12}$$

$$-\frac{\mathrm{d}}{\mathrm{d}r}\left(\frac{1}{r}\frac{\mathrm{d}}{\mathrm{d}r}(r|A|)\right) + |\phi|^2 \left(|A|e^2 - \frac{e}{r}\right) = 0 . \tag{2.13}$$

The exact solution of these two equations has so far not been obtained analytically. We shall be content with a solution of the type where

$$|\phi| \simeq \text{const (for large } r) . \tag{2.14}$$

If we treat $|\phi|$ as a constant, eq. (2.13) can be solved without further approximations. One finds, with c a constant of integration,

$$|A| = \frac{1}{er} + \frac{c}{e} K_1(e|\phi|r) \underset{r\to\infty}{\longrightarrow} \frac{1}{er} +$$

$$\frac{c}{e}\sqrt{\frac{\pi}{2e|\phi|r}}\, \mathrm{e}^{-e|\phi|r} + \text{lower order terms.} \tag{2.15}$$

Eq. (2.11) then gives

$$|H| = c|\phi| K_0(e|\phi|r) \underset{r\to\infty}{\longrightarrow} \frac{c}{e}\sqrt{\frac{\pi|\phi|}{2er}}\, \mathrm{e}^{-e|\phi|r} + \text{lower order} . \tag{2.16}$$

Eq. (2.12) is then approximately satisfied if

$$|\phi| \simeq \sqrt{\frac{c_2}{2c_4}}\,, \tag{2.17}$$

with c_2 and c_4 being large so as to take care of deviations of $|A|$ from $1/er$.

We next define the characteristic length λ (called the penetration lenฺgth in superconductivity),

$$\lambda = \frac{1}{e|\phi|} = \sqrt{\frac{2c_4}{e^2 c_2}}\,. \tag{2.18}$$

λ thus measures (see eq. (2.16)) the region over which the field H is appreciably different from zero.

To estimate the variation of $|\phi|$ we notice that eq. (2.17) gives the minimum of the potential, i.e.,

$$|\phi| = \phi_0 = \sqrt{\frac{c_2}{2c_4}}\,, \tag{2.19}$$

is the vacuum-value of the field $|\phi|$. Let us write

$$|\phi| = \phi_0 + \rho(x)\,,$$

where $\rho(x)$ give the fluctuations around the vacuum. The first derivatives of the potential

$$\tfrac{1}{2}(-c_2\phi\phi^* + c_4\phi^2\phi^{*2})\,, \tag{2.20}$$

vanish, whereas the second derivative is

$$2c_4|\phi_0|^2 = 2c_2\,, \tag{2.21}$$

which is the mass square of the scalar particle in Higg's mechanism. Then the oscillations in the potential are

$$2c_2\,\rho(x)^2 \tag{2.22}$$

leading to a solution for ρ of the Yukawa-type,

$$\rho(x) \sim e^{-\sqrt{2c_2}\,r}\,. \tag{2.23}$$

We then define a new characteristic length ξ.

$$\xi = \frac{1}{\sqrt{2c_2}}\,. \tag{2.24}$$

Thus, ξ measures the distance that is takes before the field $|\phi|$ reaches it vacuum value.

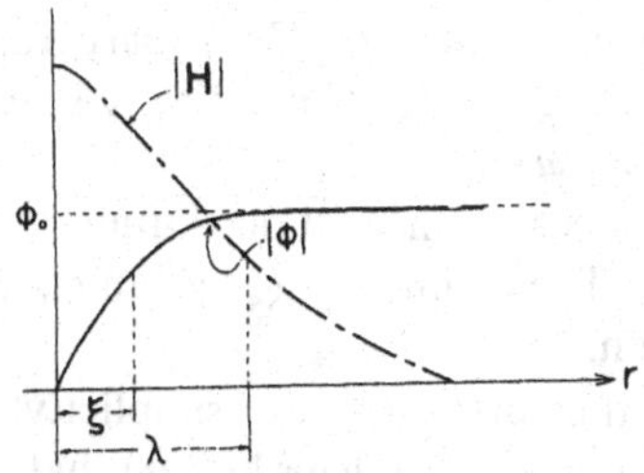

Fig. 1. An example of the behaviour of the fields $|H|$ and $|\phi|$ for a vortex solution.

In fig. 1 we have illustrated the behaviour of the fields $|H|$ and $|\phi|$. It is seen that if ξ and λ are of the same order of magnitude, then we have a well-defined vortex line, or a well defined string. The vacuum state is described by $H = 0$, and $|\phi| = \sqrt{\frac{1}{2}c_2/c_4}$, and the extension of the string is given by $\xi \sim \lambda$. The main point of this section is thus that the Higgs type of Lagrangian (2.1) allows a string-like solution. This is simply due to the fact that the Higgs Lagrangian is a relativistic generalization of the Ginzburg-Landau Lagrangian, which is well known to have vortex solutions.

The constant of integration c introduced in eq. (2.15) is determined by the requirement that the flux $\Phi(r) = 2\pi r|A(r)|$ shall go to zero for $\xi \ll r \ll \lambda$. Now, for $0 < e|\phi|r \ll 1$ we have

$$K_1(e|\phi|r) \approx \frac{1}{e|\phi|r} \tag{2.25}$$

and consequently

$$c = -e|\phi|. \tag{2.26}$$

3. The Nambu Lagrangian from vortex-line structure

In the preceeding section we pointed out that the Higgs-type of Lagrangian

$$\mathcal{L} = -\frac{1}{4}F_{\mu\nu}F^{\mu\nu} + \frac{1}{2}|(\partial_\mu + ieA_\mu)\phi|^2 + c_2|\phi|^2 - c_4|\phi|^4, \tag{3.1}$$

has solutions of the vortex type. By a suitable choice of parameters we can arrange that the width of the vortex-line is much smaller than the radius of curvature of the vortex-line. Of course, in addition to the vortex solution, (3.1) certainly has other solutions. In the next section we shall discuss how to handle some of the other solutions.

In this section we shall concentrate on the vortex contribution to the Lagrangian (3.1), $\mathcal{L}_{\text{vortex}}$ say. We shall assume that the other solutions can be effectively decoupled from the vortex solution, so that it has a meaning to separate out the special

In the last section we saw that by choosing the characteristic length λ sufficiently small, the field $F_{\mu\nu}(x)$ is non-vanishing only in a small region, whereas the field $|\phi|$ is nearly always equal to its vacuum value $\sqrt{c_2/2c_4}$, apart from the region in space where $F_{\mu\nu}$ is non-vanishing. Taking $\xi \sim \lambda$, $|\phi|$ practically speaking vanishes when $F_{\mu\nu}$ is non-vanishing, and *vice versa*.

The field $F_{\mu\nu}(x)$ therefore acts as a smeared out δ-function, which is only non-vanishing along the vortex-line. The quantity $-(\frac{1}{4})F^2$ in eq. (3.1) therefore also acts as a smeared out δ-function.

In fig. 2 we have illustrated the vortex-line. For simplicity we shall assume that the end-points are at spatial infinity, or (perhaps better) that the vortex is a closed loop. Since the vortex-contribution to the Lagrangian (3.1) is a smeared out δ-function, which is non-vanishing only along the vortex-line, $\mathcal{L}$ is itself a smeared out δ-function. Of course, $\mathcal{L}$ is also relativistically invariant. It therefore follows that $\mathcal{L}_{\text{vortex}}$ is Lorentz-contracted in the transverse direction, i.e.

$$\mathcal{L}_{\text{vortex}} \propto \sqrt{1 - v_\perp^2}\,, \tag{3.2}$$

due to the motion of the vortex-line in the transverse direction. Let $\mathrm{d}s$ be the element of length along the vortex-line. Since the "transverse length" of the string, $\lambda \sim \xi$, is a constant, it follows that the action is given by (we ignore constants of proportionality)

$$S_{\text{vortex}} = \int \mathrm{d}^4x\ \mathcal{L}_{\text{vortex}} \propto \int \mathrm{d}t\ \mathrm{d}s \sqrt{1 - v_\perp^2}\,. \tag{3.3}$$

Now

$$v_\perp = \partial \boldsymbol{x}/\partial t - \partial \boldsymbol{x}/\partial s(\partial \boldsymbol{x}/\partial t \cdot \partial \boldsymbol{x}/\partial s) \tag{3.4}$$

and hence

$$S_{\text{vortex}} \propto \int \mathrm{d}t\ \mathrm{d}s\ \sqrt{1 - \left(\frac{\partial \boldsymbol{x}}{\partial t}\right)^2 + \left(\frac{\partial \boldsymbol{x}}{\partial t}\ \frac{\partial \boldsymbol{x}}{\partial s}\right)^2}\,. \tag{3.5}$$

Introducing a different parametrization of the vortex, $\boldsymbol{x}(s, t) = \boldsymbol{x}(\sigma, t)$ this leads to the Nambu action [2] (discussed in detail by Goddard, Goldstone, Rebi and Thorn [4])

$$S_{\text{vortex}} \propto \int_{-\infty}^{+\infty} \mathrm{d}t \int_{\sigma_1}^{\sigma_2} \mathrm{d}\sigma \frac{\mathrm{d}s}{\mathrm{d}\sigma} \sqrt{1 - v_\perp^2}\,, \tag{3.6}$$

provided one chooses a frame [4] where the parameter τ is identified with the time t.

It is rather clear that the arguments presented above do not depend on the details of the Lagrangian (3.1). All we need is that a vortex solution exists.

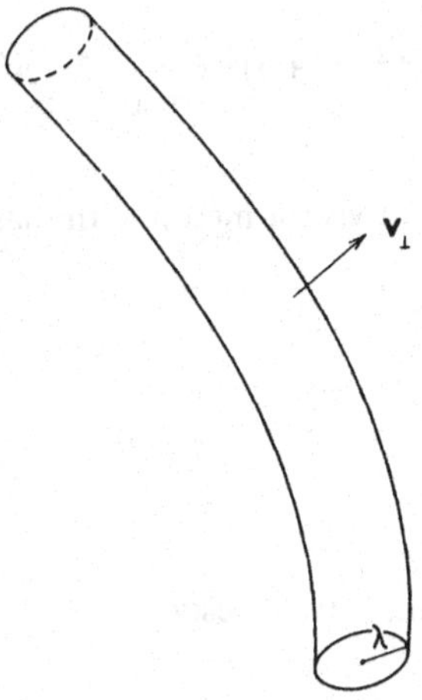

Fig. 2. Lorentz contraction of the vortex in the transverse direction.

Thus we have obtained the following rather remarkable result: the action of the vortex is proportional to the area of the surface swept out by the vortex in space and time.

This result forms the basis of our identification of the vortex solution of the Higgs-type Lagrangian (which is, of course, only a special example) with the dual string, described by the Nambu Lagrangian. It must be emphasized that our considerations are entirely classical, and so far no quantum effects are included.

4. The strong coupling limit

Our example, namely the field theory of the Landau-Ginzburg type, is the one used by Higgs to illustrate the Higgs mechanism. He showed that the theory by an appropriate gauge choice is revealed to be a theory of a massive scalar and a massive vector meson. In fact he chooses a gauge where the phase of the charged scalar field vanishes. (This would be a very inappropriate gauge choice for a vortex state). Choosing this gauge and putting

$$\phi(x) = \phi_0 + \rho(x) , \tag{4.1}$$

where ϕ_0 is given by

$$\phi_0 = \sqrt{\frac{c_2}{2c_4}} , \tag{4.2}$$

$$\mathcal{L} = -\tfrac{1}{4}F_{\mu\nu}F^{\mu\nu} + \tfrac{1}{2}(\partial_\mu \rho)^2 + \tfrac{1}{2}A_\mu^2(\phi_0 + \rho)^2 + \frac{c_2^2}{4c_4} - 2c_2\rho^2 - 2\sqrt{2c_2c_4}\rho^3 - c_4\rho^4 \,. \tag{4.3}$$

This form is easily seen to describe a vector meson with mass

$$m_V = e\phi_0 = e\sqrt{\frac{c_2}{2c_4}}\,, \tag{4.4}$$

and a scalar meson with mass

$$m_S = \sqrt{2c_2} \tag{4.5}$$

interacting with each other.

The compton wave lengths of these two mesons

$$\lambda = m_V^{-1}\,, \tag{4.6}$$

$$\xi = m_S^{-1} \tag{4.7}$$

give the width of the string. In the case $\lambda \gg \xi$ discussed above, ξ is the radius of an inner core in the string where $|\phi|$ deviates appreciably from its vacuum ϕ_0, while λ is the radius of the string determined by the width of the flux bundle (compare with fig. 1). In order that the strings are really strings, that is to say are thin[‡], we must have the penetration depth λ and the coherence length ξ small compared to the characteristic length, which for a dual string model is $\sqrt{\alpha'}$. Here α' is the universal slope for the state of the string with two ends. According to the paper by Goldstone, Goddard, Rebbi, and Thorn [4], the energy density along the dual string is

$$\text{energy density} = \frac{\gamma_\perp}{2\pi\alpha'} = \frac{1}{2\pi\alpha'\sqrt{1-v_\perp^2}}\,, \tag{4.8}$$

where $v_\perp$ is the transverse velocity of the string.

We can thus make a classical estimate of the connection of the universal slope α' to the three parameters, c_2, c_4 and e of the Landau-Ginzburg-like Lagrangian (2.1) by calculating the energy-density at rest for the vortex solution given in sect. 2.

The magnetic energy-density along the vortex string is

$$\frac{1}{2}\int_0^\infty |H|^2\, 2\pi r\, \mathrm{d}r = \tfrac{1}{2}\phi_0^4 e^2 \int_0^\infty K_0(e\phi_0 r)^2\, 2\pi r\, \mathrm{d}r = \pi\phi_0^2 \int_0^\infty K_0(y)^2\, y\, \mathrm{d}y$$

$$= \frac{c_2}{c_4}\frac{\pi}{2}\int_0^\infty K_0(y)^2\, y\, \mathrm{d}y\,. \tag{4.9}$$

The integral $\int_0^\infty K_0(y)^2\, y\, \mathrm{d}y$ converges because $K_0(y)$ behaves no worse than a logarithm for $y \to 0$ and decays exponentially for $y \to \infty$. The integral is thus of order unity.

‡ A string is by definition a thin practically one dimensional structure.

A crude estimate of the energy per unit length of the string due to the core where $|\phi|$ no longer has its vacuum value also gives something of the order of magnitude

$$\xi^2 c_2 \phi_0^2 \approx \phi_0^2 \,, \tag{4.10}$$

since $c_2\phi_0^2$ is the energy density. Thus we can conclude that the energy-density along the string to be identified with $1/2\pi\alpha'$ in the dual model is

$$\frac{1}{2\pi\alpha'} \sim \phi_0^2 \sim \frac{c_2}{c_4} \,. \tag{4.11}$$

The exact ratio $\alpha' c_2/c_4$ can be computed numerically by solving the differential equations. What is important, however, is just that it is of order unity.

The order of magnitude of the characteristic length for the hadrons, being the lower quantum mechanical levels of the dual string, is thus

$$\sqrt{\alpha'} \approx \sqrt{\frac{c_4}{c_2}} \,. \tag{4.12}$$

In order that thin strings are a good approximation, we thus need to have

$$\sqrt{\alpha'} \gg \lambda, \xi \,, \tag{4.13}$$

which again implies

$$\frac{\sqrt{\alpha'}}{\lambda} \approx e \gg 1 \tag{4.14}$$

and

$$\frac{\sqrt{\alpha'}}{\xi} \approx \sqrt{c_4} \gg 1 \,. \tag{4.15}$$

These requirements might also be written

$$m_{\mathrm{V}}, m_{\mathrm{S}} \gg \frac{1}{\sqrt{\alpha'}} \,, \tag{4.16}$$

which means that the particles corresponding directly to the local fields have masses m_{V} and m_{S} much larger than the typical hadron masses. Thus in this limit low energy phenomena (low energy meaning energies of the order of $1/\sqrt{\alpha'}$) should be dominated by hadrons, i.e. dual strings, and not by the fundamental vector and scalar particles in the theory revealed by Higgs.

We may hope that a third kind of excitation is not going to be important at low energies, so that we may have a pure dual string theory in the low energy range of this strong coupling limit of the Landau-Ginzburg-like theory, in which the two coupling constants e and c_4 are infinitely large.

We thus would like to suggest that such a strong coupling limit is the one in which a dual string theory emerges, and so we would like to postulate that if nature were to be described by the Landau-Ginzburg-like model of ours, it would have chosen fairly large values of e and $\sqrt{c_4}$, so that we could get pure dual string properties of the theory at low energies.

The limit $e, \sqrt{c_4}$ large, make however, classical field considerations very doubtful, because it is in a certain sense a super quantum-mechanical limit, i.e. something like $\hbar \to \infty$. This is at first unclear in our formulation, because we used the quantum of action as a unit $\hbar = 1$.

To see that our limit $e, \sqrt{c_4} \to \infty$ is a super quantum mechanical one, we may just remark that the masses m_V and m_S of the fundamental particles of the field theory are typical harmonic oscillator frequencies, the classical solution of the equations revealing the particles as solutions $\phi(x) \approx \epsilon\, e^{-ipx}$, $A_\mu \approx \epsilon_\mu\, e^{-ipx}$, where the coefficients ϵ and ϵ_μ are small quantities. But the typical energy of the field theory is rather the energy of say a vortex line, with a length of the order of magnitude of its width, and that is of the order of magnitude λ/α'.

So in our limit $e, \sqrt{c_4} \to \infty$ the typical energy is much smaller than the typical frequency, since from (4.14) and (4.16)

$$\frac{\lambda}{\alpha'} \ll m_S, m_V \tag{4.17}$$

in the strong coupling limit. That means that the typical action of the theory is, using eq. (4.6) and (4.14),

$$\frac{\lambda}{\alpha' m_V} = \frac{1}{\alpha' m_V^2} \approx \frac{1}{e^2} \ll 1 \,, \tag{4.18}$$

so that from the point of view of strong coupling theory in this limit the quantum of action $\hbar = 1$ is tremendously big compared to the typical amount of action e^{-2}. So the theory is very very far from being classical, since in a classical theory the typical action, say 1 erg. sec, is very large compared to $\hbar$.

The extreme quantum mechanical nature of the theory in the strong coupling limit ($e, \sqrt{c_4} \to \infty$) has the implication that if we want to justify our classical solution of the fields around and in a vortex line in sect. 2 by estimating the fluctuations in a coherent state simulating our classical solution, we may be in a very bad shape.

5. The problem of quantization

We now mention a few words concerning the problem of quantizing the above string scheme. As mentioned near the end of sect. 4, if one calculates the characteristic action of the present theory, then it turns out to be much larger than Planck's constant. Thus we have an extreme quantum mechanical problem, where one might expect important fluctuations in a coherent state approximating our classical vortex solution. Thus we may be in serious trouble in going from the classical theory to the quantized theory. We have no real solution to offer to this problem. The only thing we can say at present is the following: suppose that one could show that in the strong coupling limit the classical theory has only one stable solution, namely the vortex solution. The arguments in sect. 3 then shows that the Nambu Lagrangian comes out from the field theory Lagrangian, i.e.,

$$\mathcal{L}\begin{pmatrix}\text{classical}\\ \text{field theory}\end{pmatrix} \to \mathcal{L}\begin{pmatrix}\text{classical}\\ \text{Nambu}\end{pmatrix} \quad \text{(strong coupling limit)}\,. \tag{5.18}$$

This limiting behaviour would of course not be correct if we had other stable solutions in addition to the vortex solution, and in spite of various attempts we have not succeeded in convincing ourselves that only the vortex solution is stable (classically). However, suppose that the limit (5.18) is correct. Then we could quantize the theory just by quantizing the Nambu Lagrangian, in which case we should obtain the usual quantized dual string.

Actually the somewhat optimistic remarks above are dubious. The point is that we have formulated our theory in four dimensions, and hence we should run into trouble with respect to quantization‡, and we should also obtain a tachyon. However, the classical Hamiltonian is certainly positive definite, and it appears as a mystery how it can generate a non-definite spectrum in the quantized version. However, it may be that the solution to this apparent paradox is that the positive definite character of a classical Hamiltonian may not carry over into quantum field theory. For example, if the theory has to be renormalized, it is not at all guaranteed that the signs of the renormalized couplings are the same as the signs of the bare couplings, and hence positive definiteness may impose different conditions on the couplings in a classical theory (where the bare couplings enter) and in a qunatized theory (where the renormalized coupling enter). A recent discussion of this possibility has been given by 't Hooft [7]. We do, however, not yet know to what extent the strong coupling limit (the vortex solution) should be renormalized.

6. Conclusions

We have seen that it is possible to make field theories that (classically) have solutions corresponding to vortex-lines that are one dimensional structures moving around e.g. according to the equation of motion of the Nambu dual string. In order that the length of the string in one of the lowest mass eigenstates (i.e. one of the lowest hadron states) should be much larger than the width of it, we had to choose a strong coupling limit e, $c_4 \to \infty$ in the Ginzburg-Landau model used as the example.

It should be stressed that the Ginzburg-Landau model is only an example. Many models could easily be proposed, that would all lead to the simple Nambu string with no extra degrees of freedom in the strong coupling limit at low energies. For

‡ Notice that in order to formulate the theory in an arbitrary number of dimensions, we have to generalize the Higgs mechanism to an arbitrary number of dimensions. This has not yet been done.

instance it is easy to add some extra fields giving particles with high masses of the order of magnitude of m_V and m_S to the Ginzburg-Landau theory discussed above, without giving further vortex-lines so that the vortex-lines would still behave like Nambu dual strings. Also, in the appendix we discuss the case of Yang-Mills fields, which does not lead to additional degrees of freedom.

One is also not restricted to work in 3+1 dimensions. In fact, probably the simplest non-trivial vortex-line model is the so-called Sine-Gordon theory in 2+1 dimensions having the Lagrangian

$$\mathcal{L} = \tfrac{1}{2}(\partial_\mu \varphi(x))^2 + c \cos (d\varphi(x)) ,$$

giving the equation of motion

$$\partial_\mu \partial^\mu \varphi(x) + cd \sin (d\varphi(x)) = 0 ,$$

i.e., the Sine-Gordon-equation. It is readily seen that this theory classically allows for a static solution

$$\varphi(x) = \frac{4}{d} \arctan \exp (\sqrt{c}\, d x_2)$$

describing a vortex line along the x_1-axis. The field $\varphi(x)$ changes by $2\pi/d$ across the vortex line, which has a width of the order of magnitude $1/\sqrt{c}d$. Analogous to the case of the Ginzburg-Landau theory, this width is equal to the Compton wavelength for the particle of the theory, as is seen by considering the Klein-Gordon equation with mass square $m^2 = cd^2$ obtained as the weak field limit of the Sine-Gordon-equation. The energy density along the vortex line is

$$\frac{1}{2\pi\alpha'} = \int_{-\infty}^{+\infty} \left[\left\{ \frac{4}{d} \frac{\partial}{\partial x} \arctan \exp (\sqrt{c} d x) \right\}^2 + 2c \sin^2 (2 \arctan \exp (\sqrt{c} d x) \right] \mathrm{d}x \approx \sqrt{c}/d ,$$

so that strings that are narrow (vortex lines) compared to the hadronic length $\sqrt{\alpha'} \approx d^{\frac{1}{2}} c^{-\frac{1}{4}}$ are obtained in the strong coupling (and super quantum mechanical) limit $m\sqrt{\alpha'} \approx c^{-\frac{1}{4}}$ are obtained in the strong coupling (and super quantum mechanical) vortex lines move as Nambu dual strings.

We have shown the equivalence of string models with a certain set of solutions of some field theories in a classical approximation, namely the vortex lines. We believe, but we have not proven that in the strong coupling limit at low energies all states of the Ginzburg-Landau field theory are states that can be described as states of some system of strings. This hope cannot be taken to be true for all theories having vortex lines, since theories can be cooked up, which have e.g. zero-dimensional structures in excess of the one dimensional one. That is to say one could make a field-theory model which also had kink like type of singularity similar to the solution in the 1+1 dimensional Sine-Gordon theory. Such theories could be built in higher dimensions too.

That we have to take a super quantum mechanical limit necessitates that, (i) the field theories to be used should be quantized and (ii) the equivalence of the field theory and the corresponding string theory be proven on a quantum mechanical level (that can only be done, if at all, in a low energy and strong coupling limit).

Also we have to understand how it can happen that tachyons appear in a model having at first a positive definite Hamiltonian density, as is easily seen to be the case for the Higgs-Ginzburg-Landau model. Further, we should like to know what the sifnificance of the critical dimension ($d = 26$ in the conventional model) is in terms of field-theory models like the ones discussed, but we have not even made the 26-dimensional classical Ginzburg-Landau model yet.

If some day one understands the quantum properties better, the possibility of constructing field-theory models for strings like the ones we discussed, might be an easier way to come across a good (possibly unitary) Veneziano-model than to make a string model directly.

First of all we want to thank Don Weingarten for pointing out how to make vortex-line Sine-Gordon models in any dimensions, and C.H. Tze for finding literature. We also thank B. Zumino for calling our attention to the fact that the equation we discussed is well known as the Ginzburg-Landau equation. Secondly we want to thank our colleagues at the Niels Bohr Institute and CERN for helpful discussions.

Appendix. Discussion of the Yang-Mills type of model

It is natural to ask whether it is possible to produce strings from Lagrangian having internal degrees of freedom, e.g. isospin or SU(3). If we can manage to keep the vortex solutions for such Lagrangians, this would indeed be very nice from the point of view of the dual string, since the string would then carry internal degrees of freedom, and would therefore perhaps lead to a more realistic spectrum of hadrons (and perhaps to a more realistic dual amplitude, provided we could solve the problem of colliding strings). This, unfortunately, does not seem to be the case.

An example of a Lagrangian with internal degrees of freedom which immediately comes to the mind, is the Yang-Mills type of Lagrangian. Here we introduce a field $B_{\mu\nu}$,

$$B_{\mu\nu} = \partial_\mu B_\nu - \partial_\nu B_\mu - g B_\mu \times B_\nu \, . \tag{A.1}$$

Defining the dual field

$$B^*_{\mu\nu} = \tfrac{1}{2} \epsilon_{\mu\nu\alpha\beta} B^{\alpha\beta} \, , \tag{A.2}$$

it is easily seen that

$$(\partial^\mu + g B^\mu \times) B^*_{\mu\nu} = 0, \tag{A.3}$$

which is the analogue of the Maxwell equation

$$\partial^\mu F^*_{\mu\nu} = 0\,, \tag{A.4}$$

which states that magnetic monopoles do not exist.

The form of eq. (A.3) poses a problem with respect to the interpretation of (magnetic) flux lines. From eq. (A.4) one can derive that

$$0 = \int_{V_\mu} \partial^\mu F^*_{\mu\nu} \mathrm{d}V^\nu = \int_{S_1} F_{\mu\nu} \mathrm{d}\sigma^{\mu\nu} - \int_{S_2} F_{\mu\nu} \mathrm{d}\sigma^{\mu\nu}\,, \tag{A.5}$$

where V_μ is a volume bounded by the two surfaces S_1 and S_2. Eq. (A.5) tells us that no magnetic flux lines can start inside the volume. As far as eq. (A.3) is concerned, a similar procedure leads to

$$-g \int_{V_\mu} B^\mu \times B^*_{\mu\nu}\, \mathrm{d}V^\nu = \int_{S_1} B_{\mu\nu} \mathrm{d}\sigma^{\mu\nu} - \int_{S_2} B_{\mu\nu} \mathrm{d}\sigma^{\mu\nu}\,. \tag{A.6}$$

Thus, at least in general flux lines can originate inside the volume. If, however, we concentrate on the static situation, we can take $B_o = 0$, and $B_{ok} = 0$, and hence we have the interpretation that the flux is given by

$$\Phi = \int B_{\mu\nu}\, \mathrm{d}\sigma^{\mu\nu} \text{ (static case)}\,. \tag{A.7}$$

Next let us consider a specific Yang-Mills Lagrangian, namely

$$\mathcal{L} = -\tfrac{1}{4} B_{\mu\nu} B^{\mu\nu} + \tfrac{1}{2} [(\partial_\mu + g B_\mu \times)\, \phi]^2 + \tfrac{1}{2} [(\partial_\mu + g B_\mu \times)\, \psi]^2$$

$$+ c_2 \phi^2 - c_4 (\phi^2)^2 + d_2 \psi^2 - d_4 (\psi^2)^2 + e_2 \phi\,\psi - e_4 (\phi\,\psi)^2\,, \tag{A.8}$$

where the fields ϕ and ψ are isovector fields. The reader may wonder why we introduce two fields ϕ and ψ and not just one field. The reason for this will turn out later, where we shall see that in order to have a vortex solution at least two isovector fields are needed.

Now we are looking for a solution which quantizes the flux (A.7) in a way similar to the Higgs (Ginzburg-Landau) Lagrangian discussed in sect. 2. Defining the current to be

$$-\partial^\mu B_{\mu\nu} + g B_{\mu\nu} \times B^\mu \equiv j_\nu\,, \tag{A.9}$$

i.e., $j_\nu = 0$ for a free Yang-Mills field, we obtain from the Lagrangian (A.8) by varying B_μ

$$j_\nu = g(\phi \times \partial_\nu \phi) + g(\psi \times \partial_\nu \psi) + g^2 (B_\nu \times \phi) \times \phi + g^2 (B_\nu \times \psi) \times \psi\,. \tag{A.10}$$

Considering now the static solution with cylindrical symmetry we see that the term $B_\mu \times B_\nu$ in eq. (A.1) is smaller than the term $\partial_\mu B_\nu - \partial_\nu B_\mu$ in $B_{\mu\nu}$ for large distances from the axis of symmetry. Thus we can write the flux (A.7) as

$$\Phi = \oint B_\mu(x)\,\mathrm{d}x^\mu \quad \text{(static case, large distance)}\,, \tag{A.11}$$

provided we integrate over a circle with large radius. For such large distances j_ν vanishes, and eq. (A.10) leads to

$$(B_\nu \times \phi)\times\phi + (B_\nu \times \psi)\times\psi = (B_\nu\phi)\phi + (B_\nu\psi)\psi - (\phi^2+\psi^2)B_\nu$$
$$= \frac{1}{g}(\phi\times\partial_\nu\phi + \psi\times\partial_\nu\psi)\,. \tag{A.12}$$

Now the ground state (the vacuum) of the Lagrangian (A.8) corresponds to

$$\phi^2 = \frac{c_2}{2c_4}\,,\qquad \psi^2 = \frac{d_2}{2d_4}\,,\qquad \phi\psi = \frac{e_2}{2e_4}\,, \tag{A.13}$$

where we assume that ϕ and ψ are not in the same (or opposite) direction, i.e.

$$\frac{c_2 d_2}{c_4 d_4} > \left(\frac{e_2}{e_4}\right)^2 . \tag{A.14}$$

Thus, in the vacuum the lenghts of the isovectors ϕ and ψ are fixed, and in addition the projection of one vector on the other is fixed. This ensures that in any frame of reference in isospace at least three components of ϕ and ψ are non-vanishing (e.g., ϕ_1, ψ_1 and ψ_2).

Now let us go to a frame of reference where $\phi_3 = \psi_3 = 0$. This corresponds to selecting the flux lines in the 3-direction. It is easily seen that the condition (A.12) for the current to vanish at large distances imposes the condition on B_ν that

$$B_\nu^1 = B_\nu^2 = 0\,,$$
$$B_\nu^3 = -\frac{1}{g}\partial_\nu\chi\,, \tag{A.15}$$

where χ is the phase of $\phi_1 + i\phi_2$. Notice that due to the last condition (A.13) the phase of $\psi_1 + i\psi_2$ is the same as χ + constant. Inserting eq. (A.15) in eq. (A.11) and using the fact that the phase is only unique modulus $2\pi n$ we get that the flux is given by

$$|\Phi| = 2\pi n/g\,, \tag{A.16}$$

i.e., the flux is quantized‡. This result only follows if we have at least two fields ϕ and ψ.

Having obtained the flux quantization we then go to the strong coupling limit, in order that the width of the vortex line is made sufficiently small. The width is given

‡ We have not convinced ourselves that there are no other quantas than $2\pi/g$.

by the order of magnitude of the Compton wave lengths of the Higgs particles. As before, we have scalar particles with masses

$$m_S = \sqrt{2c_2}\,, \qquad m'_S = \sqrt{2d_2}\,. \tag{A.17}$$

The Higgs vector particle is obtained from the seagull terms in the Lagrangian (A.8), i.e. from

$$\tfrac{1}{2}g^2(B_\mu \times \phi)^2 + \tfrac{1}{2}g^2(B_\mu \times \psi)^2 = \tfrac{1}{2}g^2\,[B_\mu^2(\phi^2+\psi^2) - (B_\mu\phi)^2 - (B_\mu\psi)^2]\,. \tag{A.18}$$

Inserting the vacuum values (A.13) it is easily seen that all vector mesons B_μ acquire a mass,

$$m_V = g\sqrt{\frac{c_2}{2c_4} + \frac{d_2}{2d_4}}\,. \tag{A.19}$$

Proceeding as in sect. 4 we can now go to the strong coupling limit $g \to \infty$, $c_4 \to \infty$, $d_4 \to \infty$ (or m_S, m'_S, $m_V \gg 1/\sqrt{\alpha'}$), which then gives us the string solution. From the very general argument in sect. 3, we know that this solution corresponds to the classical Nambu Lagrangian. However, one can easily see that due to gauge invariance no new degrees of freedom are introduced in the string Lagrangian.

Note added in proof

In addition to the term (4.9) the magnetic energy-density also contains a term coming from the seagull term in the Lagrangian. The latter term gives rise to an energy-density of the order of magnitude

$$(c_2/c_4)\log(\lambda/\xi).$$

As long as $\log(\lambda/\xi)$ is not too large, this term does not change any of the conclusions in sect. 4. See ref. [8] for example.

A preprint of L.J. Fàssie [9] with a similar idea has appeared.

References

[1] Y. Nambu, Proc. Int. Conf. on symmetries and quark models, (Wayne State University, 1969);
H.B. Nielsen, 15th Int. Conf. on high energy physics, Kiev, 1970;
L. Susskind, Nuovo Cimento 69A (1970) 457.
[2] Y. Nambu, Lectures at the Copenhagen Summer Symposium, 1970.
[3] L.N. Chang and J. Mansouri, Phys. Rev. D5 (1972) 2535.
[4] P. Goddard, J. Goldstone, C. Rebbi and C.B. Thorn, Nucl. Phys. B56 (1973) 109.
[5] H.B. Nielsen and P. Olesen, Nucl. PHys. B57 (1973) 367;
P. Olesen, Niels Bohr Institute preprint NBI-HE-73-9, Nuovo Cimento, to be published.
[6] D. Saint-James, G. Sarma and E.J. Thomas, Type II super-conductivity (Pergamon Press, 1969).
[7] G. 't Hooft, CERN preprint TH 1666 (1973).
[8] A. De Gennes, Superconductivity of metals and alloys (1966).
[9] L.J. Fassie, Lines of quantized magnetic flux and the relativistic string of the dual resonance model of hadrons, Canberra preprint (1973).

PHYSICAL REVIEW D VOLUME 10, NUMBER 8 15 OCTOBER 1974

Confinement of quarks*

Kenneth G. Wilson
Laboratory of Nuclear Studies, Cornell University, Ithaca, New York 14850
(Received 12 June 1974)

A mechanism for total confinement of quarks, similar to that of Schwinger, is defined which requires the existence of Abelian or non-Abelian gauge fields. It is shown how to quantize a gauge field theory on a discrete lattice in Euclidean space-time, preserving exact gauge invariance and treating the gauge fields as angular variables (which makes a gauge-fixing term unnecessary). The lattice gauge theory has a computable strong-coupling limit; in this limit the binding mechanism applies and there are no free quarks. There is unfortunately no Lorentz (or Euclidean) invariance in the strong-coupling limit. The strong-coupling expansion involves sums over all quark paths and sums over all surfaces (on the lattice) joining quark paths. This structure is reminiscent of relativistic string models of hadrons.

I. INTRODUCTION

The success of the quark-constituent picture both for resonances and for deep-inelastic electron and neutrino processes makes it difficult to believe quarks do not exist. The problem is that quarks have not been seen. This suggests that quarks, for some reason, cannot appear as separate particles in a final state. A number of speculations have been offered as to how this might happen.[1]

Independently of the quark problem, Schwinger observed many years ago[2] that the vector mesons of a gauge theory can have a nonzero mass if vacuum polarization totally screens the charges in a gauge theory. Schwinger illustrated this result with the exact solution of quantum electrodynamics in one space and one time dimension, where the photon acquires a mass $\sim e^2$ for any nonzero charge e [e has dimensions of $(\text{mass})^{1/2}$ in this theory]. Schwinger suggested that the same effect could occur in four dimensions for sufficiently large couplings.

Further study of the Schwinger model by Lowenstein and Swieca[3] and Casher, Kogut, and Susskind[4] has shown that the asymptotic states of the model contain only massive photons, not electrons. Nevertheless, as Casher *et al.* have shown in detail, the electrons are present in deep-inelastic processes and behave like free pointlike particles over short times and short distances. The polarization effects which prevent the appearance of electrons in the final state take place on a longer time scale (longer than $1/m_\gamma$, where m_γ is the photon mass).

A new mechanism which keeps quarks bound will be proposed in this paper. The mechanism applies to gauge theories only. The mechanism will be illustrated using the strong-coupling limit of a gauge theory in four-dimensional space-time. However, the model discussed here has a built-in ultraviolet cutoff, and in the strong-coupling limit all particle masses (including the gauge field masses) are much larger than the cutoff; in consequence the theory is far from covariant.

The confinement mechanism proposed here is soft (long-time scale). However, in the model discussed here the cutoff spoils the possibility of free pointlike behavior for the quarks.

The model discussed in this paper is a gauge theory set up on a four-dimensional Euclidean lattice. The inverse of the lattice spacing a serves as an ultraviolet cutoff. The use of a Euclidean space (i.e., imaginary instead of real times) instead of a Lorentz space is not a serious restriction; the energy eigenstates (including scattering states) of the lattice theory can be determined from the "transfer-matrix" formalism as has been discussed by suri[5] and reviewed by Wilson and Kogut.[6] A brief discussion of the transfer-matrix method is given in Sec. III.

In Schwinger's speculations about four dimensions, the photon mass would be zero for any charge e less than a critical coupling e_c; for $e > e_c$ the photon mass would be nonzero and vary with e. Figure 1 shows how a plot of m_γ vs e might look. The point e_c is a point of nonanalyticity. Similar nonanalytic points, called critical points, occur in solid-state physics at certain types of phase transitions. Consider, for instance, a ferromagnet in the absence of an external field. For any temperature above the Curie temperature T_C, the spontaneous magnetization M is 0. Below T_C, M is nonzero and a function of temperature. At T_C there may be either a first-order phase transition (in which case M is discontinuous at T_C) or a second-order phase transition (critical point) for which $M \to 0$ as $T \to T_C$ from either side.

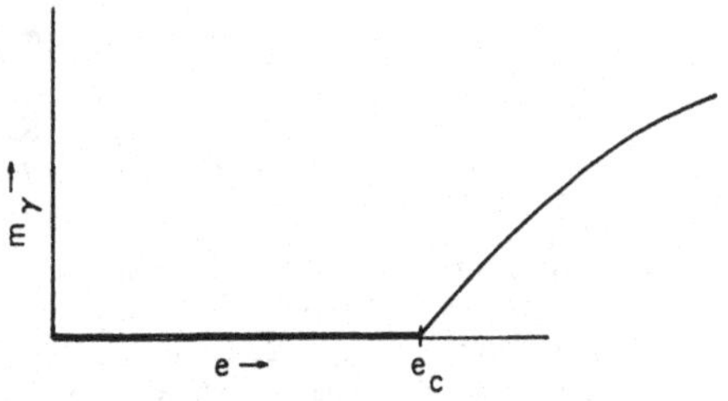

FIG. 1. Speculative plot of photon mass vs renormalized charge e, in unknown units. The transition at e_c is second-order (see text).

By analogy with the solid-state situation one can think of the transition from zero to nonzero photon mass as a change of phase: this analogy is best understood by imagining the particles of quantum electrodynamics to be the excitations of a medium (the ether). In this case it is the ether which undergoes a change of phase at e_c. There is again a question whether this change of phase is first-order (cf. Fig. 2) or second-order (Fig. 1). (Coleman and Weinberg[7] have found a nontrivial example of a first-order transition in another context.)

The model discussed in this paper is a single Abelian gauge field coupled (with strength g) to massive quarks. In weak coupling the gauge field behaves like a normal free zero-mass field (despite modifications introduced in the lattice quantization) and the quarks are unbound. In strong coupling the gauge field is massive and the quarks are bound, showing the existence of the second phase. Thus there should be a phase transition at some intermediate value of g. Nothing is known about this transition at the present time.

The quantization procedure and strong-coupling approximation described in this paper can be applied to non-Abelian gauge theories also. This will be explained briefly in Sec. III.

An extraordinary feature of the strong-coupling expansion of the lattice theory (see Sec. IV) is that it has the same general structure as the relativistic string models of hadrons.[8] The vacuum expectation values of the gauge theory involve (in the strong-coupling expansion) sums over all quark paths and sums over all surfaces connecting these paths; the surfaces are generated by the gauge field treated in strong coupling. The paths and surfaces are defined on a discrete lattice. There are geometrical difficulties in relating the surfaces on the lattice to the continuum surfaces of the string models; it is not known at present whether these difficulties can be overcome.

In Sec. II the nature of the quark binding mechanism will be discussed, qualitatively. In Sec. III the gauge theory will be formulated on a discrete lattice, both classically and quantum mechanically. In Sec. IV the strong-coupling expansion for the lattice gauge theory is explained. In Sec. V a cursory discussion of weak coupling is given. In Sec. VI there is a brief discussion of the problem of Lorentz invariance and the relation to string models.

II. QUARK BINDING MECHANISM

The binding mechanism will be explained in this section using the Feynman path-integral picture. The path-integral framework will be used in an intuitive rather than a formal way. Consider the current-current propagator

$$D_{\mu\nu}(x) = \langle\Omega|\, T J_\mu(x) J_\nu(0) \,|\Omega\rangle \ , \qquad (2.1)$$

whose Fourier transform determines the e^+-e^- annihilation cross section into hadrons. Assume that the currents $J_\mu(x)$ are built from quark fields as in the quark-parton model. Assume that the quarks interact through a single gauge field. (The restriction to one field is only for simplicity.) In the Feynman path-integral picture the propagator

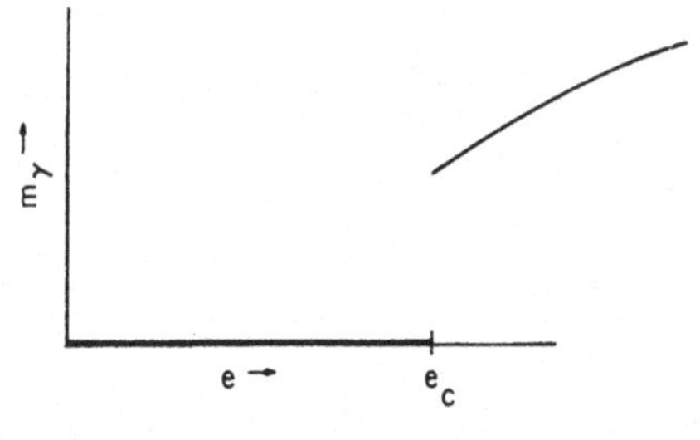

FIG. 2. Speculative plot of photon mass vs renormalized charge e if there is a first-order transition at e_c.

$D_{\mu\nu}(x)$ is given by a weighted average over all possible classical quark paths and all possible classical values of the gauge field. The currents $J_\mu(x)$ and $J_\nu(0)$ are thought of as producing a quark pair at the origin which later annihilate at the point x: One has to sum over all paths joining the points 0 and x for each of the pair of quarks.

An example of paths for the quark and antiquark are shown in Fig. 3. The vacuum can also emit and absorb quark pairs; this leads to further closed loops as illustrated in Fig. 4. All possible loops must be summed over too. It is also possible to have independent loops for the points 0 and x (Fig. 5), but this possibility will not be important here.

The weight associated with a given quark path or set of paths includes a factor of $\exp[ig\oint A_\mu(x) \times ds^\mu]$, where $A_\mu(x)$ is the gauge field. Here $\oint\cdots ds^\mu$ is a line integral or a sum over line integrals for each of the quark-antiquark loops, including the loop joining the points 0 and x. The constant g is the coupling constant of the gauge theory. There are further weight factors independent of A. Finally, independently of the quark paths there is another weight factor, namely, the exponential of the free action for the gauge field. The combined weight factor is then averaged over all quark paths and all gauge fields $A_\mu(x)$ to give the current-current propagator.

In order that quarks exist as separate final-state particles it must be possible to have quark-antiquark loops with well-separated quark and antiquark lines, at least when x and 0 are far apart. This is illustrated in Fig. 6(a). If the quark and antiquark paths are unlikely to separate beyond a fixed size, say 10^{-13} cm [see Fig. 6(b)], then clearly no detector will see a quark or antiquark in isolation.

It is assumed in this discussion that vacuum loops are not important. If vacuum loops are important enough then space will be filled with a high *density* of vacuum-produced quark-antiquark pairs. In particular, there will be many quark-antiquark pairs inside a detector of macroscopic size. The question then is whether there can be an excess of quarks over antiquarks, or vice versa, in a region of macroscopic size. This is a more difficult question to answer and will not be discussed in this paper.

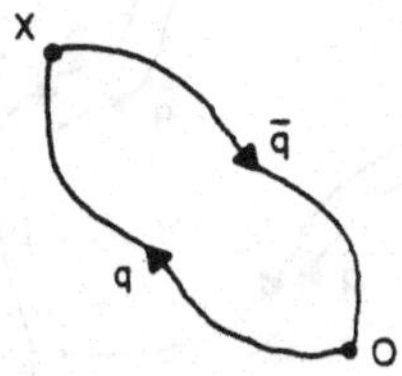

FIG. 3. An example of quark (q) and antiquark ($\bar{q}$) paths connecting the points 0 and x.

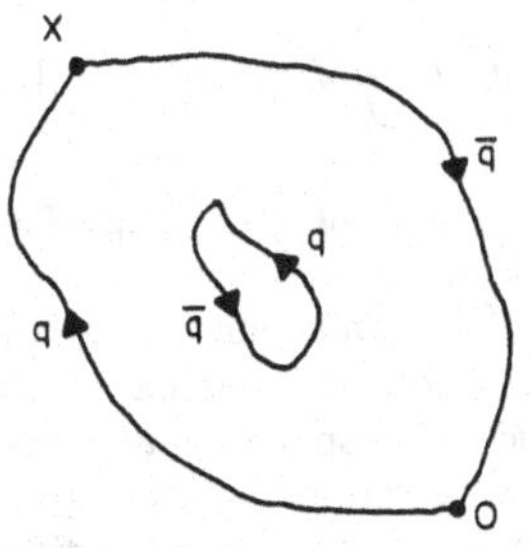

FIG. 4. Example of current loop (as in Fig. 3) with an extra vacuum loop.

Note that x must be large: If x is small there is little likelihood of finding a large size loop. This may seem a bit peculiar: One expects quarks to appear in the final state of e^+-e^- annihilation only at large virtual-photon momentum q if they appear at all, and large q means small x, not large x. The answer to this paradox has two parts. First, the important paths in the Feynman path integral bear no detailed relation to possible physical final states (the paths are paths of bare particles, not physical particles). Secondly, large x does not necessarily mean small q. In fact the study of whether well-separated quark-antiquark paths exist for large x is really a search for a quark-antiquark threshold in e^+-e^- annihilation, which would contribute a term $\sim\exp(2mi\sqrt{x^2})$ to the current-current propagator for *large* x, where m is the quark mass. (Here ~ means up to a power of x^2.) Such a term corresponds in momentum space to the singularity at the threshold $q^2 = (2m)^2$.

Suppose the gauge-field averaging is performed before the quark-paths averaging. Then one determine the average over all gauge fields of the weight factor $\exp[i\oint gA_\mu(y)ds^\mu]$ weighted further with the exponential of the free gauge-field action. For an Abelian gauge theory this average can be computed explicitly: It is

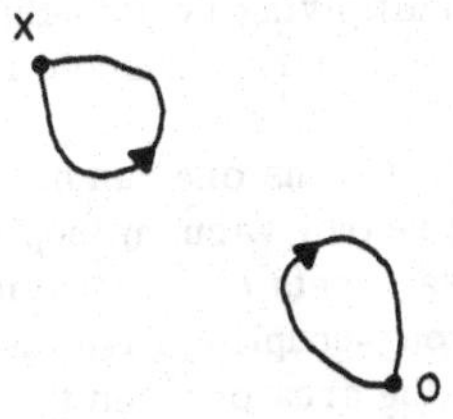

FIG. 5. Example of separate quark loops for the points 0 and x. (Integration over the gauge field produces gauge propagators which connect these loops.)

$$\exp\left[-g^2\oint ds^\mu \oint ds'^\nu D_{\mu\nu}(y-y')\right],$$

where $D_{\mu\nu}(y-y')$ is the free gauge-field propagator.

The quark binding mechanism can be seen by comparing the above expression for one space dimension and three space dimensions. In three space dimensions this calculation gives no binding (the binding occurs only with a modified gauge-field action: see Sec. IV), while there is binding in one space dimension. In three space dimensions $D_{\mu\nu}(y-y')$ behaves as $(y-y')^{-2}$. In consequence large values of $(y'-y)$ are negligible in the double line integral. Hence, the double integral is proportional to P, where P is the length of the loop. Unfortunately the integral is divergent at $y'=y$; a cutoff is needed for the integral to make sense. Since a cutoff will be introduced anyway in this paper, this is not a major concern. For simple loops, the perimeter P is roughly of order $(x^2)^{1/2}$ (ignoring the case that x is close to the light cone). Thus one has an exponential of the type one expects when free quarks are present.

In one space dimension, $D_{\mu\nu}(y-y')$ behaves as $\ln[(y-y')^2]$ for $y'-y$ large, and y' and y can freely range separately over the loop. In this case the double integral is proportional to P^2. Now the gauge-field average behaves as e^{-icP^2}, where c is a constant. In this case the contribution of large loops is heavily suppressed and there are no free quarks. [The case of nearby quark-antiquark pairs as in Fig. 6(b) is special—in this case large $y-y'$ is unimportant due to cancellation between the quark path and the nearby and oppositely directed antiquark path. In this case the double integral behaves as P, not P^2, but in this case there are no isolated quarks.]

In the strongly coupled lattice gauge theory described in later sections, the gauge-field average of $\exp[ig\oint A_\mu(x)ds^\mu]$ behaves as $\exp(ic'A)$, where A is the *enclosed area* of the loop. This heavily suppresses large loops, such as in Fig. 6(a), where A is of order P^2. One can think of one factor P as being roughly $(x^2)^{1/2}$, the other P as being analogous to a mass multiplying $(x^2)^{1/2}$. Since $P\to\infty$ as $x\to\infty$, the quark-antiquark threshold is at infinite mass.

In all these calculations one can have a large loop if there is a nearby vacuum loop (Fig. 7). In this case one always gets $e^{ic''P}$ behavior. For example, in the strong-coupling case the relevant enclosed area is the area between the two loops which is proportional to the perimeter P provided the separation of the two loops is fixed independently of P. This is in accord with Schwinger's picture. While an isolated well-separated loop

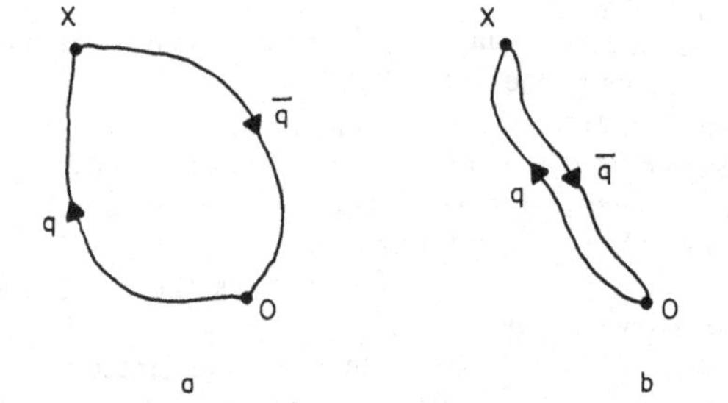

FIG. 6. (a) Loop with well-separated quark and antiquark. (b) Loop with small separation between quark and antiquark.

may be suppressed (due to P^2 or A dependence in the exponential) a loop closely shielded by a vacuum-polarization loop is always unsuppressed.

The binding mechanism proposed here is soft: The exponential damping is associated with large size loops having large areas. The behavior of small loops is irrelevant to the binding mechanism. Also for small loops both their area and perimeter are small and neither is of great importance in an exponential.

The mechanism discussed here works equally well for Dirac quarks or scalar quarks. This is in contrast to the Higgs mechanism which can wipe out the charge of scalar particles only.

III. LATTICE QUANTIZATION OF GAUGE FIELDS

A. Classical action on a lattice

In this section the gauge-field action (space-time integral of the Lagrangian) will be defined on a discrete lattice with spacing a in Euclidean space-time. The simplest way to proceed is to consider a continuum action, substitute finite-difference approximations for derivatives, and replace the space-time integral by a sum over the lattice sites. However, the result of this is an action which is not gauge-invariant for nonzero a. Because of the vagaries of renormalization this is likely to mean that the quantized theory still lacks

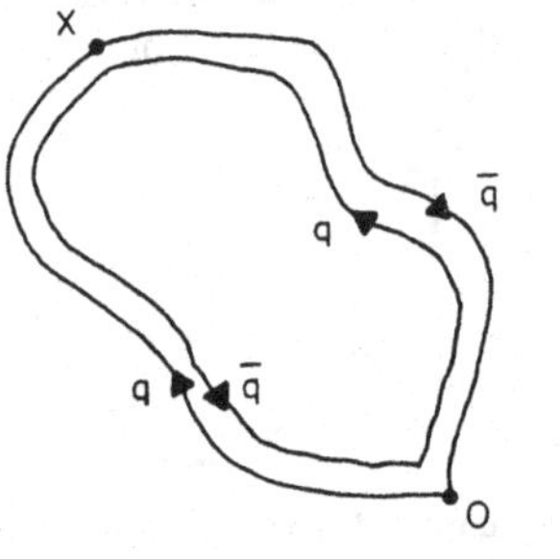

FIG. 7. Quark-antiquark loop with nearby vacuum loop.

gauge invariance in the limit $a \to 0$ (if such a limit exists). The alternative is to formulate gauge invariance for a lattice theory, and tinker with the action so that it is gauge invariant for any a. This alternative will be pursued here.

For convenience a charged Dirac field ψ coupled to a single gauge field A_μ will be discussed in detail. Generalizations to non-Abelian gauge groups will be noted later.

On the lattice the fields are ψ_n, $\bar{\psi}_n$, and $A_{n\mu}$, where n is a four-vector with integer components referring to points on a simple hypercubic lattice. A simple action on the lattice for the Dirac field (ignoring the gauge field for now) has the form

$$A_\psi = -a^4 \sum_n m_0 \bar{\psi}_n \psi_n + \tfrac{1}{2} a^3 \sum_n \sum_\mu \bar{\psi}_n \gamma_\mu (\psi_{n+\hat{\mu}} - \psi_{n-\hat{\mu}}) , \tag{3.1}$$

where m_0 is the bare mass; $\hat{\mu}$ is a unit vector along the axis μ; $a^4 \sum_n$ replaces the space-time integration of the continuum theory, and $(\psi_{n+\hat{\mu}} - \psi_{n-\hat{\mu}})/2a$ replaces $\nabla\psi$. There is no over-all factor of i due to the Euclidean metric. A gauge transformation on the lattice can be defined as follows:

$$\psi_n \to e^{i y_n g} \psi_n , \tag{3.2}$$

$$\bar{\psi}_n \to e^{-i y_n g} \bar{\psi}_n , \tag{3.3}$$

$$A_{n\mu} \to A_{n\mu} - (y_{n+\hat{\mu}} - y_n)/a , \tag{3.4}$$

where g is the coupling constant and y_n is arbitrary. The terms $\bar{\psi}_n \psi_{n+\hat{\mu}}$ and $\bar{\psi}_n \psi_{n-\hat{\mu}}$ are not invariant to this transformation; the corresponding gauge-invariant expressions are $\bar{\psi}_n \psi_{n+\hat{\mu}} \times \exp(iagA_{n\mu})$ and $\bar{\psi}_n \psi_{n-\hat{\mu}} \exp(-iagA_{n-\hat{\mu},\mu})$. Thus, a gauge-invariant form for A_ψ is

$$A_\psi = \tfrac{1}{2} a^3 \sum_n \sum_\mu (\bar{\psi}_n \gamma_\mu \psi_{n+\hat{\mu}} e^{iagA_{n\mu}} - \bar{\psi}_{n+\hat{\mu}} \gamma_\mu \psi_n e^{-iagA_{n\mu}}) - a^4 \sum_n m_0 \bar{\psi}_n \psi_n . \tag{3.5}$$

It is convenient to define

$$B_{n\mu} = agA_{n\mu} . \tag{3.6}$$

In the action A_ψ, the field $B_{n\mu}$ acts like an angular variable: A_ψ is periodic in $B_{n\mu}$ with period 2π. The free gauge-field action will be defined to preserve this property. This does not mean that $A_{n\mu}$ is an angular variable in the continuum limit. Owing to the relation (3.6), $A_{n\mu}$ becomes an infinitesimal angular variable $A_\mu(x)$ for $a \to 0$; such a variable has the range $-\infty < A_\mu(x) < \infty$ without any periodicity.

A gauge-invariant lattice approximation for $\nabla_\mu A_\nu - \nabla_\nu A_\mu$ is

$$F_{n\mu\nu} = (A_{n+\hat{\mu},\nu} - A_{n\nu} - A_{n+\hat{\nu},\mu} + A_{n\mu})/a . \tag{3.7}$$

It is convenient to define a rescaled form of $F_{n\mu\nu}$:

$$f_{n\mu\nu} = a^2 g F_{n\mu\nu} = B_{n\mu} + B_{n+\hat{\mu},\nu} - B_{n+\hat{\nu},\mu} - B_{n\nu} . \tag{3.8}$$

A simple lattice action for the gauge field which preserves periodicity is

$$A_B = \frac{1}{2g^2} \sum_{n\mu\nu} e^{if_{n\mu\nu}} . \tag{3.9}$$

In the continuum limit, $f_{n\mu\nu} \to 0$ due to the factor a^2 in the definition of $f_{n\mu\nu}$. Thus for small a, one can write

$$A_B \simeq \frac{1}{2g^2} \sum_{n\mu\nu} (1 + if_{n\mu\nu} - \tfrac{1}{2} f_{n\mu\nu}^2 - \cdots) . \tag{3.10}$$

The constant term is irrelevant. The linear term in $f_{n\mu\nu}$ is 0 because $f_{n\mu\nu}$ is odd in the indices μ and ν. The quadratic term gives

$$A_B \simeq -\tfrac{1}{4} a^4 \sum_{n\mu\nu} F_{n\mu\nu}^2 , \tag{3.11}$$

which is the conventional gauge-field action in a lattice approximation. The terms involving $f_{n\mu\nu}^3$, $f_{n\mu\nu}^4$, etc. all vanish for $a \to 0$ even after removing a factor a^4 to convert $\sum_n$ into an integral.

The full action may now be written

$$A = -c \sum_n \bar{\psi}_n \psi_n + K \sum_n \sum_\mu (\bar{\psi}_n \gamma_\mu \psi_{n+\hat{\mu}} e^{iB_{n\mu}} - \bar{\psi}_{n+\hat{\mu}} \gamma_\mu \psi_n e^{-iB_{n\mu}}) + \frac{1}{2g^2} \sum_n \sum_{\mu\nu} e^{if_{\mu\nu}} , \tag{3.12}$$

with $c = m_0 a^4$, $K = a^3/2$. This action reduces to the usual continuum action for $a \to 0$; for finite a it is gauge invariant and periodic in the gauge field. Note, however, that the continuum limit is a *classical* limit in which the lattice variables ψ_n, $\bar{\psi}_n$, and $A_{n\mu}$ approach continuum functions $\psi(x)$, $\bar{\psi}(x)$, and $A_\mu(x)$ with $x = na$. The continuum limit of the quantized theory is much harder to discuss owing to renormalization problems.

B. Quantization

The problem of principal interest here is the quantization of the gauge field. Therefore, the gauge field will be quantized by itself to start with. Later the quantization of ψ will be discussed. At the end of this section the generalizations to non-Abelian gauge theories will also be described.

The quantization of the lattice gauge theory will

be carried out in two steps. The first step will be to define a lattice version of Euclidean vacuum expectation values, starting from a lattice version of the Feynman path integral. The second step will be to define a quantum theory on the lattice, which will allow the introduction of a real time variable and the definition of particle states and scattering amplitudes. In both cases the lattice provides an ultraviolet cutoff and there is no Lorentz invariance. Lorentz invariance can only be achieved in the limit a (lattice spacing) $\to 0$, if at all, and in practice this is a difficult limit to evaluate.

As discussed in Sec. I, one would like to calculate the gauge-field average of $\exp[ig\oint A_\mu(x)ds^\mu]$ weighted with the gauge-field action. On a lattice the line integral becomes a sum over a closed path P on the lattice (see, e.g., Fig. 8). The sum has the form: $i\sum_P(\pm)B_{n\mu}$, where a particular $B_{n\mu}$ is present in the sum if the path connects the sites n and $n+\hat{\mu}$ ($-B_{n\mu}$ appears if the path goes from $n+\hat{\mu}$ to n).

On the lattice, an average over all gauge fields involves integrating over all values of the $B_{n\mu}$ for all n and μ. Normally one would have integrals over an infinite range: $-\infty < B_{n\mu} < \infty$, but because of the periodicity in $B_{n\mu}$ there is no point to integrating over more than a single period. Thus the lattice version of the gauge-field average is

$$I(P) = Z^{-1}\left(\prod_m \prod_\nu \int_{-\pi}^{\pi} dB_{m\nu}\right) \times \exp\left(i\sum_P (\pm)B_{n\mu} + \frac{1}{2g^2}\sum_{n\mu\nu} e^{if_{n\mu\nu}}\right), \tag{3.13}$$

with

$$Z = \left(\prod_m \prod_\nu \int_{-\pi}^{\pi} dB_{m\nu}\right) \exp\left(\frac{1}{2g^2}\sum_{n\mu\nu} e^{if_{n\mu\nu}}\right). \tag{3.14}$$

Note that no gauge-fixing term has been added to the action. The finite range of $B_{n\mu}$ makes a gauge-fixing term unnecessary. In continuum gauge theories where $A_\mu(x)$ has an infinite range $[-\infty < A_\mu(x) < \infty]$ a gauge-fixing term is essential to have a convergent functional integral. The reason for this is that the volume in path-integral space generated by all possible gauge transformations is infinite; the gauge-fixing term provides a convergence factor in this volume.[10] In the lattice theory the total volume of integration is finite if the lattice itself is of finite extent; no convergence factor is required. For a lattice of infinite extent there are divergences due to the infinite number of integrations (in other words, the infinite lattice volume), but these divergences are normally removed by the division by Z (this division is equivalent to removing all vacuum loops in perturbation theory).

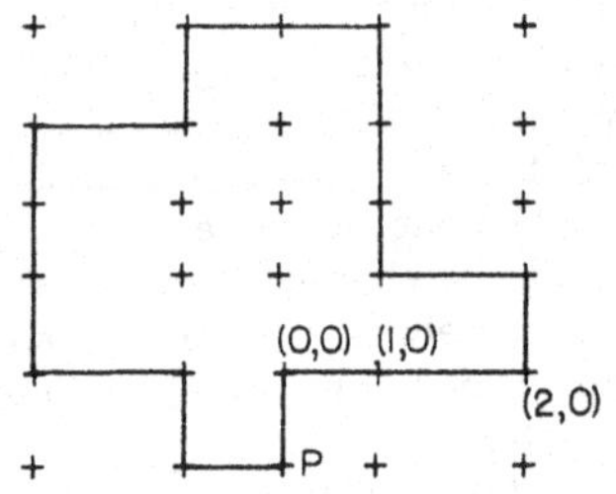

FIG. 8. Example of a lattice path P.

One can define more conventional vacuum expectation values in a similar manner. For instance, one can define a propagator. In the absence of a gauge-fixing term it is awkward to define a propagator for the gauge field $B_{n\mu}$ itself; instead one can define a gauge-invariant propagator as

$$D_{n\mu\pi,\sigma\tau} = Z^{-1}\left(\prod_m \prod_\nu \int_{-\pi}^{\pi} dB_{m\nu}\right) \times \exp\left(if_{n\mu\pi} - if_{0\sigma\tau} + \frac{1}{2g^2}\sum_{m\nu\rho} e^{if_{m\nu\rho}}\right). \tag{3.15}$$

This is a propagator for the operator $e^{if_{n\mu\nu}}$; it is defined only for the lattice points n of a Euclidean space-time lattice. If the lattice spacing is a, this means the propagator is defined only for imaginary times of the form in_0a, where n_0 is an integer.

A theory defined only for discrete imaginary values of the time leaves much to be desired. Fortunately, one can generalize the theory to define a Hamiltonian for a quantized theory. The particle eigenstates and scattering amplitudes of the theory can then be obtained, in principle, by diagonalizing the Hamiltonian. The Hamiltonian will be defined using the transfer-matrix formalism. Only a brief discussion of the transfer-matrix approach will be given here. For a review of the ideas see Wilson and Kogut.[6] A detailed discussion including approximate calculation of single-particle energies and scattering amplitudes in a simple scalar field theory is given by suri.[5]

Consider the expression for Z. Introduce finite bounds on the lattice coordinate n_0, say

$$-N \leq n_0 \leq N. \tag{3.16}$$

Introduce periodic boundary conditions (see below). Then one can write Z as the trace of a matrix V, more precisely

$$Z = \mathrm{Tr}\, V^{2N+1} . \tag{3.17}$$

This formula is made possible by the fact that each term in the action A involves no more than two adjacent values of n_0.

To set up the matrix V, one must first understand the space on which it acts. The space used here is the space of all functions $\psi(B_{\vec{n}i})$ (periodic in each $B_{\vec{n}i}$ with period 2π), where the index i runs from 1 to 3 only, and the lattice variable $\vec{n}$ has only three components (n_1, n_2, n_3). The matrix V will be defined as a function of two sets of arguments, say $B_{\vec{n}i}$ and $B'_{\vec{n}i}$, these two sets of arguments referring to the space-time fields B_{ni} for two adjacent values of n_0. Matrix multiplication of two V's involves integrations over a set of variables $\{B_{\vec{n}i}\}$. Define V as

$$V = \prod_{\vec{m}} \int_{-\pi}^{\pi} dB_{\vec{m}0} e^U . \tag{3.18}$$

The quantity U as written out below looks complicated, but all it is is that part of the action A referring to a given nearest-neighbor pair of values for n_0; terms in A referring to a single value of n_0 are divided equally between the matrices connecting n_0 to n_0+1 and n_0 to $n_0 - 1$. The result is

$$U = \frac{1}{4g^2} \sum_{\vec{n}} \sum_i \sum_j (e^{if'_{\vec{n}ij}} + e^{if_{\vec{n}ij}}) + \frac{1}{2g^2} \sum_{\vec{n}} \sum_i (e^{if_{\vec{n}i0}} + e^{if_{\vec{n}0i}}) , \tag{3.19}$$

where

$$f_{\vec{n}ij} = B_{\vec{n}i} + B_{\vec{n}+\hat{i},j} - B_{\vec{n}+\hat{j},i} - B_{\vec{n}j} \tag{3.20}$$

and

$$f_{\vec{n}i0} = B_{\vec{n}i} + B_{\vec{n}+i,0} - B'_{\vec{n}i} - B_{\vec{n}0} . \tag{3.21}$$

With the definition of V given here, the trace $\mathrm{Tr}\, V^{2N+1}$ is easily seen to reproduce all the integrations involved in the equation for Z, and the sum of the $2N+1$ exponents U reproduces the action A except for some additional terms coupling the boundary $n_0 = N$ to the boundary $n_0 = -N$ to achieve a periodic structure.

Note that the 0 components of $B_{n\mu}$ have received special treatment. This is because there are no terms in A involving B_{n0} for more than one value of n_0; this makes it possible to include integrations over B_{n0} in the definition of V rather than in the definition of matrix multiplication.

The matrix V is used to define the quantized theory. Briefly, this is accomplished as follows. V is a Hermitian matrix, i.e.,

$$V(B, B')^* = V(B', B) . \tag{3.22}$$

[This result can be verified by close examination of Eqs. (3.18)–(3.21). One must remember that the variables $B_{\vec{n}0}$ are integrated out in the definition of V. This means that in forming the complex conjugate of V one can also make the change of variable $B_{\vec{n}0} \to -B_{\vec{n}0}$.] Hence V has a complete orthogonal set of eigenstates Ψ and eigenvalues λ. The Hamiltonian H is now *defined* as follows: The eigenstates of H are the eigenstates of V; the eigenvalues of H are given by

$$E = -a^{-1} \ln\lambda , \tag{3.23}$$

where λ is the corresponding eigenvalue of V. The reason for the factor a^{-1} will be evident shortly. The reason for using the logarithm is so that the energies of multiparticle scattering states will be the sum of single-particle energies (see suri[5]).

A problem arises with this definition if V has any negative eigenvalues λ. If this were to happen, H would have complex eigenvalues. This did not happen in the case studied by suri[5]; whether it happens here the author does not know; this question must be studied further. Even if V has negative eigenvalues, they may be irrelevant in the limit $a \to 0$ if such a limit exists.

The definition (3.23) means that

$$V = e^{-aH} . \tag{3.24}$$

This means V is the operator which propagates a state through an imaginary time ia. It is a consequence of this that the propagator $D_{n\mu\pi,\sigma\tau}$ is a vacuum expectation value for imaginary time in_0a in the theory with Hamiltonian H. For proof of this statement see Refs. 5 and 6.

The lattice quantization procedure can be extended to non-Abelian gauge theories. This is done as follows. In place of the single variable $B_{n\mu}$, one has a set of variables $B^\alpha_{n\mu}$ where α is an internal index. For each n and μ, $B^\alpha_{n\mu}$ is to parametrize an element $b_{n\mu}$ of the gauge group. In place of $\exp(iB_{n\mu})$ one substitutes $U(b_{n\mu})$, where U is the unitary matrix representing $b_{n\mu}$ in the quark representation. The product $\bar{\psi}_n \gamma_\mu \psi_{n+\hat{\mu}} \exp(iB_{n\mu})$ is replaced by $\bar{\psi}_n \gamma_\mu U(b_{n\mu}) \psi_{n+\hat{\mu}}$. A gauge transformation is defined by a set of group transformations y_n; under these transformations

$$\psi_n \to U(y_n)\psi_n , \tag{3.25}$$

$$\bar{\psi}_n \to \bar{\psi}_n U^\dagger(y_n) , \tag{3.26}$$

$$b_{n\mu} \to y_n b_{n\mu} y_{n+\hat{\mu}}^{-1} , \tag{3.27}$$

$$U(b_{n\mu}) \to U(y_n) U(b_{n\mu}) U^\dagger(y_{n+\hat{\mu}}) , \tag{3.28}$$

where $y_n b_{n\mu} y_{n+\hat{\mu}}^{-1}$ is computed according to the multiplication law of the group. A simple gauge-invariant action for the gauge field is

$$A_B = \frac{1}{2g^2} \sum_{n\mu\nu} \mathrm{Tr}\, U_A(b_{n\mu} b_{n+\hat{\mu},\nu} b_{n+\hat{\nu},\mu}{}^{-1} b_{n\nu}{}^{-1}) , \tag{3.29}$$

where the unitary transformation U_A is taken from the adjoint representation of the group. (Any representation will do as well.) In the classical continuum limit, $b_{n\mu}$ becomes an infinitesimal group transformation $[B^{\alpha}_{n\mu} = gaA^{\alpha}_{\mu}(na)$ with A^{α}_{μ} fixed as $a \to 0]$; in this limit one can show, with some effort, that A_B reduces to the standard continuum Yang-Mills action. When the non-Abelian theory is quantized, the integrations over $b_{m\nu}$ are group integrations over all elements of the group. These are compact integrations (for compact groups) and no gauge-fixing term is required.

Since quark binding can be illustrated using Abelian theory, the non-Abelian theory will not be studied further.

Finally, the quantization of the Dirac field will be discussed. It is convenient initially to quantize the Dirac field in an analogous manner to the gauge field. The only problem that occurs is to define "integration" for a Fermi field. This can be done.[11]

The property of the integral that is crucial for quantization is translational invariance in the integration variable. For example, when quantizing a scalar field ϕ on a lattice the field averaging involves the integral $\int_{-\infty}^{\infty} d\phi_n$ which has the translational invariance

$$\int_{-\infty}^{\infty} d\phi_n f(\phi_n + J_n) = \int_{-\infty}^{\infty} f(\phi_n) d\phi_n \tag{3.30}$$

for any integrable function f and any constant J_n. It is this translational invariance that makes the Feynman path integral provide a realization of Schwinger's action principle (see, e.g., Ref. 12 for further discussion). Analogously one needs to define an integration over Fermi fields with the same translational invariance. Stated abstractly, one wants to define a bracket operation $\langle \cdots \rangle$ defined on functions of purely anticommuting Fermi fields ψ_n with the property

$$\langle f(\psi_n + \eta_n, \bar{\psi}_n + \bar{\eta}_n)\rangle = \langle f(\psi_n, \bar{\psi}_n)\rangle , \tag{3.31}$$

where η_n and $\bar{\eta}_n$ are anticommuting c-numbers (these have been introduced by Schwinger). The bracket operation should produce a number for every function f; it should also be a linear operation. Thus for a finite lattice it is sufficient to specify the bracket $\langle \cdots \rangle$ for all monomials in the ψ_n and $\bar{\psi}_n$. Because of the anticommutation rules, $\psi_n{}^2$ and $\bar{\psi}_n{}^2$ vanish (more correctly, $\psi_{n\alpha}{}^2$ and $\bar{\psi}_{n\alpha}{}^2$ vanish where $\psi_{n\alpha}$ and $\bar{\psi}_{n\alpha}$ are any component of ψ_n and $\bar{\psi}_n$), therefore, there are only a finite number of possible monomials. It is now easy to see that the bracket $\langle \cdots \rangle$ must vanish for all products of ψ's and $\bar{\psi}$'s, except the product containing all possible ψ's and $\bar{\psi}$'s. For example, suppose there are two lattice sites 0 and 1, and ψ_n and $\bar{\psi}_n$ are single component fields. Then the brackets must be

$$\begin{aligned} &\langle 1 \rangle = 0 , \\ &\langle \psi_0 \rangle = \langle \psi_1 \rangle = \langle \bar{\psi}_0 \rangle = \langle \bar{\psi}_1 \rangle = 0 , \\ &\langle \psi_0 \psi_1 \rangle = \langle \psi_0 \bar{\psi}_1 \rangle = \cdots = 0 , \\ &\langle \bar{\psi}_0 \psi_0 \psi_1 \rangle = \cdots = 0 , \\ &\langle \bar{\psi}_0 \psi_0 \bar{\psi}_1 \psi_1 \rangle = 1 , \end{aligned} \tag{3.32}$$

where 1 is a constant which was chosen arbitrarily. This definition of the bracket operation satisfies translational invariance: for example,

$$\langle (\bar{\psi}_0 + \bar{\eta}_0)(\psi_0 + \eta_0)(\bar{\psi}_1 + \bar{\eta}_1)(\psi_1 + \eta_1)\rangle = \langle \bar{\psi}_0 \psi_0 \bar{\psi}_1 \psi_1 \rangle = 1$$

because the terms multiplying the η's are all 0. Note also that the anticommutation rules mean that, for example,

$$\langle \psi_0 \bar{\psi}_0 \bar{\psi}_1 \psi_1 \rangle = -1 . \tag{3.33}$$

(In analogy to the scalar case, one requires $\psi_0 \bar{\psi}_0 = -\bar{\psi}_0 \psi_0$, not $\psi_0 \bar{\psi}_0 = 1 - \bar{\psi}_0 \psi_0$.)

One can now define the Feynman path integral on a lattice for the complete gauge theory including the Dirac fields. For example, the current-current propagator on the lattice is

$$D_{n\mu\nu} = Z_{\text{tot}}{}^{-1} \left(\prod_m \prod_\nu \int_{-\pi}^{\pi} dB_{m\nu} \right) \langle \bar{\psi}_n \gamma_\mu \psi_n \bar{\psi}_0 \gamma_\nu \psi_0 e^A \rangle , \tag{3.34}$$

where A is the full action of Eq. (3.12) and

$$Z_{\text{tot}} = \left(\prod_m \prod_\nu \int_{-\pi}^{\pi} dB_{m\nu} \right) \langle e^A \rangle . \tag{3.35}$$

This formulation of the path integral is different from the formulation discussed in Sec. II. However, one can easily derive a lattice form of the path integrals of Sec. II from the present expression. The procedure is to expand Eq. (3.34) in powers of K, where K is the coefficient of the nearest-neighbor coupling terms $\bar{\psi}_n \gamma_\mu \psi_{n+\hat{\mu}} e^{iB_{n\mu}}$ etc. This nearest-neighbor coupling term can be represented diagrammatically by a line from the site n to the site $n+\hat{\mu}$. The expansion is best described by studying an example of a term from the expansion of the numerator of Eq. (3.34), which will now be discussed. An example of a term in the expansion is represented diagrammatically in Fig. 9. The expression for this term is

$$\left(\prod_m \prod_\nu \int_{-\pi}^{\pi} dB_{m\nu}\right) K^4 \langle \bar{\psi}_{11}\gamma_\mu\psi_{11}\bar{\psi}_{00}\gamma_\nu\psi_{00}\bar{\psi}_{00}\gamma_0\psi_{10}e^{iB_{00,0}}\bar{\psi}_{10}\gamma_1\psi_{11}e^{iB_{10,1}}\bar{\psi}_{11}\gamma_0\psi_{01}e^{-iB_{01,0}}\bar{\psi}_{01}\gamma_1\psi_{00}e^{-iB_{00,1}}e^{A_0}\rangle = D \tag{3.36}$$

(D for diagram), where the four lattice sites involved are $(n_0, n_1) = (0, 0)$, $(1, 0)$, $(0, 1)$, and $(1, 1)$; the values of n_2 and n_3 are constant and have been suppressed in the notation of Eq. (3.36). The action A_0 omits the K term, and is

$$A_0 = -c\sum_n \bar{\psi}_n\psi_n + \frac{1}{2g^2}\sum_n \sum_{\mu\nu} e^{if_{n\mu\nu}} . \tag{3.37}$$

The calculation of D has two parts, one being the integration over all gauge fields $B_{m\mu}$, the other being the calculation of the $\psi, \bar{\psi}$ bracket. These are independent calculations, i.e., D factors into $D_\psi D_B$. The quantity D_B is

$$D_B = \left(\prod_m \prod_\nu \int_{-\pi}^{\pi} dB_{m\nu}\right) \exp\left[i(B_{00,0} + B_{10,1} - B_{01,0} - B_{00,1}) + \frac{1}{2g^2}\sum_{n\mu\nu} e^{if_{n\mu\nu}}\right] . \tag{3.38}$$

This is an example of a guage-field average of the exponential of a line integral over a closed loop, the loop being the loop of Fig. 9. The ψ bracket calculation can be factorized further into separate bracket calculations for each lattice site, since A_0 contains no terms involving ψ or $\bar{\psi}$ and coupling different lattice sites. Consider only the four lattice sites on the loop, for simplicity. By moving the ψ's around some (using the anticommuting rule) the bracket becomes

$$\langle \bar{\psi}_{00}\gamma_0\psi_{10}e^{-c\bar{\psi}_{10}\psi_{10}}\bar{\psi}_{10}\gamma_1\psi_{11}e^{-c\bar{\psi}_{11}\psi_{11}}\bar{\psi}_{11}\gamma_\mu\psi_{11}\bar{\psi}_{11}\gamma_0\psi_{01}e^{-c\bar{\psi}_{01}\psi_{01}}\bar{\psi}_{01}\gamma_1\psi_{00}e^{-c\bar{\psi}_{00}\psi_{00}}\bar{\psi}_{00}\gamma_\nu\psi_{00}\rangle .$$

To make a product of all possible ψ's and $\bar{\psi}$'s means one must have products of all possible ψ_{00}'s and $\bar{\psi}_{00}$'s, all possible ψ_{01}'s and $\bar{\psi}_{01}$'s, etc. In summary the complete bracket may be written as a product of four separate brackets. Define

$$D_\psi^1 = \langle \psi_{10}e^{-c\bar{\psi}_{10}\psi_{10}}\bar{\psi}_{10}\rangle , \tag{3.39}$$

$$D_{\psi\mu} = \langle \psi_{11}e^{-c\bar{\psi}_{11}\psi_{11}}\bar{\psi}_{11}\gamma_\mu\psi_{11}\bar{\psi}_{11}\rangle . \tag{3.40}$$

Both D_ψ^1 and $D_{\psi\mu}$ are matrices in spin space due to the spinor indices implied for ψ_{10}, $\bar{\psi}_{10}$, ψ_{11}, and $\bar{\psi}_{11}$. The full bracket is simply

$$D_\psi = -K^4\,\mathrm{Tr}(D_\psi^1\gamma_1 D_{\psi\mu}\gamma_0 D_\psi^1\gamma_1 D_{\psi\nu}\gamma_0) . \tag{3.41}$$

The matrices D_ψ^1 and $D_{\psi\mu}$ are easily determined. For example, D_ψ^1 explicitly is

$$D^1_{\psi\alpha\beta} = \left\langle \psi_{10\alpha}\exp\left(-c\sum_\gamma \bar{\psi}_{10\gamma}\psi_{10\gamma}\right)\bar{\psi}_{10\beta}\right\rangle . \tag{3.42}$$

The exponential can be expanded in powers of c; assuming the spinors have four components only the c^3 term can produce a product of all four ψ_{10}'s times all four $\bar{\psi}_{10}$'s; the result is

$$D^1_{\psi\alpha\beta} = -\delta_{\alpha\beta}(-c)^3 \tag{3.43}$$

(the minus sign comes from the convention that the bracket is positive when $\bar{\psi}_{10\beta}$ appears to the left of $\psi_{10\beta}$). A similar calculation gives

$$D_{\psi\mu} = c^2\gamma_\mu . \tag{3.44}$$

The results of this example are easily generalized. A term of general order K^l is nonzero only if the nearest-neighbor couplings combine to form closed loops (the lattice site at the endpoint of an open line would have an extra ψ_n or $\bar{\psi}_n$ so the bracket at n would give 0). The bracket calculation for a closed loop gives a trace involving K times a γ matrix for each line in the loop and D's for each lattice site in the loop (except the points n and 0 where there are currents). The average over gauge fields involves an exponential of a sum of $B_{n\mu}$'s around each loop. There can be any number of loops.

IV. STRONG-COUPLING APPROXIMATION

The gauge field average $I(P)$ which determines whether quarks are bound was defined on a lattice in Sec. III (Eq. 3.13). There are two limits in which this average can be calculated. The most interesting limit is the strong-coupling limit $g\to\infty$. This is the limit which exhibits quark binding. A strong-coupling expansion will be derived in this section.

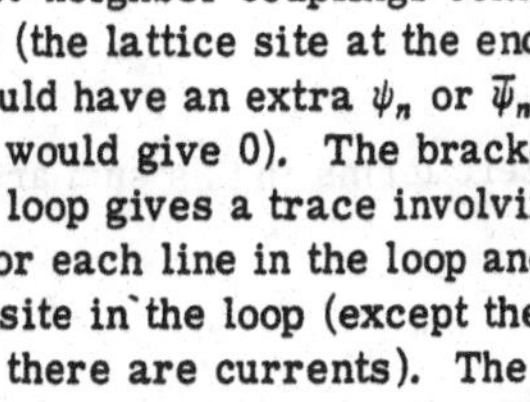

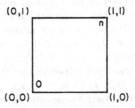

FIG. 9. Elementary square on the lattice.

The strong-coupling expansion will be the basis for a reformulation of the gauge-field theory as a string model. This will also be explained in this section.

Consider specifically the numerator of Eq. (3.13), to be denoted $I_N(P)$:

$$I_N(P)=\left(\prod_m\prod_\nu\int_{-\pi}^{\pi}dB_{m\nu}\right)\times\exp\left[i\sum_P(\pm)B_{n\mu}+\frac{1}{2g^2}\sum_{n\mu\nu}e^{if_{n\mu\nu}}\right]. \qquad (4.1)$$

Expanding in powers of $1/g^2$, the zeroth-order term is

$$I_N^{(0)}(P)=\left(\prod_m\prod_\nu\int_{-\pi}^{\pi}dB_{m\nu}\right)\exp\left[i\sum_P(\pm)B_{n\mu}\right]. \qquad (4.2)$$

This term vanishes, since for any $B_{n\mu}$ which appears in $\sum_P$, there is an integral $\int_{-\pi}^{\pi}dB_{n\mu}\times\exp(\pm iB_{n\mu})$ which is zero. Thus one must seek higher-order terms in g^{-2} which cancel the $B_{n\mu}$ in the line integral. The first-order term is

$$I_N^{(1)}(P)=\frac{1}{2g^2}\left(\prod_m\prod_\nu\int_{-\pi}^{\pi}dB_{m\nu}\right)\sum_{l\pi\sigma}\exp\left[i\sum_P(\pm)B_{n\mu}+if_{l\pi\sigma}\right]. \qquad (4.3)$$

The quantity $f_{l\pi\sigma}$ is itself a line integral of the gauge field; it is the line integral around a square originating at the lattice site l of size a (unit square). The integral for $I_N^{(1)}(P)$ will vanish unless it is possible to find a unit square such that $f_{l\pi\sigma}$ cancels completely the line integral $\sum_P(\pm)B_{n\mu}$. This is possible only if the path P is itself a unit square. Otherwise the first-order term vanishes and one must study the terms of order g^{-4} or higher.

The term of order g^{-2k} has the form

$$I_N^{(k)}(P)=\frac{1}{k!}\left(\frac{1}{2g^2}\right)^k\left(\prod_m\prod_\nu\int_{-\pi}^{\pi}dB_{m\nu}\right)\sum_{l_1\pi_1\sigma_1}\cdots\sum_{l_k\pi_k\sigma_k}\exp\left[i\sum_P(\pm)B_{n\mu}+if_{l_1\pi_1\sigma_1}+\cdots+if_{l_k\pi_k\sigma_k}\right]. \qquad (4.4)$$

The only nonzero terms in this sum are those for which

$$\sum_P(\pm)B_{n\mu}+f_{l_1\pi_1\sigma_1}+\cdots+f_{l_k\pi_k\sigma_k}=0. \qquad (4.5)$$

[See Eq. (3.8) for the definition of $f_{l\pi\sigma}$ in terms of $B_{n\mu}$.] This equation can be understood geometrically. Each $f_{l\pi\sigma}$ corresponds to a square of size a on the lattice. For this sum to vanish the set of squares defined by $f_{l_1\pi_1\sigma_1}\cdots f_{l_k\pi_k\sigma_k}$ must combine to make a surface with boundary P. (To be precise, each $f_{l\pi\sigma}$ corresponds to a line integral around a square, and when these squares are joined to make a surface the line integrals must cancel along all internal lines of the surface. The line integrals along the boundary P of the surface must run in the opposite direction to the original path P.) See Fig. 10.

For a given path P the lowest nonzero order in $I_N(P)$ is determined by the minimal area A enclosed by P, the area A being the area of any surface built of unit squares on the lattice with boundary P. Then $I_N(P)\sim(g^2)^{-A/a^2}$, apart from a numerical factor.

This is the result promised in Sec. II: The gauge-field average $I_N(P)$ behaves as $\exp[-A(\ln g^2)/a^2]$, i.e., exponentially in the area enclosed by P. Hence, according to the arguments of Sec. II, quark paths will not separate macroscopically, and there will be no quarks among the final-state particles.

Consider higher-order terms in the expansion of $I_N(P)$ for given P. There are many such terms because there are many surfaces with boundary P. In particular, there are many ways to combine subsets of f's to add to zero so such subsets can be added to any minimal sum of f's which forms a surface with boundary P. The simplest example of a set of f's which add to zero are the set of f's corresponding to the six faces of a unit cube. Written out, this gives

$$0=f_{n\mu\nu}-f_{n+\hat{\pi},\mu\nu}+f_{n\nu\pi}-f_{n+\hat{\mu},\nu\pi}+f_{n\pi\mu}-f_{n+\hat{\nu},\pi\mu}, \qquad (4.6)$$

which is easily checked using Eq. (3.8). [Equation

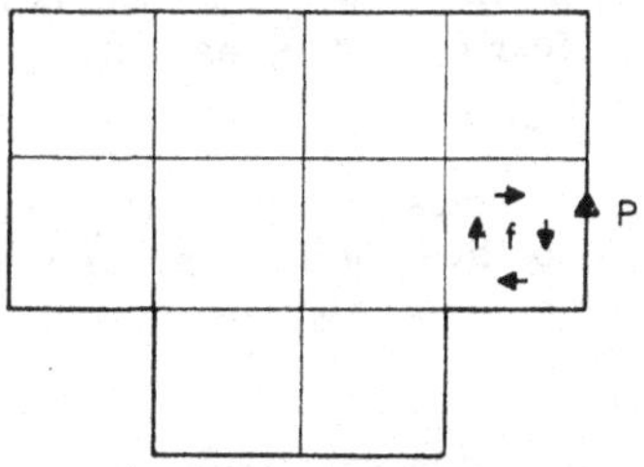

FIG. 10. Filling of enclosed area of path P by elementary squares.

(4.6) is the lattice analog of the equation $\epsilon^{\mu\nu\pi\sigma}\nabla_\pi F_{\mu\nu}(x)=0$.]

Let $A(P)$ be the minimal area as defined above enclosed by P. Since one can place a unit cube anywhere on this minimal area, it means that there are roughly $A(P)/a^2$ more terms of order $g^{-12}(g^2)^{-A(P)/a^2}$ in the expansion of $I_N(P)$ than there are terms of order $(g^2)^{-A(P)/a^2}$. [One can place unit cubes anywhere in space, not just on the minimal surface; but when one divides $I_N(P)$ by Z all disconnected terms cancel, as usual.] This suggests that the $1/g^2$ expansion is not very useful in the limit $A(P)\to\infty$, which is the limit of interest for quark binding. However, experience with related problems suggests that $I_N(P)$ is not the appropriate quantity to expand; instead one should try writing

$$I(P)=Z^{-1}I_N(P)$$

$$=(g^2)^{-A(P)}e^{-c(P,g^2)}\,, \tag{4.7}$$

and expand $c(P,g^2)$ in powers of g^{-2} instead. One would expect $c(P,g^2)$ to be dominated by a term proportional to $A(P)$, say

$$c(P,g^2)=A(P)f(g^{-2})+O(P) \tag{4.8}$$

(where P is the length of the path P). The crucial question is the nature of the series for $f(g^{-2})$. Past experience with similar types of expansions (namely, the high-temperature expansions of statistical mechanics: see, e.g., Ref. 13) suggests that $f(g^{-2})$ will have a convergent expansion at least for g^{-2} less than a critical value g_c^{-2}. However, no calculations have been done in the gauge-field theory for $f(g^{-2})$ as yet.

Consider the complete expansion of $I(P)$. Each nonzero term in the expansion corresponds to a surface with perimeter P. The complete expansion corresponds to a sum over all possible surfaces with given perimeter P. "All possible" surfaces include surfaces which intersect themselves (to take into account terms where a given $f_{n\mu\nu}$ appears several times in the sum $f_{l_1\pi_1\sigma_1}+\cdots+f_{l_k\pi_k\sigma_k}$). There is a weight factor for each surface, aside from the power of g^{-2} determined by the area of the surface. For a simple surface, the weight factor is 1; the weight is more complicated for self-intersecting surfaces.

Thus, the strong-coupling expansion for the current-current propagator has the same general structure as in string models of hadrons. One is actually dealing here with a double expansion. An expansion in the coefficient K (appearing in the Dirac field action) was needed to define quark loops on the lattice; the sum of the K expansion is a sum over all possible quark loops. The g^{-2} expansion is needed to define surfaces filling in the quark loops. The sum of the g^{-2} expansion is a sum over all such surfaces. This is precisely the structure appearing in string models: combined sums over quark loops and interpolating surfaces. However, the loops and surfaces of the gauge-field theory are defined on a lattice whereas the loops and surfaces of the string models are defined on a continuum. It may not be easy to derive quantitative relations between the two types of surfaces.

V. WEAK-COUPLING APPROXIMATION

The weak-coupling approximation will be discussed briefly, leaving many questions open. Only the pure gauge field will be discussed. Consider again the expression

$$I_N(P)=\left(\prod_m\prod_\nu\int_{-\pi}^{\pi}dB_{m\nu}\right)\times\exp\left[i\sum_P(\pm)B_{n\mu}+\frac{1}{2g^2}\sum_{n\mu\nu}e^{if_{n\mu\nu}}\right]. \tag{5.1}$$

Suppose the integration variables were $f_{n\mu\nu}$ rather than $B_{m\nu}$. For small g, only small values of $f_{n\mu\nu}$ would be important in the integral, in order that $\mathrm{Re}\,e^{if_{n\mu\nu}}$ be near its maximum value 1. One would then expand:

$$\frac{1}{2g^2}\sum_{n\mu\nu}e^{if_{n\mu\nu}}\simeq\frac{1}{2g^2}\sum_{n\mu\nu}(1-\tfrac{1}{2}f_{n\mu\nu}{}^2)\,. \tag{5.2}$$

With this approximation one could extend the limits of integration on $f_{n\mu\nu}$ from $\pm\pi$ to $\pm\infty$, with negligible error; one would then have a set of Gaussian integrals to evaluate.

In practice the integration variables are the $B_{m\nu}$, not the $f_{n\mu\nu}$. However, one can make a change of variable from the $B_{m\nu}$ to the $f_{n\mu\nu}$. It is not possible to eliminate all the $B_{m\nu}$ by this transformation, and not all the variables $f_{n\mu\nu}$ are independent. Nevertheless, the transformation is sufficient to make $I_N(P)$ calculable for small g^2.

To make the change of variables precise, consider a system of finite size ($1\le n_i\le N$) with periodic boundary conditions. Then one can change variables from the $B_{m\nu}$ to a subset of the $f_{n\mu\nu}$ plus some gauge transformation variables ϕ_n, plus four extra variables ζ_μ, as follows:

(i) For $n_0\ne N$, n_1, n_2, n_3 arbitrary, $B_{n\mu}$ ($\mu=1,2,3$) is replaced by $f_{n\mu0}$. For B_{n0}, one writes

$$B_{n0}=\phi_{n+\hat{0}}-\phi_n \tag{5.3}$$

and replaces B_{n0} by ϕ_n. This is the essence of the transformation from $B_{n\mu}$ to $f_{n\mu0}$ for $\mu\ne0$ and from B_{n0} to ϕ_n. To complete the transformation one must discuss the surface $n_0=N$.

(ii) For $n_0=N$, $n_1\neq N$, and n_2 and n_3 arbitrary, $B_{n\mu}$ ($\mu\neq 1$) is replaced by $f_{n\mu 1}$. B_{n1} is replaced by ϕ_n, with

$$B_{n1}=\phi_{n+\hat{1}}-\phi_n \,. \tag{5.4}$$

(iii) For $n_0=n_1=N$, $n_2\neq N$, and n_3 arbitrary, $B_{n\mu}$ ($\mu\neq 2$) is replaced by $f_{n\mu 2}$. B_{n2} is replaced by ϕ_n, with $B_{n2}=\phi_{n+\hat{2}}-\phi_n$.

(iv) For $n_0=n_1=n_2=N$, and $n_3\neq N$, $B_{n\mu}$ ($\mu\neq 3$) is replaced by $f_{n\mu 3}$, B_{n3} by ϕ_n, where $B_{n3}=\phi_{n+\hat{3}}-\phi_n$.

(v) For $n_0=n_1=n_2=n_3=N$ one writes $B_{n\mu}=\zeta_\mu$ with ζ_μ being the new variables. One also sets $\phi_n=0$ for this value of n.

The variables $f_{n\mu\nu}$ which are not integration variables (for example, $f_{n\mu\nu}$ with $\mu\neq 0$ and $\nu\neq 0$, for $n_0\neq N$) can be expressed in terms of the independent $f_{n\mu\nu}$ variables using Eq. (4.6); neither the ϕ_n nor ζ_μ appear in these expressions.

When an arbitrary $B_{m\nu}$ is expressed in terms of the new independent variables, one finds (ζ_ν is present only if $m_\nu=N$):

$$B_{m\nu}=\zeta_\nu+\phi_{m+\hat{\nu}}-\phi_m + (\text{linear combination of } f_{n\pi\sigma}) \,, \tag{5.5}$$

i.e., the ϕ variables define a gauge transformation and ζ_ν represents a translation of some of the B's. It is easily verified that the integrand of $I_N(P)$ involves only the f's: It is independent of both the ϕ's and ζ_ν (the latter does not appear because any closed path P has as many $-B_{n\nu}$ terms as $+B_{n\nu}$ terms on the sublattice $n_\nu=N$). Hence the ϕ_n and ζ_ν integrations can be computed trivially. The f integrations are nontrivial because of the constraint (4.6).

What one wants to accomplish is to reduce the lattice theory for small g to something like a conventional free gauge-field theory. This means restoring the $B_{m\nu}$ as the integration variables, but with infinite limits of integration, and with a gauge-fixing term included. Suppose, for example, one starts with

$$I_N'(P)=\left(\prod_m\prod_\nu\int_{-\infty}^{\infty}dB_{m\nu}\right)\exp\left(i\sum_P(\pm)B_{n\mu}-\frac{\alpha}{2}\sum_n\left[\sum_\mu(B_{n\mu}-B_{n-\hat{\mu},\mu})\right]^2-\frac{1}{4g^2}\sum_{n\mu\nu}f_{n\mu\nu}^{\,2}\right)\,, \tag{5.6}$$

where the α term is a lattice version of a $(\nabla_\mu A_\mu)^2$ gauge-fixing term. This integral can be computed by explicit Gaussian integration methods rather more easily than the $f_{n\mu\nu}$ integrations for $I_N(P)$. In addition, $I_N(P)$ can also be reduced to an integral over a subset of the $f_{n\mu\nu}$, using the same change of variables as for $I_N(P)$. The result is different in this case due to the α term which couples the ϕ's to the f's; also there is no convergence factor for the ζ integration. To make $I_N'(P)$ well defined and equal to $I_N(P)$, one must (a) put in a convergence factor for the ζ integral, i.e., a term $-\frac{1}{2}\beta(\sum_n B_{n\mu})^2$ and (b) add a quadratic form in the f's to compensate for the result of the ϕ integration of the gauge-fixing term. The author has not carried through this calculation; but since the net result is still that $I_N(P)=I_N'(P)$ is a Gaussian integration in the B's, the result will presumably be similar to the conventional free-field calculation reported in Sec. II.

VI. PHASE TRANSITIONS

In the strong-coupling limit ($g\to\infty$, $K\to 0$) the gauge theory is far from being Lorentz-invariant. More precisely, since the action was defined on a Euclidean metric, it is Euclidean invariance that is missing. In the strong-coupling limit, vacuum expectation values decrease rapidly at separations of only a few lattice sites (there is a factor g^{-2} or K or both for each unit lattice spacing of separation). This corresponds to the existence of masses much larger than the cutoff. [The usual rule is that if a propagator falls as $e^{-x/\xi}$ for x large then the lowest mass intermediate state contributing to the propagator has mass $1/\xi$. If the propagator behaves as g^{-2n} for distances $x=na$, then the corresponding mass is $2(\ln g)/a$. This is larger than the cutoff momentum π/a if g is large.]

Thus, one is interested in practice in values of g and K such that the correlation length ξ is much larger than the lattice spacing a, in order that the corresponding mass is much less than the cutoff. One knows from statistical mechanics that large correlation lengths are associated with second-order phase transitions (critical points). Thus one seeks special values g_c and K_c for g and K at which there is a phase transition.[14]

It has already been argued that there are two distinct phases for the gauge field, a strong-coupling phase for large g which binds quarks, and a weak-coupling phase for small g which does not bind quarks. The arguments given neglected quark vacuum loops, which is reasonable if K is small. There should be a transition between these two phases which would occur at a critical value g_c for any g and any small value of K. This is one possible phase transition; it is this transition which was discussed in Sec. II. But, as will be argued below, this is probably a first-order transition rather than second order.

Suppose one wishes to construct a model of strong interactions using the lattice theory of this paper with the gauge group separate from ordinary SU(3)×SU(3) symmetry. Then the gauge fields would all be SU(3)×SU(3) singlets, while the quark fields would carry SU(3)×SU(3) quantum numbers as well as gauge-group quantum numbers. A little thought shows that in the strong-coupling limit ($g\to\infty$, $K\to 0$), SU(3)×SU(3) is an exact symmetry rather than a spontaneously broken symmetry. Varying g does not change this situation; so one must hope that by increasing K one can change the exact SU(3)×SU(3) into broken SU(3)×SU(3). If this does not work one is free to introduce additional terms into the quark field action in hopes of forcing a spontaneous breaking of SU(3)×SU(3). Suppose, for simplicity, that SU(3)×SU(3) can be broken by increasing K. Then there will be a phase transition at a critical value K_c for K where one changes from exact SU(3)×SU(3) to spontaneously broken SU(3)×SU(3). If this transition is a second-order transition then there will be a large correlation length for K near K_c; in this case the theory might be a realistic model of broken SU(3)×SU(3) for K slightly greater than K_c (with g large enough to maintain quark binding).

In summary, the transition of real interest is a transition in K (or some other parameter introduced into the quark action) rather than g.

Apart from special limits ($g\to\infty$ and $K\to 0$, or g small) it is very difficult to solve the lattice theory. It is especially difficult to solve the lattice theory near a critical point with a large correlation length. Various methods have been developed by statistical mechanicians to deal with this problem. In the remainder of this section these methods will be discussed briefly. There are essentially three approaches to consider: (1) mean-field techniques, (2) series expansions, and (3) the renormalization-group approach.

Mean-field techniques[15] are the simplest and crudest methods for studying a critical point; invariably they are the methods one uses first in studying a new situation. They are used to determine if there is a phase transition, whether it is first or second order, and to give rough estimates of the behavior near the transition. None of the results of a mean-field calculation are entirely trustworthy. Examples of mean-field calculations will be given later.

An example of a series expansion would be the expansion of the current-current propagator for small momentum (momentum $\ll 1/a$) in powers of g^{-2} and K, to high order in g^{-2} and K. One then uses Padé-approximant techniques to look for singularities in either g or K that would be associated with a mass approaching 0. In simple statistical-mechanical problems one can generate 12 terms or so in analogous expansions. The expansion for the lattice theory of this paper is more complicated, but one could hope to generate maybe 6 or 7 orders with some practice. Series expansions require considerably more effort than mean-field calculations; they apply mainly to propagators, being very awkward to perform on three- and four-point functions, and one must have a clear idea of what one is trying to learn before attempting such calculations. See Ref. 16 for one of the best series-expansion formalisms; see Ref. 13 for a general review.

The renormalization-group approach is potentially the most powerful and accurate method for studying lattice theories near a critical point, but at present the renormalization-group techniques are too limited in scope to be applicable to the present problem. See Refs. 6 and 17.

Return to mean-field ideas.[15] The prototype mean-field calculation is a calculation of the magnetization as a function of the external field for an Ising ferromagnet. Let s_n be the spin at site n with values ±1 only; let the interaction be

$$\frac{-H}{kT} = K\sum_n \sum_\mu s_n s_{n+\hat{\mu}} + h\sum_n s_n \,, \tag{6.1}$$

where K is related to the spin-spin coupling and h is proportional to the external field. Then

$$M = Z^{-1}\left\langle s_0 \exp\left(K\sum_n\sum_\mu s_n s_{n+\hat{\mu}} + h\sum_n s_n\right)\right\rangle \,, \tag{6.2}$$

where $\langle\cdots\rangle$ means a sum over all configurations of all spins, and

$$Z = \left\langle \exp\left(K\sum_n\sum_\mu s_n s_{n+\hat{\mu}} + h\sum_n s_n\right)\right\rangle \;. \tag{6.3}$$

In the mean-field approximation, one assumes that the spins $s_{\hat{\mu}}$ coupled to s_0 can be replaced by their average value M. As a result, the formula for M simplifies to a sum over s_0 only, namely

$$M = Z_0^{-1}\sum_{s_0=\pm 1} s_0 e^{(2dKM+h)s_0} \,, \tag{6.4}$$

with d being the dimensionality (3 usually) and

$$Z_0 = \sum_{s_0=\pm 1} e^{(2dKM+h)s_0} \;. \tag{6.5}$$

The result is

$$M = \tanh(2dKM + h) \;. \tag{6.6}$$

If $2dK<1$ this equation has a unique solution for M as a function of h; in particular, $M=0$ for $h=0$. For $2dK>1$ the solution is multiple-valued; stability considerations show that one must choose a solution with $M\neq 0$ when $h=0$.

In this approximation one has actually replaced $\sum_\mu s_{\hat{\mu}}$ by $2dM$, which is a good approximation if d is large. This is generally true of mean-field theories.

An analogous mean-field calculation can be performed for the lattice gauge theory. In this case a simple external field term has the form

$$h\sum_{n\mu}(e^{iB_{n\mu}}+e^{-iB_{n\mu}})$$

(this is to be added to the gauge-field action), and M can be defined to be the expectation value of $e^{iB_{0\mu}}$. The question is whether M is zero in the limit $h\to 0$. In the gauge-field theory $e^{iB_{0\mu}}$ couples to a product of three other exponentials; as a mean-field approximation one replaces this product by M^3. The result of this is that

$$M=Z_0^{-1}\int_{-\pi}^{\pi}dB_{0\mu}e^{iB_{0\mu}} \times\exp\left[\left(\frac{2(d-1)}{2g^2}M^3+h\right)(e^{iB_{0\mu}}+e^{-iB_{0\mu}})\right], \tag{6.7}$$

where d is the space-time dimensionality, and

$$Z_0=\int_{-\pi}^{\pi}\exp\left[\left(\frac{2(d-1)}{g^2}M^3+h\right)(e^{iB_{0\mu}}+e^{-iB_{0\mu}})\right]. \tag{6.8}$$

The result of this is that

$$M=f\left(\frac{(d-1)}{g^2}M^3+h\right), \tag{6.9}$$

where f is a ratio of Bessel's functions. If g is large the solution to this equation is unique and $M=0$ for $h=0$. If g is small then there are solutions with $M\neq 0$ for $h=0$, and stability considerations show again that the $M\neq 0$ solutions are preferred.

In the magnetic case, one finds that the spontaneous magnetization M goes to zero for $2dk\to 1$ [from Eq. (6.6)]. However, the gauge-field case never has a solution for $h=0$ with M small but nonzero. Thus there is a first-order transition at the value of g for which M changes from zero to being nonzero.

A nonzero value of M in the limit $h\to 0$ means one has spontaneous breaking of the gauge-field symmetry. So for small g the theory shows spontaneous breaking.

A much more thorough discussion of the mean-field approximation has been given by Balian, Drouffe, and Itzykson.[18] A Hamiltonian formulation of the lattice gauge theory has been given by Kogut and Susskind.[19] A clear review of quark confinement in the lattice theory is given in Ref. 20. Another formulation of the connection between strongly coupled gauge theories and string models is given in Ref. 21.

ACKNOWLEDGMENTS

The author has benefited from conversations with many people in developing and especially in understanding the lattice gauge theory. Persons I am indebted to include J. Bjorken, R. P. Feynman, M. E. Fisher, D. Gross, J. Kogut, L. Susskind, T.-M. Yan, and members of a seminar at Orsay.

*Work supported in part by the National Science Foundation.

[1]See, e.g., J. M. Cornwall and R. E. Norton, Phys. Rev. D 8, 3338 (1973); R. Jackiw and K. Johnson, *ibid.* 8, 2386 (1973); J. Kogut and L. Susskind, *ibid.* 9, 3501 (1974); D. Amati and M. Testa, Phys. Lett. 48B, 227 (1974); G. 't Hooft, Nucl. Phys. B (to be published); P. Olesen, Phys. Lett. 50B, 255 (1974); A. Chodos, R. L. Jaffe, K. Johnson, C. B. Thorn, and V. F. Weisskopf, Phys. Rev. D 9, 3471 (1974).

[2]J. Schwinger, Phys. Rev. 125, 397 (1962); 128, 2425 (1962).

[3]J. H. Lowenstein and J. A. Swieca, Ann. Phys. (N.Y.) 68, 172 (1971).

[4]A. Casher, J. Kogut, and L. Susskind, Phys. Rev. Lett. 31, 792 (1973); Phys. Rev. D 10, 732 (1974).

[5]A. suri, Ph.D. thesis, Cornell University, 1969 (unpublished).

[6]K. G. Wilson and J. Kogut, Phys. Rep. 12C, 75 (1974). Sec. X.

[7]S. Coleman and E. Weinberg, Phys. Rev. D 7, 1888 (1973).

[8]Y. Nambu, in *Symmetries and Quark Models*, proceedings of the International Conference on Symmetries and Quark Models, Wayne State Univ., 1969, edited by Ramesh Chand (Gordon and Breach, New York, 1970); L. Susskind, Nuovo Cimento 69A, 457 (1970); G. Konisi, Prog. Theor. Phys. 48, 2008 (1972); P. Goddard, J. Goldstone, C. Rebbi, and C. B. Thorn, Nucl. Phys. B56, 109 (1972); J.-L. Gervais and B. Sakita, Phys. Rev. Lett. 30, 716 (1973); S. Mandelstam, Nucl. Phys. B64, 205 (1973).

[9]R. P. Feynman, Phys. Rev. 80, 440 (1950).

[10]See, e.g., E. S. Abers and B. W. Lee, Phys. Rep. 9C, 1 (1973).

[11]See, e.g., F. A. Berezin, *The Method of Second Quantization* (Academic, New York, 1966), pp. 52ff.

[12]K. G. Wilson, Phys. Rev. D 6, 419 (1972).

[13]M. E. Fisher, Rep. Prog. Phys. 30, 615 (1967).

[14]For the relation of relativistic field theory to critical

points see Ref. 6, Sec. X, and references cited therein.

[15]For a review see R. Brout, *Phase Transitions* (Benjamin, New York, 1965), p. 8.

[16]D. Jasnow and M. Wortis, Phys. Rev. 176, 739 (1968).

[17]K. G. Wilson, Cargèse (1973) Lecture Notes, in preparation.

[18]R. Balian, J. M. Drouffe, and C. Itzykson, Phys. Rev. D (to be published).

[19]J. Kogut and L. Susskind, Phys. Rev. D (to be published).

[20]K. Wilson, Cornell Report No. CLNS-271 (to be published in the proceedings of the conference on Yang-Mills Fields, Marseille, 1974).

[21]H. B. Nielsen and P. Olesen, Nucl. Phys. B61, 45 (1973); L. J. Tassie, Phys. Lett. 46B, 397 (1973).

Volume 59B, number 1 PHYSICS LETTERS 13 October 1975

COMPACT GAUGE FIELDS AND THE INFRARED CATASTROPHE

A.M. POLYAKOV
Landau Institute for Theoretical Physics, Moscow, USSR

Received 19 August 1975

It is shown that infrared phenomena in the gauge theories are guided by certain classical solutions of the Yang-Mills equations. The existence of such solutions can lead to a finite correlation length which stops infrared catastrophe. In the present paper we deal only with theories with a compact but abelian gauge group. In this case the problems of correlation length and charge confinement are completely solved.

It was pointed out by different authors [1] several years ago that the infrared phenomena, occurring with a gauge field, might provide a natural explanation for the confinement of quarks. At the same time there exist no methods for analyzing the interaction of gauge fields in the deep infrared region. It is the purpose of the present paper to work out a formalism which permits, at least partly, to take into account the infrared effects in gauge-field interactions. Our main idea is that the system of gauge fields acquires a finite correlation length through the following phenomenon.

Imagine that we are calculating a certain correlation function in the euclidean formulation of the gauge theory. This means averaging over all possible fields A_μ with the weight equal to:

$$\exp\{-S(A)\} = \exp\left\{-\frac{1}{4g^2}\,\mathrm{Sp}\int F_{\mu\nu}^2\, \mathrm{d}^4x\right\} \tag{1}$$

$$F_{\mu\nu} = \partial_\mu A_\nu - \partial_\nu A_\mu + [A_\mu, A_\nu]\,.$$

Assume that the charge $g^2 \ll 1$; then the leading role in the averaging will be played by the fields close to that defined by the equation:

$$\frac{\delta S}{\delta \bar{A}_\mu(x)} = 0; \quad S[\bar{A}] < \infty. \tag{2}$$

Usually one takes into account only the trivial minima of S, i.e. $A_\mu = 0$, and developes the perturbation theory as a small deviation from this. For the correlation function with the distance R the parameter of the perturbation expansion is $g^2 \log R/a$ where a is the inverse cut-off. Hence, for very large R, perturbation theory is not applicable and another $\bar{A}$ might become essential. Indeed, though the weight with which non-trivial minima enters the averaging is small being proportional to

$$\exp\{-S(\bar{A})\} = \exp(-E/g^2) \tag{3}$$

(where E is certain constant) their influence on the cor relation is large if the classical field $\bar{A}$ is long ranged. (In fact, the contribution to the correlation will be shown to be proportional to $\exp\{-E/g^2\}R^4$).

Now assume that the fields $\bar{A}_\mu$ are such as if they were produced by certain "particles" in the four-dimensional euclidean space. In other words there exist the "one-particle" minima of S, the "two-particle" and so on. Of course, the "energy" E depends on the number of the above mentioned pseudo-particles. The average density of pseudo-particles in our system is very small, being proportional to $\exp(-E/g^2)$. However, their existence creates long range random fields in our system. Due to these random fields, the correlation length becomes finite. This is precisely the phenomena we are going to investigate.

The above discussion was based on the crucial assumption that there exists pseudo-particle solutions of the gauge field equations. It will be proved in the second paper of this series that such solutions indeed exist for every compact nonabelian gauge group.

In this first paper we confine ourselves to the problem of realizing the above program in the case of compact but abelian gauge fields. The purpose of this consideration is two fold. First, it is a good and simple model for trying our program on. Second, the compactness of quantum electrodynamics seems to be an attractive hypothesis and our results may have physical applications. For example we shall prove the existence of a certain critical charge in QED.

Volume 59B, number 1 PHYSICS LETTERS 13 October 1975

The definition of the theory is as follows. Let us introduce a lattice in the four dimensional space, necessary in the definition of functional integrals. Generally, the action should have the form:

$$S = \sum_{x,\mu,\nu} f(F_{x,\mu\nu}) \qquad (3)$$

$$F_{x,\mu\nu} = A_{x,\mu} + A_{x+a_\mu,\nu} - A_{x+a_\nu,\mu} - A_{x,\nu}$$

where a_μ is lattice vector,

$$f(x) \underset{x\to 0}{\approx} \frac{1}{4g^2} x^2.$$

The hypothesis of the compactness of the gauge group means that $A_{x,\mu}$ are the angular variables, and the group is the circle and not the line. This is equivalent to the hypothesis that:

$$f(x + 2\pi) = f(x). \qquad (4)$$

Gauge theories on the lattice have been considered earlier by Wilson [2] and the present author (unpublished). See also [3].

The immediate consequence of the periodicity of $f(x)$ is that the nearest neighbours $A_{x+a_\lambda,\mu}$ and $A_{x,\mu}$ can be different by $2\pi N$ (where N is integer) without producing large action. Hence, in the continuous limit $F_{\mu\nu}$ may have the following singularities:

$$F_{\mu\nu}(x) = F_{\mu\nu}^{\text{reg}} + 2\pi \sum_i N_{i\mu\nu}\, \delta^{(S_i)}(x) \qquad (5)$$

where $\delta^{(S)}(x)$ is the surface δ-function. The second term in (5) will not contribute to the action, due to the periodicity.

It will be convenient for us to analyze first the three dimensional theory. In this case there exists quasiparticle solutions of Maxwell equations which simply coincide with the Dirac monopole solution. If we introduce the field:

$$F_\alpha \equiv \tfrac{1}{2}\epsilon_{\alpha\beta\gamma} F_{\beta\gamma} \qquad (6)$$

then the general pseudo-particle solution will be given by:

$$F_\alpha = \sum_a \frac{q_a}{2} \cdot \frac{(x - x_a)_\alpha}{|x - x_a|^3} - 2\pi\delta_{\alpha 3} \sum q_a \theta(x_3 - x_{3a})\, \delta(x - x_{1a})\, \delta(x - x_{2a}). \qquad (7)$$

If $\{q_a\}$ are integers then the singularities in (7) are just of the permitted type.

The action is given by:

$$S(\bar{A}) = E/g^2$$
$$E = \frac{\pi}{2} \sum_{a\neq b} \frac{q_a q_b}{|x_a - x_b|} + \epsilon \sum q_a^2 \qquad (8)$$

(the value of the constant ϵ depends on the lattice type and is not essential for us).

Now let us analyze the correlation function introduced in [2] which is most convenient in the confinement problem:

$$F(C) \equiv \exp\{-W(C)\} = \langle \exp\{i \oint_C A_\mu\, dx_\mu\}\rangle \qquad (9)$$

(here C is some large contour).

For the evaluation of (9) let us substitute $A_\mu = \bar{A}_\mu + a_\mu$. Since the integral over a_μ is gaussian we get:

$$F(C) = F_0(C) \frac{\Sigma \exp\{-S(\bar{A})\} \exp\{i \oint \bar{A}_\mu\, dx_\mu\}}{\Sigma \exp\{-S(\bar{A})\}} \qquad (10)$$

(Here F_0 is the contribution of $\bar{A} = 0$).

The sum in (10) goes over all possible configurations of pseudo-particles. Now, let us use the formula:

$$\exp\{i \oint A_\mu\, dx_\mu\} = \exp\{i \int F_\alpha\, d\sigma_\alpha\} \qquad (11)$$

in which, due to the periodicity of the exponent, only the first term from (7) should be substituted.

The problem is reduced now to the calculation of the free energy of the monopoles plasma with the "temperature" g^2 in the external field:

$$\varphi^c(x) = i \frac{\partial}{\partial x_\alpha} \int \frac{d\sigma_\alpha}{|x - y|}. \qquad (12)$$

This problem was solved by using Debye method which is correct for sufficiently small g^2. The result is two fold. First, there exist the Debye correlation length and the corresponding photon mass m equal to:

$$m^2 = \exp\{-\epsilon/g^2\} \qquad (13)$$

(in the units of the inverse lattice length).

Secondly:

$$W[C] = \text{const}(g^2 mA) \qquad (14)$$

where A is the area of the contour C. Eq. (14) was derived for arbitrary planar contour. According to

Wilson [1] this result means "charge confinement" in the three dimensional QED with the compact gauge group.

In the case of the four dimensional QED it can be shown that the only classical solutions with finite action are closed rings. This follows from the fact that singular points in this case should form lines. To prove this, assume that it is not so, and consider the pseudo particle solution with $x = 0$. Consider the cube K with $x_4 = 0$. Then it should be:

$$\oint F_{\mu\nu}\, \mathrm{d}\sigma_{\mu\nu} = 2\pi q \qquad (15)$$

But, after small variation of the x_4, our pseudo-particle will be outside the cube, and this contradicts (15).

Since the closed rings produce only dipole forces their influence on the correlation are rather weak. We showed that in this case the correlation length remains infinite and that

$$W[C] = \text{const} \exp(-B/g^2)\cdot L \qquad (16)$$

where L is the length of the contour C, and B is some constant. This result means the absence of the charge confinement for small g^2. Since it was proved in [2] that for large g^2 the charge confinement exist there are some critical charge g_c^2 at which the phase transition occurs. It is not clear now whether this critical charge is connected with the fine structure constant.

The extension of the above ideas on the nonabelian theory will be presented in the other papers of this series.

References

[1] S. Weinberg, D. Gross and F. Wilezek, Phys. Rev. D8 (1973) 3633; Phys. Rev. Lett. 31 (1973) 494.
[2] K. Wilson, Phys. Rev. D10 (1974) 2445.
[3] R. Balian et al., Phys. Rev. D10 (1974) 3376.

Nuclear Physics B72 (1974) 461–473. North-Holland Publishing Company

A PLANAR DIAGRAM THEORY FOR STRONG INTERACTIONS

G. 't HOOFT
CERN, Geneva

Received 21 December 1973

Abstract: A gauge theory with colour gauge group U(N) and quarks having a colour index running from one to N is considered in the limit $N \to \infty$, g^2N fixed. It is shown that only planar diagrams with the quarks at the edges dominate; the topological structure of the perturbation series in $1/N$ is identical to that of the dual models, such that the number $1/N$ corresponds to the dual coupling constant. For hadrons N is probably equal to three. A mathematical framework is proposed to link these concepts of planar diagrams with the functional integrals of Gervais, Sakita and Mandelstam for the dual string.

1. Introduction

The question we ask ourselves in this paper is how to construct a field theory of strong interactions in which quarks form inseparable bound states. We do not claim to have a satisfactory solution to that problem, but we do wish to point out some remarkable features of certain (gauge) field theories that make them an interesting candidate for such a theory.

First we have the singular infra-red behaviour of massless gauge theories [1] that makes it impossible to describe their spectra of physical particles by means of a perturbation expansion with respect to the coupling constant. It is not inconceivable that in an infra-red unstable theory long range forces will accumulate to form infinite potential wells for single quarks in hadrons.

The Han-Nambu quark theory [2] gives a qualitative picture of such forces between quarks: a very high, or infinite, energy might be required to create a physical state with non-zero "colour" quantum number. It is natural to take the symmetry corresponding to this quantum number to be a local gauge symmetry of some group SU(N). In that case, a formal argument in terms of functional integrals has been given by Amati and Testa [3] that supports the conjecture that "coloured" states have infinite energy.

In this paper we put the emphasis on an interesting coincidence. If we consider the parameter N of the colour gauge group SU(N) as a free parameter, then an expansion of the amplitudes at $N \to \infty$ arranges the Feynman diagrams into sets which have exactly the topology of the quantized dual string with quarks at its ends. The analogy with the string can be pursued one step further by writing the planar diagrams in the light cone reference frame. In sect. 6, we write down a Hamiltonian that generates all planar diagrams, in a Hilbert space of a fixed number of quarks. The quarks are inseparable if and only if the spectrum of this Hamiltonian becomes discrete in the presence of the interactions.

2. U (N) gauge theory

In order to show that the set of planar diagrams may play a leading rôle if certain physical parameters have certain values, we first formulate a possible gauge theory for strong interactions in which the parameters N and g have arbitrary values.

The quarks p_i, n_i and λ_i form three representations of the group U(N); $i = 1, \ldots, N$. Let us assume that an observer can distinguish between p, n and λ, but that he cannot distinguish the different colour components (see also sect. 3) *.

There is an anti-Hermitian gauge (vector) field

$$A_i{}^j{}_\mu(x) = -A^*{}_j{}^i{}_\mu(x), \tag{2.1}$$

and the Lagrangian is

$$\mathcal{L} = \tfrac{1}{4}\, G_{\mu\nu i}{}^j\, G_{\mu\nu j}{}^i - \bar{q}^{ai}(\gamma_\mu D_\mu + m_{(a)})\, q^a{}_i\,, \tag{2.2}$$

where

$$\begin{aligned} G_{\mu\nu i}{}^j &= \partial_\mu A_i{}^j{}_\nu - \partial_\nu A_i{}^j{}_\mu + g\,[A_\mu, A_\nu]_i{}^j\,;\\ D_\mu q^a{}_i &= \partial_\mu q^a{}_i + g\,A_i{}^j{}_\mu\, q^a{}_j\,. \end{aligned} \tag{2.3}$$

The index a runs from one to three

$$q^1 = p;\quad q^2 = n;\quad q^3 = \lambda. \tag{2.4}$$

For sake of simplicity we do not make the restriction that the trace of the gauge field, $A^i_{i\mu}$ should vanish, and so we will have a photon corresponding to the Abelian subgroup U(1) of U(N), and coupling to baryon number. Of course we could dispose of it, either by replacing U(N) by SU(N), or when we switch on weak and electromagnetic interactions through the Higgs mechanism. But for the time being it is there and we must keep it in mind when we finally interpret the results of our calculations.

The Feynman rules [4, 5] may be formulated as usual in any suitable gauge. Let us take the Feynman gauge. We add to the Lagrangrian

$$\tfrac{1}{2}\,\partial_\mu A_i{}^j{}_\mu\,\partial_\nu A_j{}^i{}_\nu - \partial_\mu\,\phi_j^{*i}\,(\partial_\mu \phi_i{}^j + g\,[A_\mu, \phi]_i{}^j), \tag{2.5}$$

* We do not know whether this assumption is really essential for the theory, but it does simplify the arguments in sect. 3.

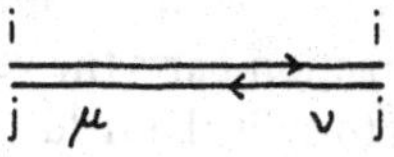

$-\dfrac{\delta_{\mu\nu}}{k^2 - i\epsilon}$ (vector field)

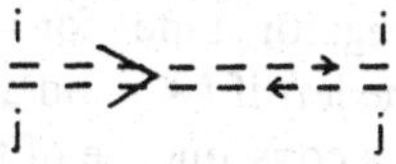

$\dfrac{1}{k^2 - i\epsilon}$ (F.P. ghost)

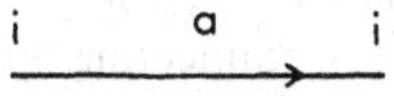

$\dfrac{1}{m_{(a)} + i\gamma k - i\epsilon}$ (quark q_i^a)

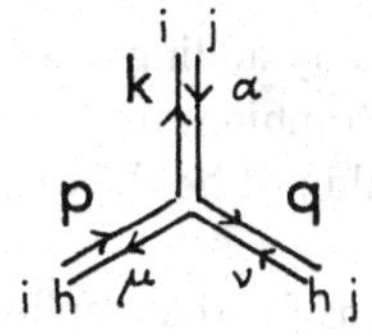

$ig\{\delta_{\alpha\nu}(k-q)_\mu + \delta_{\alpha\mu}(p-k)_\nu + \delta_{\mu\nu}(q-p)_\alpha\}$

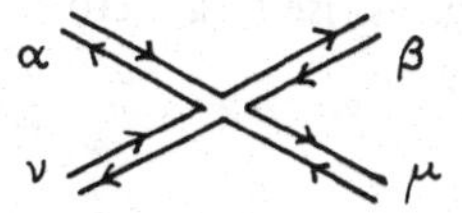

$g^2\{2\delta_{\alpha\mu}\delta_{\beta\nu} - \delta_{\alpha\beta}\delta_{\mu\nu} - \delta_{\beta\mu}\delta_{\alpha\nu}\}$

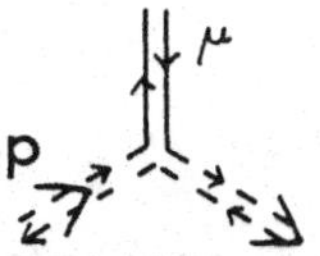

igp_μ

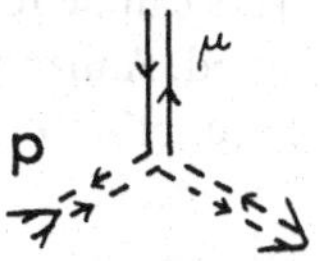

$-igp_\mu$

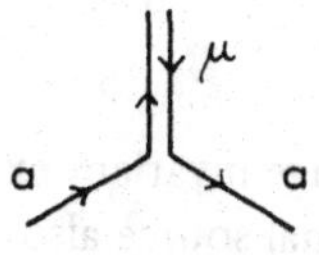

$-g\gamma_\mu$

-1 (Fermi statistics)

Fig. 1. Feynman rules for U(N) gauge theory in Feynman gauge.

where ϕ is the Feynman-DeWitt-Faddeev-Popov ghost field. Now the bilinear parts of the Lagrangian generate the propagators and the interaction parts the vertices.

In order to keep track of the indices, it is convenient to split the fields $A_{\mu i}{}^j$ into complex fields for $i > j$ and real fields for $i = j$. One can then denote an upper index by an incoming arrow, and a lower index by an outgoing arrow. The propagator is then denoted by a double line. In fig. 1, the vector propagator stands for an $A_{\mu i}{}^j$ propagator to the right if $i > j$; an $A_{\mu j}{}^i$ propagator to the left if $i < j$ and a real propagator if $i = j$. The extra minus sign in this propagator is a consequence of the anti-Hermiticity of the field A (eq. (2.1)). The ghost fields satisfy no Hermiticity condition and therefore their propagators have an additional arrow (fig. 1).

The vertices always consist of Kronecker delta functions connecting upper and lower indices, and thus connect ingoing with outgoing arrows. The quark propagators consist of a single line.

As usual, amplitudes and Green functions are obtained by adding all possible (planar and non-planar) diagrams with their appropriate combinatory factors. Note now that the number N does not enter in fig. 1 (this would not be the case if we would try to remove the photon).

But, of course, the number N will enter into expressions for the amplitudes, and that is when an index-line closes. Such an index loop gives rise to a factor

$$\sum_i \delta_i{}^i = N.$$

3. The $N \to \infty$ limit

In sect. 2 we assumed that the observer is colour-blind. This can be formulated more precisely: only gauge-invariant quantities can be measured. A measuring apparatus can formally be represented by a c number source function $J(x)$ which is coupled to a gauge invariant current, for instance

$$\sum_i \bar{p}^i n_i . \tag{3.1}$$

We observe from fig. 2 that index lines never stop at a gauge invariant external source, but they continue. "Index loops" going through an external source also obtain a factor N, because of the summation in (3.1).

We are now in the position that we can classify the diagrams with gauge invariant sources according to their power of g and their power of N. Let there be given a connected diagram. First we consider the two-dimensional structure obtained by attaching little surfaces to each index loop. We get a big surface, with edges formed by the quark lines, and which is in general multiply connected (contains "worm holes"). We close the surface by also attaching little surfaces to the quark loops separately.

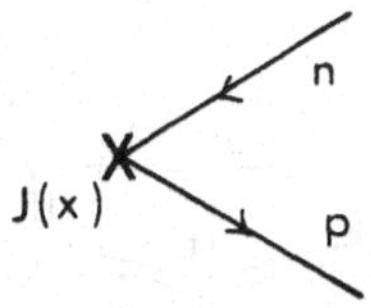

Fig. 2. Gauge invariant source function.

Let that surface have F faces, P internal lines or propagators, and V vertices. Here $F = L + I$, where L is the number of quark loops and I the number of index loops; and $V = \Sigma_n V_n$, where V_n is the number of n-point vertices. The diagram is associated with a factor

$$r = g^{V_3 + 2V_4} N^I . \qquad (3.2)$$

By drawing a dot at each end of each internal line, we find that the number of dots is

$$2P = \sum_n n V_n , \qquad (3.3)$$

and eq. (3.2) can be written as

$$r = g^{2P - 2V} N^{F-L} . \qquad (3.4)$$

Now we apply a well-known theorem of Euler:

$$F - P + V = 2 - 2H, \qquad (3.5)$$

where H counts the number of "holes" in the surface and is therefore always positive (a sphere has $H = 0$, a torus $H = 1$, etc.). And so,

$$r = (g^2 N)^{\frac{1}{2} V_3 + V_4} N^{2-2H-L} . \qquad (3.6)$$

Suppose we take the limit

$$N \to \infty, \quad g \to 0, \quad g^2 N = g_0^2 \text{ (fixed)}. \qquad (3.7)$$

If the sources are coupled to quarks, then there must be at least one quark loop: $L \geqslant 1$. The leading diagrams in this limit have $H = 0$ and $L = 1$, they are the planar diagrams with the quark line at the edges (fig. 3).

Note, however, that the above arguments not only apply to gauge fields but also to theories with a global U(N) symmetry containing fields with two U(N) indices, but from the introduction, it will be clear why we concentrate mainly on gauge fields.

It is interesting to compare our result with that of Wilson [6], who considers gauge fields on a dense lattice and also finds structures with the topology of a two-

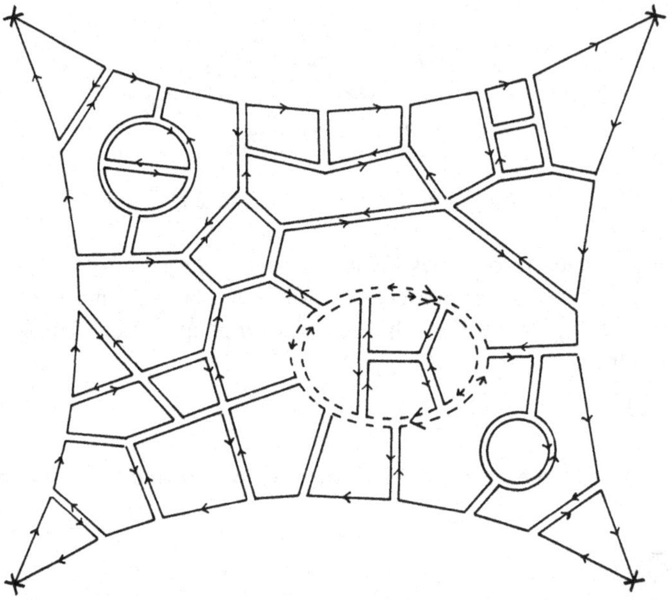

Fig. 3. One of the leading diagrams for the four-point function.

dimensional surface. It is not difficult to show that also Wilson's surfaces are associated with factors $1/N^2$ and $1/N$ for each worm hole or fermion loop, respectively.

The dual topology of the set of planar diagrams has been noted before [7]. Here we see that the analogy with dual models goes even further; the expansion in powers of $1/N$ corresponds to the expansion with respect to the dual coupling constant in

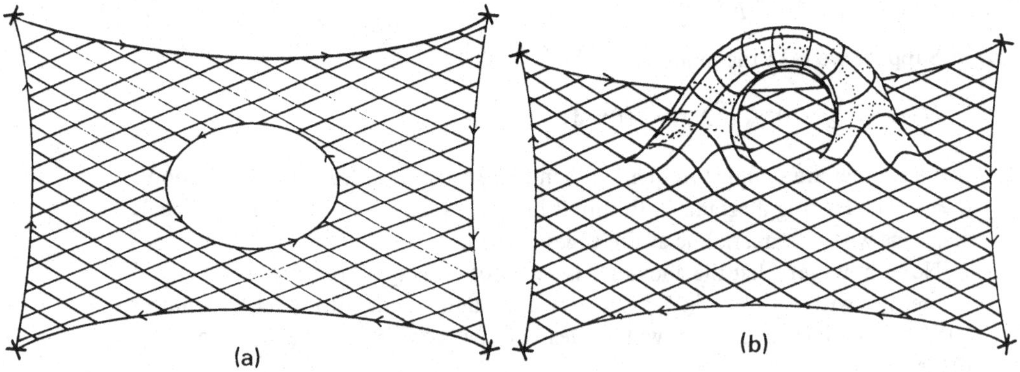

Fig. 4. Two diagrams of higher order in $1/N$: (a) obtain a factor $1/N$, (b) obtain a factor $1/N^2$, as compared with the lowest-order graphs of the previous figure.

dual models. If we adopt the Han-Nambu picture of hadrons [2] then N is very likely to be three. This seems to give a reasonable order of magnitude for the dual coupling constant.

Let us now formulate our theory more precisely. We assume that there is a local gauge group of the type U(3), (or SU(3)) for which no preferred reference frame in the form of a Higgs field exists. Such a theory is infra-red unstable [1] which implies that infra-red divergences accummulate instead of cancel, and the physical spectrum is governed by long range forces. A simple-minded perturbation expansion with respect to g_0 in eq. (3.7) does not describe the spectrum and the S-matrix. But the $1/N$ expansion may be a reasonable perturbation expansion, in spite of the fact that N is not very big.

4. Planar diagrams in the light-cone frame

The theory implies that we have to sum all planar diagrams in order to get the leading contributions to the amplitudes. Attempts to calculate certain large planar diagrams are known in the literature [7] but it seems to us that the choice of diagrams there is rather arbitrary, and the replacement of a propagator by Gaussian expressions seems to be a bad approximation. We believe that a more careful study of this problem is necessary.

Let us consider any large planar diagram (fig. 5). For a moment we shall abandon the rather complicated Feynman rules of fig. 1, replacing the vertices by simple local ϕ^3 or ϕ^4 interactions.

We immediately face two problems:

(i) how to find a convenient parametrization scheme to indicate a point of the graph in the plane, in terms of two parameters σ and τ;

(ii) how to arrive at Gaussian integrands, in order to be able to do the integrations.

These two problems can be solved simultaneously by going to light-cone co-ordinates: we write [8]

$$p^{\pm} = \frac{1}{\sqrt{2}}(p^3 \pm p^0), \quad \tilde{p} = (p^1, p^2),$$

$$x^{\pm} = \frac{1}{\sqrt{2}}(x^3 \pm x^0), \quad \tilde{x} = (x^1, x^2). \tag{4.1}$$

Although the gauge particles are massless, we shall consider the slightly more general case of arbitrary masses. The propagators are then

$$\frac{1}{(2\pi)^4 i} \frac{1}{(\tilde{p}^2 + 2p^+p^- + m^2 - i\epsilon)}, \tag{4.2}$$

(for sign conventions, see ref. [5]). We go over to a mixed momentum coordinate representation: at each vertex $V_{(\alpha)}$ we perform an integration over its time co-ordi-

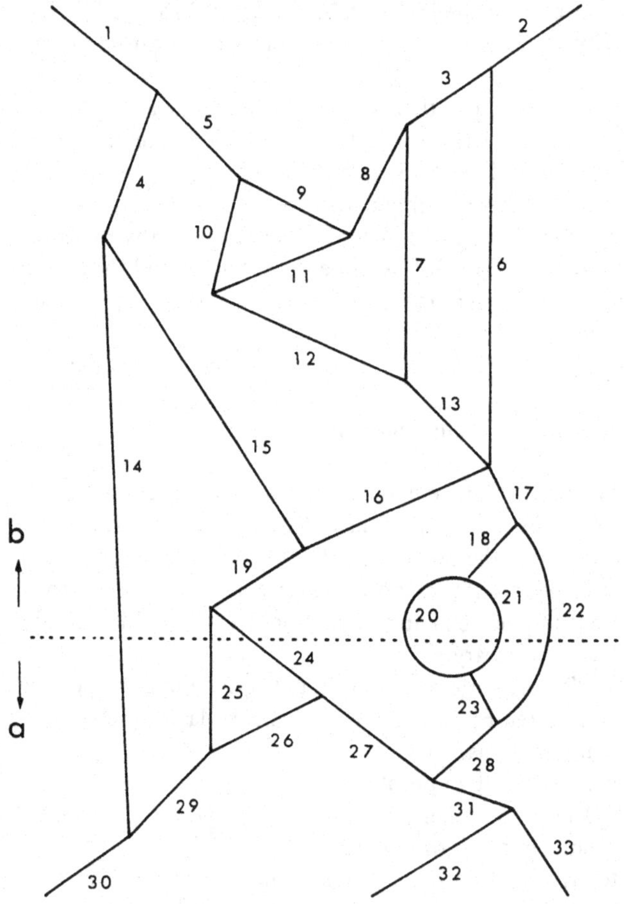

Fig. 5. Example of a planar diagram, divided into two regions a and b (see text).

nate $x^+_{(\alpha)}$, and to each window $F^{(i)}$ of the graph corresponds an integration over the momenta $\tilde{p}_{(i)}$ and $p^+_{(i)}$ (always directed anti-clockwise). In terms of these variables the propagator is the Fourier transform of (4.2) with respect to p^-,

$$\frac{1}{(2\pi)^3 2|p^+|}\,\theta\,(x^+p^+)\exp - i\,\frac{x^+}{2p^+}\,(m^2+\tilde{p}^2)\,. \tag{4.3}$$

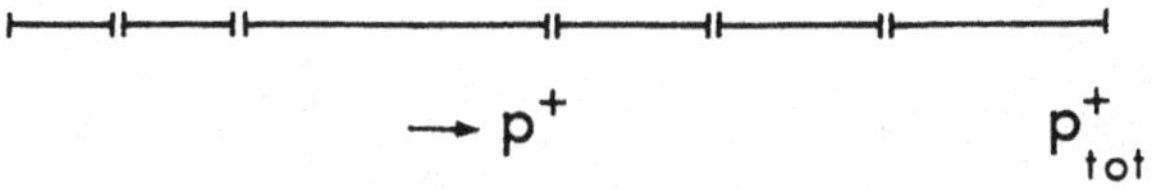

Fig. 6. The components p^+ of the momenta of the propagators that cross the dotted line in fig. 5.

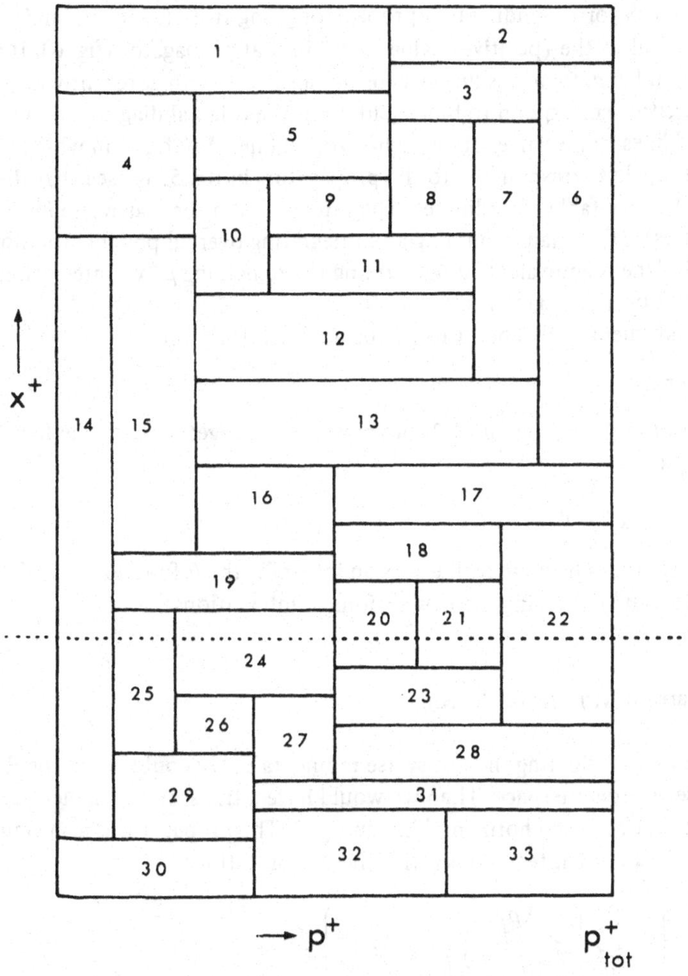

Fig. 7. A new representation of the same diagram. The blocks here correspond to the propagators in fig. 5 and have been numbered accordingly.

Here x^+ is the time difference between the two end points of the line, and $(\tilde{p}, p^+)$ are the difference of the momenta $(\tilde{p}_{(i)}, p^+_{(i)})$ circulating in the windows at both sides of the line. Note that the propagator (4.3) is Gaussian in the transverse momenta $\tilde{p}$.

The parametrization problem can be solved by exploiting the famous θ function in (4.3). For simplicity we shall assume that all external lines with positive p^+ lie next to each other in the plane†. If we divide the set of vertices into: (a) those with $x^+_{(\alpha)} < a$, and (b) those with $x^+_{(\alpha)} > a$, then all lines * going from (a) to (b) have

† If this condition is not fulfilled the resulting plane of fig. 7 will get several "sheets".

* If we want to keep the diagram planar while dividing it into blobs (a) and (b), then we must expect lines going from (a) through (b) back to (a), etc. But it is easy to convince oneself that in those cases the diagram is zero as a consequence of the θ functions.

positive p^+. Now imagine a horizontal line with length p^+_{total} and divide it into segments, each corresponding to a propagator going from (a) to (b). and with a length equal to the (positive) value of p^+ in that propagator (fig. 6). If we now vary the number a, then this line sweeps out a surface with constant width, in which the propagators correspond to blocks; loops in the original diagrams now correspond to vertical lines, and vertices are now horizontal lines. See fig. 7 in which we numbered the blocks corresponding to the propagators in fig. 5. We see that the variables p^+ and x^+ are suitable co-ordinates. The integration in the transverse momenta (or co-ordinates) is Gaussian. Summing and integrating over all possible topologies in the p^+x^+ plane is equivalent to performing the remaining p^+x^+ integrations and the summations over the diagrams.

It is convenient at this point to perform a Wick rotation.

$$i x^+ = \tau . \tag{4.4}$$

The factor i in the exponent (4.3) now disappears, together with the factors i at each vertex:

$$(2\pi)^3 \, i \, \lambda \, \mathrm{d} x^+ \to (2\pi)^3 \, \lambda \, \mathrm{d} \tau , \tag{4.5}$$

and all amplitudes become real, Gaussian integrals (the θ function in (4.3) now becomes $\theta\,(\tau p^+)$, defining the new regions of integrations).

5. Comparison with the dual string

Instead of considering the transverse momenta $\tilde{p}$, we could study the diagrams in transverse coordinate space. Then we would have a transverse variable $\tilde{x}$ at each vertex of fig. 5, or at each horizontal line in fig. 7. The propagator is also Gaussian in terms of the $\tilde{x}$. The integrand is (after the Wick rotation)

$$C \exp \left\{ - \sum_{ij} \left| \frac{\Delta p^+_{ij}}{2\,(\tau_i - \tau_j)} \right| (\tilde{x}_i - \tilde{x}_j)^2 \right\} . \tag{5.1}$$

where C is independent of the transverse variables, and the summation is performed over all pairs of adjacent horizontal lines in fig. 7. Δp^+_{ij} stands for the width of the block between i and j.

This is to be compared with (the essential part of) the functional integrand for the quantized string:

$$C \exp - \int \mathrm{d}\sigma \, \mathrm{d}\tau \left[\left(\frac{\partial \tilde{x}}{\partial \sigma} \right)^2 + \left(\frac{\partial \tilde{x}}{\partial \tau} \right)^2 \right] , \tag{5.2}$$

where $\tilde{x}\,(\sigma, \tau)$ is now a continuous variable on a similar rectangular surface [9]. The difference between (5.1) and (5.2) is profound. The first difference is that in eq. (5.1) we have a partition of the dual surface into meshes, and secondly in (5.1) one must also integrate over all longitudinal variables and sum over all diagrams. This integration and summation together correspond to the summation over all partitions into meshes. The detailed structure of the meshes will depend on the initial Feynman rules, and from those it will probably depend whether (5.1) can be approximated by (5.2) in any way. If so, then the dual string will be an approximate solution of the dynamical equations of our gauge model.

6. A Hamiltonian formalism

Attempts to attain more understanding of the peculiarities of planar diagram field theory have failed until now. There exists, however, a Hamiltonian for this system that might be useful. For simplicity, we confine ourselves to the planar diagrams of ϕ^3 theory (again defined by means of a certain $N \to \infty$ limit). A representation of states $|\psi\rangle$ in a Hilbert space is defined as a set of structures like in fig. 6: a number of "particles" is sitting on a line segment with length p^+_{total}. They have coordinates p^+_i, $i = 1, \ldots, r; r = 0, 1, \ldots$. A transverse loop integration momentum $\tilde{p}_i$ is assigned to each particle (the particles in fig. 6 actually correspond to loops in the original diagram). We put $\tilde{p}_{\text{total}} = 0$, so that $\tilde{p} = 0$ on the boundaries at the left and at the right.

A Wick rotation is not necessary here, so we can take x^+ to be real. The x^+ axis is divided into small segments $x^+_0, x^+_1, \ldots, x^+_n$, with

$$x^+_{k+1} - x^+_k = \epsilon. \tag{6.1}$$

Now we write the amplitude formally as

$$A = {}_{x^+_n}\langle \text{out} \,|\, \mathrm{e}^{-i\epsilon H} \,|\, \psi\rangle_{x^+_{n-1}} \; \langle \psi \,|\, \mathrm{e}^{-i\epsilon H} \,|\, \psi\rangle_{x^+_{n-2}} \; \langle \psi \,|\, \ldots \qquad \ldots |\text{in}\rangle_{x^+_0}\,, \tag{6.2}$$

where summation and integration over the intermediate states is understood. We now construct the Hamiltonian H that will yield the sum of all planar diagrams. Expand

$$\mathrm{e}^{-i\epsilon H} = 1 - i\epsilon H = 1 - i\epsilon\,(H_0 + H_1), \tag{6.3}$$

where H_0 will be taken to be diagonal in the above-defined representation. If no vertex occurs between x^+_k and x^+_{k+1} then only H_0 contributes to

$${}_{x^+_{k+1}}\langle \psi | \mathrm{e}^{-i\epsilon H} | \psi \rangle_{x^+_k}\,.$$

Taking

$$H_0 = \sum_i \frac{m_i^2 + (\tilde{p}_i - \tilde{p}_{i-1})^2}{2\,(p^+_i - p^+_{i-1})}\,, \tag{6.4}$$

we get the correct exponential parts of the propagators (compare (4.3)).

At the vertices (horizontal lines in fig. 7), our particles are created or annihilated. Here H_1 is in action. Let us define operators $a^\dagger(\tilde{p}, p^+)$ and $a(\tilde{p}, p^+)$, with

$$[a^\dagger(\tilde{p}, p^+), a(\tilde{k}, k^+)] = \delta^2(\tilde{p} - \tilde{k})\,\delta(p^+ - k^+), \tag{6.5}$$

creating respectively annihilating particles. We can then take $H_1 = V + V^\dagger$, with

$$V = -(2\pi)^3 \lambda (16\pi^3)^{-\frac{3}{2}} \int dp^+ \int d^2\tilde{p}\, \frac{a(\tilde{p}, p^+)}{\sqrt{(p_r^+ - p_l^+)(p_r^+ - p^+)(p^+ - p_l^+)}} \tag{6.6}$$

where λ is the coupling constant; p_r^+ and p_l^+ are the coordinates of the closest neighbours at the right and at the left of the point p^+.

Substituting this interaction Hamiltonian into (6.3) and (6.2), we find exactly the Feynman rules for planar diagrams: the square root of the width of each block in fig. 7 always occurs twice, thus giving rise to the required factor $1/p^+$ in the propagator (4.3).

In our gauge theory model, a similar Hamiltonian will describe one quark and one antiquark in interaction. If our theory is to describe hadrons, then its spectrum should come out to be discrete.

7. Conclusion

We are still far away from a satisfactory theory for bound quarks. But, guided by the topological structure of the dual theories, we are led to the planar diagram field theory, in terms of which our problem can easily be formulated: if the eigenstates of a certain Hamiltonian crystallize into a discrete spectrum, despite the fact that the zeroth order Hamiltonian is continuous, then the original particles will condensate into a string that keeps quarks together.

As for baryons, the situation is even more complicated. The Han-Nambu theory clearly suggests $N = 3$. In that case we can raise or lower indices in the following way:

$$\lambda_i \to \lambda^{ij} = \epsilon^{ijk} \lambda_k = -\lambda^{ji}. \tag{7.1}$$

Taking p_i, n_i and λ^{ij} as our elementary fermions we can again consider the $N \to \infty$ limit. The λ quark will then sit in the middle of a string with p and/or n quarks at its ends: we have a string with Σ or Λ baryons! Similarly protons, neutrons and all other baryons can be constructed.

It will be clear that in the case of baryons the $1/N$ expansion is extremely delicate.

If calculations will be possible at all in this theory, then the dual coupling constant will be calculable and of order $\frac{1}{3}$.

References

[1] G. 't Hooft, Marseille Conf., 19–23 June 1972, unpublished;
H.D. Politzer, Phys. Rev. Letters 30 (1973) 1346;
D.J. Gross and F. Wilczek, Phys. Rev. Letters 30 (1973) 1343;
D.J. Gross and F. Wilczek, NAL preprint, Princeton preprint (1973);
S. Coleman and D.J. Gross, Princeton preprint (1973);
G. 't Hooft, Nucl. Phys. B61 (1973) 455; B62 (1973) 444.
[2] M.Y. Han and Y. Nambu, Phys. Rev. 139B (1965) 1006.
[3] D. Amati and M. Testa, CERN preprint TH. 1770 (1973).
[4] G. 't Hooft and M. Veltman, Nucl. Phys. B50 (1972) 318.
[5] G. 't Hooft and M. Veltman, "DIAGRAMMAR", CERN report 73–9 (1973).
[6] K.G. Wilson, Lecture given in Orsay (August 1973), unpublished.
[7] H.B. Nielsen and P. Olesen, Phys. Letters 32B (1970) 203;
B. Sakita and M.A. Virasoro, Phys. Rev. Letters 24 (1970) 1146.
[8] S.-J. Chang and S.-K. Ma, Phys. Rev. 180 (1969) 1506.
[9] J.-L. Gervais and B. Sakita, Phys. Rev. Letters 30 (1973) 716;
S. Mandelstam, Nucl. Phys. B64 (1973) 205.

Volume 59B, number 1 PHYSICS LETTERS 13 October 1975

PSEUDOPARTICLE SOLUTIONS OF THE YANG-MILLS EQUATIONS

A.A. BELAVIN, A.M. POLYAKOV, A.S. SCHWARTZ and Yu.S. TYUPKIN
Landau Institute for Theoretical Physics, Academy of Sciences, Moscow, USSR

Received 19 August 1975

We find regular solutions of the four dimensional euclidean Yang-Mills equations. The solutions minimize locally the action integrals which is finite in this case. The topological nature of the solutions is discussed.

In the previous paper by one of the authors [1] the importance of the pseudoparticle solutions of the gauge field equations for the infrared problems was shown. By "pseudoparticle" solutions we mean the long range fields A_μ which minimize locally the Yang-Mills actions S and for which $S(A) < \infty$. The space is euclidean and four-dimensional. In the present paper we shall find such a solution. Let us start from the topological consideration which shows the existence of the desired solutions.

All fields we are interested in satisfy the condition:

$$F_{\mu\nu} = \partial_\mu A_\nu - \partial_\nu A_\mu + [A_\mu, A_\nu] \underset{x\to\infty}{\to} 0\,. \tag{1}$$

Consider a very large sphere S^3 in our 4-dimensional space. The sphere itself is of course 3-dimensional. From (1) it follows that

$$A_\mu|_{S^3} \approx g^{-1}(x) \left.\frac{\partial g(x)}{\partial x_\mu}\right|_{S^3} \tag{2}$$

where $g(x)$ are matrices of the gauge group. Hence every field $A_\mu(x)$ produce a certain mapping of the sphere S^3 onto the gauge group G. It is clear that if two such mappings belong to different homotopy classes then the corresponding fields $A_\mu^{(1)}$ and $A_\mu^{(2)}$ cannot be continuously deformed one into another. It is well known [2] that there exists an infinite number of different classes of mappings of $S^3 \to G$ if G is a nonabelian simple Lie group. Hence, the phase space of the Yang-Mills fields are divided into an infinite number of components, each of which is characterized by some value of q, where q is a certain integer.

Our idea is to search for the absolute minimum of the given component of the phase space. In order to do this we need the formula expressing the integer q through the field A_μ ‡. It is easy to check that

$$q = \frac{1}{8\pi^2}\, \epsilon_{\mu\nu\lambda\gamma} \operatorname{Sp} \int F_{\mu\nu} F_{\lambda\gamma}\, d^4x. \tag{3}$$

To prove this let us use the identity:

$$\epsilon_{\mu\nu\lambda\gamma} \operatorname{Sp} F_{\mu\nu} F_{\lambda\gamma} = \partial_\alpha J_\alpha$$
$$J_\alpha = \epsilon_{\alpha\beta\gamma\delta} \operatorname{Sp}(A_\beta(\partial_\gamma A_\delta + \tfrac{2}{3} A_\gamma A_\delta)). \tag{4}$$

From (4) follows:

$$q = \frac{1}{8\pi^2} \oint_{S^3} J_\alpha\, d^3\sigma^\alpha$$
$$= \frac{1}{8\pi^2}\, \tfrac{4}{3}\, \epsilon_{\alpha\beta\gamma\delta} \oint \operatorname{Sp}(A_\beta A_\gamma A_\delta)\, d^3\sigma_\alpha \tag{5}$$

where

$$A_\mu = g^{-1}(x)\partial g/\partial x_\mu \tag{6}$$

Now consider the case $G = SU(2)$. In this case it is clear that:

$$d\mu(g) = \operatorname{Sp}(g^{-1}dg \times g^{-1}dg \times g^{-1}dg) \tag{7}$$

is just the invariant measure on this group, since it is the invariant differential form of the appropriate dimension. The meaning of the notation in (7) is as follows. Let $g(\xi_1 \xi_2 \xi_3)$ be some parametrization of SU(2), say, through the Euler angles. Then the invariant measure will be:

‡ Formulas like (3) are known in topology by the name of "Pontryagin class".

$$d\mu = \mathrm{Sp}\left(g^{-1}\frac{\partial g}{\partial \xi_1}\, g^{-1}\frac{\partial g}{\partial \xi_2}\, g^{-1}\frac{\partial g}{\partial \xi_3}\right) d\xi_1 \dots d\xi_3 \,. \quad (8)$$

Comparing (8) with (5) we see that the integrand in (5) is precisely the Jacobian of the mapping of S^3 on SU(2). Hence q is the number of times the SU(2) is covered under this mapping. It is just the definition of the mapping degree. In the case of the arbitrary group G one should consider the mapping of S^3 on its SU(2) subgroup and repeat the above. There exists an important inequality which will be extensively used below. Consider the following relation:

$$\mathrm{Sp}\int (F_{\mu\nu} - \widetilde{F}_{\mu\nu})^2 d^4x \geqslant 0 \quad (9)$$

where $\widetilde{F}_{\mu\nu} = \frac{1}{2}\epsilon_{\mu\nu\lambda\gamma}F_{\lambda\gamma}$. From (9) and (3) it follows that:

$$E \geqslant 2\pi^2 |q| \quad (10)$$

where

$$S(A) \equiv E(A)/g^2$$

and g^2 is a coupling constant.

The formula (10) gives the lower bound for the energy of the quasiparticles in each homotopy class. We shall show now that for $q = 1$ this bound can be saturated. In other words one can search the solution of the equation, which replace the usual Yang-Mills one:

$$F_{\alpha\beta} = \pm\tfrac{1}{2}\epsilon_{\alpha\beta\gamma\delta}F_{\gamma\delta}$$
$$F_{\alpha\beta} = \partial_\alpha A_\beta - \partial_\beta A_\alpha + [A_\alpha A_\beta] \,. \quad (11)$$

Again it is sufficient to consider the case $G = \mathrm{SU}(2)$. In this case it is convenient though not necessary to extend this group up to $\mathrm{SU}(2) \times \mathrm{SU}(2) \approx \mathrm{O}(4)$. The gauge fields for O(4) are $A_\mu^{\alpha\beta}$ where A_μ are antisymmetric on $\alpha\beta$. The SU(2) gauge field are connected with $A_\mu^{\alpha\beta}$ by the formulas:

$$\pm A_\mu^i = \tfrac{1}{2}(A_\mu^{oi} \pm \tfrac{1}{2}\epsilon_{ikl}A_\mu^{kl}). \quad (12)$$

Now, two equations:

$$\pm F_{\mu\nu}^i = \pm\tfrac{1}{2}\epsilon_{\mu\nu\lambda\gamma} \pm F_{\lambda\gamma}^i$$

are equivalent to the following one:

$$\epsilon_{\alpha\beta\gamma\delta}F_{\mu\nu}^{\gamma\delta} = \epsilon_{\mu\nu\lambda\gamma}F_{\lambda\gamma}^{\alpha\beta}. \quad (13)$$

Let us search the solution of (13) which is invariant under simultaneous rotations of space and isotopic space. The only possibility is:

$$A_\mu^{\alpha\beta} = f(\tau)(x_\alpha\delta_{\mu\beta} - x_\beta\delta_{\mu\alpha}). \quad (14)$$

It is easy to calculate F:

$$F_{\mu\nu}^{\alpha\beta} = (2f - \tau^2 f^2)(\delta_{\mu\alpha}\delta_{\nu\beta} - \delta_{\mu\beta}\delta_{\nu\alpha})$$
$$+ (f'/\tau + f^2)(x_\alpha x_\mu \delta_{\nu\beta} - x_\alpha x_\nu \delta_{\mu\beta} \quad (15)$$
$$+ x_\beta x_\nu \delta_{\mu\alpha} - x_\beta x_\mu \delta_{\nu\alpha}).$$

It is evident that the first tensor structure (15) satisfies the equation (13) and the second does not. Hence we are to choose:

$$f'/\tau + f^2 = 0, \quad f(\tau) = \frac{2}{\tau^2 + \lambda^2} \quad (16)$$

where λ is an arbitrary scale. The quasi-energy E is given by

$$E = \tfrac{1}{4}\mathrm{Sp}\int_\pm F_{\mu\nu}^2 d^4x$$
$$= \tfrac{1}{32}\mathrm{Sp}\int (F_{\mu\nu}^{\alpha\beta})^2 d^4x = 2\pi^2. \quad (17)$$

Comparison of (17) and (10) shows that we find absolute minimum for $q = 1$.

Another representation for the solution (14) is given by the formulas:

$$A_\mu = \frac{\tau^2}{\tau^2 + \lambda^2}\, g^{-1}(x)\frac{\partial g(x)}{\partial x_\mu}$$
$$g(x) = (x_4 + i\boldsymbol{x}\cdot\boldsymbol{\sigma})(x_4^2 + \boldsymbol{x}^2)^{-1/2} \quad (18)$$
$$g^\dagger g = 1, \quad \tau^2 = x_4^2 + \boldsymbol{x}^2$$

($\boldsymbol{\sigma}$ are Pauli matrixes).

For arbitrary group G one should consider its subgroup SU(2) for which A_μ is given by (18) and all other matrix elements of A_μ let be zero.

Our solution, as is evident from the scale invariance, contains the arbitrary scale λ. Hence these fields are long range and are essential in the infrared problems.

We do not know whether any solutions of (13) exist with $q > 1$. One may consider of course several

Volume 59B, number 1 PHYSICS LETTERS 13 October 1975

pseudoparticles with $q = 1$. However, we do not know whether they are attracted to each other and form the pseudoparticle with $q > 1$ or whether there exists repulsion and no stable pseudoparticle.

One of us (A.M.P.) is indebted to S.P. Novikov for explanation of some topological ideas.

References

[1] A.M. Polyakov, Phys. Lett. 59B (1975) 82.

VOLUME 37, NUMBER 3 PHYSICAL REVIEW LETTERS 19 JULY 1976

COMMENTS

Vacuum Periodicity in a Yang-Mills Quantum Theory*

R. Jackiw and C. Rebbi
Laboratory for Nuclear Science and Physics Department, Massachusetts Institute of Technology, Cambridge, Massachusetts
(Received 1 June 1976)

We propose a description of the vacuum in Yang-Mills theory and arrive at a physical interpretation of the pseudoparticle solution and the attendant violation of symmetries. The existence of topologically inequivalent classical gauge fields gives rise to a family of quantum mechanical vacua, parametrized by a CP-nonconserving angle. The requirement of vacuum stability against gauge transformations renders the vacua chirally noninvariant.

A classical pseudoparticle solution to the SU(2) Yang-Mills theory in Euclidean four-dimensional space has been given by Belavin, Polyakov, Schwartz, and Tyupkin,[1] with the suggestion that it be used to dominate the functional integral which describes a quantum field theory continued to Euclidean space. 't Hooft[2] has shown that these nontrivial minima of the action give nonvanishing contributions to amplitudes which would be zero in the ordinary sector. Specifically in a theory of fermions coupled to Yang-Mills fields, with chiral U(1) and CP symmetries, symmetry-nonconserving effects are found through the pressence of the axial-vector-current anomaly.[3] Thus he provides a possible resolution of the long-standing U(1) problem[4] and an intriguing suggestion for the origin of CP nonconservation. The phenomena are $O(\exp(-8\pi^2/g^2))$, where g is the gauge coupling constant; they are nonperturbative.

The fact that the classical field configuration which is responsible for the new results is in Euclidean four-dimensional space, i.e., imaginary time, leads one to suspect that the pseudoparticle is associated with quantum-mechanical tunneling by which field configurations in the ordinary three-dimensional space are joined in the course of the (real-time) evolution through the penetration of an energy barrier.[5] Also the exponentially small magnitude is indicative of tunneling. Here we wish to present a further explanation of this point, which we hope, will clarify the physical interpretation of the pseudoparticle solution and will supplement 't Hooft's more formal computations. Our considerations lead to a description of the quantum mechanical vacuum state of a Yang-Mills theory which is unexpectedly rich.

In the quantum field theory, a state of the system can be represented by a wave functional $\Psi[\vec{A}]$ of the field configuration. Having in mind a Yang-Mills theory, we have taken the potentials $\vec{A}(\vec{x})$ (anti-Hermitian matrices in the space of the infinitesimal group generators) as argument of the functional, excluding the time components $A^0(\vec{x})$, because they are dependent variables. In defining scalar products and matrix elements of observables one must avoid infinities associated with the volume of the gauge group. Without repeating details of the well-kown gauge-fixing procedure, let us only recall that it removes from the functional integral over $\vec{A}$ configurations of the fields which can be joined by a *continuous* gauge transformation to configurations already counted. In particular, one does not integrate over potentials of the form

$$\vec{A}(\vec{x})=g^{-1}(\vec{x})\nabla g(\vec{x}), \tag{1}$$

where g is the unitary matrix of a gauge transformation that can be joined to the identity through a one-parameter continuous family of transformations $g(\vec{x},\alpha)$:

$$g(\vec{x},1)=g(\vec{x}); \quad g(\vec{x},0)=I. \tag{2}$$

The potentials of Eq. (1) are of course gauge equivalent to $\vec{A}=0$.

But it is important to realize that there are values of $\vec{A}$ that can be obtained from each other by

VOLUME 37, NUMBER 3 PHYSICAL REVIEW LETTERS 19 JULY 1976

gauge transformations which cannot be continuously joined with the identity transformation. For instance, we may consider

$$g_1(\vec{x}) = \frac{\vec{x}^2 - \lambda^2}{\vec{x}^2 + \lambda^2} - \frac{2i\lambda\vec{\sigma}\cdot\vec{x}}{\vec{x}^2 + \lambda^2} \qquad (3)$$

which gives origin to

$$\vec{A}(\vec{x}) = g_1^{-1}(\vec{x})\nabla g_1(\vec{x}) = \frac{2i\lambda}{(\vec{x}^2+\lambda^2)^2}[\sigma(\lambda^2 - \vec{x}^2) + 2\vec{x}(\sigma\cdot\vec{x}) + 2\lambda\vec{x}\times\vec{\sigma}] \qquad (4)$$

and of course to vanishing field strengths F_{ij}. Values of the potentials like those of Eq. (4), although gauge equivalent to $\vec{A}=0$, should *not* be removed from the integrations over the field configurations by the gauge fixing procedure, and indeed we shall argue that physical effects are associated with them.

Before proceeding, let us characterize the classes of gauge-equivalent, but not continuously gauge-equivalent, potentials. We study effects which are local in space and therefore, when we consider a gauge transformation g, we require

$$g(\vec{x}) \xrightarrow[|\vec{x}|\to\infty]{} I. \qquad (5)$$

Thus g defines a mapping of the three-dimensional space, *with all the directions at* ∞ *identified,* into the group space. From the topological point of view, the Euclidean space E^3 with points at ∞ identified is equivalent (homeomorphic) to a three-dimensional sphere S^3; but the manifold of SU(2) is also homeomorphic to S^3, so that g defines a mapping

$$S^3 \xrightarrow[g]{} S^3. \qquad (6)$$

It is known that these mappings fall into homotopy classes (mappings belonging to different classes cannot be continuously distorted into each other) classified by an integer n,

$$g_n(\vec{x}) = [g_1(\vec{x})]^n \qquad (7)$$

with g_1 given in Eq. (3) being a representative of the nth class.

We can make contact now with the pseudoparticle solution.[1] Observe that the field configuration of Eq. (4) has zero potential energy, and that there is no energy-conserving evolution of the system which adiabatically connects that configuration with $\vec{A}=0$. Such an evolution should be a continuous gauge transformation; but this is impossible because g_1 and the identity belong to different homotopy classes. All paths joining the two field configurations in real time must go over an energy barrier. To exemplify this, let us multiply the potentials of Eq. (4) by $\frac{1}{2}-\alpha$ and increase α adiabatically from $-\frac{1}{2}$ to $+\frac{1}{2}$. Now the field strength is nonvanishing, but proportional to $\frac{1}{4} - \alpha^2$. The energy, $-\frac{1}{8}\int d^3x \,\mathrm{Tr} F_{ij}F^{ij} \geq 0$, becomes proportional to $(\alpha^2 - \frac{1}{4})^2$ and exhibits a barrier shape as α varies from $-\frac{1}{2}$ to $\frac{1}{2}$.

In the quantum theory, tunneling will occur across this barrier. It is well known that a semiclassical description of tunneling can be given by solving the classical equations of motion with *imaginary* time, thus achieving an evolution which would be classically forbidden for real time.[5] The pseudoparticle solution[1] serves precisely this purpose: It carries zero energy (the Euclidean stress tensor vanishes); it can be arranged to connect $g=g_1$ at $x_4 = -it = -\infty$ with $g=I$ at $x_4=\infty$. The physical implication of the pseudoparticle solution is that the quantal description of the vacuum state cannot be limited to fluctuations around any definite classical configuration of zero energy.

Let us now describe in greater detail the nature of the vacuum wave functional. Consider any of the field configurations

$$\vec{A}_n(\vec{x}) = g_n^{-1}(\vec{x})\nabla g_n(\vec{x}) \qquad (8)$$

with vanishing F^{ij}. Neglecting tunneling effects we might expect the vacuum to be of the form

$$\psi_n[\vec{A}] = \varphi[\vec{A} - \vec{A}_n], \qquad (9)$$

where the wave functional φ is peaked about zero and has a spread due to quantum fluctuations and any $\vec{A}_n$ can be chosen as representative of the classical vacuum, i.e., the classical zero-energy configuration.

But the pseudoparticle solution connects $\vec{A}_n$ with $\vec{A}_{n+1}$, giving origin to tunneling between the different ψ_n. The true quantal vacuum state will therefore be a superposition of the form

$$\Psi[\vec{A}] = \sum_n c_n \psi_n[\vec{A}] + O(\exp(-8\pi^2/g^2)). \qquad (10)$$

To determine the coefficients c_n in this equation let us observe that the finite gauge transformation g_1 changes ψ_n into ψ_{n+1}. Requiring the vacuum state to be stable against gauge transformations determines the coefficients to be

$$c_n = e^{in\theta}. \quad (11)$$

Thus we find a family of vacua, parametrized by an angle θ, where under the gauge transformation g_1

$$\Psi_\theta[\vec{A}] \underset{g_1}{\to} e^{-i\theta}\Psi_\theta[\vec{A}]. \quad (12)$$

The occurrence of multiple vacua is intriguing and is reminiscent of the situation encountered in the Schwinger model.[6]

The significance of the phase θ in Eqs. (10) and (11) becomes apparent when massless fermions are coupled to the Yang-Mills fields. One may then introduce the U(1) axial-vector current

$$J_5^{\mu} = i\bar{\psi}\gamma^{\mu}\gamma^5\psi$$

which however is not conserved because of the anomaly.[3] A conserved, but gauge-variant, current is given by

$$\tilde{J}_5^{\mu} = J_5^{\mu} - 4\pi^{-2}\epsilon^{\mu\nu\alpha\beta} \times \mathrm{Tr}(A_\tau\partial_\alpha A_\beta + \tfrac{2}{3}A_\nu A_\alpha A_\beta) \quad (13a)$$

and the conserved axial charge is

$$\tilde{Q}_5 = \int d^3x\, \tilde{J}_5^{0}. \quad (13b)$$

To exhibit the gauge dependence of $\tilde{Q}_5$, we perform a finite gauge transformation g, with $g(\vec{x}) \underset{|\vec{x}|\to\infty}{\longrightarrow} I$, and find

$$\Delta\tilde{Q}_5 = \frac{1}{12\pi^2}\int d^3x\, \mathrm{Tr}\epsilon_{ijk}(g^{-1}\partial_i g)(g^{-1}\partial_j g)(g^{-1}\partial_k g) = \frac{1}{12\pi^2}\int d\mu(g) = 2n, \quad (14)$$

where $d\mu(g)$ is the invariant measure of the group and n is the integer which characterizes the homotopy class of g. (g belongs to the nth homotopy class when it is continuously deformable to g_n.) The fact that $\tilde{Q}_5$ commutes with the Hamiltonian, and that it changes by two units under the gauge transformation g_1, together with Eq. (11), implies that

$$\exp(-\tfrac{1}{2}i\theta'\tilde{Q}_5)\Psi_\theta[\vec{A}] = \Psi_{\theta+\theta'}[\vec{A}] \quad (15)$$

which in turn shows that all the vacua are degenerate in energy and define the same theory. Equation (15) also demonstrates the possibility of symmetry breaking without Goldstone bosons: This may provide a solution to the U(1) problem.[2,4]

An explanation of the nonconservation of the gauge-invariant fermionic axial charge

$$Q_5(t) = \int d^3x\, J_5^{0}(t,\vec{x}) \quad (16)$$

may be given. The tunneling process between adjacent vacuum components $\psi_n[\vec{A}]$ and $\psi_{n+1}[\vec{A}]$ is equivalent to a gauge transformation g_1 which changes $\tilde{Q}_5 - Q_5$ by two units; see Eq. (14). But $\tilde{Q}_5$ is conserved, and therefore $\Delta Q_5 = -2$.

Finally, we remark that, whereas in the massless case conservation of $\tilde{Q}_5$ renders all vacua degenerate, we expect that if the fermions are massive, so that $\tilde{Q}_5$ is no longer conserved, different values of θ define nonequivalent theories as in the Schwinger model.[6] A nonzero value of θ could describe *CP* nonconservation,[2] but in the theory as developed thus far there is no indication how to compute θ.

We are happy to acknowledge our indebtedness to G. 't Hooft, whose calculations made our observations possible. We also thank S. Coleman and L. Susskind for discussions.

Added note.—After completion of this manuscript, we received a paper by C. Callan, R. Dashen, and D. Gross (to be published) who arrive at conclusions similar to ours.

---

*This work is supported in part through funds provided by the U. S. Energy Research and Development Administration under Contract No. E(11-1)-3069.

[1]A. Belavin, A. Polyakov, A. Schwartz, and Y. Tyupkin, Phys. Lett. 59B, 85 (1975); A. Polyakov, Phys. Lett. 59B, 82 (1975).

[2]G. 't Hooft, Phys. Rev. Lett. 37, 8 (1976), and lectures at various universities.

[3]S. L. Adler, in *Lectures on Elementary Particles and Quantum Field Theory*, edited by S. Deser, M. Grisaru, and H. Pendleton (The MIT Press, Cambridge, Mass., 1970); R. Jackiw, in *Lectures on Current Algebra and Its Applications*, edited by S. Treiman, R. Jackiw, and D. Gross (Princeton Univ. Press, Princeton, N. J., 1972).

[4]H. Pagels, Phys. Rev. D 13, 343 (1976).

[5]Use of imaginary time in discussions of tunneling is well known; see, e.g., K. Freed, J. Chem. Phys. 56, 692 (1972); D. McLaughlin, J. Math. Phys. (N.Y.) 13, 1099 (1972).

[6]J. Schwinger, Phys. Rev. 128, 2425 (1962); J. Lowenstein and A. Swieca, Ann. Phys. (N.Y.) 68, 172 (1971); S. Coleman, R. Jackiw, and L. Susskind, Ann. Phys. (N.Y.) 93, 267 (1975). (This analogy was developed in conversations with Coleman and Susskind.) There is no pseudoparticle solution for *spinor* electrodynamics in two space-time dimensions. However, by adding to the model charged spinless fields, with a Higgs potential and minimal electromagnetic coupling, a pseudoparticle solution exists in Euclidean two-dimension space—it is just the Nielsen-Olesen string

[Nucl. Phys. B61, 45 (1973)]. The topologically conserved object is $\int d^2x\epsilon_{\mu\nu}F^{\mu\nu}$, and $\epsilon_{\mu\nu}F^{\mu\nu}$ is also proportional to the anomalous divergence of the axial vector current; K. Johnson, Phys. Lett. 5, 253 (1963); R. Jackiw, in *Laws of Hadronic Matter*, edited by A. Zichichi (Academic, New York, 1975).

Volume 63B, number 3 PHYSICS LETTERS 2 August 1976

THE STRUCTURE OF THE GAUGE THEORY VACUUM*

C.G. CALLAN, Jr.
Joseph Henry Laboratories, Princeton University, Prinecton, New Jersey 08540, USA

R.F. DASHEN*
Institute for Advanced Study, Princeton, New Jersey 08540, USA

and

D.J. GROSS
Joseph Henry Laboratories, Princeton University, Princeton, New Jersey 08540, USA

Received 20 May 1976

The finite action Euclidean solutions of gauge theories are shown to indicate the existence of tunneling between topologically distinct vacuum configurations. Diagonalization of the Hamiltonian then leads to a continuum of vacua. The construction and properties of these vacua are analyzed. In non-abelian theories of the strong interactions one finds spontaneous symmetry breaking of axial baryon number without the generation of a Goldstone boson, a mechanism for chiral SU(N) symmetry breaking and a possible source of T violation.

Polyakov [1] has recently pointed out that the Euclidean classical equations of motion of gauge theories have soliton-like solutions and has suggested that when properly included in the Euclidean functional integral they may have a bearing on the dynamics of confinement. The physical interpretation of these solutions has, however, been obscure since they are localized in time as well as space. In this letter we shall show that Euclidean gauge solitons describe events in which topologically distinct realizations of the gauge vacuum *tunnel* into one another and that this process radically changes the nature of the vacuum state. In fact, we find a continuum of vacua, each one of which is a superposition of the vacua with difinite topology and stable under the tunnelling process. The new vacua are the ground states of independent, and in general, inequivalent worlds (most striking, P and T are spontaneously violated in some of them!). When massless fermions are present, the vacuum tunnelling process forces a redefinition of the fermion vacuum as well and leads directly to spontaneous breakdown of chiral invariance without generating a "ninth" Goldstone boson.

We have, in effect, shown that the vacuum "seizes" as suggested by Kogut and Susskind [2], and identified the mechanism by which it does so. Our primary aim in this letter will be to give arguments for the existence of the new vacuum structure and to present the correct form of the functional integral appropriate to studying the properties of a particular vacuum. In the spirit of displaying qualitative consequence of the new vacuum structure we shall also briefly summarize results obtained from rather crude approximations to the functional integral.

To explore the structure of the vacuum we study the Euclidean functional integral

$$\langle 0|\exp(-Ht)|0\rangle \underset{t\to\infty}{\longrightarrow} \int [DA_\mu D\psi] \times \exp\{-\int d^dx[\mathcal{L}(A_\mu,\psi_i...)+\mathcal{L}_{gf}]\} \qquad (1)$$

where d is the dimension of space time, $\mathcal{L}$ is the Langrange density of the theory, $\mathcal{L}_{gf}$ is a gauge-fixing term and the integration is to be done over all fields that approach vacuum values ($F_{\mu\nu} = 0$) at infinity. Now since $F_{\mu\nu} = 0$ implies $A_\mu = g^{-1}(x)\partial_\mu g(x)$ takes on values in the gauge group, G, any gauge field included in the functional integration defines a map of the sphere at Euclidean infinity into G. As pointed out by

* Research supported by the National Science Foundation under Grant Number MPS 75-22514.
* Research sponsored in part by the ERDA under Grant No. E(11-1)-2220.

Volume 63B, number 3 PHYSICS LETTERS 2 August 1976

Belavin et al. [3] these maps fall into homotopy classes corresponding to elements of the homotopy group, $\Pi_{d-1}(G)$. For most non-Abelian groups in four dimensional space time and U(1) in two dimensional spacetime, this homopoty group is Z. In these theories (the only ones we consider) the gauge fields integrated over in eq. (1) fall into discrete classes indexed by an integer ν running from $-\infty$ to $+\infty$ (we shall use the notation $[DA_\mu]_\nu$ to denote functional integration over the νth class). Thus there is actually a discrete infinity of funtional integrals and one must ask which, if any, is the "right" one.

One can clearly see what is going on by working in the gauge $A_0 = 0$ and requiring $F_{\mu\nu}$ to vanish outside a large, but finite, spacetime volume, V (this boundary condition is, of course, gauge invariant). The dynamical variables are now just the space components, A_i, for the vector potential and at large negative and positive times they must take on time independent vacuum values, $A_i(\boldsymbol{x}) = g^{-1}(\boldsymbol{x})\partial_i g(\boldsymbol{x})$. The topological quantum number, ν, associated with any particular Euclidean gauge field time history may be written as a gauge invariant volume integral

$$\begin{aligned}\nu &= \frac{1}{8\pi^2}\int d^4x \, \mathrm{tr}\,(F_{\mu\nu}\tilde{F}^{\mu\nu}), \quad d=4 \\ &= \frac{1}{4\pi}\int d^2x \, \epsilon_{\mu\nu}F^{\mu\nu} \quad , \quad d=2.\end{aligned} \tag{2}$$

In both cases, the integrand is a total divergence and $A_0 = 0$ gauge ν may be rewritten as a surface integral, $\nu = n(t=+\infty) - n(t=-\infty)$, where

$$\begin{aligned}n &= \frac{1}{6\pi^2}\epsilon_{ijk}\int d^3x \, \mathrm{tr}\,(A_i A_j A_k), d=4 \\ &= \frac{1}{2\pi}\int dx A_1 \, , \qquad d=2.\end{aligned} \tag{3}$$

With no loss of generality (we have the freedom of making time independent gauge transformations) we may choose $n(t=-\infty)$ to be an integer. Then since ν is integral the gauge vacuum configuration at $t=-\infty$ must also have integral winding number $n(+\infty) = n(-\infty) + \nu$.

Therefore we must admit the existence of a discrete infinity of vacuum states, $|n\rangle$, labelled by a winding number taking on integral values from $-\infty$ to $+\infty$. The interpretation of the multiplicity of Euclidean functional integrals corresponding to different ν-classes, is then straightforward:

$$\begin{aligned}\langle n|\exp(-Ht)|m\rangle \underset{t\to\infty}{\longrightarrow} &\int [DA_\mu ...]_{(n-m)} \\ &\times \exp\{-\int d^d x[\mathcal{L}(A_\mu) + \mathcal{L}_{\mathrm{gf}}]\}\,.\end{aligned} \tag{4}$$

The functional integral over homotopy class ν describes a vacuum-to-vacuum transition in which the vacuum winding number changes by ν! Now the minimum action for $\nu = 0$ is zero (corresponding to $A_\mu \equiv 0$) so that in the WKB sense the $|n\rangle \to |n\rangle$ amplitude is O(1). In the $\nu \neq 0$ sectors the minimum action is in general non-zero – for $\nu = 1$ in four dimensions it corresponds to the Belavin et al. instanton [3], whose action is $8\pi^2/g^2$. Thus in the same WKB sense the $|n\rangle \to |n+1\rangle$ amplitude is $O(\exp(-8\pi^2/g^2))$. This is a typical "tunnelling" amplitude, vanishing exponentially for small coupling and unseen by standard perturbation theory. Indeed, perturbation treatments of gauge theories expand about $A_\mu = 0$ and pretend that the vacuum $|n=0\rangle$ is true vacuum. Because of vacuum tunnelling, this is completely wrong and causes perturbation theory to miss qualitatively significant effects.

What then is the true vacuum? A convenient way of constructing it is to consider the generators of time independent gauge transformations characterized by a gauge function $\lambda^a(\boldsymbol{x})$:

$$Q_\lambda = \int d^{d-1}x[F^a_{0i}D_i\lambda^a + gJ^a_0\lambda^a]$$

where D_i is the covariant derivative and J^a_0 is the gauge source of fields other than the gauge field itself. In order to satisfy Gauss' law, $D_iF^a_{0i} = gJ^a_0$, it is sufficient to restrict the state space by $Q_\lambda|\psi\rangle = 0$ for all gauge functions λ which vanish at infinity. In particular, all our vacuum states $|n\rangle$ are annihilated by such local gauge transformations. There also exist gauge functions which do not vanish at infinity and generate gauge transformations, T, which change the vacuum topology. One can easily construct a unitary T effecting such a non-local gauge transformation: $T = \exp(iG_\infty)$, with $G_\infty\,(2\pi/g)[E(\infty) + E(-\infty)]$ for the two-dimensional abelian theory or $G_\infty = (2\pi/g)\int d^2S_iE^a_i\hat{x}^a$ for the four-dimensional non-Abelian theory, and T satisfies $T|n\rangle = |n+1\rangle$.

Since T is a gauge transformation, the hamiltonian

Volume 63B, number 3 PHYSICS LETTERS 2 August 1976

commutes with it and energy eigenstates must be T eigenstates. Since T is unitary, its eigenvalues are $e^{i\theta}$, $0 \leqslant \theta \leqslant 2\pi$, and the eigenstates are $|\theta\rangle = \Sigma e^{in\theta}|n\rangle$. This diagonalization of H is obviously unaffected by including in $\mathcal{L}$ sources coupled to gauge invariant densities. Thus, each $|\theta\rangle$ vacuum is the ground state of an independent and in general physically inequivalent sector within which we may study the propagation of gauge invariant disturbances. Since the different θ-worlds do not communicate with each other, there is no a-priori way of deciding which world is the right one. It is gratifying that this multiplicity of worlds is known to exist in the Schwinger model, corresponding there to different values of background electric field [4].

Finally, we must express the functional integral, eq. (4), in θ basis:

$$\langle\theta'|\exp(-Ht)|\theta\rangle \xrightarrow[t\to\infty]{} \delta(\theta-\theta')I(\theta)$$

$$I(\theta) = \sum_\nu \exp(-i\nu\theta)$$

$$\times \int [DA_\mu \ldots]_\nu \exp\{-\int d^dx[\mathcal{L}(A_\mu \ldots) + \mathcal{L}_{gf}]\}$$

$$= \int [DA_\mu \ldots] \exp\{-\int d^dx[\mathcal{L}_{gf} + \mathcal{L}_\theta]\} \qquad (5)$$

where $\mathcal{L}_\theta = (i\theta/8\pi^2)\,\mathrm{tr}\,(F_{\mu\nu}\tilde{F}^{\mu\nu})$ for $d = 4$, $\mathcal{L}_\theta = (i\theta/4\pi)\,\epsilon_{\mu\nu}F_{\mu\nu}$ for $d = 2$ and in the second expression for $I(\theta)$ all gauge field topologies are summed over. $I(\theta)$ contains all possible information about physics in the θ-world and requires no further modification. The second form for $I(\theta)$ makes manifest one of the peculiar ways in which the θ-worlds differ from one another. In four dimensional pure Yang-Mills theory, re-expressed in Minkowski coordinates, the effective Lagrangian is $\mathrm{tr}[F_{\mu\nu}F^{\mu\nu} + (\theta/8\pi^2 F_{\mu\nu}\tilde{F}^{\mu\nu}]$. This clearly breaks P and T invariance (except for $\theta = 0$) and we must in general expect spontaneous breaking of space-time symmetries in all but a few special θ-worlds!

As a concrete illustration of these general remarks we should like to present an approximate evaluation of $I(\theta)$ in two-dimensional charged scalar electrodynamics. In the sector with $\nu = \pm 1$ the field configuration with minimum action is just the Nielson-Olesen vortex [5] in which there is a localized region of non-zero field of flux $\pm 2\pi/g$, radius μ^{-1} (μ is the heavy photon mass), arbitrary location and total action, S_0, proportional to μ^2/g^2. We shall construct the sectors with topological quantum number ν by superposing n_+ $\nu = +1$ vortices and n_- $\nu = -1$ vortices with $n_+ - n_- = \nu$, neglecting any interactions between vortices (since fields decrease exponentially this is not too bad for low vortex density, which turns out to mean small g). In this "dilute gas" approximation, the functional integral is

$$\langle\theta'|\exp(-Ht)|\theta\rangle \sim \delta(\theta'-\theta)\sum_{n_+,n_-=0}^{\infty} \exp\{-(n_+ + n_-)S_i\}$$

$$\times \frac{\exp\{i\theta(n_+ - n_-)\}}{n_+!\,n_-!}\left(\frac{V}{V_0}\right)^{n_+ + n_-} \qquad (6)$$

where the factors of V come from integrating over vortex locations and V_0 is a normalization factor which can be calculated from the quantum corrections to this basically semiclassical approximation. The sum is trivial and yields $\exp[2(V/V_0)e^{-S_0}\cos\theta)$. We have normalized the energy so that the naive perturbation theory vacuum energy is zero. By contrast, the θ vacua have an energy per unit volume equal to $-2V_0^{-1}\cos\theta\; e^{-S_0}$. Because $S_0 \propto 1/g^2$, this energy difference is a non-perturbative effect (a tunnelling effect) but potentially important nonetheless. Although the $\theta = 0$ vacuum has lowest energy (and no parity violation) we can't conclude that it is *the* vacuum since the other θ-vacua, though higher in energy are stable to gauge invariant perturbations. Having constructed a vacuum one can then calculate Green' functions of gauge invariant operators perturbatively. In the path integral this corresponds to performing ordinary perturbation theory about the appropriate classical solution for each topologically distinct sector and summing.

If we try the above sort of approximation on the non-abelian theory in four dimensions, there is a problem. The classical theory is scale invariant and the basic $\nu = 1$ solution (instanton) has an arbitrary scale parameter, λ, as well as an arbitrary position. The integration over λ need not diverge since scale invariance is broken by quantum corrections. Indeed, the renormalization group should tell us whether the integral converges at the short distance end. In the dilute gas approximation one finds for the vacuum energy density

Volume 63B, number 3 PHYSICS LETTERS 2 August 1976

$$\mathcal{E}_\theta = -2\cos\theta \int_0^\infty \lambda^3\, \mathrm{d}\lambda\, \exp\left[-\frac{8\pi^2}{\bar{g}(\lambda/\mu)^2} + \mathcal{F}(\bar{g}(\lambda/\mu))\right] \tag{7}$$

where $\bar{g}(\lambda/\mu)$ is the usual effective coupling, normalized so that $\bar{g}(\lambda = \mu) = g$, μ is arbitrary, and $\mathcal{F}$ summarizing the effect of loop corrections, can be computed perturbatively. If the theory is asymptotically free and there are not too many quark multiplets, $\bar{g}$ can vanish rapidly enough for the integral to converge in the limit of large λ (small instanton size). This condition is met for any pure SU(N) gauge theory and for SU(3) with no more than *ten* flavors of quark.

On the other hand, in the limit of large instanton size, one is driven to large coupling (unless β has a small infrared fixed point) and the dilute gas approximation breaks down (instanton overlap and have long range interactions). Thus the attempt to construct the vacuum may run into an essential strong coupling problem because the quantum corrections to vacuum tunnelling will be large for large instanton size. In fact, there may not be a sensible way of perturbatively calculating even Green's functions of gauge invariant operators, no matter how small one makes g. This phenomenon is typical of a theory with no inherent mass scale which produces masses dynamically. If one sets the renormalization mass scale, μ, equal to some physical mass (e.g. $4\sqrt{\mathcal{E}_\theta}$), then g is determined (dimensional transmutation) and typically of order 1.

These problems should not, however, affect the standard applications of asymptotic freedom which rely on one's ability to compute operator product expansion coefficients at short distances. Precisely because of asymptotic freedom, vacuum tunnelling is suppressed at arbitrarily small scales and leading short distance behavior will agree with conventional calculations. There will, however, be calculated non-leading terms suppressed by powers of momentum, which reflect the mass scale set non-perturbatively by the tunnelling phenomenon.

The arguments presented above require some modification when massless fermions are present. We again confine non-zero $F_{\mu\nu}$ to a large but finite space-time volume, V, and again encounter a discrete infinity, $\{|n\rangle\}$, of vacuum states characterized by a vacuum gauge field with winding number n and a standard fermi vacuum with all negative energy states filled. In principle we must allow for transitions between vacuum, and evaluate $\langle n|\mathrm{e}^{-Ht}|m\rangle$ for general n and m. In fact, for massless quarks, $\langle n|\mathrm{e}^{-Ht}|m\rangle \propto \delta_{nm}$!

The reason for this is that, because of the anomaly, the *conserved* axial charge

$$Q_5 = \int \mathrm{d}\bar{x} J^5_{\mathrm{c}},$$

$$J^5_\mu = \sum_{\text{flavor,color}} \bar{\psi}\gamma_\mu\gamma_5\psi - \mathrm{tr}\left\{\frac{g^2N}{32\pi^2}\epsilon_{\mu\nu\lambda\sigma}A_\nu(\partial_\lambda A_\sigma + \tfrac{2}{3}A_\lambda A_\sigma)\right\} \tag{8}$$

while invariant under local gauge transformations, is not invariant under global gauge transformations. In particular, one has $TQ_5T^{-1} = Q_5 - 2N$ where T is the global gauge transformation, introduced earlier, which changes gauge field winding number by one unit and N is the number of flavors. If the vacuum states of different topology are defined by $|n\rangle = T^n|0\rangle$, with $Q_5|0\rangle = 0$ one finds that $Q_5|n\rangle = 2N\cdot n|n\rangle$. However, Q_5 is *conserved*, so that it must be true that $\langle n|\mathrm{e}^{-Ht}|m\rangle \propto \delta_{n,m}$. In general, we must find $\langle n|\mathrm{e}^{-Ht}D_\nu|m\rangle \propto \delta_{n-m,\nu}$ where D_ν is any operator of chirality $2N\nu$ (D_ν may stand for multiple insertions of local operators at different times – all that matters is net chirality). Therefore, we may replace eq. (4) by

$$\langle n|\mathrm{e}^{-Ht}|m\rangle \to \delta_{nm}\int [DA_\mu ...]_0 \exp\{-\int \mathrm{d}^dx[\mathcal{L}(A_\mu ...) + \mathcal{L}_{\mathrm{gf}}]\}$$

and

$$\langle n|\mathrm{e}^{-Ht}D_\nu|m\rangle \to \delta_{n+\nu,m}\int [DA_\mu ...]_\nu D_\nu \times \exp\{-\int \mathrm{d}^dx[\mathcal{L}(A_\mu ...) + \mathcal{L}_{\mathrm{gf}}]\} \tag{10}$$

with the same meaning still attached to D_ν. The restriction on the topology of the gauge field histories would actually have emerged directly from a mindless application of eq. (4): Doing the fermion integrations for fixed A_μ yields $[\det(\partial\!\!\!/ - A\!\!\!/)]^{+1}$. This determinant vanishes whenever $(\partial\!\!\!/ - A\!\!\!/)$ has a zero eigenvalue. 't Hooft [6] has noted that if A_μ is taken equal to the $\nu = +1$ or $\nu = -1$ instanton there is a zero eigenvalue, and our argument is just telling us that whenever A_μ belongs to a $\nu \neq 0$ class, $(\partial\!\!\!/ - A\!\!\!/)$ has a zero eigenvalue, eliminating the $\nu \neq 0$ sectors from the integration.

Though vacuum tunnelling is now suppressed, the

$|n\rangle$ vacua are not acceptable because they violate cluster decomposition for operators of non-zero chirality . Consider an operator D of chirality $2N$. The arguments of the preceeding paragraph show that $\langle n|D|n\rangle = 0$, $\langle n+1|D|n\rangle \neq 0$. Then $\langle n|D^{+}(x)D(y)|n\rangle$ will not vanish for large $|x-y|$ as required by cluster decomposition and the vanishing of the "vacuum" expectation $\langle n|D|n\rangle$: it obviously approached $\langle n|D^{+}|n+1\rangle\langle n+1|D|n\rangle$. The solution to this problem is obvious (it was solved in the Schwinger model years ago!): The proper vacuum states are the $|\theta\rangle$ vacua, in which basis the functional integrals have the form

$$\langle\theta'|e^{-Ht}|\theta\rangle \rightarrow \delta(\theta'-\theta)\int[DA_{\mu}...]_{(0)}$$

$$\times \exp\{-\int d^{d}x[\mathcal{L}(A_{\mu}...)+\mathcal{L}_{gf}]\}$$

$$\langle\theta'|e^{-Ht}D_{\nu}|\theta\rangle \rightarrow \delta(\theta'-\theta)\int[DA_{\mu}...]_{(\nu)}$$

$$\times \exp\{-\int d^{d}x[\mathcal{L}(A_{\mu}...)+\mathcal{L}_{gf}]\}D_{\nu}\,. \qquad (11)$$

The cluster problem is resolved by the non-vanishing vacuum expectation value of D in the true vacuum state. The fact that only one topological class of gauge field history contributes to each functional integral makes physical quantities have a trivial dependence on θ: The vacuum energy density, while non-zero, is independent of θ. The variation of the vacuum energy with respect to θ is just $\langle \mathrm{tr}\, F_{\mu\nu}\tilde{F}^{\mu\nu}\rangle$, the quantity whose non-zero value is the signal for P and T violation. In the massless fermion case, P and T appear not to be spontaneously violated and, indeed, all the $|\theta\rangle$ vacua are physically equivalent. Finally, the axial baryon number invariance of the original Lagrangian is violated by a vacuum expectation value of operators with non-zero chirality. It should be said that at this stage we have in the N-flavor case, only broken axial U(1) and *not* axial SU(N). Actually, since the axial charge rotates θ, a discrete subgroup of order $2N$ of U(1) is left unbroken, consisting of those elements which rotate θ by a multiple of 2π. There is no associated Goldstone boson because the conserved, but gauge-variant, U(1) charge takes one out of a given $|\theta\rangle$ sector ($e^{i\alpha Q_5}|\theta\rangle = |\theta+2N\alpha\rangle$) while $\mathrm{tr}(F\tilde{F})$, the divergence of the gauge invariant axial current, has non-vanishing matrix elements. That Q_5 causes transitions between different vacua is characteristic of the "vacuum seizing" mechanism postulated by Kogut and Susskind [2] while the non-vanishing of $F\tilde{F}$ in instanton solutions as a possible escape from the U(1) problem was noted by G. 't Hooft [7].

Although the presence of zero mass fermions suppresses vacuum tunnelling in the strict asymptotic sense, tunnelling does have a profound effect on the vacuum energy and other physically relevant quantities. When the vacuum tunnels, fermion pairs are produced. Although the pair must ultimately be absorbed by an anti-tunnelling, since the fermions are massless the pair may live for a long time and tunnelling occurs freely in intermediate states. To get some notion of what goes on it is instructive to attempt a crude calculation of the basic functional integrals of eq. (11) in the case of a single flavor.

We shall assume that the integral over A_{μ} is dominated by configurations of widely separated instantons (n_{+} in number) and anti-instantons (n_{-} in number). To compute the vacuum energy we must set $n_{+} = n_{-}$ (configurations with $\nu = 0$), sum over n_{+} and integrate over instanton locations. We will ignore the integration over instanton sizes. For a given gauge field configuration, the integration over fermi fields yields $\det(\not{\partial} - \not{A})$. This determinant must also be approximated.

Now, as 't Hooft has pointed out [7], individual instantons have a zero-energy eigenfunction $\psi_0^{\pm}(x, x_{\pm})$ ($x_{\pm}$ is the instanton location and the $\pm$ label distinguishes instanton from anti-instanton). Since the interesting physical effects arise precisely from these zero energy solutions, we shall compute the determinant of $(\not{\partial} - \not{A})$ in the subspace spanned by the $2n_{+}$ functions $\psi_0^{+}(x, x_i^{+})$, $\psi_0^{-}(x, x_i^{-})$. In the widely separated instanton approximation, these functions are orthonormal and one has to compute the determinant of the matrix

$$M_{ij} = (\psi_{0,i}^{*}|(\not{\partial} - \not{A})|\psi_{0,j}^{-}).$$

In this approximation A_{μ} differs from a gauge transformation only in the neighborhood of each instanton and M_{ij} can be approximated by

$$M_{ij} \approx (\phi_{0,i}^{+}|(-\not{\partial})^{-1}|\phi_{0,j}^{-})$$

where $\phi_0 = \not{\partial}\psi_0$. The γ_5 structure of the ψ_0 forbids i and j to be both instantons or both anti-instantons. The ψ_0 fall off at large distances exactly like the free fermi propagator, which is why we introduce the ϕ_0 's which are localized as well as the instanton itself. The cycle expansion of the determinant then gives a sum of terms with a graphical interpretation in terms of closed fermion loops. For instance, if $n_+ = n_- = 1$ we get

$$\mathrm{Det} \approx (\phi_0^+|(\not{\partial})^{-1}|\phi_0^-)(\phi_0^-|(\not{\partial})^{-1}|\phi_0^+)\,,$$

which has the obvious interpretation of massless quark propagators connecting non-local vertices, $V^{\pm} = \phi_{0\pm}^*(x)\phi_{0\pm}(x')$, associated with the instantons. The functional integral weights each vertex with a factor proportional to $e^{-S_{\mathrm{cl}}}$, where S_{cl} is the instanton classical action. The γ_5 structure of $\phi_0^{\pm}$ is such that $V^{\pm}$ is like $\bar{\psi}(1 \pm \gamma_5)\psi$ in its Dirac matrix structure, which is to say that $V^{\pm}$ looks like a non-local, or momentum-dependent mass term. Summing over numbers and locations of instantons simply completes the vacuum fermion loop analogy by providing all possible insertions of the pseudo mass terms, $V^{\pm}$, on massless quark loops. Then calculations of physical quantities proceed in a perfectly conventional way so long as we remember to add the mass term $\sim e^{-S_{\mathrm{cl}}}(V_+ + V_-)$ to the massless quark propagator. Anything which directly depends on the mass term, such as the vacuum expectation of $\bar{\psi}\psi$, will be proportional to $e^{-S_{\mathrm{cl}}}$ and will have only the by now familiar dependence on g characteristic of a tunnelling process. In the "scale invariant" four dimensional theory $\langle\bar{\psi}\psi\rangle \neq 0$ implies spontaneous generation of mass and dimensional transmutation as before.

If the fermion is given a bare mass tunnelling is allowed and one is driven directly to a θ vacuum (whose energy depends on θ). The limit of zero mass is smooth. If the bare mass is small compared to the spontaneously generated mass, it acts as a small perturbation on the $m_0 = 0$ theory.

If there are N flavors, the above discussion is modified in an important way: the effective instanton-quark interaction is no longer billinear, but $2N$-linear. Indeed, the instanton (anti-instanton) vertex has the structure $V_+ = \Pi_{i=1}^N \bar{\Psi}_i(1+\gamma_5)\psi_i$ ($V_- = \Pi_{i=1}^N \bar{\Psi}_i(1-\gamma_5)\psi_i$). As a result, summing over instantons in the dilute gas approximation will not just produce a mass term in the quark propagator, but does something more complicated. To produce a quark mass term, one must break the global chiral SU(N), while we have argued that the vacuum tunnelling phenomenon is only guaranteed to violate the chiral U(1) symmetry. On the other hand, the effective interactions between quarks of different helicity generated by the instanton provide new ways of identifying sums of graphs which can lead to the desired symmetry breakdown. We have constructed a simple Hartree–Fock type argument for $N = 2$ which has a chance of being correct in a weak coupling theory and which seems, on superificial examination, to generate quark masses. The inevitable Goldstone bosons arise in this case from iterated bubble graphs generated by the four-fermion interactions, $V_{\pm}$. We do not wish to make too much of these crude arguments other than to suggest that the new interactions generated by vacuum tunnelling are likely to play a key role in generation of quark masses and Goldstone bosons.

In terms of the picture presented here Polyakov's ideas about confinement appear as follows. For an isolated quark located at x the tunnelling amplitude $|n, x\rangle \to |n+1, x\rangle$ will be reduced relative to the vacuum to vacuum amplitude. A quark state will then have more energy than the vacuum, as it should. In the dilute gas approximation the energy difference is proportional to the integral over all instantons which overlap the quark. Integrating over instanton locations and then over the scale size λ^{-1} leads to an integral which tends to diverge at small λ. For the large instantons (small λ), however, the dilute gas approximation is not valid, and one is again confronted with a strong coupling problem.

Obviously, much remains to be done to fully exploit the phenomena we have found. The major difficulty, of course, is that the theory we are really interested in, quantum chromodynamics, is basically a strong coupling theory and reliable calculations are difficult, if not impossible. However, one may hope that a new understanding of the qualitative physics will suggest new methods of calculation. We are especially encouraged by the appearance, already in semiclassical approximations, of a vacuum that breaks chiral symmetry and sets a dynamical mass scale. We are also intrigued by the natural appearance of spontaneous violation of P and T invariance but have so far not seen how to understand why these effects are small in the real world or how to exploit them to explain observed violations of these symmetries. Perhaps super-unified theories will shed some light on these questions.

One of us (D.J. Gross) would like to acknowledge V.N. Gribov for stimulating conversations and in particular for the suggestion that the Euclidean solitons might be relevant to the structure of the vacuum.

References

[1] A.M. Polyakov, Phys. Lett. 59B (1975) 82.
[2] J. Kogut and L. Susskind, Phys. Rev. D11 (1975) 3594.
[3] A.A. Belavin, A.M. Polyakov, A.S. Schwartz and Yu. S. Tyupkin, Phys. Lett. 59B (1975) 85.
[4] J. Lowenstein and A. Swieca, Ann. Phys. 68 (1971) 172. S. Coleman, Harvard preprint (1975).
[5] H. Nielsen and P. Olesen, Nucl. Phys. B61 (1973) 45.
[6] T.S. Bell and R. Jackiw, Nuovo Cimento 60 (1969) 47; S. Adler, Phys. Rev. 177 (1969) 2426.
[7] G. 't Hooft, Harvard preprint (1976).

VOLUME 52, NUMBER 1 PHYSICAL REVIEW LETTERS 2 JANUARY 1984

Gauge Noninvariance and Parity Nonconservation of Three-Dimensional Fermions

A. N. Redlich
Center for Theoretical Physics, Laboratory for Nuclear Science and Department of Physics, Massachusetts Institute of Technology, Cambridge, Massachusetts 02139

(Received 30 September 1983)

The effective gauge field action due to an odd number of fermion species in three-dimensional SU(N) gauge theories is shown to change by $\pm\pi|n|$ under a homotopically nontrivial gauge transformation with winding number n. Gauge invariance can be restored by use of Pauli-Villars regularization, which, however, introduces parity nonconservation in the form of a parity-nonconserving, topological term in the effective action.

PACS numbers: 11.15.Tk, 11.30.Er

The action, $I[A,\psi]$, for an odd number of massless fermions coupled to SU(N) gauge fields in odd dimensions is invariant under space-time reflection (which we shall call parity). I establish here, however, that the ground-state current, $\langle J_a{}^\mu\rangle$, a physical quantity, must violate this symmetry: parity is "spontaneously broken." The calculations will be limited to three dimensions, although they can be easily generalized to higher odd dimensions.

To demonstrate that parity conservation must be violated in this theory, I make use of an intriguing connection between parity nonconservation and the noninvariance of the action under homotopically nontrivial gauge transformations. The argument goes as follows: While the action, $I[A,\psi]$, is invariant under local SU(N) gauge transformations, the effective action, $I_{\rm eff}[A]$, may not be—$I_{\rm eff}[A]$ is obtained by integrating out fermionic degrees of freedom. In a functional integral formulation of the theory, we require $\exp iI_{\rm eff}[A]$ to be gauge invariant. If under a gauge transformation, $I_{\rm eff}[A]$ changes by a number, the parameters of the theory must be quantized so that number is an integral multiple of 2π.[1-4] I find that there are two ways to regulate the ultraviolet divergences in the calculation of $I_{\rm eff}[A]$. The first way maintains parity as a good symmetry, but does not maintain gauge invariance: $I_{\rm eff}[A]$ is found to change by an odd multiple of π under a homotopically nontrivial gauge transformation. The second way introduces a heavy Pauli-Villars regulator field and subtracts $\lim_{M\to\infty} I_{\rm eff}[A,M]$ from $I_{\rm eff}[A]$, thus canceling the gauge noninvariance in $I_{\rm eff}[A]$, but introducing parity nonconservation through $I_{\rm eff}[A,M]$—a three-dimensional mass term $M\bar\psi\psi$ violates parity conservation and for $M\neq 0$ the fermions have parity-nonconserving spin equal to[2] $\frac{1}{2}M/|M| = \pm\frac{1}{2}$. I find that $\lim_{M\to\infty} I_{\rm eff}[A,M]$ contains a parity-nonconserving topological term, $\pm\pi W[A]$—$W[A]$ is the Chern-Simons secondary characteristic class[2,5]—which changes by $\pm\pi n$ under a homotopically nontrivial gauge transformation, U_n, with winding number n. Since $\langle J_a{}^\mu\rangle$ is equal to $\delta I_{\rm eff}{}^R/\delta A_{\mu a}$ ($I_{\rm eff}{}^R \equiv I_{\rm eff}[A] - \lim_{M\to\infty} I_{\rm eff}[A,M]$), $\langle J_a{}^\mu\rangle$ contains a topological term, due to $I_{\rm eff}[A,M]$, which violates parity conservation.

As an alternative to introducing a heavy Pauli-Villars regulator to restore gauge invariance, one may simply add the topological term $\pm\pi W[A]$ to the gauge field action. Another way to restore gauge invariance is to work with an even number of fermion species, so that the effective action changes by $2\pi n$ under a large gauge transformation. In this case, parity conservation need not be violated, which is not surprising, since an even number of fermions in three dimensions can be paired to form Dirac fermions with parity-conserving mass terms.

The nonconservation of parity in odd dimensions is analogous to the nonconservation of the axial current in two and four dimensions where Pauli-Villars regularization introduces a mass which violates axial symmetry. We therefore complete the program begun by generalizing the axial anomaly to higher even dimensions[6] by establishing the existence of a similar phenomenon in odd dimensions.[7] The "anomaly" in odd dimensions appears as a parity-nonconserving topological term in the ground-state current $\langle J_a{}^\mu\rangle$, rather than as a topological term ($\sim{}^*FF$) in $\partial_\mu\langle J_a{}^\mu\rangle$ (there exists no axial current in odd dimensions, i.e., no γ_5). In both cases, the "anomaly" causes a "physical" ground-state current to violate a symmetry of the original action, $I[A,\psi]$.

To show that fermions induce a gauge-noninvariant term in the action, we begin with the functional integral

$$Z = \int d\bar\psi d\psi dA \exp\{i\int[\operatorname{tr}F^2/2 + i\bar\psi(\not\partial + A)\psi]d^3x\} \quad (1)$$

and integrate over the fermion fields:

$$Z = \int dA \exp\{i(\int \mathrm{tr}(F^2/2)d^3x + I_{\mathrm{eff}}[A])\}, \qquad (2)$$

where

$$I_{\mathrm{eff}}[A] = -i \ln\det(\not\partial + \not A). \qquad (3)$$

I use the usual matrix notation $A = gT^a A_\mu{}^a$, where T^a are anti-Hermitian generators of the group. For definiteness, we work with SU(2) and a doublet of fermions—where $T^a = \sigma^a/2i$, and σ^a are the Pauli matrices—but the results hold in any group for which Π_3 is the additive group of integers, Z, and the fermions are in the fundamental representation. The Dirac matrices in three dimensions are Pauli matrices $(\sigma_3, i\sigma_2, i\sigma_1)$.

I now demonstrate that det $(\not\partial + \not A)$ and, by (3), the effective action $I_{\mathrm{eff}}[A]$ are not gauge invariant. More precisely, I show that

$$\det(\not\partial + \not A) \to (-1)^{|n|}\det(\not\partial + \not A) \qquad (4)$$

under a homotopically nontrivial gauge transformation, U_n, with winding number n. By (3) this is equivalent to $U_n: I_{\mathrm{eff}}[A] \to I_{\mathrm{eff}}[A] \pm \pi|n|$.

To prove (4), we follow closely the analogous calculations performed by Witten in four dimensions.[3] We use here a Euclidean formulation of the three-dimensional theory, and we consider gauge transformations which approach the identity at large distances; hence the base manifold is S_3 rather than R_3.

To begin, we observe that det $i\sigma_\mu(\partial_\mu + A_\mu)$, $\mu = 1, 2, 3$, may be written $\det^{1/2} i\not D_4$, where $\not D_4 = \gamma_\mu(\partial_\mu + A_\mu)$ and

$$\gamma_\mu = \begin{pmatrix} \sigma_\mu & 0 \\ 0 & -\sigma_\mu \end{pmatrix}. \qquad (5)$$

Since there exists a 4×4 matrix which anticommutes with $i\not D_4$, the spectrum of $i\not D_4$ is symmetric about zero (and is real because is $i\not D_4$ is Hermitian). Therefore, for a particular gauge field, A_μ, we may define the square root as the product of the positive eigenvalues of $i\not D_4[A_\mu]$. (It is of course assumed that there is no zero eigenvalue.) Since det$i\not D_4$ can be regulated by the introduction of a parity-invariant *Dirac* mass,[8] this procedure maintains parity as a good symmetry.

We now vary the gauge field along a continuous path, parametrized by τ, from $A_\mu(x^\mu, \tau) = 0$ at $\tau = -\infty$ to the pure gauge $A_\mu(x^\mu, \tau) = U_n^{-1}\partial_\mu U_n$ at $\tau = +\infty$, where U_n belongs to the nth homotopy class (U_n has winding number n). The spectrum of $i\not D_4[A]$ at $\tau = -\infty$ is identical to the spectrum at $\tau = +\infty$. However, as τ is varied from $-\infty$ to $+\infty$, the gauge field must pass through configurations in field space which are not pure gauge, because U_n is not continuously deformable to the identity. Therefore, the eigenvalues of $i\not D_4$ may become rearranged as τ goes from $-\infty$ to $+\infty$. In particular, one or more eigenvalues which are positive at $\tau = -\infty$ may cross zero and become negative at $\tau = +\infty$ (see Fig. 1). The square root of the determinant, $\det^{1/2} i\not D[A]$, defined as the product of the positive eigenvalues of $i\not D_4$ at $\tau = -\infty$, will therefore change sign if the number of eigenvalues which flow from positive to negative values (or vice versa) is odd.

We now recognize that the family of vector potentials, $A_\mu(x^\mu, \tau)$, is equivalent to an instanton-like four-dimensional gauge field, A^i, in the gauge $A^4 = 0$ [the space $x^i = (x^\mu, \tau)$, $i = 1, 2, 3, 4$, is the cylinder $S_3 \times R$]. The remaining components of $A^i(x^\mu, \tau)$ vary *adiabatically* as a function of $\tau = x^4$ along the path considered above.

The number of zero crossings of the eigenvalues of $i\not D_4[A^\mu(\tau)]$ is related to the number of normalizable zero modes of the four-dimensional operator

$$\not D = \gamma_i(\partial_i + A_i), \quad i = 1, 2, 3, 4, \qquad (6)$$

with

$$\gamma^4 = \gamma^\tau = i\begin{pmatrix} 0 & I \\ -I & 0 \end{pmatrix}.$$

To see this, we write the Dirac equation $\not D\psi = 0$ as

$$d\psi/d\tau = -\gamma^\tau \not D_4 \psi. \qquad (7)$$

Equation (7) is soluble in the adiabatic approximation; we choose $\psi(x^\mu, \tau) = f(\tau)\varphi^\tau(x^\mu)$, where $\varphi^\tau(x^\mu)$ satisfies the eigenvalue equation

$$\gamma^\tau \not D_4 \varphi^\tau(x^\mu) = \lambda(\tau)\varphi^\tau(x^\mu). \qquad (8)$$

Since the spectrum of $\gamma^\tau \not D_4$ is equal to the spectrum of $i\not D_4$, the eigenvalues $\lambda(\tau)$ vary continuous-

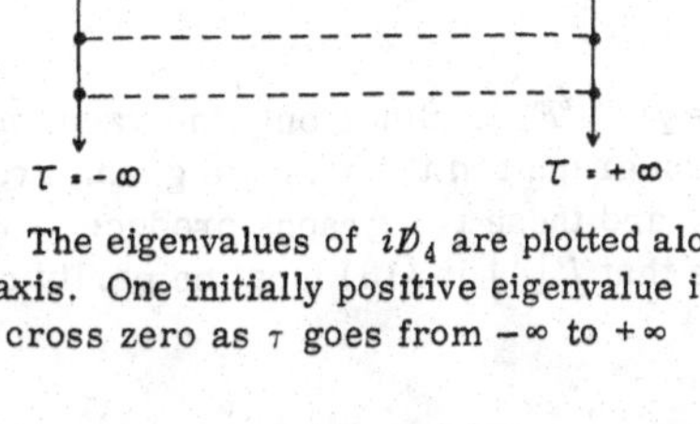

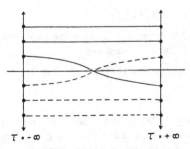

FIG. 1. The eigenvalues of $i\not D_4$ are plotted along the vertical axis. One initially positive eigenvalue is shown to cross zero as τ goes from $-\infty$ to $+\infty$

ly along the curves of Fig. 1. In the adiabatic approximation, Eq. (7) becomes

$$df/d\tau = -\lambda(\tau)f(\tau) \quad (9)$$

which has the solution

$$f(\tau) = f(0)\exp[-\int_0^\tau d\tau'\lambda(\tau')]. \quad (10)$$

Only if λ is positive for $\tau = +\infty$ and negative for $\tau = -\infty$ is this solution normalizable.

Therefore, there exists a one to one correspondence between the number of normalizable zero modes of $\not{D}[A^i]$ and the number of eigenvalues of $i\not{D}_4[A^\mu(\tau)]$ which pass from negative to positive (or from positive to negative) values as τ is varied from $-\infty$ to $+\infty$. The number of zero modes of $\not{D}[A^i]$ is well known from instanton studies.[9] If we choose the eigenfunctions of $\not{D}\psi_\pm = 0$ to be eigenfunctions of γ_5, $\gamma_5\psi_\pm = \pm\psi_\pm$, and denote the number of zero modes of ψ_+ (ψ_-) by n_+ (n_-), then

$$n_- - n_+ = n, \quad (11)$$

where n is the instanton number (the winding number of U_n). Therefore, the total number of zero modes, N_T, of $\not{D}$ is

$$N_T = 2n_+ + n, \quad (12)$$

which is odd if the winding number n is odd. This completes the proof of (4): The determinant changes sign under a large gauge transformation with odd winding number.

To show that gauge-invariant regularization procedures, such as Pauli-Villars regularization, restore gauge invariance at the cost of introducing parity nonconservation, I have performed two types of calculations—details will be given elsewhere.

First, one may calculate $\lim_{M\to\infty} I_{\rm eff}[A, M]$ in perturbation theory[2,10]; I find

$$I_{\rm eff}{}^R \equiv I_{\rm eff}[A] - \lim_{M\to\infty} I_{\rm eff}[A, M] = \pm\pi W[A] + I'[A], \quad (13)$$

where $I_{\rm eff}{}^R$ is finite and $W[A]$ is the parity-nonconserving Chern-Simons term

$$W[A] = (1/8\pi^2)\int d^3x\, {\rm tr}({}^*F_\mu A^\mu - \tfrac{1}{3}A^\mu A^\nu A^\alpha \epsilon_{\mu\nu\alpha}), \quad (14)$$

with ${}^*F^\mu = \frac{1}{2}\epsilon^{\mu\nu\alpha}F_{\nu\alpha}$. Since only the vacuum polarization graph and the triangle graph are divergent, and these two graphs produce $\pm\pi W[A]$, I conclude that $I'[A]$ in (13) must be parity conserving.

Second, in the special case of gauge fields with constant field strength tensor, $F^{\mu\nu}$, the effective action (3) can be calculated exactly. The calculational procedure is the three-dimensional analog of the Euler-Heisenberg calculation of the effective action for constant field strength, as presented by Schwinger.[11] I find

$$I_{\rm eff} = \pm W[A] + I_{\rm NA}[A] \quad (15)$$

where $I_{\rm NA}[A]$ is parity conserving, but is nonanalytic in the gauge field. In performing this calculation, it is necessary to introduce a parity-nonconserving fermion mass, M, to regulate divergences in a gauge-invariant manner.[2] When the mass is set equal to zero at the end of the calculation, however, the parity-nonconserving term, $\pm\pi W[A]$, does not vanish. (The sign depends upon the sign of the regulator mass.) The existence of the nonanalytic expression, $I_{\rm NA}[A]$, is characteristic of three-dimensional gauge theories coupled to massless fermions, where such terms are known to appear in partial sums of infrared-divergent Feynman diagrams.[12] For the even more special case of an Abelian, constant field strength, the ground-state current is given exactly by

$$\langle J_a\rangle = (g^2/8\pi)\,{}^*F_a{}^\mu. \quad (16)$$

This current is identically conserved, but it violates parity conservation explicitly since ${}^*F^\mu$ is a pseudovector.

The discovery of anomalous parity nonconservation in three dimensions has wide ranging consequences. First, the violation of space-time reflection in three dimensions may have direct, measurable consequences in condensed matter physics, where models of vortex-particle interactions in superconductors are mathematically equivalent to three-dimensional QED—while the discussion here is limited to SU(N) theories, parity nonconservation occurs in $\langle J^\mu\rangle$ in QED as well. The relativistic Dirac equation used here appears also in a nonrelativistic system as is seen in the discovery of crystals which model the four-dimensional anomaly.[12] At high temperatures, four-dimensional field theories effectively reduce to three-dimensional theories. Although fermions are known to decouple at high temperatures, the effective theory in three dimensions, obtained by "integrating out" the fermions, may still be relevant in the study of high-temperature four-dimensional theories. The induced topological term $\pm W[A]$ in $I_{\rm eff}[A]$ is known to produce a mass for the gauge fields.[2,13] Not only must parity conservation be violated in odd-dimensional theories with an odd number of fermions, but the gauge fields *must* become massive as well. Since the gauge fields acquire a mass without the need for Higgs fields, this implies that the analogous phenomenon in solid state physics, the Meissner effect, may be possible without the need for a condensate of fermion pairs. Above all, the discovery of anomalous parity nonconservation in three dimensions provides a different and simpler laboratory in which to study the subtle interplay between "anomalous" violation of symmetries and global topological properties of gauge theories.

Results similar to those presented in this paper have been obtained independently by L. Alvarez-Gaumé and E. Witten (to be published).

Recently, E. D'Hoker and E. Farhi have shown how to compensate for the gauge noninvariance of the fermion determinant in four-dimensional theories by the addition of bosonic terms to the Lagrangian.[14] Also, Niemi and Semenoff have shown how to derive the results presented here using the anomaly in two dimensions.[7]

I would like to thank R. Jackiw for suggesting this problem and for his continued assistance, as well as A. Guth and G. W. Semenoff for helpful conversations. I am also grateful to E. Witten for communicating to me his results, and emphasizing the necessity of parity nonconservation. This work was supported in part through funds provided by the U. S. Department of Energy under Contract No. DE-AC02-76ERO3069.

[1]The Dirac monopole is the oldest example of this.

[2]S. Deser, R. Jackiw, and S. Templeton, Phys. Rev. Lett. 48, 975 (1982), and Ann. Phys. (N.Y.) 140, 372 (1982); R. Jackiw, in *Asymptotic Realms of Physics*, edited by A. Guth, K. Wan, and R. Jaffe (Massachusetts Institute of Technology Press, Cambridge, Mass., 1983).

[3]E. Witten, Phys. Lett. 117B, 324 (1982).

[4]E. Witten, Nucl. Phys. B223, 422 (1983).

[5]R. Jackiw and C. Rebbi, Phys. Rev. Lett. 37, 172 (1976).

[6]P. H. Frampton and T. W. Kephart, Phys. Rev. Lett. 50, 1343 (1983).

[7]A. T. Niemi and G. W. Semenoff, Phys. Rev. Lett. 51, 2077 (1983).

[8]This argument is due to E. Witten.

[9]For a review, see R. Jackiw, C. Nohl, and C. Rebbi, in *Particles and Fields*, edited by D. H. Boal and A. W. Kamal (Plenum, New York, 1978).

[10]I. Affleck, J. Harvey, and E. Witten, Nucl. Phys. B206, 413 (1982).

[11]J. Schwinger, Phys. Rev. 82, 664 (1951).

[12]S. Templeton, Phys. Lett. 103B, 132 (1981), and Phys. Rev. D 24, 3134 (1981); E. I. Guendelman and Z. M. Radulovic, Phys. Rev. D 27, 357 (1983).

[13]W. Siegel, Nucl. Phys. B156, 135 (1979); R. Jackiw and S. Templeton, Phys. Rev. D 23, 2291 (1981); J. Schonfeld, Nucl. Phys. B185, 157 (1981).

[14]E. D'Hoker and E. Farhi, Massachusetts Institute of Technology Report No. CTP 1111 (to be published).

PHYSICAL REVIEW D VOLUME 12, NUMBER 12 15 DECEMBER 1975

Concept of nonintegrable phase factors and global formulation of gauge fields

Tai Tsun Wu*
Gordon McKay Laboratory, Harvard University, Cambridge, Massachusetts 02138

Chen Ning Yang†
Institute for Theoretical Physics, State University of New York, Stony Brook, New York 11794
(Received 8 September 1975)

Through an examination of the Bohm-Aharonov experiment an intrinsic and complete description of electromagnetism in a space-time region is formulated in terms of a nonintegrable phase factor. This concept, in its global ramifications, is studied through an examination of Dirac's magnetic monopole field. Generalizations to non-Abelian groups are carried out, and result in identification with the mathematical concept of connections on principal fiber bundles.

I. MOTIVATION AND INTRODUCTION

The concept of the electromagnetic field was conceived by Faraday and Maxwell to describe electromagnetic effects in a space-time region. According to this concept, the field strenght $f_{\mu\nu}$ describes electromagnetism. It was later realized,[1] however, that $f_{\mu\nu}$ by itself does not, in quantum theory, completely describe all electromagnetic effects on the wave function of the electron. The famous Bohm-Aharonov experiment, first beautifully performed by Chambers,[2] showed that in a multiply connected region where $f_{\mu\nu}=0$ everywhere there are physical experiments for which the outcome depends on the loop integral

$$\frac{e}{\hbar c} \oint A_\mu dx^\mu \tag{1}$$

around an unshrinkable loop. This raises the question of what constitutes *an intrinsic and complete description* of electromagnetism. In the present paper we wish to discuss this question and also its generalization to non-Abelian gauge fields.

An examination of the Bohm-Aharonov experiment indicates that in fact only *the phase factor*

$$\exp\left(\frac{ie}{\hbar c} \oint A_\mu dx^\mu\right), \tag{2}$$

and *not the phase* (1), is physically meaningful. In other words, the phase (1) contains more information than the phase factor (2). But the additional information is not measurable. This simple point, probably implicitly recognized by many authors, is discussed in Sec. II. It leads to the concept of nonintegrable (i.e., path-dependent) phase factor as the basis of a description of electromagnetism.

This concept has been taken[3] as the basis of the definition of a gauge field. The discussions in Ref. 3, however, centered only on the local properties of gauge fields. To extend the concept to global problems we analyze in Sec. III the field produced by a magnetic monopole. We demonstrate how the quantization of the pole strength, a striking result due to Dirac,[4] is understood in this concept of electromagnetism. The demonstration is closely related to that in the original Dirac paper. Dirac discussed the phase factor of the wave function of an electron (which, among other things, depends on the electron energy). Our emphasis is on the nonintegrable electromagnetic phase factor (which does not depend on such quantities as the energy of the electron).

The monopole discussion leads to the recognition that in general the phase factor (and indeed the vector potential A_μ) can only be properly defined in each of many overlapping regions of space-time. In the overlap of any two regions there exists a gauge transformation relating the phase factors defined for the two regions. This discussion is made more precise in Sec. IV. It leads to the definition of global gauges and global gauge transformations.

In Sec. V generalizations to non-Abelian gauge groups are made. The special cases of SU_2 and SO_3 gauge fields are discussed in Secs. VI and VII. A surprising result is that the monopole types are quite different for SU_2 and SO_3 gauge fields and for electromagnetism.

The mathematics of these results is in fact well known to the mathematicians in *fiber bundle theory*. An identification table of terminologies is given in Sec. V. We should emphasize that our interest in this paper does not lie in the beautiful, deep, and general mathematical development in fiber bundle theory. Rather we are concerned with the necessary *concepts to describe the physics of gauge theories*. It is remarkable that these concepts have already been intensively studied as mathematical constructs.

Section VII discusses a "*gedanken*" generalized Bohm-Aharonov experiment for SU_2 gauge fields.

Unfortunately, the experiment is not feasible unless the mass of the gauge particle vanishes. In the last section we make several remarks.

II. DESCRIPTION OF ELECTROMAGNETISM

The Bohm-Aharonov experiment explores the electromagnetic effect on an electron beam (Fig. 1) in a doubly connected region where the electromagnetic field is zero. As predicted[1] by Aharonov and Bohm, the fringe shift is dependent on the phase factor (2), which is equal to

$$\exp\left(\frac{-ie}{\hbar c}\Omega\right),$$

where Ω is the magnetic flux in the cylinder. Thus two cases a and b for which

$$\Omega_a - \Omega_b = \text{integer} \times (hc/e) \tag{3}$$

give the same interference fringes in the experiment. This we shall state and prove as follows.

Theorem 1: If (3) is satisfied, no experiment outside of the cylinder can differentiate between cases a and b.

Consider first an electron outside of the cylinder. We look for a gauge transformation on the electron wave function ψ_a and the vector potential $(A_\mu)_a$ for case a, which changes them into the corresponding quantities for case b, i.e. we try to find $S = e^{-i\alpha}$ such that

$$S = S_{ab} = (S_{ba})^{-1},$$

$$\psi_b = S^{-1}\psi_a, \quad \text{or } \psi_b = e^{i\alpha}\psi_a, \tag{4}$$

$$(A_\mu)_b = (A_\mu)_a - \frac{i\hbar c}{e} S \frac{\partial S^{-1}}{\partial x^\mu}, \quad \text{or } (A_\mu)_b = (A_\mu)_a + \frac{\hbar c}{e}\frac{\partial\alpha}{\partial x^\mu}. \tag{5}$$

For this gauge transformation to be definable, S must be *single-valued*, but α itself need not be. Now $(A_\mu)_b - (A_\mu)_a$ is curlless; hence (5) can always be solved for α. But it is multiple-valued with an increment of

$$\Delta\alpha = \frac{e}{\hbar c}\oint [(A_\mu)_b - (A_\mu)_a]\, dx^\mu$$

$$= \frac{e}{\hbar c}(\Omega_b - \Omega_a) \tag{6}$$

every time one goes around the cylinder. If (3) is satisfied, $\Delta\alpha = 2\pi \times \text{integer}$ and S is single-valued. Case a and case b outside of the cylinder are then gauge-transformable into each other, and no physically observable effects would differentiate them. The same argument obviously holds if one studies the wave function of an interacting system of particles provided the charges of the particles are all integral multiples of e. Thus we have shown the validity of Theorem 1.

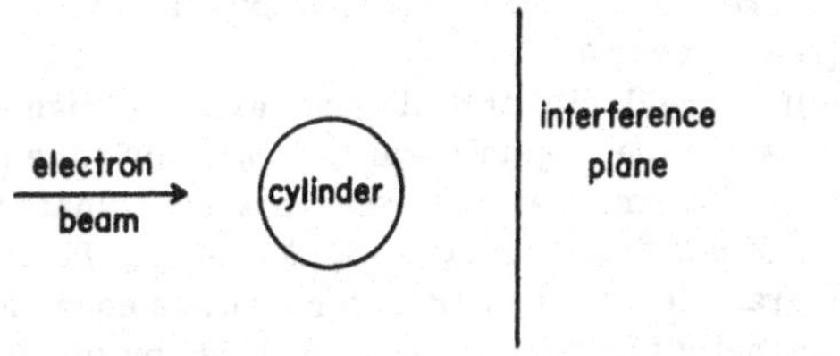

FIG. 1. Bohm-Aharonov experiment (Refs. 1, 2). A magnetic flux is in the cylinder. Outside of the cylinder the field strength $f_{\mu\nu} = 0$.

We conclude: (a) The field strength $f_{\mu\nu}$ underdescribes electromagnetism, i.e., different physical situations in a region may have the same $f_{\mu\nu}$. (b) The phase (1) overdescribes electromagnetism, i.e., different phases in a region may describe the same physical situation. What provides a complete description that is neither too much nor too little is the phase factor (2).

Expression (2) is less easy to use (especially when one makes generalizations to non-Abelian groups) as a fundamental concept than the concept of a phase factor for any path from P to Q

$$\Phi_{QP} = \exp\left(\frac{ie}{\hbar c}\int_P^Q A_\mu dx^\mu\right) \tag{7}$$

provided that an arbitrary gauge transformation

$$\exp\left(\frac{ie}{\hbar c}\int_P^Q A_\mu dx^\mu\right)$$

$$\to \exp\left(\frac{ie}{\hbar c}a(Q)\right)\exp\left(\frac{ie}{\hbar c}\int_P^Q A_\mu dx^\mu\right)\exp\left(\frac{-ie}{\hbar c}a(P)\right) \tag{8}$$

does not change the prediction of the outcome of any physical measurements. Following Ref. 3, we shall call the phase factor (7) a nonintegrable (i.e., path-dependent) phase factor.

Electromagnetism is thus the gauge-invariant manifestation of a nonintegrable phase factor. We shall develop this theme further in the next section.

III. FIELD DUE TO A MAGNETIC MONOPOLE

The definition of a nonintegrable phase factor (7) in a general case may present problems. To illustrate the problem, let us study the magnetic monopole field of Dirac.[4] Consider a static magnetic monopole of strength $g \neq 0$ at the origin $\vec{r} = 0$ and take the region R of space-time under consideration to be all space-time minus the origin $\vec{r} = 0$. We shall now show the following:

Theorem 2: There does not exist a singularity-free A_μ over all R.

If a singularity-free A_μ does exist throughout R, consider the loop integral $\oint A_\mu dx^\mu$ for time $t=0$ around a circle at fixed spherical coordinates r and θ with azimuthal angle $\phi = 0 \to 2\pi$. This integral, denoted by $\Omega(r,\theta)$ for $r>0$, is equal to the magnetic flux through a cap bounded by the loop, or more explicitly $\Omega(r,\theta) = 2\pi g(1-\cos\theta)$. At $\theta=0$, $\Omega(r,0)=0$. Increasing θ leads to a continuous increase in Ω till one approaches $\theta=\pi$, at which

$$\Omega(r,\pi) = 4\pi g\,. \tag{9}$$

But at $\theta=\pi$ the loop shrinks to a point. Therefore $\Omega(r,\pi)=0$ since A_μ has no singularity. We have thus reached a contradiction and Theorem 2 is proved.

With an A_μ which has singularities, the nonintegrable phase factor becomes undefined if the path goes through a singularity. This difficulty *must* be resolved in order to use a nonintegrable phase factor as a fundamental concept to describe electromagnetism. It can be resolved in the following way. Let us seek to divide R into two overlapping regions R_a and R_b and to define $(A_\mu)_a$ and $(A_\mu)_b$, each singularity-free in their respective regions, so that (i) their curls are equal to the magnetic field and (ii) in the overlapping region $(A_\mu)_a$ and $(A_\mu)_b$ are related by a gauge transformation. One possible choice is to take the regions to be

$$\begin{aligned} R_a&:\ 0\le\theta<\pi/2+\delta \quad 0<r,\quad 0\le\phi<2\pi,\ \text{all } t\\ R_b&:\ \pi/2-\delta<\theta\le\pi \quad 0<r,\quad 0\le\phi<2\pi,\ \text{all } t \end{aligned} \tag{10}$$

with an overlap extending throughout $\pi/2-\delta<\theta<\pi/2+\delta$. (We assume $0<\delta\le\pi/2$.) Take

$$\begin{aligned} (A_t)_a=(A_r)_a=(A_\theta)_a=0\,,\quad (A_\phi)_a=\frac{g}{r\sin\theta}(1-\cos\theta),\\ (A_t)_b=(A_r)_b=(A_\theta)_b=0\,,\quad (A_\phi)_b=\frac{-g}{r\sin\theta}(1+\cos\theta)\,. \end{aligned} \tag{11}$$

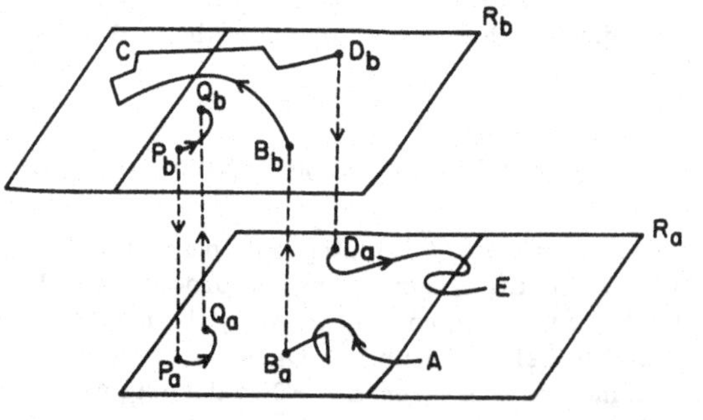

FIG. 2. Schematic diagram illustrating the relationship between R_a and R_b.

The gauge transformation in the overlap of the two regions is

$$S=S_{ab}=\exp(-i\alpha)=\exp\left(\frac{2ige}{\hbar c}\phi\right). \tag{12}$$

This is an allowed gauge transformation if and only if S is single-valued, i.e.,

$$\frac{2ge}{\hbar c}=\text{integer}=D\,, \tag{13}$$

which is Dirac's quantization. With (13) we have

$$S_{ab}=\exp(iD\phi)\,. \tag{12'}$$

To define the phase factor for a path we refer to Fig. 2, where a point in the overlapping region, such as point P, is regarded as two points P_a and P_b. If a path is entirely within region a or b, we define Φ along the path by (7) with $(A_\mu)_a$ or $(A_\mu)_b$ in the integrand in the exponent. If the path $Q\leftarrow P$ is entirely within the overlapping region we have then two possible phase factors $\Phi_{Q_aP_a}$ and $\Phi_{Q_bP_b}$. It is easy to prove that

$$\Phi_{Q_bP_b}=S^{-1}(Q)\Phi_{Q_aP_a}S(P)\,, \tag{14}$$

i.e.,

$$\Phi_{Q_aP_a}S(P)=S(Q)\Phi_{Q_bP_b}\,, \tag{14'}$$

which merely states that $(A_\mu)_a$ and $(A_\mu)_b$ are related by a gauge transformation with the transformation factor (12).

For a path that crisscrosses in and out of the overlapping region, such as $A\to B\to C\to D\to E$ in Fig. 2, the definition of Φ is

$$\Phi_{EDCBA}=\Phi_{ED_a}S_{ab}(D)\Phi_{D_bCB_b}S_{ba}(B)\Phi_{B_aA}\,. \tag{15}$$

Notice that fixing the path but sliding the points B and D along it does not change Φ_{EDCBA} [because of formulas like (14′)] so long as B and D remain in the overlapping region.

The phase factor so defined satisfies the group property, e.g.,

$$\begin{aligned}\Phi_{EDCBA}&=\Phi_{ED_a}\Phi_{D_aCBA}\\ &=\Phi_{ED_b}\Phi_{D_bCBA}\\ &=\Phi_{EDC}\Phi_{CBA},\ \text{etc.}\end{aligned} \tag{16}$$

The relationship between the electromagnetic field and the phase factor around a loop is the same as usual. One only has to be careful that if the starting and terminating point A is in the overlapping region, the phase factor is taken to be $\Phi_{A_aBA_a}=\Phi_{A_bBA_b}$, and not $\Phi_{A_aBA_b}$ or $\Phi_{A_bBA_a}$. The phase factor around the loop is then equal to

$$\exp\left(\frac{ie}{\hbar c}\right)\Omega\,,$$

where Ω is the magnetic flux through a cap bordered by the loop. Notice that because of Dirac's quantization condition, the phase factor is the same whichever way one chooses the cap provided it does not pass through the point $\vec{r}=0$ (any t).

We have satisfactorily resolved the difficulty mentioned at the beginning of this section, provided Dirac's quantization condition (13) is satisfied. We shall now prove the following.

Theorem 3: If (13) is not satisfied (the above method of resolving the difficulty would not work since) there exists no division of R into overlapping regions $R_a, R_b, R_c, \ldots$ so that condition (i) and (ii) stated above, properly generalized to the case of more than two regions, would hold.

To prove this statement, observe that if such a division is possible, one could generalize (15) and arrive at a satisfactory definition of the phase factor. The phase factor around a loop is then a continuous function of the loop. Take the loop to be a parallel on the sphere r fixed, $t=0$, θ fixed, $\phi=0-2\pi$. The phase factor defined by the generalization of (15) is equal to

$$\exp\left[\frac{ie}{\hbar c}\Omega(r,\theta)\right]=\exp\left[\frac{ie}{\hbar c}2\pi g(1-\cos\theta)\right]. \quad (17)$$

This is not equal to unity when $\theta=\pi$, since (13) is assumed to be invalid. Thus we have a contradiction.

Theorem 3 shows that if Dirac's quantization condition (13) is not satisfied, then the field of a magnetic monopole of strength g cannot be taken as a realizable physical situation in R. (Of course, if one excludes the half-line $x=y=0$, $z<0$, or any half-line starting from $\vec{r}=0$ leading to infinity, then it is possible to have any value for g.) This conclusion is the same as Dirac's, but viewed from a somewhat different point of emphasis.

IV. GENERAL DEFINITION OF GAUGE AND GLOBAL GAUGE TRANSFORMATION

Assuming that (13) holds, to round out our concept of a nonintegrable phase factor the question of the flexibility in the choice of the overlapping regions and the flexibility in the choice of A_μ in the regions must be faced. Both of these questions are related to gauge transformations.

Consider a gauge transformation ξ in R_b (ξ will be assumed to be many times differentiable, but not necessarily analytic), resulting in a new potential $(A_\mu)'_b$. We shall illustrate schematically the transformation by "elevating" the region b in Figure 3(a).

One could extend the region b. One could also contract it, provided the whole R remain covered.

One could create a new region by considering a subregion of b as an additional region R_c [Figure 3(b)], and define the gauge transformation connecting them as the identity transformation so that $(A_\mu)_c=(A_\mu)_b$. One can then "elevate" R_c and contract R_b, which results in Fig. 3(c).

Through operations of the kind mentioned in the last three paragraphs, which we shall call *distortions*, we arrive at a large number of possibilities, each with a particular choice of overlapping regions and with a particular choice of gauge transformation from the original $(A_\mu)_a$ or $(A_\mu)_b$ to the new A_μ in each region. Each of such possibilities will be called a *gauge* (or *global gauge*). This definition is a natural generalization of the usual concept, extended to deal with the intricacies of the field of a magnetic monopole.

For each choice of gauge there is a definition of a nonintegrable phase factor for every path. The group condition $\Phi_{C_cBA_a}=\Phi_{C_cB_b}\Phi_{B_bA_a}$ is always satisfied.

Notice that the original gauge we started with was characterized by (a) specifying [in (10)] the regions [R_a and R_b] and (b) specifying the gauge transformation factor (12′) in the overlap (between R_a and R_b). *It does not refer to any specific A_μ.* [A distortion may of course lead to no changes in characterizations (a) and (b). Thus two different gauges may share the same characterizations (a) and (b).] In the case of the monopole field, we had chosen the vector potential to be given by (11). But, in fact, we can attach to this gauge any $(A_\mu)_a$ and $(A_\mu)_b$ provided they are gauge-transformed into each other by (12′) in the region of overlap. (The resultant $f_{\mu\nu}$ is, of course, not a monopole field in general.) *Thus a gauge is a concept not tied to any specific vector potential.* We shall call the process of distortion leading from one gauge to another a *global gauge transformation.* It is also a concept not tied to any specific vector potential. It is a natural generalization of the usual gauge transformation.

The collection of gauges that can be globally gauge-transformed into each other will be said to

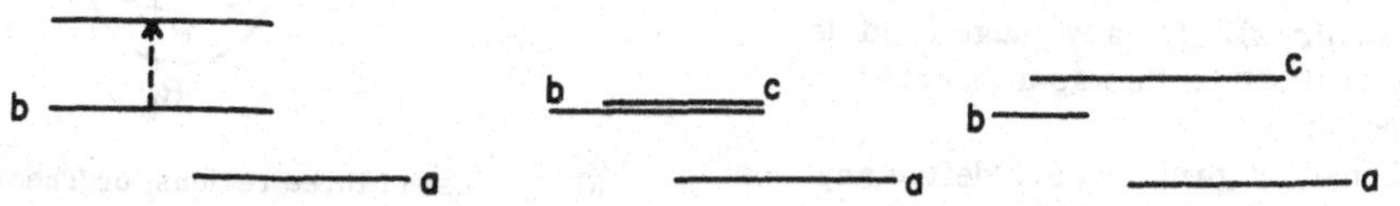

FIG. 3. Distortions allowed in gauge transformation.

belong to the same *gauge type*.

The phase factor around a loop starts and ends at the same point in the same region. Thus it does not change under any global gauge transformation, i.e. we have, for Abelian gauge fields, the following.

Theorem 4a: The phase factor around any loop is invariant under a global gauge transformation.

It follows trivially from this, by taking an infinitesimal loop, that

Theorem 5a: The field strength $f_{\mu\nu}$ is invariant under a global gauge transformation.

For a given value of D, the gauge defined by (10) and (12) will be denoted by $\mathcal{G}_D$. For $D \neq D'$, the relationship, or rather the lack of relationship, between $\mathcal{G}_D$ and $\mathcal{G}_{D'}$ is shown by Theorem 6.

Theorem 6: For $D \neq D'$, $\mathcal{G}_D$ and $\mathcal{G}_{D'}$ are not related by a global gauge transformation, i.e., they are not of the same gauge type.

To prove this theorem we use Theorem 7.

Theorem 7: Between two gauge fields defined on the same gauge there exists a continuous interpolating gauge field defined on the same gauge.

To prove Theorem 7, we simply make a linear interpolation between the two original gauge fields which we shall denote by $(A_\mu)^{(\alpha)}$ and $(A_\mu)^{(\beta)}$:

$$A^{(\gamma)} = t(A_\mu)^{(\alpha)} + (1-t)(A_\mu)^{(\beta)}, \quad 0 \leq t \leq 1. \tag{18}$$

In an overlap between regions a and b this interpolating vector potential assumes values $(A_\mu)_a^{(\gamma)}$ and $(A_\mu)_b^{(\gamma)}$ which are related by the proper gauge transformation belonging to this overlap. Thus we have proved Theorem 7.

Now go back to Theorem 6 and assume it to be invalid. Then we can gauge-transform the vector potential belonging to the monopole of strength $D'\hbar c/2e$ to the gauge $\mathcal{G}_D$. For this gauge we have then two monopole fields of different pole strengths. Using Theorem 7 we interpolate between them and obtain unquantized magnetic monopoles, which contradict Theorem 3.

Notice that although in this proof of Theorem 6 we have used two specific gauge fields, the theorem itself does not refer to any specific gauge fields at all.

By the same argument as used in the proof of Theorem 7, any gauge field defined on $\mathcal{G}_D$ must have a magnetic monopole of strength $D\hbar c/2e$ at the excluded point $\vec{r}=0$, in addition to possible fields produced by electric charges and currents. Thus the total magnetic flux around the origin $\vec{r}=0$ is equal to $(2\pi\hbar c/e)D$ for any gauge field defined on $\mathcal{G}_D$. We shall state this as a theorem and give another proof of it.

Theorem 8: Consider gauge $\mathcal{G}_D$ and define any gauge field on it. The total magnetic flux through a sphere around the origin $\vec{r}=0$ is *independent of the gauge field* and only depends on the gauge:

$$\oiint f_{\mu\nu}\, dx^\mu dx^\nu = \frac{-i\hbar c}{e} \oint \frac{\partial}{\partial x^\mu}(\ln S_{ab})\, dx^\mu, \tag{19}$$

where S is the gauge transformation defined by (12) for the gauge $\mathcal{G}_D$ in question, and the integral is taken around any loop around the origin $\vec{r}=0$ in the overlap between R_a and R_b, such as the equator on a sphere $r=1$.

To prove this theorem we observe that the flux through the upper half of the sphere $r=1$ is equal to the following integral around the equator:

$$\oint (A_\mu)_a dx^\mu. \tag{20a}$$

The flux through the lower half is equal to a similar integral around the equator:

$$-\oint (A_\mu)_b dx^\mu. \tag{20b}$$

Hence

$$\text{total flux} = \oint [(A_\mu)_a - (A_\mu)_b]\, dx^\mu$$

$$= \frac{-i\hbar c}{e} \oint \frac{\partial}{\partial x^\mu}(\ln S_{ab})\, dx^\mu, \tag{21}$$

which completes the proof. Using (13) and (12), the right-hand side of (21) is equal to $4\pi g$, as expected.

If one starts with any gauge which is of the same gauge type as $\mathcal{G}_D$, and makes a global gauge transformation on it, the total flux is not changed by Theorem 5a. Thus (19), which depends only on the gauge, is in fact the same for all gauges of the same type. Notice that if there are more regions in a gauge than two, (19) should be replaced by a sum of line integrals along paths that are in the various overlaps between the regions. For a case of three regions there are three paths, which are illustrated in Fig. 4. Along each path the integral is of the form (19) with S denoting the gauge transformation factor, such as (12), between the two regions containing the path. To prove Theorem 8 in this case one need only add three loop integrals to-

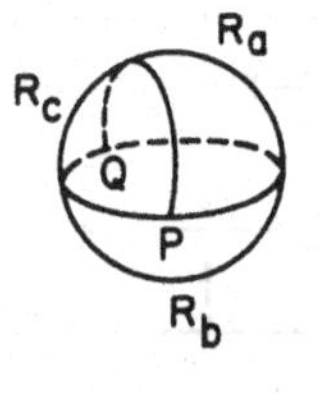

FIG. 4. Case of three regions for Theorem 8. The three paths from P to Q are in the three overlapping regions between (R_a, R_b), (R_b, R_c), and (R_c, R_a).

gether, each of the form of (20a) and (20b), and notice that along each path the integrand is always the difference of the vector potential A_μ between two regions, very much as in (21).

The first proof we gave above of Theorem 8 is easy and is "obvious" to a physicist. The second proof is more involved but is more intrinsic. The theorem is a special case of the Chern-Well theorem which evolved from the famous Gauss-Bonnet-Allendoerfer-Weil-Chern theorem, a seminal development in contemporary mathematics.[5] We want to emphasize two consequences of the theorem. (i) The right-hand side of (19) is independent of the gauge field, and only depends on the gauge type. (ii) The right-hand side of (19) has as integrand the gradient of $\ln S$. Since S is single-valued, the integral must be equal to an *integral multiple* of a constant (in this case $2\pi i$). A remarkable fact is that these consequences remain valid in the general mathematical theorem, which is very deep.

V. GENERALIZATION TO NON-ABELIAN GAUGE FIELD

So far we have only considered electromagnetism and described it in terms of an Abelian gauge field that corresponds to the group U_1, or equivalently SO_2. On the basis of the discussions in the preceding section, the generalization to the non-Abelian case can be carried out without much difficulty. For a local region this has been done in Ref. 3. Extension to global considerations is our present focus of interest.

A gauge is defined by (a) a particular choice of overlapping regions and (b) a particular choice of *single-valued* gauge transformations S_{ab} in the overlapping regions. The choice of gauge transformations clearly must satisfy the following two conditions.

(1) In the overlapping region $R_a \cap R_b$, the gauge transformations S_{ba} from a to b and S_{ab} from b to a are related by

$$S_{ab}S_{ba}=1\,,$$

where 1 is the identity element of the gauge group.

(2) If three regions R_a, R_b, and R_c overlap, then there are gauge transformations $S_{ab}, S_{ba}, S_{ac}, S_{ca}, S_{bc}, S_{cb}$ so that

$$S_{ab}S_{bc}S_{ca}=1\,,\ \text{etc.}$$

in $R_a \cap R_b \cap R_c$.

As in the case of electromagnetism, both the concept of a gauge and the concept of a global gauge transformation are not tied to any specific gauge potentials, denoted in general by b^k_μ.

The *nonintegrable phase factor* for a given path is now an element of the gauge group. We shall still call it a phase factor. Since these phase factors do not in general commute with each other, Theorems 4a and 5a for the Abelian case need to be modified as follows.

Theorem 4: Under a global gauge transformation, the phase factor around any loop remains in the same class. The class does not depend on which point is taken as the starting point around the loop.

Theorem 5: The field strength $f^k_{\mu\nu}$ is covariant under a global gauge transformation.

Only theorem 4 is not immediately transparent. For a loop $ABCA$, under a gauge transformation[3]

$$\Phi_{ABCA} \to \Phi'_{ABCA} = \xi(A)\Phi_{ABCA}\xi^{-1}(A)\,.$$

Thus Φ'_{ABCA} and Φ_{ABCA} are in the same class. Also around the same loop if we change the starting point from A to C,

$$\Phi_{CABC} = \Phi_{CA}\Phi_{ABCA}\Phi_{AC}\,.$$

Hence changing the starting point does not change the class.

Theorem 4 defines *the class of a loop.* This concept is the generalization of the phase factor for electromagnetism around a loop with the magnetic flux as the exponent. It is a gauge-invariant concept.

These concepts have been extensively studied by the mathematicians in the framework of more general[6] mathematical constructs. A translation of terminology is given in Table I.

VI. CASE OF SU_2 GAUGE FIELD

For the SU_2 case we take the infinitesimal generators X_k to satisfy

$$X_1X_2 - X_2X_1 = X_3\,,\ \text{etc.} \tag{22}$$

and define the phase factor, as a generalization of (7), by[7]

$$\Phi_{QP} = \left[\exp\left(\int_P^Q \frac{-e}{\hbar c} b^k_\mu X_k dx^\mu\right)\right]_{\text{ordered}}, \tag{23}$$

i.e., we make the replacement

$$ieA_\mu \to -eb^k_\mu X_k\,, \tag{24}$$

or

$$A_\mu \to ib^k_\mu X_k\,. \tag{25}$$

[The subscript "ordered" means that, in the definition of the exponential in terms of a power series, the factors $b^k_\mu X_k$ are ordered along the path from P to Q with the factor $b^k_\mu(P)X_k$ at the right end of the product.] The algebraic operators X_k can be thought of as the collection of all irreducible representations of (22). The eigenvalue of iX_k with the

TABLE I. Translation of terminology.

Gauge field terminology	Bundle terminology
gauge (or global gauge)	principal coordinate bundle
gauge type	principal fiber bundle
gauge potential b_μ^k	connection on a principal fiber bundle
S_{ba} (see Sec. V)	transition function
phase factor Φ_{QP}	parallel displacement
field strength $f_{\mu\nu}^k$	curvature
source [a] J_μ^K	?
electromagnetism	connection on a $U_1(1)$ bundle
isotopic spin gauge field	connection on a SU_2 bundle
Dirac's monopole quantization	classification of $U_1(1)$ bundle according to first Chern class
electromagnetism without monopole	connection on a trivial $U_1(1)$ bundle
electromagnetism with monopole	connection on a nontrivial $U_1(1)$ bundle

[a] I.e., electric source. This is the generalization (see Ref. 3) of the concept of electric charges and currents.

minimum absolute value is $\pm\frac{1}{2}$. Therefore the minimum "charge" of all physical states can be read off from (24) by taking the 2×2 irreducible representation of X_k:

$$X_k = -\frac{i\sigma_k}{2}, \qquad (26)$$

where σ_k are the Pauli matrices. Thus

$$\text{minimum "charge"} = \frac{e}{2}. \qquad (27)$$

The particle of the gauge field belongs to the adjoint representation. Its "charges" are e, 0, and $-e$. Thus

$$\frac{\text{"charge" of gauge particle}}{\text{minimum "charge"}} = 2 \text{ for } SU_2. \qquad (28)$$

We shall now try to define a Dirac monopole field as a special SU_2 field along only one isospin direction $k=3$, i.e., we define

$$b_\mu^1 = b_\mu^2 = 0, \quad b_\mu^3 = A_\mu, \qquad (29)$$

where A_μ is given in the two regions (10) by (11). In the overlapping region, transformation factor S of (12) and (14) now becomes

$$S_{ab} = \exp\left(-\frac{2ge}{\hbar c}\phi X_3\right) \qquad (30)$$

by replacement (25). This is single-valued if and only if the quantization condition

$$\frac{eg}{\hbar c} = \text{integer} = D \qquad (31)$$

is satisfied because for SU_2

$$\exp(4\pi X_3) = 1, \quad \exp(2\pi X_3) \neq 1,$$

which follows from the existence of half-integral representations such as (26).

The phase factor (30) describes a great circle, wound D times, on the manifold of SU_2 when ϕ varies from $0 \to 2\pi$. Such a circle can be continuously shrunk to the identity element, in contrast with the situation for electromagnetism. Thus, by a global gauge transformation S may be changed to $S'=1$, and the two regions a and b after the global gauge transformation can be *fused into one single region*. The gauge potential b_μ^k is then defined *everywhere* in R as a single region. Thus we have the following theorem.

Theorem 9: For the SU_2 gauge group, the gauges $\mathcal{G}_D$ for different D can be transformed into each other by global gauge transformations. The different monopole fields are therefore of the same type.

We shall only exhibit the global transformation for the case $\mathcal{G}_{-1}$ for which

$$S_{ba} = \exp(-2\phi X_3), \qquad (32)$$

$$\frac{e}{\hbar c} = \frac{-1}{g}. \qquad (33)$$

The gauge transformations we shall seek are illustrated in Fig. 5. We shall choose

$$\xi = \exp[\theta(X_1 \sin\phi - X_2\cos\phi)], \qquad (34)$$

$$\eta = \exp[(\pi-\theta)(X_1\sin\phi - X_2\cos\phi)]\exp(\pi X_2). \qquad (35)$$

It is easy to see that ξ is analytic in the coordinates x^μ at all points in R_a. (One only has to verify this statement at $\theta = 0$, which is easily done.) Similarly η is analytic in R_b. ξ and η are therefore allowed gauge transformations in, respectively, R_a and R_b.

Now one can prove after some algebra that[8]

$$S'_{ba}=\eta S_{ba}\xi=1\,.$$

Thus after the gauge transformations ξ and η, which together form a global gauge transformation, regions R_a and R_b are related by the identity gauge transformation in their overlap, i.e., the two regions can be fused into one. To calculate the gauge potentials $b^{k'}_\mu$ after the global gauge transformation we use

$$\xi(Q)\left[1+\frac{1}{g}(b^k_\mu)'_a X_k dx^\mu\right]\xi^{-1}(P)=1+\frac{1}{g}(b^k_\mu)_a X_k dx^\mu =1+\frac{1}{g}(A_\mu)_a X_3\, dx^\mu\,, \tag{36}$$

where A_μ is given by (11) and $Q=P+dx$. By choosing dx^μ to be along the t and r directions, one obtains $b^{k'}_t=b^{k'}_r=0$. By choosing dx^μ to be along the θ direction, one obtains

$$b^{1'}_\theta=\frac{-g}{r}\sin\phi\,,\quad b^{2'}_\theta=\frac{g}{r}\cos\phi\,,\quad b^{3'}_\theta=0\,. \tag{37}$$

Now take dx^μ to be along the ϕ direction. We obtain, to order $d\phi$,

$$1+\frac{1}{g}(b^k_\phi)'X_k r\sin\theta\, d\phi =\xi^{-1}(Q)\xi(P)+\frac{1}{g}(A_\phi)_a r\sin\theta\, d\phi\, \xi^{-1}(P)X_3\xi(P)\,. \tag{38}$$

The first term on the right-hand side can be[8] computed in a straightforward manner:

$$\xi^{-1}(Q)\,\xi(P)=1-\sin\theta\, d\phi[(X_1\cos\phi+X_2\sin\phi)-X_3\tan\tfrac{1}{2}\theta]\,.$$

The second term also can be easily computed since

$$\xi^{-1}(P)\,X_3\xi(P)=\sin\theta(X_1\cos\phi+X_2\sin\phi)+X_3\cos\theta$$

and $(A_\phi)_a$ was given by (11). Finally one arrives at

$$b^{1'}_\phi=\frac{-g}{r}\cos\theta\cos\phi\,,\qquad b^{2'}_\phi=\frac{-g}{r}\cos\theta\sin\phi\,,\qquad b^{3'}_\phi=\frac{g}{r}\sin\theta\,. \tag{39}$$

Combining these results and remembering (33), we obtain

$$\frac{e}{\hbar c}b^{k'}_\mu X_k dx^\mu=\frac{-1}{r^2}\epsilon_{ikj}\,x^j dx^i X_k\,,$$

i.e.,

$$b^{k'}_4=0\,,\quad \frac{e}{\hbar c}b^{k'}_i=-\frac{1}{r^2}\epsilon_{ikj}\,x^j\,. \tag{40}$$

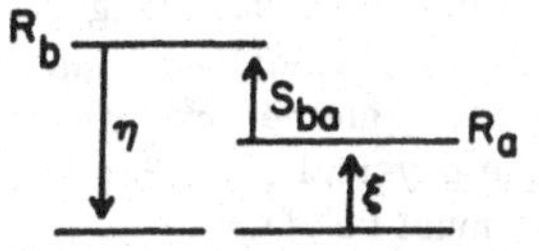

FIG. 5. A global transformation after which R_a and R_b can be fused.

Thus the new potential $b^{k'}_\mu$ is analytic in R_a. Because $\eta S_{ba}\xi=1$ the new potential (in the overlapping region) for R_b must be the same as (40). By analyticity (40) is seen to be valid throughout R. Notice that (40) is the same potential as one of the solutions [solution (12a)], for a sourceless gauge field, in Ref. 9.

The global gauge transformation that transforms $\mathcal{G}_D$ into $\mathcal{G}_0$ for $D\neq -1$ can be obtained by slightly modifying (34) and (35).

We shall discuss Theorem 9 further in the next section.

VII. CASE OF SO_3 GAUGE FIELD

We turn to SO_3, which is locally the same as SU_2, but for which

$$e^{2\pi X_k}=1\,. \tag{41}$$

Equations (22) to (25) remain unaltered. The minimum "charge" of all physical states is now

$$\text{minimum "charge"}=e\,, \tag{42}$$

giving

$$\frac{\text{"charge" of gauge particle}}{\text{minimum "charge"}}=1 \quad \text{for } SO_3\,. \tag{43}$$

This last formula differentiates physically the SO_3 case from the SU_2 case.

We emphasize here a point already made in the literature[10] for electromagnetism: The local character of the gauge group is of course determined by the interactions (which determine the conservation laws). We want to ask what determines the global character. The global character (compact or noncompact in the case of electromagnetism, SU_2 or SO_3 in the isospin case) *is determined by* the representations for all states which *physically* exist. For example, in electromagnetism, if all charges are integral multiples of a single unit, the gauge group is compact,[10] because the group is *physically defined* as the simultaneous local phase factor change of *all* charge fields. There is then no physically definable meaning to the noncompact group. In the case of SU_2 or SO_3, if (43) is satisfied, then all representations of X_k physically realizable are integral representations.

Thus the simultaneous local changes of isospin phase factor of all physical systems *cannot differentiate* the group element $e^{2\pi X_k}$ from the identity. *Therefore*, the physical definition of $e^{2\pi X_k}$ is unity and the group must be SO_3.

Turning now to the monopole field for SO_3 we find that (30) is still correct. Equation (41) then leads to the quantization condition

$$\frac{2eg}{\hbar c} = \text{integer} = D \tag{44}$$

in order that S (as an element of SO_3) be a single-valued function of the coordinates x^μ in R.

As ϕ increases from 0 to 2π, the phase factor (30) describes a closed circuit in the group space of SO_3, starting from the identity element and returning to it. If one continuously traced the corresponding element of the group SU_2, one would have started from the identity and ended with the element that corresponds to

$$\begin{pmatrix} -1 & \\ & -1 \end{pmatrix}$$

in the 2×2 representation of SU_2 when D is odd. In such a case, no distortion of the closed circuit in SO_3 described by the phase factor (30) can shrink it to the identity element. This means that the gauge type for even D is not the same as that for odd D. By constructing explicit gauge transformations like (34) and (35) one can then complete the proof of the following theorem.

Theorem 10: For SO_3, all gauges $\mathcal{G}_D$ for D = even are of one type, and all gauges $\mathcal{G}_D$ for D = odd are of one type. These two types are different.

Summarizing the situation for U_1, SU_2, and SO_3 we find that in each case the "magnetic" monopole fields have quantized strengths. They belong to, respectively, infinitely many types for U_1 gauge group (electromagnetism), one type for the SU_2 gauge group, and two types for SO_3 gauge groups.

The physical meaning of these statements are as follows. In the SU_2 case, all magnetic monopole fields can be continuously changed into each other by the process of continuous changes[19] of "electric" sources. For example, starting with the "magnetic monopole" field for $\mathcal{G}_{-1}$ of Theorem 9 we can, by a gauge transformation, obtain the potentials b' [on $\mathcal{G}_0$] given in (40). We can then consider the potential (on $\mathcal{G}_0$): $b'' = \alpha b$, where $0 \le \alpha \le 1$. The gauge field for b'' is no longer electrically sourceless outside of the origin, but is magnetically sourceless except at the orgin, where it is not sourceless either magnetically or electrically. As α changes from 1 to 0 we thus have a continuous change of the original magnetic monopole field to empty space through a process during which there are continuous changes of electric charge-current distributions. Such a process is not possible for electromagnetism, by Theorem 6. (In the SO_3 case it is also not possible, although it is possible to change the magnetic monopole strength by two units by a similar process.) Thus the meaning of a magnetic monopole field in the non-Abelian case is quite different from that in electromagnetism.

It is not really surprising that in the case of electromagnetism one cannot change the magnetic monopole strength by changing electric sources: In the region R there are no magnetic monopoles. The continuity of magnetic lines of forces in R is guaranteed by the equation $\nabla \cdot \vec{H} = 0$. No continuous movement of magnetic lines of force could therefore increase or decrease the net total flux around the origin. That this state of affairs does not obtain for SU_2 and SO_3 is due to the fact that in general $\nabla \cdot \vec{H}^k \neq 0$ in the non-Abelian case, so that one cannot define the magnetic flux through a loop. However, we had seen before (Theorem 4) that in the case of a non-Abelian gauge field what takes the place of the magnetic flux is the *phase factor of a loop*. One may then ask what takes the place of the total magnetic flux outwards from a sphere around the origin $\vec{r} = 0$. To answer this question consider the loop

$$r = 1\,, \quad \theta = \text{fixed}\,, \quad \phi = 0 \to 2\pi\,. \tag{45}$$

As θ changes from 0 to π the phase factor of the loop changes and it describes a continuous circuit (in the space of the group) starting from and ending at the identity element. Clearly any other way of "looping" over the sphere only leads to a distortion of this circuit, without changing the starting and ending point. We shall call this circuit the *total circuit* for the gauge field around the origin $\vec{r} = 0$. It is a concept that replaces the total magnetic flux around $\vec{r} = 0$ in electromagnetism.

We can now prove the following generalization of Theorem 8.

Theorem 11: Consider region R and the group SU_2 or SO_3. Consider a gauge $\mathcal{G}$ and define any gauge field on it. The total circuit for the gauge field around the origin $\vec{r} = 0$ is independent of the gauge field and only depends on the gauge type of $\mathcal{G}$. For the case of $\mathcal{G}_D$,

total circuit of the gauge field

$$\simeq [S_{ba}(\phi) \text{ for } \phi = 2\pi \to 0]\,, \tag{46}$$

where $\simeq$ means "can be continuously distorted into."

This last formula is the generalization of (19).

To prove Theorem 11, consider first the loop (45). The phase factor in R_a and R_b will be denoted by $\Phi^a(\theta)$ and $\Phi^b(\theta)$. They are related in the overlap by

$$\Phi^b(\theta)=\Phi^a(\theta) \tag{47}$$

since $S(\phi=0)=I$. [One uses a generalization of (14).] Next consider the loop $L(\theta)$ which lies on the sphere $r=1$ with its projection onto the x-y plane given in Fig. 6. It consists of a first part $(BA)_1$ around the equator and a second part $(AB)_2$ not on the equator except for points A and B. It is clear that

$$[\text{loop(45) for } \theta=0\to\pi/2]\simeq[L(\theta) \text{ for } \theta=0\to\pi/2]$$

because both sides "loop over" the upper hemisphere. Thus

$$[\Phi^a(\theta) \text{ for } \theta=0\to\pi/2]\simeq[\Phi^a_{L(\theta)} \text{ for } \theta=0\to\pi/2]$$
$$=[\Phi^a_{(AB)_2}\Phi^a_{(BA)_1} \text{ for } \theta=0\to\pi/2].$$

$\Phi^a_{(AB)_2}$ is continuous in θ, and

$$[\Phi^a_{(AB)_2} \text{ for } \theta=0\to\pi/2]\simeq \text{identity element}.$$

Thus

$$[\Phi^a(\theta) \text{ for } \theta=0\to\pi/2]\simeq[\Phi^a_{(BA)_1} \text{ for } \theta=0\to\pi/2]. \tag{48}$$

Similarly

$$[\Phi^b(\theta) \text{ for } \theta=\pi/2\to\pi]\simeq[\Phi^b_{(BA)_1} \text{ for } \theta=\pi/2\to\pi]. \tag{49}$$

At $\theta=\pi/2$, the left-hand sides of (48) and (49) match because of (47). Also the right-hand sides match. Thus we can take (48) and (49) in tandem, obtaining

$$\text{total circuit of gauge field}\simeq[\Phi^a_{(BA)_1} \text{ for } \theta=0\to\pi/2 \text{ followed by } S_B\Phi^a_{(BA)_1} \text{ for } \theta=\pi/2\to\pi], \tag{50}$$

where we have used

$$\Phi^b_{(BA)_1}=S_B\Phi^a_{(BA)_1}S^{-1}{}_A=S_B\Phi^a_{(BA)_1}. \tag{51}$$

Now

$$[\Phi^a_{(BA)_1} \text{ for } \theta=0\to\pi/2 \text{ followed by } \Phi^a_{(BA)_1} \text{ for } \theta=\pi/2\to\pi] \tag{52}$$

is a loop *that doubles back* on itself, i.e., (52) can be distorted to the identity element. Applying this fact to (50) one obtains

$$\text{total circuit of gauge field}\simeq[I \text{ for } \theta=0\to\pi/2 \text{ followed by } S_B \text{ for } \theta=\pi/2\to\pi]. \tag{53}$$

Now $S_B=S_{ba}(\phi=4\pi-4\theta)$. As $\theta=\pi/2\to\pi$, $S_B=S_{ba}(\phi)$ for $\phi=2\pi\to0$. Substitution into (53) leads to (46).

To complete the proof of Theorem 11 we need the generalization of (46) to gauges that contain more than two regions. This can be done without much difficulty, e.g., for the case that region b is further divided into regions c and d, as schematically illustrated in Fig. 7(a), (46) should be replaced by

$$\text{total circuit of gauge field}\simeq[S_{dc}(B)S_{ca}(x) \text{ for } x=A\to A \text{ along direction of arrow},$$
$$\text{followed by } S_{dc}(y)S_{ca}(A) \text{ for } y=B\to B \text{ along direction of arrow}] \tag{54}$$

For the case that $\mathcal{G}$ has four regions a,b,c,d as illustrated in Fig. 7(b), (46) should be replaced by

$$\text{total circuit of gauge field}\simeq[S_{da}(x) \text{ for } x=A\to B,$$
$$\text{followed by } S_{db}(y)S_{ba}(x) \text{ for } x=B\to C,\ y=B\to D,$$
$$\text{followed by } S_{dc}(D)S_{cb}(x)S_{ba}(C) \text{ for } x=D\to C,$$
$$\text{followed by } S_{dc}(y)S_{ca}(x) \text{ for } x=C\to E,\ y=D\to E,$$
$$\text{followed by } S_{da}(x) \text{ for } x=E\to A]. \tag{55}$$

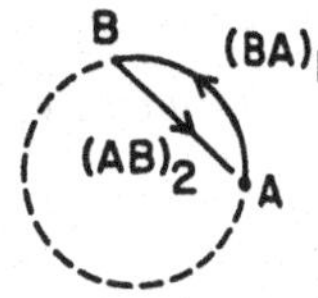

FIG. 6. Projection onto x-y plane of loop $L(\theta)$. The loop lies entirely on sphere $r=1$, and is in the upper (lower) hemisphere for $0\le\theta\le\pi/2$ ($\pi/2<\theta\le\pi$). The portion $(BA)_1$ lies on the equator. Coordinates for A: $r=1$, $\theta=\pi/2$, $\phi=0$. Coordinates for B: $r=1$, $\theta=\pi/2$, $\phi=h(\theta)$, where $h(\theta)=4\theta$ for $0\le\theta\le\pi/2$ and $h(\theta)=4\pi-4\theta$ for $\pi/2<\theta\le\pi$.

Notice that the right-hand sides of (54) and (55) are dependent only on the gauge type, and not on the specific gauge field.

VII. GENERALIZED BOHM-AHARONOV EXPERIMENT

The concept of an SU_2 gauge field was first discussed in 1954. In recent years many theorists, perhaps a majority, believe that SU_2 gauge fields do exist. However, so far there is *no experimental proof* of this theoretical idea, since conservation of isotopic spin only suggests, and does not require, the existence of an isotopic spin gauge field. What kind of experiment would be a definitive test of the existence of an isotopic spin gauge field? A generalized Bohm-Aharonov experiment would be.

If the gauge particle for isospin group SU_2 is massless, it is possible to design a *gedanken* generalized Bohm-Aharonov experiment as illustrated in Fig. 1. One constructs the cylinder of material for which the total I_z spin is not zero, e.g., a cylinder made of heavy elements with a neutron excess. One spins the cylinder around its axis, setting up a "magnetic" flux inside the cylinder, along the I_z "direction." If one scatters a proton beam around the cylinder, the fringe shift would be in the opposite direction from the corresponding shift observed with a neutron beam. To be more specific, imagine that one spins the cylinder clockwise. The magnetic flux would be emerging from the diagram towards the reader, since the cylinder has a net negative value for I_z. This means that for a proton (neutron) beam, the flux produces an increment (decrement) of path length counterclockwise around the cylinder. This increment (decrement) produces a net downward (upward) shift of the fringes, i.e., a shift toward the bottom (top) of the diagram.

If one scatters a coherent mixture of neutron and proton in a pure state, in the interference plane one would observe not only fluctuations of nucleon intensity, but also fluctuations of the neutron-proton mixing ratio. A variation of this phenomenon

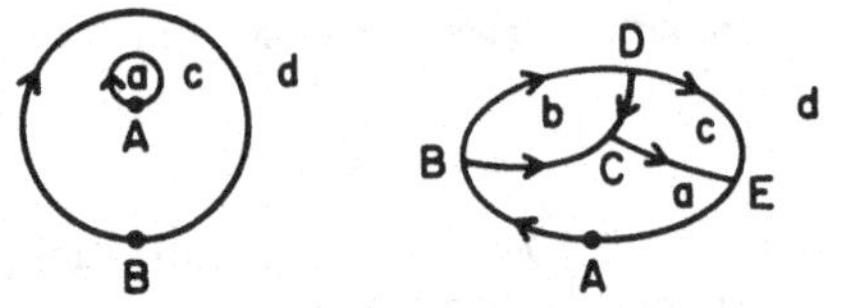

FIG. 7. Schematic diagrams for division lines in overlaps between three or more regions. The drawings are projections from the sphere $r=1$. The projection is from the south pole of the sphere onto the tangent plane at the north pole. The south pole is underneath the plane of the paper.

obtains if one imagines rotating a cylinder which has an average $\langle\vec{I}\rangle$ which is not zero, and is not in the I_z direction. A magnetic flux would then be set up which is in a "direction" other than I_z. Scattering a beam of protons would then produce some neutrons as well as protons in the interference plane. This implies, of course, that there is electric charge transfer between the beam and the cylinder together with the gauge field around it.

If the gauge particle has a finite mass $m>0$, the experiment becomes difficult because the return flux would hug the outside surface of the cylinder, to a distance $\sim\hbar/mc$. Unless the fringe plane lies within this distance of the cylinder, the effect of the flux will be negligible.

IX. REMARKS

(a) From the viewpoint of the present paper, the electric charge and the magnetic charge play completely unsymmetrical roles. This matter deserves further comments. In the non-Abelian case it was in fact already pointed out[11] that the dual of an unquantized sourceless gauge field is not necessarily a gauge field. Thus the asymmetry between electric and magnetic phenomena is not due to the formalism, but is of an intrinsic nature in the non-Abelian case. In contrast, in the Abelian case the asymmetry is only formal since the electric and magnetic charges interact with the electromagnetic field in entirely symmetrical ways. In other words, one can use the phase factor Φ^M associated with the magnetic charges to describe the electromagnetic field, rather than the phase factor Φ discussed in the present paper. The mathematical relationship between these two kinds of phase factors (or between the associated vector potentials A_μ^M and A_μ) remains to be explored. So does the corresponding question in any second-quantized theory[12] of all the fields.

(b) In the proof of Theorem 9 we had shown explicitly how a magnetic monopole field for the SU_2 gauge group can be gauge-transformed into the solution (12a) of Ref. 9. Now a magnetic monopole field is not a gauge field at the origin $\vec{r}=0$ since it does not satisfy the Bianchi identity[3] at the origin. Thus, although solution (12a) of Ref. 9 is (electrically) sourceless at all points, including the origin, it is not a proper gauge field at the origin, a fact we did not realize before. All three solutions, (12a), (12d), and (12e), are, of course, of the same gauge type.

(c) In Sec. II it was emphasized that $f_{\mu\nu}$ underdescribes electromagnetism because of the Bohm-Aharonov experiment which involves a doubly connected space region. For non-Abelian cases, the field strength $f^k_{\mu\nu}$ underdescribes the gauge field even in a singly connected region. An example of this underdescription was given in Ref. 13.

(d) For the region of space-time outside of the cylinder of Fig. 1 there is only one gauge type. All electromagnetic fields in the region can be continuously distorted into each other by the movement of electric charges and currents inside and outside the cylinder.

(e) The phase factor for the group U_1 is the phase factor of the algebra of complex numbers. It is perhaps not accidental that such a phase factor provides the basis for the description of a physically realized gauge field—electromagnetism. Now the only possible more complicated division algebra is the *algebra of quaternions*. The phase factors of the quaternions form the group SO_3. It is tempting to speculate that such a phase factor provides the basis for the description of a physically realized gauge field—the SU_2 gauge field. Speculation about the possible relationship between quaternions and isospin has been made before.[14] Such speculations were, however, not made with reference to gauge fields. If one believes that gauge fields give the underlying basis for strong and/or weak interactions, then the fact that gauge fields are fundamentally *phase factors* adds weight to the speculation that quaternion algebra is the real basis of isospin invariance.

(f) It is a widely held view among mathematicians that the fiber bundle is a natural geometrical concept.[15] Since gauge fields, including in particular the electromagnetic field, are fiber bundles, *all gauge fields are thus based on geometry*.[16] To us it is remarkable that a geometrical concept formulated without reference to physics should turn out to be exactly the basis of one, and indeed maybe all, of the fundamental interactions of the physical world.

ACKNOWLEDGMENTS

It is a pleasure to thank Professor Shiing-shen Chern for correspondence and discussions. We are especially indebted to Professor J. Simons, whose lectures and patient explanations have revealed to us glimpses of the beauty of the mathematics of fiber bundles.

While we were making corrections on the draft of this paper, a report on the experimental discovery of a magnetic monopole[17] reached us.

Additional references to fiber bundles, monopoles and quaternions are given in footnote 18.

---

*Work supported in part by the U. S. ERDA under Contract No. AT(11-1)-3227.

†Work supported in part by the National Science Foundation under Grant No. MPS74-13208 A01.

[1]Y. Aharonov and D. Bohm, Phys. Rev. 115, 485 (1959). See also W. Ehrenberg and R. E. Siday, Proc. Phys. Soc. London B62, 8 (1949).

[2]R. G. Chambers, Phys. Rev. Lett. 5, 3 (1960).

[3]Chen Ning Yang, Phys. Rev. Lett. 33, 445 (1974). This paper introduced the formulation of gauge fields in terms of the concept of nonintegrable phase factors. The differential formulation of gauge fields for Abelian groups was first discussed by H. Weyl, Z. Phys. 56, 330 (1929); for non-Abelian groups it was first discussed by Chen Ning Yang and Robert L. Mills, Phys. Rev. 96, 191 (1954). See also S. Mandelstam, Ann. Phys. (N.Y.) 19, 1 (1962); 19, 25 (1962); I. Białynicki-Birula, Bull. Acad. Pol. Sci., Ser. Sci. Math. Astron. Phys. 11, 135 (1963); N. Cabibbo and E. Ferrari, Nuovo Cimento 23, 1146 (1962); R. J. Finkelstein, Rev. Mod. Phys. 36, 632 (1964); N. Christ, Phys. Rev. Lett. 34, 355 (1975); and A. Trautman, in *The Physicist's Conception of Nature*, edited by J. Mehra (Reidel, Boston, 1973), p. 179.

[4]P. A. M. Dirac, Proc. R. Soc. London A133, 60 (1931). Since this brilliant work of Dirac, there have been several hundred papers on the magnetic monopole. For a listing of papers until 1970, see the bibliography by D. M. Stevens, Virginia Polytechnic Institute Report No. VPI-EPP-70-6, 1970 (unpublished).

[5]See J. Milnor and J. Stasheff, *Characteristic Classes* (Princeton Univ. Press, Princeton, N.J., 1974); C. B. Allendoerfer and A. Weil, Trans. Am. Math. Soc. 53, 101 (1943); Shiing-shen Chern, Ann. Math. 45, 747 (1944). See also H. Weyl, Amer. J. Math. 61, 461 (1939), and Ref. 6 below.

[6]There are many books on fiber bundles. See e.g., N. Steenrod, *The Topology of Fibre Bundles* (Princeton Univ. Press, Princeton, N. J., 1951). For connection, see, e.g., S. Kobayashi and K. Nomizu, *Foundations of Differential Geometry* (Interscience, New York, Vol. I-1963, Vol. II-1969).

[7]The notation here is the same as that in Ref. 3, except for the normalization factor $e/\hbar c$ which was absorbed into b in Ref. 3. To avoid confusion with the azimuthal angle, we write Φ for the ϕ of Ref. 3. Notice that

$$\Phi_{(A+dx)A} = I - \frac{e}{\hbar c} b^k_\mu(x) X_k dx^\mu .$$

All formulas are the same as in Ref. 3, but the name for Φ_{QP} will now be "the phase factor from P to Q." (In Ref. 3 the same name applied to Φ_{PQ}.) The new name is in accordance with the usual convention of time ordering.

[8]Since the 2×2 representation is faithful, it is sufficient for computational purposes to use the representation (26) for X_k. This makes the algebra quite simple since one can apply the formula $e^{i\theta\sigma_k} = \cos\theta + i\,\sigma_k \sin\theta$.

[9]Tai Tsun Wu and Chen Ning Yang, in *Properties of Matter under Unusual Conditions*, edited by H. Mark and S. Fernbach (Wiley, New York, 1969), p. 349.

[10]Chen Ning Yang, Phys. Rev. D 1, 2360 (1970).

[11]Gu Chao-hao and Chen Ning Yang, Sci. Sin. 18, 483 (1975).

[12]P. A. M. Dirac, Phys. Rev. 74, 817 (1948); J. Schwinger, *Particles, Sources and Fields* (Addison-Wesley, Reading, Mass., Vol. 1-1970, Vol. 2-1973).

[13]Tai Tsun Wu and Chen Ning Yang, preceding paper, Phys. Rev. D 12, 3843 (1975).

[14]Cheng Ning Yang, comments after J. Tiomno's talk, session 9, *Proceedings of the Seventh Annual Rochester Conference on High-Energy Nuclear Physics, 1957* (Interscience, New York, 1957); Chen Ning Yang, in *The Physicist's Conception of Nature*, edited by J. Mehra (Reidel, Boston, 1973), p. 447.

[15]See, e.g., Shiing-shen Chern, Geometry of Characteristic Classes, Proceedings of the 13th Biennial Seminar, Canadian Mathematics Congress, 1972, p. 1.

[16]This is in sharp contrast with an interaction (if it exists), which is not related to gauge concepts.

[17]P. B. Price, E. K. Shirk, W. Z. Osborne, and L. S. Pinsky, Phys. Rev. Lett. 35, 487 (1975).

[18]There have been many papers on fiber bundles, monopoles, and quaternions in the physics literature. The following is only a partial list: Elihu Lubkin, Ann. Phys. (N.Y.) 23, 233 (1963); J. Math. Phys. 5, 1603 (1964); D. Finkelstein, J. M. Jauch, S. Schiminovich, and D. Speiser, J. Math. Phys. 4, 788 (1963); articles by J. A. Wheeler, B. S. DeWitt, A. Lichnerowicz, and C. W. Misner, in *Relativity, Groups and Topology*, edited by C. DeWitt and B. S. DeWitt (Gordon and Breach, New York, 1963); also C. Misner, K. Thorne, and J. A. Wheeler, *Gravitation* (Freeman, San Francisco, 1973); A. Trautman, Rep. Math. Phys. 1, 29 (1970); G. 't Hooft, Nucl. Phys. B79, 276 (1974); Hendricus G. Loos, Phys. Rev. D 10, 4032 (1974); J. Arafune, P. G. O. Freund, and C. J. Goebel, J. Math. Phys. 16, 433 (1975); B. Julia and A. Zee, Phys. Rev. D 11, 2227 (1975); M. K. Prasad and Charles M. Sommerfield, Phys. Rev. Lett. 35, 760 (1975).

[19]*Footnote added in proof.* Professor A. Lenard has raised an interesting question in this connection: In the continuous changes of "electric" sources, are sources quantized according to (44)? The answer to this question is "no," and requires explanations. The "electric" charges play two separate roles. They act as sources of gauge fields, and they also act as responders to gauge fields. In a physical situation the two roles are of course interrelated. In the discussion here, however, we separate the two roles. We therefore do not require quantization of electric charges as sources, but require quantization of electric charges as responders.